Geometric and Algorithmic Aspects of Computer–Aided Design and Manufacturing

DIMACS

Series in Discrete Mathematics
and Theoretical Computer Science

Volume 67

Geometric and Algorithmic Aspects of Computer–Aided Design and Manufacturing

DIMACS Workshop
Computer Aided Design and Manufacturing
October 7–9, 2003
Piscataway, New Jersey

Ravi Janardan
Michiel Smid
Debasish Dutta
Editors

Center for Discrete Mathematics
and Theoretical Computer Science
A consortium of Rutgers University, Princeton University,
AT&T Labs–Research, Bell Labs (Lucent Technologies),
NEC Laboratories America, and Telcordia Technologies
(with partners at Avaya Labs, HP Labs, IBM Research,
Microsoft Research, and Stevens Institute of Technology)

American Mathematical Society

This DIMACS volume presents the proceedings from the workshop Computer Aided Design and Manufacturing held at the DIMACS Center, Rutgers University, October 7–9, 2003.

2000 *Mathematics Subject Classification.* Primary 12Y05, 52C45, 65D17, 65D18, 68U05, 68U07, 68W30, 68W40.

Library of Congress Cataloging-in-Publication Data

DIMACS Workshop Computer Aided Design and Manufacturing (2003 : Piscataway, N.J.)

Geometric and algorithmic aspects of computer-aided design and manufacturing : DIMACS Workshop Computer Aided Design and Manufacturing, October 7–9, 2003, Piscataway, New Jersey / Ravi Janardan, Michiel Smid, Debasish Dutta, editors.

p. cm. — (DIMACS series in discrete mathematics and theoretical computer science ; v. 67)

Includes bibliographical references.

ISBN 0-8218-3628-5 (alk. paper)

1. Engineering design. 2. Manufacturing processes. 3. Computer-aided design. I. Janardan, Ravi. II. Smid, Michiel. III. Dutta, D. (Debasish) IV. Title. V. Series.

TA174.D498 2003
620′.0042′0285—dc22 2004062274

---

10 9 8 7 6 5 4 3 2 1 10 09 08 07 06 05

Contents

Foreword

A workshop on Computer-Aided Design and Manufacturing was held on October 7-9, 2003 at Rutgers University. We would like to express our appreciation to Debasish Dutta, Ravi Janardan and Michiel Smid for their efforts to organize and plan this successful conference.

The workshop was part of the 2002-2005 Special Focus on Computational Geometry and Applications. We extend our thanks to Steven Fortune, Bernard Chazelle and Bill Steiger for their work as special focus organizers.

The workshop brought together theoreticians and practitioners working in the field of Computer-Aided Design and Manufacturing (CAD/CAM) and those working in Computational Geometry (CG), for the purpose of promoting and exploring links between these areas.

Topics addressed by the workshop included computational metrology and tolerancing, geometric constraint solving, geometric modeling related to manufacturing, numerically-controlled machining, part immobilization, rapid prototyping and layered manufacturing, and surface modeling and reconstruction. These and many other topics are all major themes of current interest in both CAD/CAM and CG.

DIMACS gratefully acknowledges the generous support that makes these programs possible. Special thanks go to the National Science Foundation, the New Jersey Commission on Science and Technology, and to DIMACS partners at Rutgers, Princeton, AT&T Labs - Research, Bell Labs, NEC Laboratories America, and Telcordia Technologies, and affiliate partners Avaya Labs, HP Labs, IBM Research, Microsoft Research, and Stevens Institute of Technology.

Fred S. Roberts
Director

Robert Tarjan
Co-Director for Princeton

Preface

Computer-Aided Design and Manufacturing (CAD/CAM) is concerned with all aspects of the process of designing, prototyping, manufacturing, inspecting, and maintaining complex geometric objects under computer control. As such, there is a natural synergy between this field and Computational Geometry (CG), which involves the design, analysis, implementation, and testing of efficient algorithms and data representation techniques for geometric entities such as points, polygons, polyhedra, curves, and surfaces. On the one hand, CG can bring about significant performance improvements in CAD/CAM, while, on the other hand, CAD/CAM can be a rich source of interesting problems that can spur new research in CG. Indeed, such two-way interaction has already been witnessed in recent years in areas such as numerically-controlled machining, casting and injection molding, rapid prototyping and layered manufacturing, metrology, and mechanism/linkage design, to name just a few.

To further promote the interaction between the two fields, we organized a Workshop on *"Computer-Aided Design and Manufacturing"* at the DIMACS Center in Piscataway, NJ, from Oct. 7-9, 2003. The Workshop, which was attended by about fifty-five individuals from academia, research laboratories, and industry, was organized around a series of invited talks, contributed presentations, and informal discussions geared towards fostering collaborative research. Details about the Workshop can be found at `http://dimacs.rutgers.edu/Workshops/CompAided/`.

Subsequently, authors of papers presented at the Workshop were invited to submit expanded and polished versions of their research to this Volume. The intent was to provide, in one place, a set of self-contained papers that could serve as the starting point for further research. All submissions were refereed. The papers in this Volume are an outgrowth of this process and cover topics such as geometric modeling, computational topology, computational metrology, geometric constraint solving, part immobilization, geometric aspects of machining, layered manufacturing, and algebraic methods.

We thank Fred Roberts (DIMACS Director) and Steve Fortune (Member, DIMACS Council and Chair of the DIMACS Special Focus on Computational Geometry and Applications) for their support and encouragement, DIMACS and AMS staff members for organizational help, the referees for their help with the reviews, and, of course, the authors for their research contributions. We are grateful to the National Science Foundation for financial support for the Workshop.

Ravi Janardan (Minneapolis, MN)
Michiel Smid (Ottawa, Canada)
Debasish Dutta (Ann Arbor, MI)
July 2004

DIMACS Series in Discrete Mathematics
and Theoretical Computer Science
Volume **67**, 2005

A Survey of Subdivision-Based Tools for Surface Modeling

Ioana Boier-Martin, Denis Zorin, and Fausto Bernardini

This paper is dedicated to Henning Biermann, in memoriam.

ABSTRACT. Subdivision surfaces have emerged as a powerful representation for surface modeling and design. They address important limitations of traditional spline-based methods, such as the ability to handle arbitrary topologies and to support multiscale editing operations. In this paper we survey existing subdivision-based modeling methods with emphasis on interactive tools for styling and decoration of 3D models.

1. Introduction

Subdivision surfaces offer several advantages over both irregular meshes and spline patches, two of the most commonly used surface representations today. Subdivision offers a compact way to represent geometry with minimal connectivity information. It generalizes the classical spline patch approach to arbitrary topology, it naturally accommodates multiple levels of detail, and produces meshes with well-shaped elements arranged in almost regular structures, suitable for digital processing. When combined with multiresolution analysis, subdivision offers a powerful modeling tool, allowing for complex editing operations to be applied efficiently at different resolutions.

In recent years, the set of tools available for manipulating subdivision surfaces has been growing steadily. Algorithms for direct evaluation [**Sta98, ZK02**], editing [**BKZ01, BMBZ02, BMZB02, BLZ00**], texturing [**PB00**], and conversion to other popular representations [**Pet00**] have been devised and hardware support for rendering of subdivision surfaces has been proposed [**BAD⁺01, BKS00, PS96**].

This survey focuses on the use of subdivision-based representations for styling and conceptual design. We explore various methods for manipulating subdivision surfaces and, whenever possible, we illustrate the evolution of such methods from related representations. We pay particular attention to interactive tools which are suitable for design as they allow the designer to instantaneously evaluate results. While we are trying to provide a thorough overview of the area and include the most relevant methods, we realize that the volume of published work goes well beyond

Key words and phrases. Subdivision surfaces, geometric modeling.

that covered in this survey which is by no means exhaustive (see also [**DL02, Sab02**] for additional surveys). Many of the topics presented relate to issues we have addressed in our own work which we hope will provide some insights to those pursueing similar interests.

2. Background

The basic idea of using subdivision to produce smooth curves and later, smooth surfaces, has been around for many years (see [**ZSD$^+$00**] for a brief incursion into the history of subdivision). However, it is only recently that powerful design tools based on this representation have emerged. This is partly due to the recent advent of multiresolution techniques that facilitate capturing of non-trivial shapes and partly due to even more recent advances in subdivision theory and methods for direct and efficient evaluation of subdivision surfaces. For the purpose of this survey, we provide a brief review of the basic concepts pertaining to subdivision surfaces. For additional details we refer the reader to [**ZSD$^+$00, WW01**].

Subdivision defines a smooth surface recursively as the limit of a sequence of meshes (see Figure 1). Each finer mesh is obtained from a coarse mesh by using a set of refinement rules which define a *subdivision scheme*. Many schemes have been proposed in the literature. Examples include Doo-Sabin [**DS78**], Catmull-Clark [**CC78**], Loop [**Loo87**], Butterfly [**DLG90, ZSS96**], Kobbelt [**Kob96a**], Midedge [**PR97**]. Different schemes lead to limit surfaces with different smoothness characteristics. For design purposes, the Catmull-Clark [**CC78**], Loop [**Loo87**] schemes are most often employed as they are closely related to splines (a de-facto standard in modeling today) and generate C^2-continuous surfaces over arbitrary meshes.

FIGURE 1. Subdivision defines a smooth surface recursively as the limit of a sequence of meshes.

Multiresolution subdivision extends the concept of subdivision by allowing *detail vectors* to be introduced at each level. Hence, a finer mesh is computed by adding detail offsets to the subdivided coarse mesh. Given a *semi-regular mesh*, i.e., a mesh with subdivision connectivity, it can be easily converted to a multiresolution surface by defining a smoothing operation to compute a coarse level from a finer level. The details are then computed as differences between levels. This representation was introduced by several authors in different forms [**LDW97, PL97, ZSS97**]. Figure 2 illustrates the power of multiresolution in capturing complex shapes.

A close connection exists between multiresolutin subdivision and wavelets [**SDS96**]. In particular, two operations known as *Synthesis* and *Analysis* can be defined to propagate data from coarse to fine and in reverse throughout the subdivision hierarchy, similar to wavelet transforms. *Analysis* computes positions of control points

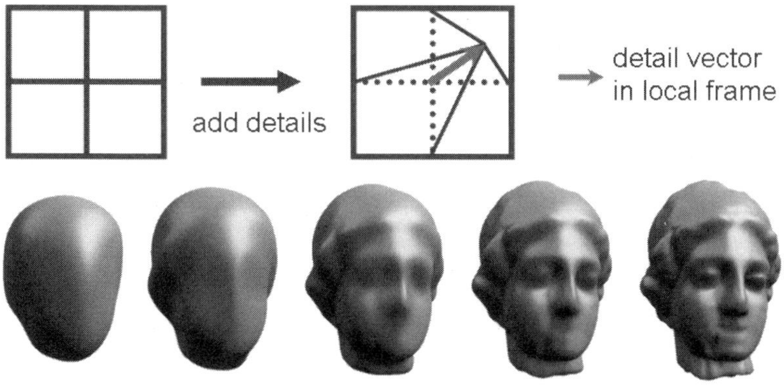

FIGURE 2. Top: multiresolution subdivision extends the concept of subdivision by introducing detail vectors at each level. Bottom: surfaces obtained by subdivision of the same coarse mesh look very different depending on the amount of detail introduced and the level at which it is introduced. From left to right: no details to progressively more details added on finer levels.

on a coarse level $i-1$ by applying a smoothing filter to points on level i. Multiresolution details on level i are computed as differences between the two levels. Conversely, *Synthesis* reconstructs the data on level i by subdividing the control mesh of level $i-1$ and adding the details [**ZSS97**].

An important property of subdivision surfaces is that they can be naturally interpreted as functions on the domain defined by the base mesh (see Figure 3). This parametric interpretation is useful in many circumstances related to design, from derivation of differential quantities to dealing with constraints along arbitrary curves as described in section 3. Figure 3 illustrates this natural parameterization.

3. Surface Modeling Tools

3.1. Free-Form Editing. Free-form manipulation of 3D models is a popular method for modifying existing shapes which attempts to mimic to a certain extent the process of modeling or sculpting a physical object by hand. The applications are numerous, from animated character creation, to virtual restorations, to industrial design.

The sculpting metaphor for geometric modeling has its roots in the parametric surface works of Sabin [**Sab71**] and Bezier [**Béz74**] which contain early mentions of surface deformations. Subsequent work has spanned more than three decades and continues to be investigated in the context of modern systems and surface representations (e.g., [**Bar84, SP86, Coq90, HKD93, CR94, MJ96, SF98, Kob96b, ZSS97, PL97, QMV98, Tak98, WW98, MQ00, TO02, GS01, BMRB04**].

The basic idea of free-form modeling is to introduce a degree of transparency between the designer and the mathematical model of the surface being shaped. Instead of controlling the shape through a set of non-intuitive surface parameters, free-form deformations allow the shape to be controlled through intuitive

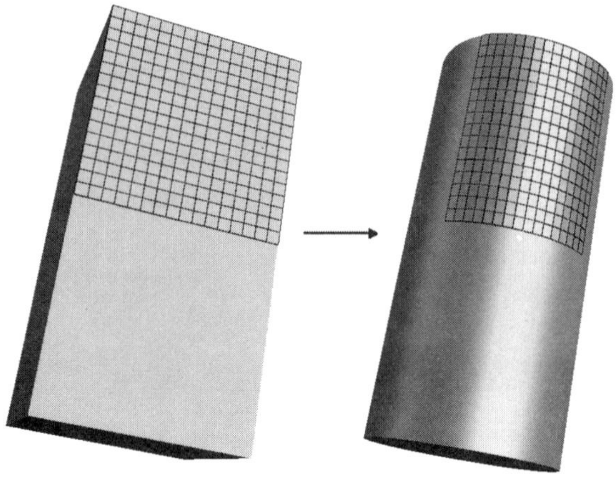

FIGURE 3. Natural parameterization of a subdivision surface. Each time we apply the subdivision rules to compute the finer control mesh we also apply midpoint subdivision to a copy of the initial control mesh. A mapping from a denser and denser subset of the control polyhedron (left) to the control points of a finer and finer control mesh (right) is obtained through repeated subdivision. In the limit, a map from the control polyhedron to the surface is obtained.

manipulation of the surface itself or the space surrounding it. The main challenge is to perform the manipulation through a limited set of controls and to define natural deformations of the surface away from the control positions. Different variations of this paradigm have been developed, including axial deformations [**Bar84, CST94, LCJ94**] which alter the axis of a shape to induce its deformation, lattice deformations [**SP86, Coq90, MJ96**] which operate on the cells of a space lattice to deform the volume inside the lattice, manipulations on scalar field embeddings [**HQ03**], control mesh editing methods which shape parameterically-defined surfaces by imposing constraints on their control meshes [**ZSS97**], and variational methods which operate by optimizing an energy functional over the surface under constraints [**Tak98, BMRB04**].

We focus our attention on methods that take advantage of subdivision representations and among these, we emphasize those that support interactive multiscale modeling. Subdivision representations are particularly suitable for free-form editing due to their hierarchical nature which easily accommodates multiscale edits, as well as their efficiency in terms of storage and access. For a survey of deformable models based on other representations see [**GM97**].

3.1.1. *Control mesh manipulations.* Manipulating control meshes offers a straightforward interface which supports interactive shape deformations. This approach has been extensively employed in spline-based modeling [**CRE01**] and can be naturally extended to subdivision surfaces. Collections of control mesh vertices, edges, and faces are re-positioned so as to induce modifications of the resulting limit surface. In addition, control points can be added and edges and faces can be split to increase the complexity of the shape as editing progresses. This type of manipulation is very common and can be found at the basis of commercial modeling packages with support for subdivision surfaces. It is routinely used for animated character design (e.g., in Discreet's 3D Studio Max [**dsm**], in Alias' Maya [**may**]) and is becoming increasingly popular for industrial modeling (e.g., in Dassault Systèmes' Catia [**cat**]). Figure 4 illustrates examples of shape modeling through control point manipulation.

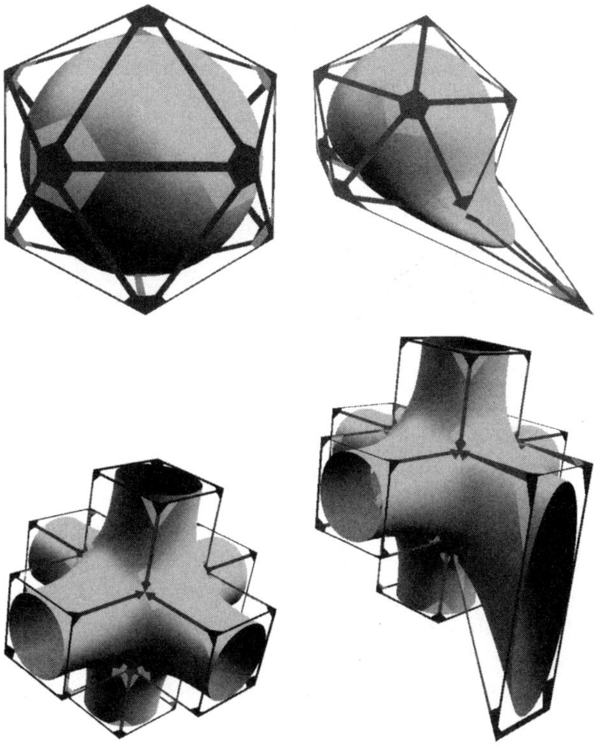

FIGURE 4. Shape modeling through control point manipulation: Loop subdivision surface (top), Catmull-Clark subdivision surface (bottom).

Single resolution control mesh manipulations offer only limited flexibility in designing shapes: only coarse shape deformations can be accommodated. Multiresolution subdivision surfaces are a much more powerful representation which lends itself very naturally to multiscale editing. Depending on the level at which the editing occurs, either a global deformation (coarse level) or a local deformation (fine level) is induced. This idea was exploited, for instance, in [**ZSS97, PL97**] for interactive multiresolution editing of Loop surfaces and in [**DKT98**] for Catmull-Clark ones. Using a combination of subdivision (i.e., transforming a coarse mesh into a finer one) and smoothing (i.e., transforming a fine mesh into a coarser one), edits performed at different levels of subdivision can be propagated through the hierarchy while keeping the magnitude of multiresolution details under control. Figure 5 illustrates edits at various scales performed on the Armadillo model.

Variations of this approach include modeling with *displaced subdivision surfaces* [**LMH00**] and subdivision surface *fitting* [**STKK99, LLS01a, MZ00**]. The

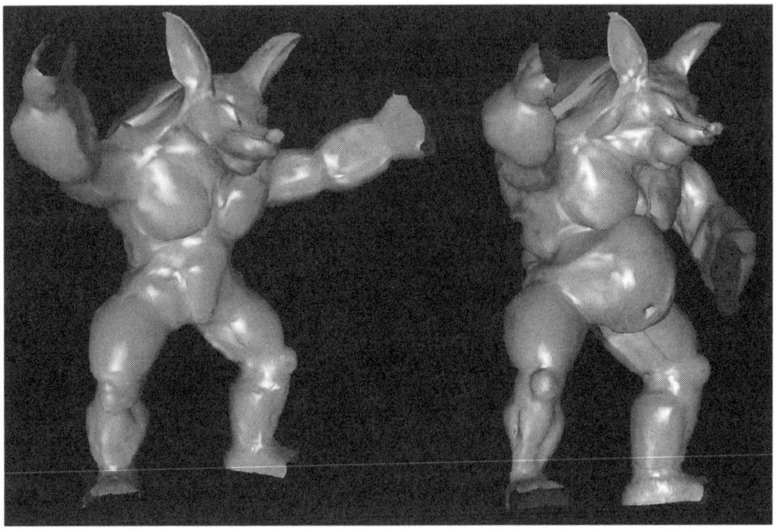

FIGURE 5. Multiresolution editing according to [**ZSS97**]: left – input model; right – editing result. Note the large-scale edit of the belly and the fine-scale edit around the chin.

displaced representation can be viewed as a restricted form of multiresolution subdivision consisting of a control mesh and a single level of scalar details. A domain surface is generated from the control mesh using Loop subdivision [**Loo87**]. A displacement map computed from the scalar displacement is then applied over the domain to generate the final surface. The displacements can be edited to create fine-level features on the surface, while control mesh edits lead to global shape alterations. In surface fitting a surface is deformed to conform to the shape of another given data set (e.g., points, curves, another surface). This approach is somewhat different than those discussed so far in that it is less suitable for interactive manipulation. Typically some optimization of the surface being fitted is performed in order to determine optimal control point positions which lead to a best fit between the surface and the target. The accuracy of the fit is controlled through a threshold parameter that bounds the error between the target and the fitted surface.

3.1.2. *Variational design.* Variational surface design operates on the principle of modifying a shape so that its *fairness* is optimized. Surface fairness is typically measured in terms of its energy and the idea is to find a minimum-energy state which, in turn, corresponds to the fairest possible shape. In Computer Graphics, energy-minimizing surfaces became popular in the context of simulating physical properties of materials [**Bar84, TF88, WW92**]. Celniker and Gossard [**CG99**] and later Welch and Witkin [**WW92**] pointed out the relationship between fair surface design and energy minimization.

Most commonly, fairness is expressed as an integral of a physical parameter associated with a real object bearing the shape of the surface [**Hal96**]. A widely used measure of fairness is the combination of *stretching* and *bending* energies:

$$Energy(S) = \alpha \int ||I||^2 dS + \beta \int ||II||^2 dS \tag{3.1}$$

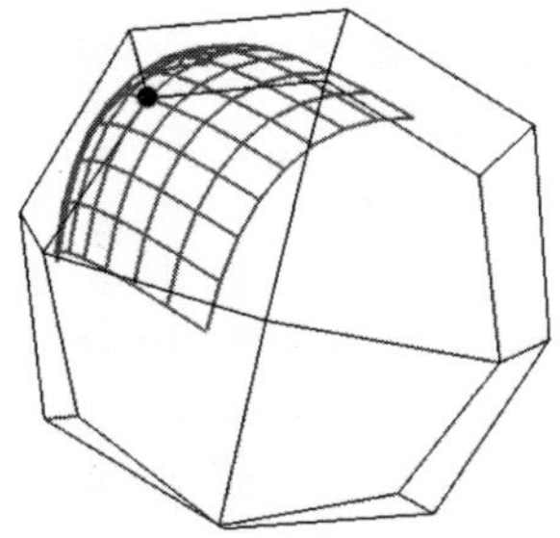

FIGURE 6. Local quadratic interpolant used to approximate first and second order derivatives [**BMRB04**].

where I and II denote the first and second fundamental forms of the surface and $||\cdot||$ is a suitably chosen matrix norm [**TPBF87**].

For practical purposes, discretized linear forms of equation (3.1) using parametric derivatives are typically employed:

$$E_{stretch} \approx \int_\Omega \left(\frac{\partial S}{\partial u}\right)^2 + \left(\frac{\partial S}{\partial v}\right)^2 dudv \tag{3.2}$$

$$E_{bend} \approx \int_\Omega \left(\frac{\partial^2 S}{\partial u^2}\right)^2 + 2\left(\frac{\partial^2 S}{\partial u \partial v}\right)^2 + \left(\frac{\partial^2 S}{\partial v^2}\right)^2 dudv \tag{3.3}$$

where Ω denotes the parametric domain of the surface S. Most variational approaches take advantage of these expressions, although alternative approaches have been proposed (e.g., [**CSA$^+$02**]). The main differences are in the types of parameterizations used to derive the differential quantities. For example, Greiner [**Gre94**] and later Kobbelt [**Kob96a**] suggested a discrete exponential map for local parameterizations (see Figure 6) such that each vertex P_0 has coordinates $(0,0)$ and its 1-ring neighbors $P_i \in R(P_0)$ are assigned coordinates:

$$(u_i, v_i) = e_i\left(\cos\left(\sum_{j\in R(P_0)} \alpha_j\right), \sin\left(\sum_{j\in R(P_0)} \alpha_j\right)\right) \tag{3.4}$$

where

$$\alpha_j = \frac{2\pi\angle(P_j^l P_0^l P_{j+1}^l)}{\sum_{j\in R(0)} \angle(P_j^l P_0^l P_{j+1}^l)}. \tag{3.5}$$

In the context of subdivision surfaces, Halstead et al. [**HKD93**] were among the first to describe a method for interpolating a given shape with a Catmull-Clark surface while minimizing surface fairness. Given the lack of a "natural" parameterization near extraordinary points, they re-formulated the stretch and thin-plate energy definitions in terms of the control meshes at different subdivision levels (rather than the limit surface). In their method subdivision is used to isolate extraordinary vertices and bi-cubic spline evaluation is used to evaluate the fairness norm away from such vertices.

Kobbelt [**Kob96b**] introduced the concept of *variational subdivision* to create interpolatory subdivision rules that place newly inserted vertices so as to minimize a global energy functional. Using a similar idea, Weimer and Warren [**WW98**] propose two schemes for variational subdivision of thin-plate splines. One scheme provides an exact solution to the variational problem, but the subdivision matrix has to be recomputed at every subdivision level. The other scheme is only approximate, but has the advantage that rules can be precomputed. Both schemes are restricted to rectilinear grids. Another method which connects subdivision with fairing and cascading multigrid methods was proposed in [**DMR02**]. The basic idea in this case is to interpret the evolution of the surface under curvature motion as a filtering proces.

Later on, Friedel et al. [**FMS03**] proposed using the characteristic map parametrization to construct first order data-dependent energies. This leads to a nonlinear minimization problem which is solved by re-writing the surface energy as a linear combination of precomputed stiffness matrices.

Constraints play an important role in variational design methods. In their absence, the optimization problem has a trivial solution, which usually leads to the collapse of the surface to a single point (an exception is the method of Boier-Martin et al. [**BMRB04**] in which the trivial solution corresponds to the input surface). We distinguish between two classes of constraints [**WW92**]:

- *Finite-dimensional*: involve point and normal constraints at discrete locations on the surface. These are the most commonly used. Point constraints are used to enforce spatial interpolation conditions. For subdivision surfaces such constraints typically correspond to control points and are easy to implement by solving linear systems. Normal constraints are used to enforce surface normals at certain points on a surface. Different approaches can be used to constrain normals: expressing the fact that two tangent vectors must be perpendicular to the prescribed normal, enforcing the positions of the vertices of a given face so that the face normal coincides with the prescribed one, or constraining tangent vectors rather than normals (the last two tend to over-constrain the problem).
- *Transfinite*: involve one or two-dimensional surface entities such as embedded curves and patches. Curve constraints are among the most common in this category. Enforcing such constraints involves solving an integral over the entity. For example, to constrain a surface curve $C(t) = S(u(t), v(t))$ along a given space curve $C_0(t)$, the following must be satisfied:

$$\int (C - C_0)^2 = 0 \tag{3.6}$$

Such constraints are usually discretized and enforced either by using a least-squares approach [**WW92**] or by reparameterizing the surface to align control points or edges with constraints [**BMRB04**] (see also Algorithm 1 in section 3.2). An alternative approach is to evaluate the curves and to incorporate the result of the evaluation into the subdivision rules to produce a limit surface that interpolates the curves. This is the object of *combined subdivision schemes* [**Lev99**] (see also [**Nas00, NA02, SWZ04**]).

Figure 7 illustrates the result of modeling with various types of constraints.

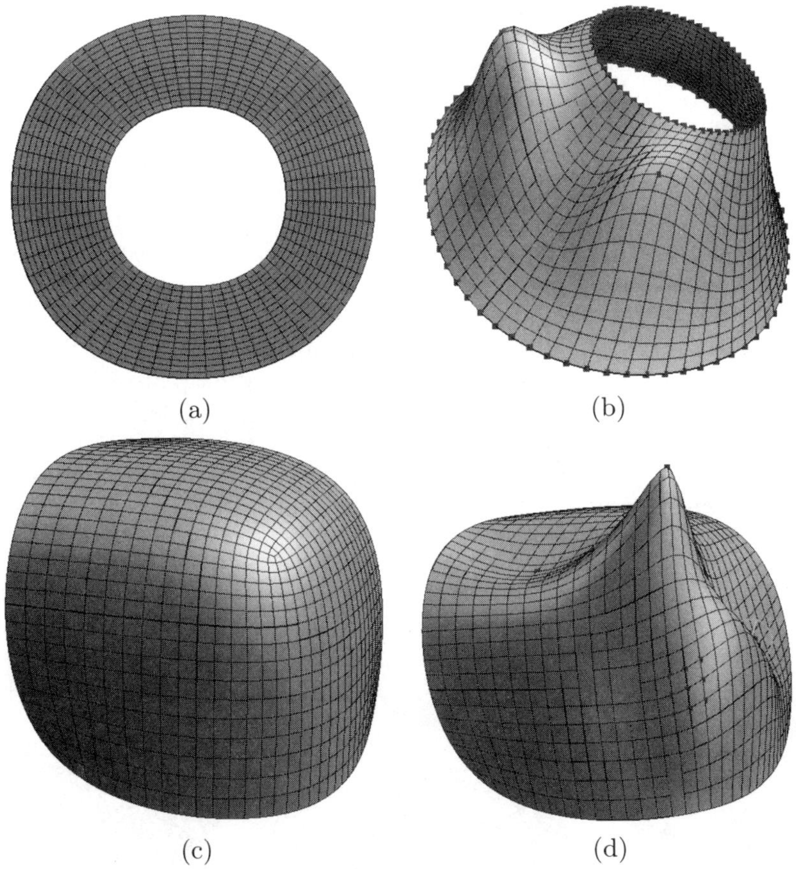

FIGURE 7. Constraint types: (a)-(b) point and discretized curve constraints; (c)-(d) normal constraints.

Another important consideration in dealing with constraints is the *region of influence* of a constraint. It is defined as the portion of the surface affected by the constraint. The region of influence can be explicitly enforced [**Kob00**] by letting the designer encircle an area on the surface. This generates boundary constraints between the surface inside the area of influence and the rest of the surface. Alternatively, in the case of hierarchical representations such as subdivision hierarchies, the region of influence can be controlled indirectly through the levels at which constraints are defined. For example, Takahashi et al. [**Tak98**] impose constraints at various scales using a wavelet framework. Constraints are being propagated from finer to coarser scales, however, the region of influence of each constraint is not controlled in any way. In [**BMRB04**] the influence of a constraint is explicitly enforced by the coarse level at which the constraint is propagated. Thus, more global or local edits can be performed depending on the level to which the constraint is restricted: a coarser level will induce a more global deformation, whereas a finer level will produce a more local edit (see Figure 8).

A related issue is that of detail preservation. When a global shape change occurs, it is often expected that the high frequency details are preserved over the modified surface. The face of Venus in Figure 9 is represented as a multiresolution subdivision surface in which non-trivial detail vectors capture the organic shape of the model. If a shape deformation is performed by pulling on a single point at the tip of the nose, a naive energy optimization approach leads to a fair shape that satisfies the constraints, but all the details of the face are lost (note that boundary constraints must also be imposed in this case to avoid the collapse of the surface to a single point). One solution is to separate the high-frequency information before optimization and to "re-apply" it to the new shape [**Kob00**]. This introduces an overhead related to saving and restoring surface details. To avoid this overhead, Boier-Martin et al. [**BMRB04**] propose to define a vector field of deformations over the surface and to optimize the energy of this vector field rather than the energy of the surface itself. Initially all deformation vectors are null. When an edit occurs, the corresponding deformation vector (i.e., at the tip of the nose) becomes non-null. The optimization procedure tries to smooth the deformation field under the constraints defined by the non-null vectors. Since the deformations are defined with respect to the detailed shape, the details are preserved during deformation. Note that, in this case, boundary constraints are not necessary as the rest shape in the absence of constraints is the input shape.

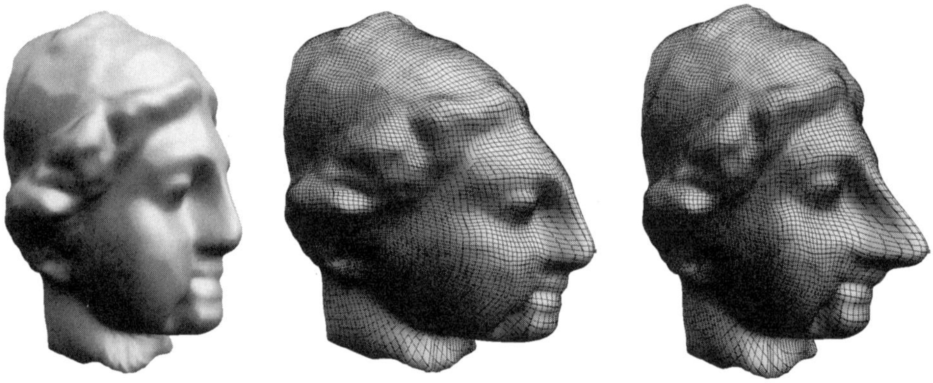

FIGURE 8. Region of influence of a multiresolution constraint: left – input model; middle – constraint is propagated to the coarsest subdivision level, inducing a global deformation of the head by pulling a single point on the nose; right – constraint is propagated only two levels coarser inducing a more localized edit.

An added advantage of subdivision hierarchies is that they facilitate the use of multigrid methods [**Bri87**] to solve the constrained minimization problem. In the presence of many constraints, however, even multigrid solvers may be too slow to yield results at interactive rates. A possible solution [**BMRB04**] is to aim for an approximate solution during interaction and a more accurate (non-interactive) result after the interaction has stopped. Figure 10 illustrates the differences between

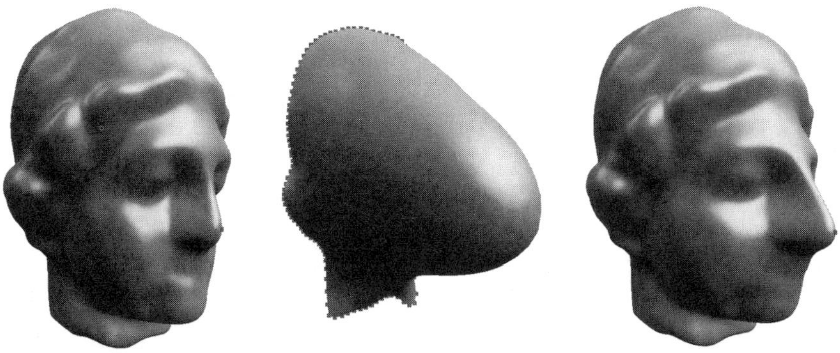

FIGURE 9. Energy optimization with constraints: left – input multiresolution subdivision surface with details; middle – optimization without detail preservation; right – optimization with detail preservation.

a Catmull-Clark approximation obtained at interactive rates and a more accurate multigrid minimization.

FIGURE 10. Computing a solution to the energy minimization problem with different accuracies: left - input model; middle - Catmull-Clark solution obtained interactively; right - multigrid solution.

For completeness, we mention the fact that the evolution of energy over time has also been considered to derive *dynamic surface models* [**TQ94, QT96**]. Dynamic models based on subdivision surfaces have been proposed by Qin et al. [**QMV98**]. Such models are typically too complex to support interactive design operations.

Topology modifications. The free-form modeling methods discussed so far operate by deforming the input surface without changing its topology. Some applications, however, may require topological modifications, such as creating handles and tunnels. An interactive sculpting environment which supports this type of edits was proposed in [**GOP99**]. The *Localized hierarchy Surface Splines* allow adding handles and punching holes, while maintaining C^1 continuity across the surface which is represented explicitly in piecewise polynomial or spline form. The main idea behind localized hierarchies is to allow local edits on locally refined mesh fragments based solely on coarser level data. Direct manipulation is performed by interacting directly with the surface rather than with control mesh. The types of operations supported include fillets, blends, semi-sharp features, extrusions, holes, and bridges.

Using meshes as an underlying representation, Guskov et al. [**GKSS02**] propose a user-driven procedure for inducing topological modifications in a semi-regular setting. The so-called *hybrid meshes* are multiresolution surface representations which enhance subdivision-based refinement operations with irregular operations that support changes in topology and approximate detailed features at multiple scales. In [**GKSS02**], hybrid meshes are defined as quadrilateral meshes on which regular $1-4$ face splits are combined with irregular operations through which groups of quads are removed and/or replaced.

3.2. Boolean Operations. Boolean operations provide a straightforward approach to creating complex models from simpler ones using intuitive combinations. Addition, subtraction, and intersection can be packaged into editing tools for modeling solids bounded by subdivision surfaces.

3.2.1. *Mesh-Based Approximations.* Traditionally, Boolean operations on boundary representations (B-reps) of solids have required intersecting parametric surfaces, removing the unwanted parts, and building new surfaces from the remaining ones. This approach presents a number of challenges, as intersections are difficult to perform for high-order B-reps and often lead to increasingly complex intersection curves. Exact matching of surfaces bordering such curves is also problematic, as it is not easy to ensure that curves in different parametric domains coincide in 3D. Consequently, subsequent editing of the resulting models may lead to unwanted artifacts in the surface (e.g., cracks) which require special handling.

A substantially simpler approach, proposed by Linsen [**Lin00**] is to use the control meshes corresponding to the parametric parts being combined, rather than the surfaces themselves. This implies that the intersections between solids are only approximately computed. At the same time, the problem of intersecting arbitrary surfaces translates into the much simpler one of intersecting arbitrary meshes. The meshes are first triangulated to avoid difficulties posed by handling of non-planar faces. Two approaches to building a combined control mesh are discussed: clipping triangles along the intersection boundaries and connecting intersection points and removing faces along the intersection curves and remeshing the resulting gaps. The latter has the advantage that it produces a more visually pleasing result. The main drawbacks in both approaches lie in the inefficiency of computing triangle-mesh intersections and robustness issues associated with such computations as well as gap filling for arbitrary gap topologies (see also [**LFKN03**] for variations on the topic of computing intersection curves for subdivision surfaces).

Using a similar control-mesh based approach, Biermann et al. [**BKZ01**] propose an approximate scheme for computing Boolean operations which deals with several important issues: matching the topology and the geometry of the intersection curve, fitting the resulting surface to the original data, and accurately capturing and representing sharp features in the result. The method uses piecewise-smooth multiresolution Loop [**Loo87**] subdivision surfaces to represent surfaces being combined. The algorithm assumes that each part being used in a Boolean operation is bounded by a closed orientable surface. It follows several steps:

(1) Compute intersection curves.
(2) Build resulting control mesh and compute an initial parameterization of the resulting surface over this mesh.
(3) Optimize the parameterization from the previous step.
(4) Use multiresolution fitting to approximate the input data as closely as possible.

For the first step, the authors improve on both the efficiency and the robustness of the naive mesh-mesh intersection approach by using bounding box hierarchies to accelerate computations and a perturbation scheme [**Sei98**] to increase robustness.

After determining the topology of the intersection, control meshes are merged with special consideration for several issues: preserving the topology of the cut, inserting a minimal number of new vertices, and keeping their valence small. The input control meshes are cut along intersection curves and a new control mesh is combined from the remaining pieces. The cutting process takes advantage of the natural parameterization of subdivision surfaces over their control meshes (see section 2) to approximate the intersection curve by alternating so-called *Snapping* and *Refinement* steps:

Algorithm 1 (snapping and refinement):

Given a domain mesh M and an intersection curve $c(t)$ in M
Repeat
    For each vertex v of a triangle intersected by c do
        1. Find $\alpha \in c$ closest to v
        2. Snap v to α if possible
    Adaptively refine parameterization
until (curve adequately approximated)

Snapping is performed between points of the curve and parametric mesh vertices, if they are sufficiently close. While optional, this step considerably reduces the complexity of the resulting domain (fewer faces). The role of the *refinement* is to increase the accuracy with which intersection curves are approximated. It is typically performed by midpoint subdivision of triangles which are intersected by curves multiple times or which fully contain curves. Figure 11 illustrates this process. The output of this step consists of piecewise linear approximations of the intersection curves, either along input edges or along newly introduced edges obtained by splitting triangles.

After cutting, the portions of the control meshes not required in the Boolean operation are removed and the meshes are joined along their boundaries. This is also done in two steps: vertices along one boundary are paired to corresponding vertices along the other boundary. When correspondences do not exist, triangles along the boundary are refined so as to introduce new vertices. Paired vertices close to one-another are merged together. During merging, intersection curves are also tagged with sharp feature tags (see also section 3.3).

By construction, the resulting merged control mesh constitutes a parameterization domain with the property that every one of its vertices belongs to one of

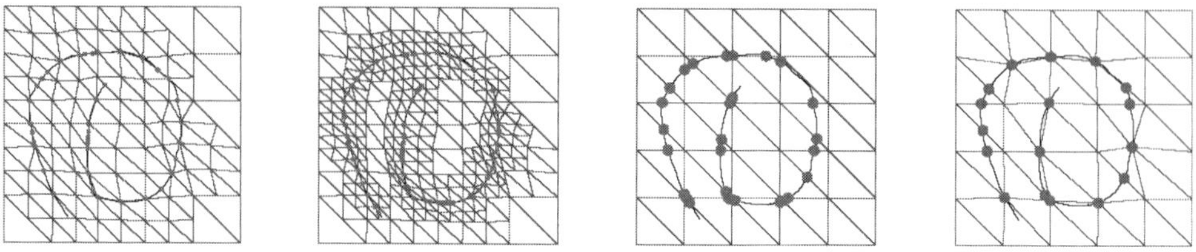

FIGURE 11. Refinement and snapping: two steps of refinement are shown on the left. The image of the curve in parameter space and vertex snapping are shown on the right (see [**BKZ01**]).

the original domains. However, the initial parameterizations of the parts of the input models corresponding to the Boolean operation may not be optimally parameterized over the new domain. An optimization procedure is used to reduce the distortion of the resulting surface over the new domain.

The last step of this method computes optimal positions of control points given the previously computed parameterization. The merged control mesh is subdivided a number of times and the resulting mesh is fitted to the original data in least-squares sense. Results of Boolean operations obtained with this method are shown in Figure 12.

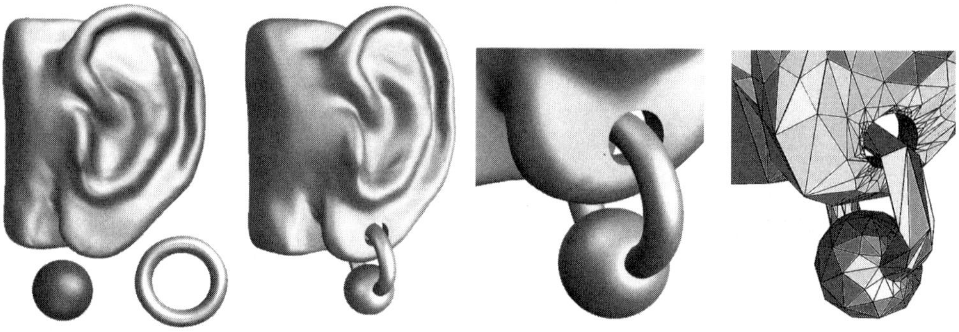

FIGURE 12. Boolean operations on multireolution subdivision surfaces [**BKZ01**].

3.2.2. *Surface Cut-and-Paste.* Surface pasting can be viewed as an instance of a Boolean operation. The basic paradigm implies creating new models by combining pieces of existing models. In its most basic form, a cut-and-paste operation involves selecting and transferring a *feature* of interest from a *source surface* to a *target surface.* There are several fundamental steps involved such an operation:

(1) Feature selection
(2) Separation of surfaces into base and detail parts
(3) Transferring the feature onto the target surface

The idea of pasting surfaces was first introduced in the context of hierarchical splines [**BBF94, CMB97**]. In this case a tensor-product B-spline surface is designated as the feature to be attached to another surface. Steps (1), (2) are assumed to have been performed in a pre-processing stage and (3) is achieved by representing tensor-product B-splines as Greville displacement B-splines [**BBF94**] and applying a mapping that takes into account the topology of the target surface and the Greville displacement representation of the feature [**BBF94**]. The main restriction is that there are no smoothness guarantees at the boundary between the feature and the target surface (not even C^0 continuity). One solution is to refine the feature surface so that its boundary better approximates the target. However, this amounts to introducing unnecessary control points over the entire feature (rather than only along boundaries), making subsequent processing of the feature very inefficient. An alternative solution was proposed by [**CM00**] and makes use of quasi-interpolation [**dF73**] to improve the result of pasting. In this case, interior feature control points are pasted using Greville displacements, while boundary points are pasted using quasi-interpolation. This leads to a composite surface which still exhibits discontinuities along the pasting boundary, however, less severe than in the original approach. In addition to the lack of continuity, the types of features that can be pasted are also limited by the underlying surface representation. Performance is also an issue due to expensive evaluations. An interactive spline-based interface was developed in [**Ma00**]. Due to performance limitations, the feature is not positioned directly onto the target surface, but rather is floating in its vicinity and the user is presented with a rough outline of the contour of the feature on the target. Once a position is decided upon, the actual pasting occurs.

Biermann et al. [**BMBZ02**] describe a more general procedure for cutting and pasting portions of existing surfaces using an intuitive approach, similar to those commonly used for 2D image cut-and-paste. The user initiates a cutting operation by selecting a feature of interest on an existing surface (termed the *source* surface (see Figure 13 (a)). She also specifies a position on a *target* surface where the source feature is to be pasted (see Figure 13 (b)). The actual pasting is performed in a sequence of steps (Figures 13(c)-(g)) which take advantage of the underlying semi-regular representation to achieve interactive rates. A discussion of the main steps follows.

Feature selection is performed interactively by the user who selects a region of interest on the source surface. A free-form closed space curve is used to outline the selection. The portion of the surface inside the curve constitutes the feature(s) of interest. The curve can have an arbitrary shape and does not have to be aligned with underlying mesh edges. The portion of the surface inside the curve must have disk topology, but it does not have to be a height field (see Figure 14).

Base / detail separation must be performed on both the source and target surfaces to define what constitutes feature detail as opposed to the larger-scale surface shape that should be ignored. Since this is largely dependent on the semantics of the operation, it is best left to the user. In [**BMBZ02**], the authors propose a continuum of base surface choices controlled by a *flatness* parameter. The base surfaces are obtained by smoothing the original surface to various degrees or by simple energy minimization within the feature boundary. Figure 15 illustrates the different effects obtained using different base surfaces for the source and target models.

Feature transfer is a complex process as it generally involves finding a mapping between two arbitrary surfaces. The solution proposed in [**BMBZ02**] is to build the mapping as a composition of maps to an auxiliary plane. The advantage of this approach is that it no longer requires parameterizing an arbitrary (source) surface onto another arbitrary (target) surface. Instead, flattening methods which

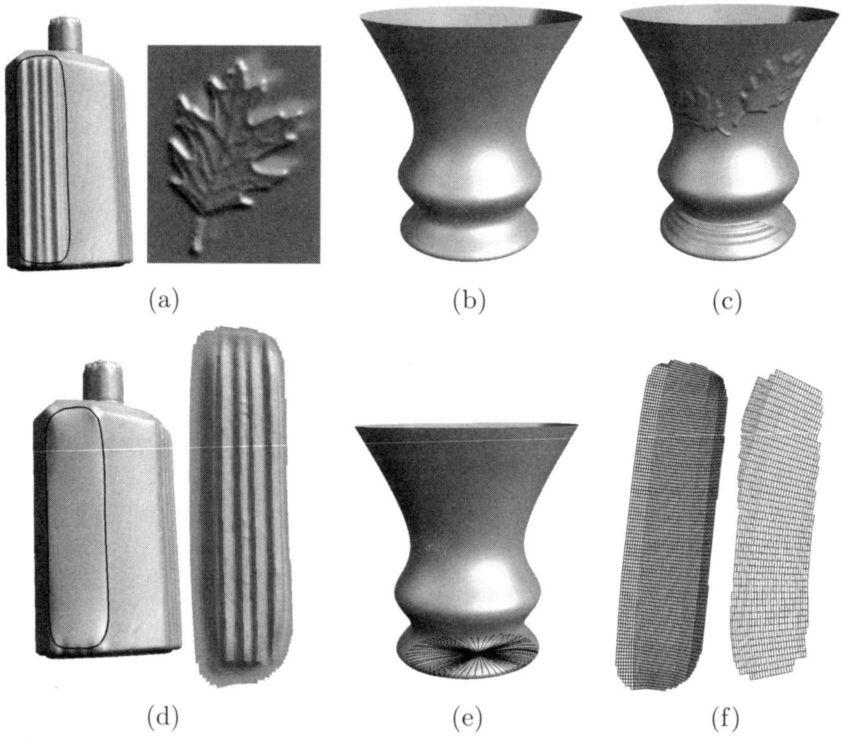

FIGURE 13. Feature-based design of an ornate vase: (a) input (source) surfaces; (b) target surface; (c) result after multiple pasting operations; Steps of a cut-and-paste sequence according to [**BMBZ02**] shown for the bottle pattern: (d) source feature selection; (e) finding a target region around a user-selected position on the target surface; (f) parameterization of source (left) and target (right) regions onto a common plane (shown displaced for illustration).

have been much more extensively researched [**SdS01, DMA02**] are needed. For the source surface the flattening is relatively easy to perform as the feature is already selected and homeomorphic to a disk. For the target surface the problem is more complicated as the surface can have any shape and can be quite large. In order to avoid flattening the entire target surface, the authors propose a method for approximating the portion of the target surface that is actually involved in pasting. As the user specifies a target location where the feature is to be pasted, the goal is to find a region that resembles the source feature in shape and size. To identify such a region, a generalized radial parameterization of the feature boundary is used [**BMBZ02**] (see also Figure 16). Once such a region is found, the transfer of the source feature onto the target model is done by aligning the planar parameterizations of the source and target regions followed by resampling the source feature onto the target connectivity.

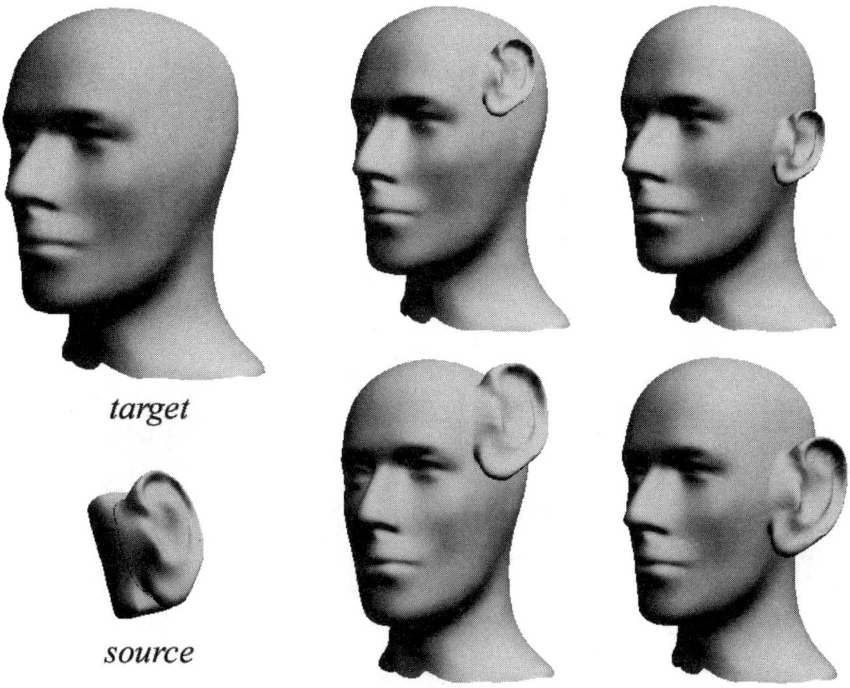

FIGURE 14. Pasting and interactively placing a complex feature: a digitized model of a clay ear constitutes the feature to be pasted onto the mannequin head. The ear can be interactively scaled, rotated, and translated on the surface of the head.

The pasting method of Biermann et al. [**BMBZ02**] provides a robust and efficient pipeline for interactive surface pasting. Its main constraints are related to self-intersections that may appear when features are pasted onto highly curved surfaces and to topological constraints on the features that can be pasted (i.e., only features with disk topology are handled). The former can be solved using a hierarchical pasting approach, in which the feature to be pasted is decomposed into frequency bands and the pasting is performed progressively, by pasting low-frequency details first, and high-frequency ones on top. The latter problem is more complex and requires more careful handling. A possible solution, albeit outside the subdivision framework, has been recently proposed in [**FMMY03**]. In this case, a volumetric approach is used to parameterize the feature and B-spline fitting is used to separate base from details. The advantage lies in the generality of features that are handled, including higher genus ones and the ability to paste them on highly curved areas. The main drawbacks are related to B-spline fitting and the need to introduce a large number of points in order to obtain a good fit. The result is not a seamless representation, but rather a composite one consisting of the original and the pasted part. In addition, the feature cannot be interactively dragged on the target surface.

3.2.3. *Surface Trimming.* Trimming, i.e., cutting holes in the surface of an object along specified curves, can also be considered as an instance of a Boolean

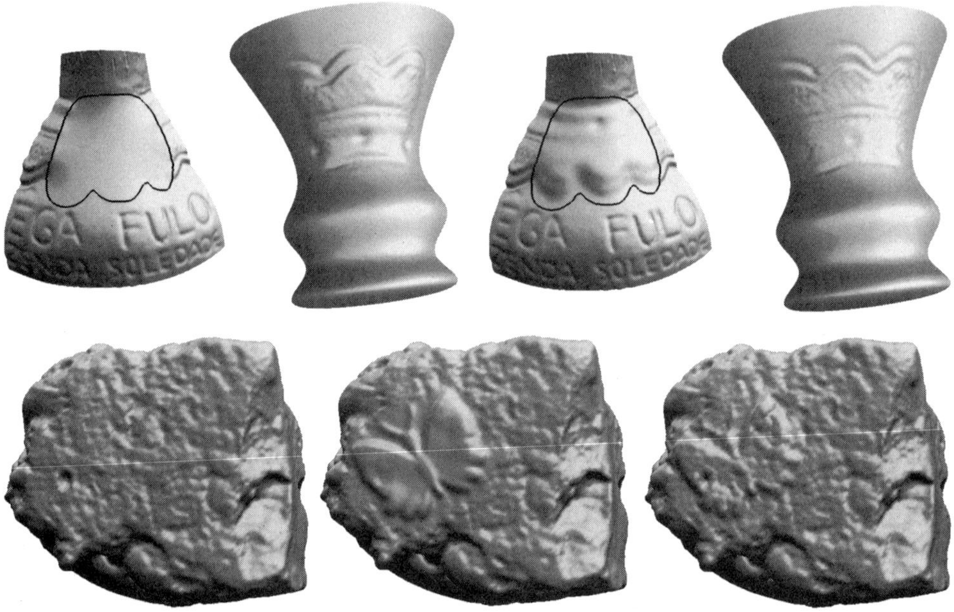

FIGURE 15. The effects of changing the base surface on the result of pasting: (top) digitized bottle detail appears on the vase differently, depending on the choice of base surface; (bottom) the butterfly feature is pasted on a rock model with and without preservation of target detail.

operation. Since this type of operation requires special subdivision rules along the trim boundary, we classify it as a special case of a non-smooth feature and we discuss it in section 3.3.

3.3. Non-Smooth Features. Subdivision surfaces can be naturally used to model smooth surfaces of arbitrary topological type. Many real objects, however, exhibit non-smooth features, such as sharp edges and boundaries, corners, and darts. While multiresolution detail vectors may be used to approximate sharp features (see Figure 17), a different setting is required to represent such features exactly. It entails altering the subdivision rules to produce limit surfaces that are only piecewise smooth, i.e., consist of smooth patches joined together along possibly sharp boundaries. To represent piecewise smooth surfaces, control mesh edges and vertices are typically *tagged* for special handling (see Figure 18). Special subdivision rules are employed in the vicinity of tagged mesh elements so as to avoid smoothing them. An edge can be tagged as a *crease edge* and vertices incident to crease edges may be tagged as one of the following (see Figure 18):

- *crease vertex*: exactly two crease edges join smoothly at this vertex
- *corner vertex*: two or more crease edges join non-smoothly at this vertex
- *dart vertex*: exactly one crease edge is adjacent to this vertex

Early mention of special rules for surfaces with boundaries appeared in the work of Doo [**Doo78**] and Nasri [**Nas91**], however accompanied only by partial analyses of the resulting surfaces. The first rules leading to provably C^1-continuous

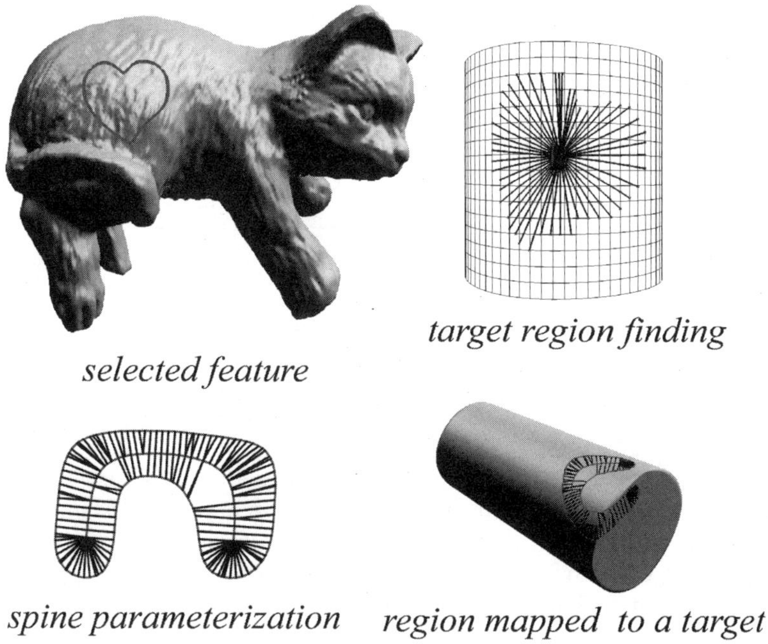

FIGURE 16. Finding a target region through radial parameterization of the feature outline.

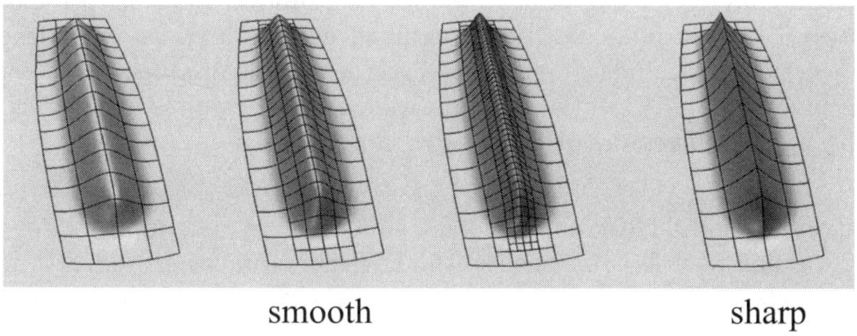

FIGURE 17. Multiresolution details are required to approximate sharp features using smooth surface representations (left three images). A piecewise smooth surface representation (right) allows sharp features to be modeled with detail vectors.

surfaces were defined in [**HDD$^+$94**] as a generalization of the Loop subdivision rules [**Loo87**]. The analysis of the resulting surfaces can be found in [**Sch96**]. As pointed out in [**BLZ00**], the rules introduced in [**HDD$^+$94**] have two main drawbacks: they are not suitable for modeling concave corners and the shape of the generated surface boundaries depends on the number of interior control points adjacent to each boundary point. The latter leads to undesirable gaps between surfaces joined along such a boundary. Both problems were handled by Biermann et al. [**BLZ00**] for the Loop [**Loo87**] and Catmull-Clark [**CC78**] subdivision schemes.

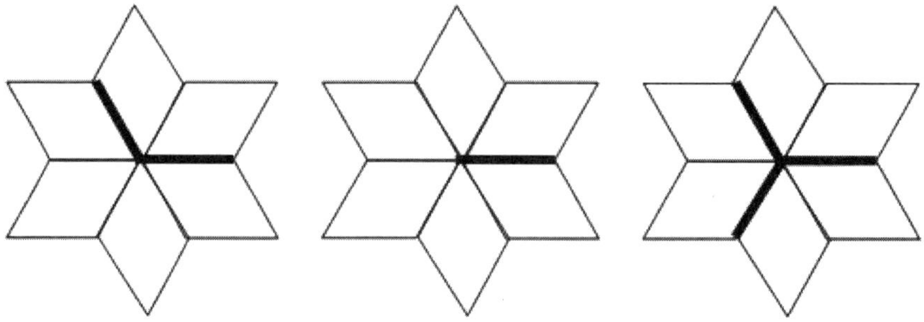

FIGURE 18. Mesh tags corresponding to (from left to right): crease, dart, and corner sharp features.

Modifications are sometimes applied to subdivision rules to achieve different effects. For example, deRose et al. [**DKT98**] propose an *edge sharpness* parameter s to vary sharpness along an edge and to allow for different degrees of sharpness. The parameter is used to blend between the positions p^{smooth} of a control point obtained with the smooth subdivision rules and a point p^{sharp} obtained with sharp subdivision rules:

$$p^{new} = (1-s)p^{smooth} + sp^{sharp},\ s \in [0,1]$$

Biermann et al. [**BLZ00**] propose a *flatness* parameter f and a *normal modification*. The flatness parameter control the speed at which control points in a neighborhood converge to the tangent plane. The subdivision rules are modified to blend between control point positions obtained without flatness modification and points in the tangent plane (p denotes the vector of control points in the neighborhood of a point,a_0, a_1, a_2 are the limit position and tangents at that point, and x^i denote the right eigenvectors of the subdivision matrix):

$$p^{new} = (1-f)p + f(a_0x^0 + a_1x^1 + a_2x^2),\ f \in [0,1]$$

The normal modification is somewhat similar, in that it interpolates between the control point position obtained without the modification and positions in a prescribed tangent plane (a normal n is prescribed through a pair of tangent vectors computed as $a'_i = (a_in)n$):

$$p^{new} = p + t((a'_1 - a_1)x^1 + (a'_2 - a_2)x^2),\ t \in [0,1]$$

Examples of sharp features modeled as proposed by Biermann et al. [**BLZ00**] are illustrated in Figure 19. A software library for piecewise smooth subdivision based on these rules is freely available from [**BZ99**].

A generalization of the subdivision concept that accommodates sharp features was developed by Sederberg et al. [**SZSS98**]. By drawing an analogy between recursive subdivision schemes and knot insertion for B-splines, the authors propose non-uniform versions of the Doo-Sabin [**Doo78**] and Catmull-Clark [**CC78**] subdivision schemes (under the general denomination of *non-uniform recursive subdivision surfaces* or NURSS). Each edge in a non-uniform Catmull-Clark control mesh (each control point in a Doo-Sabin mesh) is assigned a *knot spacing*. When all knot spacings are equal, the standard schemes are obtained. Two types of subdivision rules have to be considered for NURSS: the usual refinement scheme for the geometric positions of control points and an additional refinement scheme for knot spacings. Sharp features can be generated by setting certain knot spacings to zero.

The methods described so far require sharp features to be aligned with control mesh edges. Moreover, they provide little control over the profile of the resulting features. To address these limitations, Khodakovsky et al. [**KS99**] propose a curve-based feature editing approach. Feature curves are defined directly on the model surface through user interaction and can follow arbitrary paths, unconstrained by the connectivity of the underlying mesh. Features are obtained by perturbing the surface in the vicinity of feature curves. The curves can exist on multiple levels of a subdivision hierarchy. At each level, perturbations are computed with respect to local frames, so any coarse level modifications of the surface are carried through to finer levels. Several parameters are used to control the profile of a feature. In particular, sharp features can be obtained by specifying different normal directions for the profile on either side of the curve. This method brings forth a number of significant contributions with respect to previous approaches: it takes advantage of the multiresolution setting to define features through detail vectors at different levels, it does not impose any restrictions on the location of the feature curves on the surface or on their topology (curves can intersect or self intersect), and varying profiles allow both smooth and sharp features to be represented. The main drawback is that it does not preserve the input representation: after editing, the result is no longer a pure multiresolution subdivision surface, but rather a combined representation, consisting of a surface and a curve. This means that other subdivision-based tools that require as input a pure multireolution representation cannot be directly applied to the result of an editing operation performed with this method.

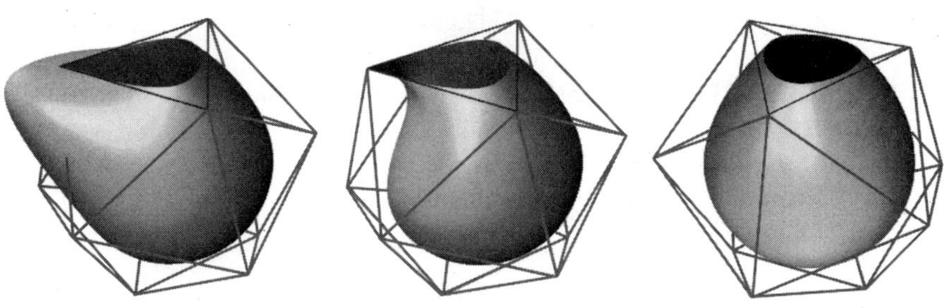

FIGURE 19. Sharp features generated with the rules proposed in [**BLZ00**]. From left to right: concave corner, convex corner, and smooth crease.

This problem is solved in [**BMZB02**] which uses the reparameterization idea described in section 3.2.1 to align the parameterization of the surface with the feature curves. Subsequently, sharp subdivision rules can be used along such curves. Figure 20 illustrates this process. An arbitrary feature curve is first projected onto the control mesh at some subdivision level (typically a coarse level which is subsequently refined). A piecewise linear approximation of the curve image in the parametric domain is computed by alternating *Snapping* and *Refinement* steps similar to those of Algorithm 1. The *Snapping* step moves mesh vertices onto the curve if they are sufficiently close, while the *Refinement* step subdivides the parameterization linearly. If $c : [0, 1] \to X$ denotes the image of a feature curve in the parameter domain X of the goal is to reparameterize the domain X such that c passes through the vertices of X. This reduces to finding a one-to-one mapping $\Pi : X \to X$ which maps vertices of X to curve points: $\Pi(v_i) = c(t_i)$, for some vertices $\{v_0, v_1, \dots\}$ and curve parameters $\{t_0, t_1, \dots\}$. After a finite number of iterations of snapping and refinement, the resulting curve $[v_0, v_1, \dots]$ is guaranteed to have the same topology as c and to follow along mesh edges (and / or diagonals in the case of the Catmull-Clark scheme). After reparameterization, the input surface is resampled according to the new parameterization. Intuitively, this moves the control mesh on the surface and places mesh vertices on the feature curve. Subsequently, the actual feature can be created by tagging the appropriate mesh edges and applying sharp subdivision rules.

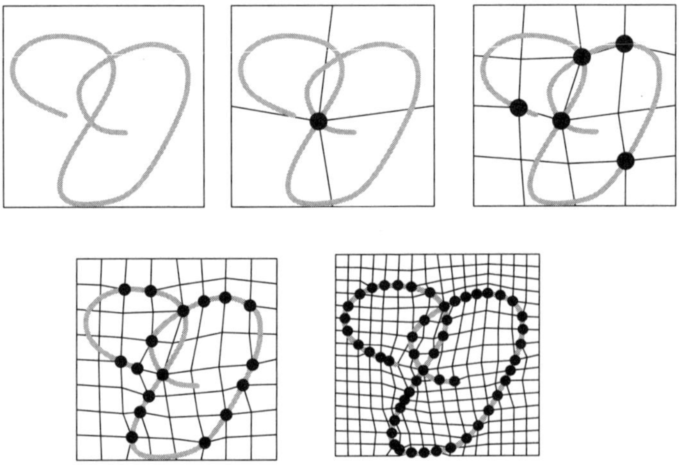

FIGURE 20. Reparameterization for approximating a feature curve: quads in parameter domain are recursively split and vertices are snapped to the curve. After several subdivision steps the curve is approximated by a sequence of vertices and follows along quad edges or diagonals.

In the case of Catmull-Clark meshes, there is an additional complication: the feature curve may pass through mesh diagonals after reparameterization (see Figure 21) and the standard crease rules do not support this situation. Biermann et al. [**BMZB02**] introduce new subdivision rules to deal with creases along quad diagonals. Sample results obtained with this method are shown in Figures 22. Note that the output surface is a multiresolution subdivision surface which can be manipulated with other tools designed to operate on such a representation. In addition, the framework is suitable not only for creating interior sharp features with various profiles (e.g., engravings, embossings), but also to create boundaries, i.e., to trim the input surface along the feature curves. An example of a trimmed surface is shown in Figure 22.

For completeness, we also mention the trimming method proposed by Litke et al. [**LLS01b**] which is complementary to that of Biermann et al. [**BMZB02**]. In this case quasi-interpolation is used to approximate a trimmed surface with a combined subdivision surface [**Lev99**].

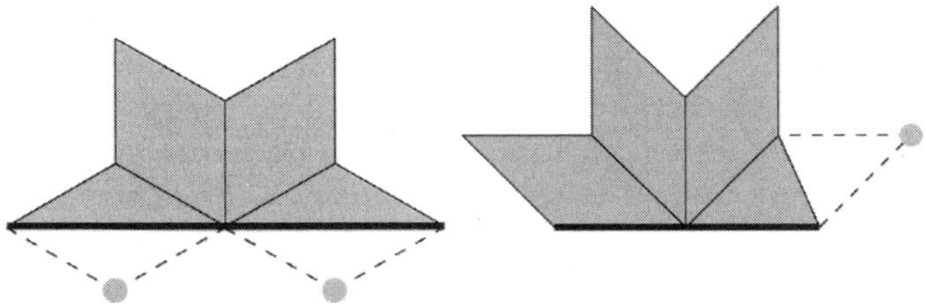

FIGURE 21. Standard sharp rules do not cover cases when the sharp edge (thick line) passes through a quad diagonal. New rules are necessary for such cases. Dotted lines and circles indicate vertices obtained by reflection used to define subdivision rules for such cases (see [**BMZB02**] for details).

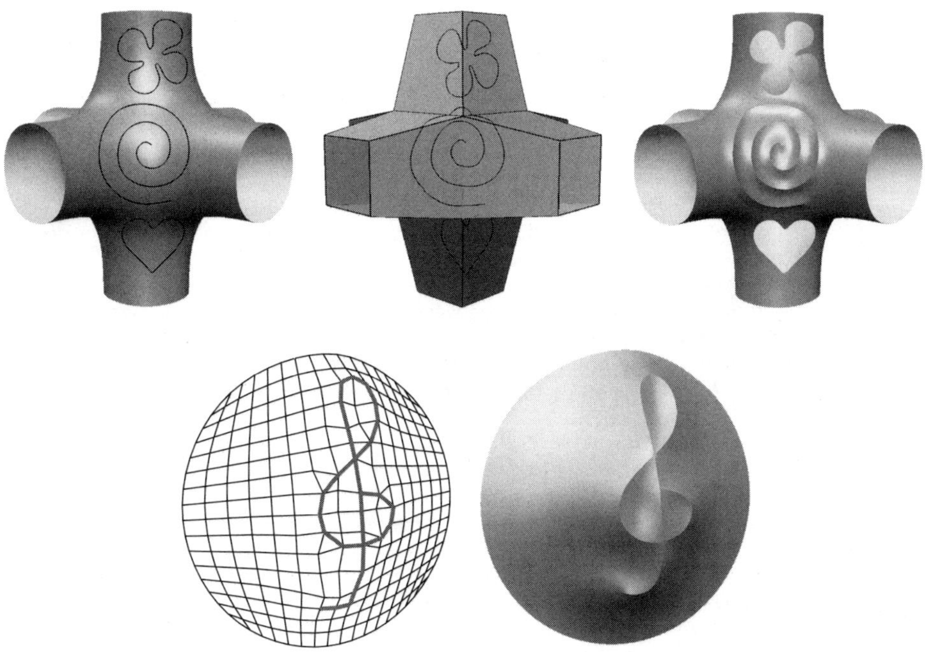

FIGURE 22. Surfaces obtained after trimming and embossing with sharp features using the method described in [**BMZB02**]. Top: input curves are shown on the surface (left) and projected into parameter space (middle). The surface obtained is shown on the right. Bottom: a self-intersecting feature.

4. Conclusions

In this paper we presented an overview of subdivision-based tools for surface design. We considered three major design categories: free-form deformations, Boolean operations, and surface decorations with sharp features. Such tools have multiple applications to animated character creation, virtual reconstructions and restorations, industrial design, shape modeling for medical and scientific simulations, etc. While modeling with subdivision surfaces has received a significant boost in recent years, there are a number of open issues still to be resolved. Among them are physical validation and interactive collision detection, guaranteed error bounds for surface fitting, and better integration between subdivision-based tools and methods operating on other representations.

References

[BAD+01] M. Boo, M. Amor, M. Doggett, J. Hirche, and Wolfgang Strasser, *Hardware support for adaptive subdivision surface rendering*, Proceedings of the ACM SIGGRAPH/EUROGRAPHICS Workshop on Graphics Hardware, 2001.

[Bar84] A. H. Barr, *Global and local deformations of solid primitives*, Proc. of SIGGRAPH 84 (1984), 21–30.

[BBF94] C. Barghiel, R. Bartels, and D. Forsey, *Pasting spline surfaces*, Mathematical Methods for Curves and Surfaces: Ulvik, Norway, Vanderbilt University Press, 1994, Available at ftp://cgl.uwaterloo.ca/pub/users/rhbartel/Paste.ps.gz, pp. 31–40.

[Béz74] P. Bézier, *Mathematical and practical possibilities of UNISURF*, CAD (1974), 127–152.

[BKS00] S. Bischoff, L. P. Kobbelt, and H.-P. Seidel, *Towards hardware implementation of loop subdivision*, Proc. of the ACM SIGGRAPH/EUROGRAPHICS workshop on Graphics Hardware, 2000, pp. 41–50.

[BKZ01] H. Biermann, D. Kristjansson, and D. Zorin, *Approximate boolean operations on free-form solids*, Proceedings of SIGGRAPH 01, August 2001, pp. 185–194.

[BLZ00] H. Biermann, A. Levin, and D. Zorin, *Piecewise smooth subdivision surfaces with normal control*, Proceedings of SIGGRAPH 00, 2000, pp. 113–120.

[BMBZ02] H. Biermann, I. Martin, F. Bernardini, and D. Zorin, *Cut-and-paste editing of multiresolution surfaces*, ACM TOG. Special issue for SIGGRAPH conference **21** (2002), no. 3, 312–321.

[BMRB04] I. Boier-Martin, R. Ronfard, and F. Bernardini, *Detail-preserving variational surface design with multiresolution constraints*, Proc. Shape Modeling International, SMI'04, 2004.

[BMZB02] H. Biermann, I. Martin, D. Zorin, and F. Bernardini, *Sharp features on multiresolution subdivision surfaces*, Graphical Models **64** (2002), no. 2, 61–77.

[Bri87] W. L. Briggs, *Multigrid tutorial*, SIAM, 1987.

[BZ99] H. Biermann and D. Zorin, *Subdivide 2.0 software*, 1999.

[cat] http://www.catia.ibm.com.

[CC78] E. Catmull and J. Clark, *Recursively generated B-spline surfaces on arbitrary topological meshes*, CAD **10** (1978), no. 6, 350–355.

[CG99] G. Celniker and D. Gossard, *Energy-based models for free-form surface shape design*, Proc. ASME Design Automation Conference, 1999.

[CM00] B. Conrad and S. Mann, *Better pasting via quasi-interpolation*, Curve and Surface Design: Saint-Malo, 1999 (Nashville, TN) (P.-J. Laurent, P. Sablonnière, and L. L. Schumaker, eds.), Vanderbilt University Press, 2000, pp. 27–36.

[CMB97] L. K. Y. Chan, S. Mann, and R. Bartels, *World space surface pasting*, Proceedings of Graphics Interface (W. Davis, M. Mantei, and V. Klassen, eds.), May 1997, pp. 146–154.

[Coq90] S. Coquillart, *Extended free-form deformation: a sculpturing tool for 3d geometric modeling*, Proc. of SIGGRAPH 90, 1990, pp. 187–196.

[CR94] Y. Chang and A. P. Rockwood, *A generalized de casteljau approach to 3D Free-Form deformation*, Proc. of SIGGRAPH 94 (1994), 257–260.

[CRE01] E. Cohen, R. F. Riesenfeld, and G. Elber, *Geometric modeling with splines*, A K Peters Ltd, 2001.

[CSA+02] F. Cirak, M.J. Scott, E.K. Antonson, M. Ortiz, and P. Schröder, *Integrated modeling, finite-element analysis, and engineering design for thin-shell structures using subdivision*, Computer Aided Design **43** (2002), 137–148.

[CST94] M. M. Chang, M. I. Sezan, and A. M. Tekalp, *Adaptive bayesian segmentation of color images*, Journal of Electronic Imaging **3** (1994), 404–414.

[dF73] C. deBoor and G. J. Fix, *Spline approximtion by quasiinterpolants*, Journal of Approximation Theory **8** (1973), 19–45.

[DKT98] T. DeRose, M. Kass, and T. Truong, *Subdivision surfaces in character animation*, Proceedings of SIGGRAPH 98, 1998, pp. 85–94.

[DL02] N. Dyn and D. Levin, *Subdivision schemes in geometric modelling*, Acta Numerica **11** (2002).

[DLG90] N. Dyn, D. Levin, and J. A. Gregory, *A butterfly subdivision scheme for surface interpolation with tension control*, ACM TOG **9** (1990), no. 2, 160–169.

[DMA02] Mathieu Desbrun, Mark Meyer, and Pierre Alliez, *Intrinsic parameterizations of surface meshes*, Eurographics conference proceedings, 2002, pp. 209–218.

[DMR02] U. Diewald, S. Morigi, and M. Rumpf, *A cascadic geometric filtering approach to subdivision*, CAGD **19** (2002), no. 9, 675 – 694.

[Doo78] D. Doo, *A subdivision algorithm for smoothing down irregularly shaped polyhedrons*, Proceedings on Interactive Techniques in Computer Aided Design (Bologna), 1978, pp. 157–165.

[DS78] D. Doo and M. Sabin, *Analysis of the behaviour of recursive division surfaces near extraordinary points*, CAD **10** (1978), no. 6, 356–360.

[dsm] http://www.discreet.com.

[FMMY03] Y. Furukawa, H. Masuda, K. T. Miura, and H. Yamato, *Cut-and-paste editing based on constrained b-spline volume fitting*, Proceedings Computer Graphics International, 2003.

[FMS03] I. Friedel, P. Mullen, and P. Schroder, *Data-dependent fairing of subdivision surfacesn*, Proc. of SM 03, 2003.

[GKSS02] I. Guskov, A. Khodakovsky, P. Schröder, and W. Sweldens, *Hybrid meshes: Multiresolution using regular and irregular refinement*, Proc. Symp. Comp. Geom., 2002, pp. 264–272.

[GM97] S. F. F. Gibson and B. Mirtich, *A survey of deformable modeling in computer graphics*, Tech. Report TR-97-19, MERL, Cambridge, MA, 1997.

[GOP99] C. Gonzalez-Ochoa and J. Peters, *Localized-hierarchy surface splines (less)*, Proc. Symp. on Interactive 3D Graphics, 1999, pp. 7–15.

[Gre94] G. Greiner, *Surface construction based on variational principles*, Wavelets, Images and Surface Fitting (P. J. Laurent, A. LeMéhauté, and L. Schumaker, eds.), AK Peters, 1994, pp. 277–286.

[GS01] E. Grinspun and P. Schröder, *Normal bounds for subdivision-surface interference detection*, Proc. of IEEE Visualization 01, 2001.

[Hal96] M. A. Halstead, *Efficient techniques for surface design using constrained optimization*, Ph.D. thesis, Univ. of California at Berkeley, 1996.

[HDD$^+$94] H. Hoppe, T. DeRose, T. Duchamp, M. Halstead, H. Jin, J. McDonald, J. Schweitzer, and W. Stuetzle, *Piecewise smooth surface reconstruction*, Computer Graphics Proceedings, Annual Conference Series, ACM Siggraph, 1994, pp. 295–302.

[HKD93] M. Halstead, M. Kass, and T. DeRose, *Efficient, fair interpolation using Catmull-Clark surfaces*, Proc. SIGGRAPH 1993, 1993, pp. 35–44.

[HQ03] J. Hua and H. Qin, *Free-Form deformations via sketching and manipulating scalar fields*, Proc. ACM Symp. on Solid Modeling and Applications, 2003, pp. 328 – 333.

[Kob96a] L. Kobbelt, *Interpolatory subdivision on openquadrilateral nets with arbitrary topology*, Proc. of Eurographics 96, 1996, pp. 409–420.

[Kob96b] L. Kobbelt, *A variational approach to subdivision*, Comput. Aided Geom. Design **13** (1996), no. 8, 743–761. MR **97j**:65029

[Kob00] L. P. Kobbelt, *Discrete fairing and variational subdivision for freeform surface design*, The Visual Computer **16** (2000), no. 3-4, 142–150.

[KS99] A. Khodakovsky and P. Schröder, *Fine level feature editing for subdivision surfaces*, Proceedings of ACM Solid Modeling, 1999.

[LCJ94] F. Lazarus, S. Coquillart, and P. Jancene, *Axial deformations: an intuitive deformation technique*, CAD **26** (1994), no. 8, 607–61.

[LDW97] M. Lounsbery, T. DeRose, and J. Warren, *Multiresolution analysis for surfaces of arbitrary topological type*, ACM TOG **16** (1997), no. 1, 34–73.

[Lev99] A. Levin, *Interpolating nets of curves by smooth subdivision surfaces*, Proc. of SIGGRAPH 99 (1999), 57–64.

[LFKN03] S. Lanquetin, S. Foufou, H. Kheddouci, and M. Neveu, *Computing subdivision surface intersection*, Proc. WSCG '03, 2003.

[Lin00] L. Linsen, *Netbased modelling*, Proc. SCCG '00, 2000, pp. 259–266.

[LLS01a] N. Litke, A. Levin, and P. Schröder, *Fitting subdivision surfaces*, Proc. of IEEE Visualization 2001, October 2001, pp. 319–324.

[LLS01b] N. Litke, A. Levin, and P. Schroder, *Trimming for subdivision surfaces*, Computer Aided Geometric Design **18** (2001), no. 5, 463–481.

[LMH00] A. Lee, H. Moreton, and H. Hoppe, *Displaced subdivision surfaces*, Proc. of SIGGRAPH 00, 2000, pp. 85–94.

[Loo87] C. Loop, *Smooth subdivision surfaces based on triangles*, Master's thesis, University of Utah, Department of Mathematics, 1987.

[Ma00] M. Ma, *The direct manipulation of pasted surfaces*, Master's thesis, University of Waterloo, Waterloo, Ontario, Canada N2L 3G1, 2000, Available on WWW as ftp://cs-archive.uwaterloo.ca/cs-archive/CS-2000-15/.

[may] http://www.alias.com.

[MJ96] R. MacCracken and K. I. Joy, *Free-Form deformations with lattices of arbitrary topology*, Proc. of SIGGRAPH 96, 1996, pp. 181–188.

[MQ00] K.T. McDonnell and H. Qin, *Dynamic sculpting and animation of Free-Form subdivision solids*, Proc. of IEEE Computer Animation, 2000.

[MZ00] W. Ma and N. Zhao, *Catmull-clark surface fitting for reverse engineering applications*, Proceedings of Geometric Modeling and Processing, 2000, pp. 274–282.

[NA02] A. H. Nasri and A. Abbas, *Designing Catmull-Clark subdivision surfaces with curve interpolation constraints*, Computers and Graphics (2002).

[Nas91] A. H. Nasri, *Surface interpolation on irregular networks with normal conditions*, CAGD **8** (1991), 89–96.

[Nas00] A. Nasri, *Interpolating meshes of boundary intersecting curves by subdivision surfaces*, The Visual Computer **16** (2000).

[PB00] D. Piponi and G. Borshukov, *Seamless texture mapping of subdivision surfaces by model pelting and texture blending*, Proceedings of SIGGRAPH 00, 2000, pp. 471–478.

[Pet00] Jörg Peters, *Patching Catmull-Clark meshes*, Proceedings of SIGGRAPH 00, 2000, pp. 255–258.

[PL97] K. Pulli and M. Lounsbery, *Hierarchical editing and rendering of subdivision surfaces*, Tech. Report UW-CSE-97-04-07, Dept. of CS&E, University of Washington, Seattle, WA, 1997.

[PR97] J. Peters and U. Reif, *The simplest subdivision scheme for smoothing polyhedra*, ACM TOG **16** (1997), no. 4.

[PS96] K. Pulli and M. Segal, *Fast rendering of subdivision surfaces*, Proc. Eurographics Workshop on Rendering, 1996, pp. 61–70.

[QMV98] H. Qin, C. Mandal, and B. Vemuri, *Dynamic Catmull-Clark subdivision surfaces*, IEEE TVCG **4** (1998), no. 3, 215–229.

[QT96] H. Qin and D. Terzopoulos, *D-NURBS: A physics based framework for geometric design*, IEEE TVCG **2** (1996), no. 1, 85–96.

[Sab71] M. Sabin, *Interrogation techniques for parametric surfaces*, Advanced computer graphics - economics, techniques and applications (R. D. Parslow and R. E. Green, eds.), Plenum Press, 1971, pp. 1095–1118.

[Sab02] M. A. Sabin, *Subdivision surfaces tutorial*, SMI 02 (2002).

[Sch96] J. E. Schweitzer, *Analysis and application of subdivision surfaces*, Ph.D. thesis, University of Washington, Seattle, 1996.

[SDS96] E. J. Stollnitz, T. DeRose, and D. H. Salesin, *Wavelets for computer graphics: theory and applications*, Morgan Kaufmann, 1996.

[SdS01] A. Sheffer and E. de Sturler, *Parameterization of faceted surfaces for meshing using angle based flattening*, Engineering with Computers **17(3)** (2001), 326–337.

[Sei98] R. Seidel, *The nature and meaning of perturbations in geometric computing*, Discrete Computational Geometry **19** (1998), no. 1, 1–17.

[SF98] K. Singh and E. Fiume, *Wires: A geometric deformation technique*, Proc. of SIGGRAPH 98, 1998, pp. 405–414.

[SP86] T. W. Sederberg and S. R. Parry, *Free-form deformation of solid geometric models*, Proc. of SIGGRAPH 86, 1986, pp. 151–160.

[Sta98] J. Stam, *Exact evaluation of catmull-clark subdivision surfaces at arbitrary parameter values*, Proceedings of SIGGRAPH 98, July 1998, pp. 395–404.

[STKK99] H. Suzuki, S. Takeuchi, F. Kimura, and T. Kanai, *Subdivision surface fitting to a range of points*, Proc. Pacific Graphics, 1999.

[SWZ04] S. Schaeffer, J. Warren, and D. Zorin, *Lofting curve networks using subdivision surfaces*, submitted (2004).

[SZSS98] T. W. Sederberg, J. Zheng, D. Sewell, and M. Sabin, *Non-uniform recursive subdivision surfaces*, Proceedings of SIGGRAPH 98, 1998, pp. 387–394.

[Tak98] S. Takahashi, *Variational design of curves and surfaces using multiresolution constraints*, The Visual Computer **14(5/6)** (1998), 208–227.

[TF88] D. Terzopoulos and K. Fleischer, *Modeling inelastic deformation: Viscoelasticity, plasticity, fracture*, Proc. of SIGGRAPH '88, 1988, pp. 269–278.

[TO02] G. Turk and J. O'Brien, *Modelling with implicit surfaces that interpolate*, ACM TOG **21** (2002), no. 4, 855 – 873.

[TPBF87] D. Terzopoulos, J. Platt, A. Barr, and K. Fleischer, *Elastically deformable models*, Proc. of SIGGRAPH 87, 1987, pp. 205–214.

[TQ94] D. Terzopoulos and H. Qin, *Dynamic NURBS with geometric constraints for interactive sculpting*, ACM TOG **13** (1994), no. 2, 103–136.

[WW92] W. Welch and A. Witkin, *Variational surface modeling*, Proceedings of SIGGRAPH '92, vol. 26, 1992, pp. 157–166.

[WW98] H. Weimer and J. Warren, *Subdivision schemes for thin-plate splines*, Proc. EUROGRAPHICS '98 **17** (1998), no. 3.

[WW01] J. Warren and H. Weimer, *Subdivision methods for geometric design: A constructive approach*, Morgan Kaufmann, 2001.

[ZK02] D. Zorin and D. Kristjansson, *Evaluation of piecewise smooth subdivision surfaces*, The Visual Computer **18** (2002), no. 5–6, 299 – 315.

[ZSD+00] D. Zorin, P. Schröder, T. DeRose, L. Kobbelt, A. Levin, and W. Sweldens, *Subdivision for modeling and animation*, SIGGRAPH'00 Course Notes, 2000.

[ZSS96] D. Zorin, P. Schröder, and W. Sweldens, *Interpolating subdivision for meshes with arbitrary topology*, Proc. of SIGGRAPH 96, August 1996, pp. 189–192.

[ZSS97] D. Zorin, P. Schröder, and W. Sweldens, *Interactive multiresolution mesh editing*, Proc. of SIGGRAPH 97, 1997, pp. 259–268.

Ioana Boier-Martin, IBM T. J. Watson Research Center, Hawthorne, NY 10532
E-mail address: ioana@us.ibm.com

Denis Zorin, Media Research Lab, New York University, New York, NY 10003
E-mail address: dzorin@mrl.nyu.edu

Fausto Bernardini, IBM T. J. Watson Research Center, Hawthorne, NY 10532
E-mail address: fausto@us.ibm.com

DIMACS Series in Discrete Mathematics
and Theoretical Computer Science
Volume **67**, 2005

Sample Based Geometric Modeling

Tamal K. Dey

This paper is dedicated to my family.

ABSTRACT. A new paradigm for digital modeling of physical objects from sample points is fast emerging due to recent advances in scanning technology. Researchers are investigating many of the traditional modeling problems in this new setting. We name this new area as *sample based geometric modeling.* The purpose of this article is to expose the readers to this new modeling paradigm through three basic problems, namely surface reconstruction, medial axis approximation and shape segmentation. The algorithms for these three problems developed by the author and his co-authors are described.

1. Introduction

Many applications in science and engineering need computer modeling of physical objects for simulations, analysis and designs. Until recently, most applications modeled these objects synthetically. An alternative approach is gaining popularity where physical objects are transported directly to the computer world through scanners. Recent advances in scanning technology are making this approach more viable and attractive. These scanners can generate a set of dense sample points on the surface of an object very quickly. These points need to be processed for producing a digital copy of the object and subsequently for other modeling purposes. We call the area dealing with these problems as *sample based geometric modeling*, or SBGM in short.

Many problems in traditional geometric modeling need to be revisited to account for the new input type, namely point cloud data. Researchers are increasingly responding to this need with new algorithms and results [**2, 3, 20, 23, 45**]. In this article, we focus on three such problems, (i) surface reconstruction, (ii) medial axis approximation and (iii) shape segmentation and matching. We will mainly describe the algorithms developed by the author and his co-authors for these problems. Although this description does not circumscribe all developments in SBGM, we hope that the readers will be exposed to a snapshot of SBGM through it.

2000 *Mathematics Subject Classification.* Primary 65D17, 65D18; Secondary 68U05, 68W99.

Key words and phrases. Sample, surface reconstruction, medial axis, shape segmentation, Voronoi diagram, Delaunay triangulation.

The author was supported in part by NSF Grants DMS-0138456, DMS-0310642 and an ARO grant DAAD19-02-0347.

SURFACE RECONSTRUCTION. Given a finite set P of points sampled from a surface Σ, the problem of surface reconstruction asks for a piecewise approximation of Σ using only P. The computed surface should have the exact topology and approximate geometry of Σ. Several algorithms for the problem have been designed. Among them the earlier ones mainly concentrated on the performance and very little on the theoretical guarantees [**11, 16, 24, 36**]. Amenta and Bern [**3**] first proposed an algorithm called CRUST with theoretical guarantees. Following this development, a number of other algorithms were designed which also provide theoretical guarantees [**4, 12, 5**]. Among them is the COCONE algorithm of Amenta, Choi, Dey and Leekha which improved CRUST algorithmically and also its proof of correctness [**4**]. We will describe the COCONE and its variants in section 3.

MEDIAL AXIS APPROXIMATION. The medial axis of a surface Σ consists of the centers of all balls that are maximally empty of points from Σ. The medial axis provides a compact representation of a shape. Researchers have found its use in a number of applications such as image processing [**42, 43**], computer vision [**10**], solid modeling [**32, 48**], mesh generation [**28**], robotics [**31**] and animations [**49**]. Exact computation of the medial axis is difficult in general though some attempts for some special cases have been made [**17, 35**]. Even these algorithms have to deal with the numerical instability associated with the medial axis computations. This instability can be attributed in part to the fact that the medial axis is sensitive to tiny changes in shape [**30, 53**]. Consequently, efforts have been made to approximate them. We present one such algorithm called MEDIAL by Dey and Zhao [**23**] in section 4.

SHAPE SEGMENTATION AND MATCHING. Let $\text{In}(\Sigma)$ denote the shape bounded by a surface Σ. Mathematically, $\text{In}(\Sigma)$ is the closure of the bounded component of $\mathbb{R}^3 - \Sigma$. The shape segmentation problem seeks to partition $\text{In}(\Sigma)$ into identifiable 'features' given some form of input representation of Σ. We are interested in the case where Σ is given by a sample. Many applications including object recognition, classification, matching, tracking need shape segmentations, see for example [**1, 9, 39, 50, 51, 52**]. Different structures such as shock graphs [**46**], medial axis [**40**], Reeb graphs [**37**] and shape distribution [**44**] have been proposed for the problem. All of these techniques require an input mesh. Thus, they are not suitable to point cloud data unless a surface is reconstructed out of the input points. It is preferable to bypass the surface reconstruction step since it may introduce artifacts on the output mesh, specially when dealing with noise. We describe an algorithm by Dey, Giesen and Goswami [**20**] in section 5 which segments a shape from its sample.

The shape matching problem where a query shape is to be matched against a pool of shapes can take advantage of the shape segmentation. If two similar shapes are segmented similarly, the segments can be compared to match shapes. We describe such a shape matching algorithm based on the work of Dey, Giesen and Goswami [**20**] in section 5.

2. Preliminaries

SAMPLED SURFACE. In what follows Σ denotes a smooth, compact surface without boundary which is embedded in $\mathbb{R}^3$. The outward normal to Σ at any point $x \in \Sigma$ is denoted $\mathbf{n}_x$. A *sample* of Σ is a finite set $P \subset \Sigma$ of discrete points. Sometimes we also refer to them as *point cloud data.*

MEDIAL AXIS. Let $X \subset \mathbb{R}^3$ be the set of all points so that for each $x \in X$, there are at least two points p and q in Σ with $\|x - p\| = \|x - q\| = \min_{y \in \Sigma} \|x - y\|$. The *medial axis* M of Σ is the closure of X. To exclude the pathological cases, we assume that Σ is semi-analytic [**15, 41**]. Under this assumption, the medial axis has a finite graph structure. The points of M are the centers of balls whose interiors are maximally empty with respect to Σ. The boundaries of these balls meet Σ only tangentially. We refer to these balls as *medial balls.* Each point in Σ has two medial balls, one touching it from outside and the other touching it from inside. It follows that the line going through a point $p \in \Sigma$ and the centers of the medial balls touching Σ at p is normal to Σ at p. See Figure 1 for an illustration in 2D.

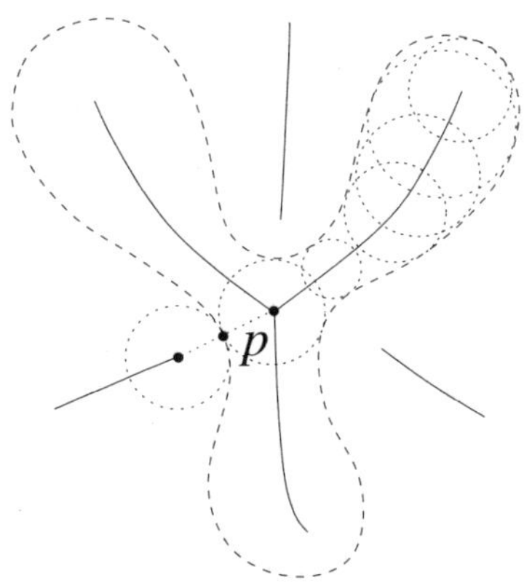

FIGURE 1. A curve in 2D (dashed), its medial axis (solid) and some medial balls (dotted).

LOCAL FEATURE SIZE AND ε-SAMPLE. Obviously, a sample P needs to be dense enough to contain information about the features of Σ if Σ is to be reconstructed from P. Following [**3**] the local feature size $f()$ is defined as a function $f : \Sigma \to \mathbb{R}$ where $f(x)$ is the distance of $x \in \Sigma$ to the medial axis M. Intuitively, $f()$ measures how complicated Σ is locally. When Σ is semi-analytic, $f()$ is positive everywhere on Σ.

A sample P of Σ is an ε-sample if each point $x \in \Sigma$ has a sample point p with $\|x - p\| \leq \varepsilon f(x)$. It is observed that $\varepsilon \leq 0.4$ is sufficient for approximating Σ with a piecewise linear surface in practice [**18**].

VORONOI DIAGRAM. For a finite set of points $P \subset \mathbb{R}^3$, the Voronoi diagram is defined as follows. The *Voronoi cell* of $p \in P$ is given as

$$V_p = \{x \in \mathbb{R}^3 \,:\, \forall q \in P - \{p\},\, \|x - p\| \leq \|x - q\|\}.$$

The sets V_p are convex polyhedra. Closed faces shared by j Voronoi cells for $1 \leq j \leq 4$ are called $(4-j)$-dimensional *Voronoi faces.* The 0-, 1-, 2-dimensional Voronoi faces are called Voronoi *vertices, edges and facets* respectively. The *Voronoi diagram* Vor P of P is the collection of all Voronoi faces of all dimensions.

DELAUNAY TRIANGULATION. The *Delaunay triangulation* of a set of points P is a dual to the Voronoi diagram of P. Assuming general position, the convex hull of $1 \leq j \leq 4$ points defines a $(j-1)$-dimensional *Delaunay simplex* if the intersection of their corresponding Voronoi cells is not empty. The 0-, 1-, 2-, 3-dimensional Delaunay simplices are called *Delaunay vertices, edges, triangles and tetrahedra* respectively. They define a decomposition of the convex hull of all points in P called the *Delaunay triangulation* Del P.

In general, the computation of the Voronoi/Delaunay diagrams takes $O(n^2)$ time and space for n points in three dimensions. However, when the points are uniformly (locally) sampled from a smooth, generic surface, the complexity goes down to $O(n \log n)$, see Attali, Boissonnat and Lieutier [**6**]. By locally uniform we mean that inter-point distances are both upper and lower bounded by a fraction of local feature size. In practice the samples are mostly locally uniform and thus the Voronoi/Delaunay diagram computations in the algorithms to be described do not take a prohibitive amount of computational resources. In fact, with the help of CGAL libraries [**54**], all algorithms have been implemented and typically, they take a minute or two on today's PCs for point cloud data containing several hundred thousands of points.

RESTRICTED DIAGRAMS. For a sample P of Σ the *restricted Voronoi diagram* $\mathrm{Vor}\, P|_\Sigma$ is the CW-complex defined by the collection of *restricted Voronoi cells* $\{V_p|_\Sigma = V_p \cap \Sigma\}$ and their lower dimensional faces. Dual to the restricted Voronoi diagram is the *restricted Delaunay triangulation* $\mathrm{Del}\, P|_\Sigma$. It is a simplicial complex where $\sigma \in \mathrm{Del}\, P|_\Sigma$ if and only if σ is the convex hull of a set of vertices $R \subseteq P$ where $\bigcap_{q \in R} V_q|_\Sigma \neq \emptyset$.

A result of Edelsbrunner and Shah [**25**] relates the topology of the restricted Delaunay triangulation to the sampled surface as stated in Theorem 2.1. We say that $\mathrm{Vor}\, P|_\Sigma$ satisfies the *topological ball property* if each Voronoi face of dimension d intersects Σ in a closed topological $(d-1)$-ball or in an empty set.

THEOREM 2.1. *The underlying space of* $\mathrm{Del}\, P|_\Sigma$ *is homeomorphic to* Σ *if* $\mathrm{Vor}\, P|_\Sigma$ *satisfies the topological ball property.*

POLES. It is an important result proved in [**3**] that the Voronoi cells are elongated along the normal direction to the surface at the sample points if the sample is sufficiently dense. The definition of *poles* plays an important role in approximating these normals [**3**].

The *pole* p^+ of a sample point p is the farthest Voronoi vertex in the Voronoi cell V_p. The *negative pole* p^- is the farthest among all Voronoi vertices that satisfies $(p^+ - p) \cdot (p^- - p) < 0$. If V_p is unbounded, p^+ is taken at infinity. The vector $\mathbf{v}_p = p^+ - p$ is called the *pole vector* for p and its direction is taken as the average of all directions of infinite edges in case V_p is unbounded, see Figure 6.

ANGLES. We will use the notation $\angle \mathbf{a}, \mathbf{b}$ to denote the acute angle between the supporting lines of two vectors $\mathbf{a}$ and $\mathbf{b}$.

3. Surface reconstruction

First, we describe the COCONE algorithm that assumes the input sample to be dense, that is, an ε-sample for sufficiently small ε. The output is a triangulated surface interpolating the input points with topological and geometric guarantees. In practice, the input point set may suffer from undersampling and noise and thereby may not be a dense sample. We will describe two other algorithms that take undersampling and noise into account.

3.1. Cocone algorithm. The COCONE algorithm developed by Amenta, Choi, Dey and Leekha [**4**] is based on some of the concepts introduced by Amenta and Bern [**3**] for the CRUST algorithm. Amenta and Bern proved that when P is a dense sample, $\angle \mathbf{v}_p, \mathbf{n}_p$ is small for any sample point $p \in P$. This is a crucial ingredient of the COCONE algorithm.

A *cocone* C_p for a point $p \in P$ is defined as the subset $\{x\} \subseteq V_p$ where $\angle(x-p), \mathbf{v}_p \geq \frac{3\pi}{8}$. In other words, it is the intersection of V_p with the complement of a double cone with apex p, axis $\mathbf{v}_p$ and an opening angle $\frac{3\pi}{8}$, see Figure 2(b). Since $\mathbf{v}_p$ approximates the normal to the surface at p, the cocone C_p (clipped within V_p) approximates a thin neighborhood around the tangent plane at p. For each point p, the algorithm then determines all Voronoi edges in V_p that are intersected by the cocone C_p. The dual Delaunay triangles of these Voronoi edges constitute the set of candidate triangles T.

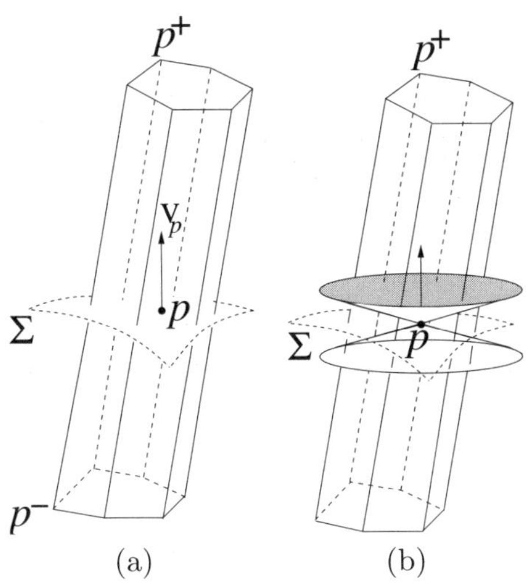

FIGURE 2. A long thin Voronoi cell V_p, the positive pole p^+, the pole vector $\mathbf{v}_p$ and the negative pole p^- (a), the cocone (b).

When the input point set P is an ε-sample of Σ for $\varepsilon \leq 0.06$, the candidate triangles enjoy some nice properties [**4**]. The circumscribing disks of the candidate triangles are small. Specifically, if $pqr \in T$ has a circumradius r, then

$$(i)\ r = O(\varepsilon) f(x) \text{ where } f(x) = \min\{f(p), f(q), f(r)\}.$$

It turns out that any triangle with such a small circumradius must lie flat to the surface, i.e., if $\mathbf{n}_{pqr}$ is the normal to a candidate triangle pqr, then

$$(ii)\ \angle \mathbf{n}_{pqr}, \mathbf{n}_x = O(\varepsilon) \text{ where } x \in \{p, q, r\}.$$

Also, it is proved that

$$(iii)\ T \text{ includes all restricted Delaunay triangles of } \operatorname{Del} P|_\Sigma.$$

The set T almost approximates Σ but may not form a surface. Nevertheless, the set T includes all restricted Delaunay triangles of $\operatorname{Del} P|_\Sigma$ according to the property (iii). Theorem 2.1 says that $\operatorname{Del} P|_\Sigma$ is homeomorphic to Σ if the topological ball property holds. Amenta and Bern show that if P is an ε-sample for $\varepsilon \le 0.06$, each Voronoi face in $\operatorname{Vor} P$ satisfies this property. This means that, if the triangles in $\operatorname{Del} P|_\Sigma$ can be extracted from T, a surface homeomorphic to S can be obtained. Unfortunately, it is impossible to detect the restricted Delaunay triangles of $\operatorname{Del} P|_\Sigma$ since Σ is unknown. However, the fact that T contains them is used in extracting a 2-manifold out of T. The algorithm for extracting a 2-manifold N from T is described in [**3**]. This 2-manifold N is output as the approximation of Σ. The following theorem holds for any 2-manifold N extracted out of T.

THEOREM 3.1. *If P is an ε-sample of a smooth, compact surface $\Sigma \subset \mathbb{R}^3$ without boundary where $\varepsilon \le 0.06$, the following properties hold:*

(i) *N is homeomorphic to Σ,*
(ii) *any point $x \in \Sigma$ is within $\varepsilon f(x)$ distance of a point in N and any point in N is within $\varepsilon f(x)$ distance of a point x in Σ.*

COCONE uses a single Voronoi diagram as opposed to two in CRUST and also eliminates a normal filtering step. Because of the Voronoi diagram computation, the COCONE runs in $O(n^2)$ time and space when there is no assumption on the distribution of sample points. Funke and Ramos [**29**] improved its complexity to $O(n \log n)$ though the resulting algorithm is not very practical. If we assume that the sample is locally uniform, the COCONE algorithm as described here runs in $O(n \log n)$ time and space thanks to the result of Attali et al. [**6**].

3.2. Undersampling. The assumption of ε-sampling for sufficiently small ε often does not hold in practice. This undersampling may be caused by inadequate attention during sampling process, machine error, or nonsmoothness. Even boundaries in a surface may be viewed as the demarkation where appropriate sampling stops and the undersampling begins. Dey and Giesen [**18**] took this unified view to detect boundaries that identify the regions of undersampling.

Given a sample P of a surface Σ, the subset $\Sigma' \subseteq \Sigma$ is called *well sampled* if each point x in Σ' has a sample point within $\varepsilon f(x)$ distance. If P undersamples Σ, the well sampled surface Σ' has boundaries. A point $p \in P$ is a *boundary sample* if V_p intersects the boundary of Σ', otherwise p is *interior*. The algorithm of Dey and Giesen [**18**] works in two phases to detect all boundary samples. In the first phase, it selects a set $R \subseteq P$ based on two conditions. The first condition requires the Voronoi cell of a point in R be long and thin and the second one requires its pole vector agree with those of all its neighbors on the surface (determined by cocones). These two conditions ensure that R consists of interior points only. In a second phase, the set R is expanded to include more points by relaxing the second condition. It is proved that under some mild assumptions on sampling, this algorithm determines all interior points and the rest of the points are correctly detected as boundary.

Once the boundary samples are detected, the COCONE algorithm is employed to filter the candidate triangles. Boundary samples are not allowed to choose any triangle. This produces the boundaries at the undersampled regions.

3.3. Water-tight Surface. Most of the surface reconstruction algorithms face difficulty while dealing with undersampled surfaces and noise. While the modified COCONE algorithm as described above can detect undersampling, it leaves holes in the surface near the vicinity of undersampling, see the second and third rows of Figure 3. Although this may be desirable for reconstructing surfaces with boundaries, many applications require that the output surface be *water-tight.* This means that the output surface bounds a solid.

Amenta, Choi and Kolluri [**5**] designed the POWER CRUST algorithm to produce water-tight surfaces. Let the circumscribing balls of the Delaunay tetrahedra be called *Delaunay balls.* The Delaunay balls with centers at poles are called *polar balls.* The set of polar balls can be partitioned into two sets, the *inner* polar balls that have centers in $\mathrm{In}(\Sigma)$, and the *outer* polar balls that have centers in the complement. Recall that $\mathrm{In}(\Sigma)$ is the closure of the bounded component of $\mathbb{R}^3 - \Sigma$. The algorithm works on the principle that the boundaries of the union of inner and outer polar balls have exact topology and approximate geometry of Σ if the sample is dense. It computes the weighted Voronoi diagrams of the inner and outer polar balls and output the Voronoi faces that are shared by both diagrams.

A drawback of the POWER CRUST is that it introduces Steiner points in the output. In some cases, where the input size is already large, this increase in the output size is undesirable. Dey and Goswami [**19**] devised a water-tight surface reconstructor called TIGHT COCONE which does not introduce any extra points. This algorithm first computes the surface with the modified COCONE. Recall that COCONE may leave some holes in the surface due to undersampling. A subsequent sculpting in the Delaunay triangulation of the input points finds the Delaunay triangles that fill up the holes. Unlike POWER CRUST, TIGHT COCONE does not add any Steiner point. Some results are shown in Figure 3.

The algorithm runs as follows. The COCONE algorithm as described in the subsection 3.2 computes a preliminary surface possibly with holes and other artifacts at the undersampled regions. The sample points in the rest of the surface have their neighborhoods well approximated. Specifically, the set of surface triangles incident to these points form a *topological disk.* We call the points whose incident surface triangles form a topological disk *good.* The rest of the points are called *poor.* The union of surface triangles incident to a good point p is denoted as U_p. After the preliminary surface computation, the algorithm goes through two phases, namely marking and peeling to fill the holes.

MARKING. In this phase the Delaunay tetrahedra of $\mathrm{Del}\, P$ are marked as *in* or *out.* This is done by a walk through the Delaunay triangulation in a depth first manner using the vertex and triangle adjacencies. It maintains a stack of pairs (p, σ) where p is a good point and σ is a tetrahedron incident to p marked *out.* Suppose the pair (p, σ) is currently popped out from the stack. The surface patch U_p locally separates the tetrahedra incident to p into two clusters, one on each side. The cluster that contains σ is marked *out* since σ is already marked *out.* The other cluster gets the marking *in.* This is done by initiating a local walk from σ that traverses all tetrahedra through triangle adjacency without ever crossing a triangle

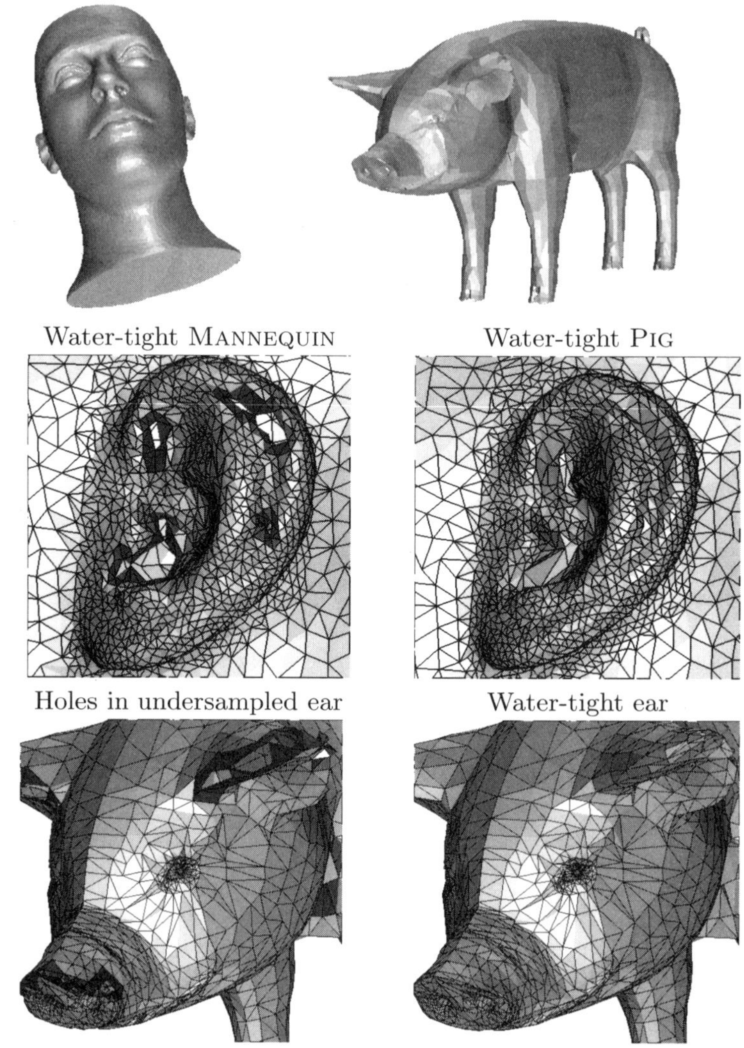

FIGURE 3. Results of TIGHT COCONE. First row shows the water-tight reconstruction of MANNEQUIN and PIG data. Second and third row show the holes in the preliminary surface computed by COCONE and how they are filled with TIGHT COCONE. Triangles bordering the holes are shaded darker.

in U_p and marking each tetrahedron as *out*. The rest of the tetrahedra that are not encountered in this walk get the *in* marking. During this local walk in the *out* cluster, when a vertex q of U_p is reached through a tetrahedron σ', the pair (q, σ') is put into the stack if q is good and is not explored yet. One should notice that some tetrahedra whose all vertices are poor are not reached by the above walk. These tetrahedra are handled with a slightly different marking step. The details are described in Dey and Goswami [**19**].

PEELING. After the marking step, a walk is initiated to peel off tetrahedra that are marked *out*. The boundary of the union of the remaining tetrahedra form the water-tight surface. This is also the surface of the union of the peeled tetrahedra.

3.4. Noise. So far, we assumed that the points in a sample always lie on the surface Σ. In practice, the samples are often noisy with points scattered around the surface. The ROBUST COCONE algorithm [**19**] is designed to handle such data sets. This algorithm has theoretical guarantees which is based on the following noise model.

NOISY (ε, κ)-SAMPLE. Suppose $\tilde{p}$ denotes the nearest point of a point $p \in \mathbb{R}^3$ on Σ. We say $P \in \mathbb{R}^3$ is a *noisy* (ε, κ)-sample of Σ for two positive constants k_1 and k_2 if the following sampling conditions hold.

(i) $\tilde{P} = \{\tilde{p} \,|\, p \in P$ is an ε-sample of $\Sigma\}$,
(ii) $\|p - \tilde{p}\| \leq k_1 \varepsilon f(\tilde{p})$,
(iii) $\|p - q\| \geq k_2 \varepsilon f(\tilde{p})$ for any two points p, q in P where q is the κth nearest sample point to p among all points in P.

The first condition says that the projection of the point set P on the surface makes a dense sample and the second says that P is close to the surface. The third condition makes the sampling locally uniform. Notice that, for $\kappa = 1$ this condition prohibits any two sample points to be arbitrarily close. But, that may be a severe restriction for samples in practice. This is why κ is introduced to make the locally uniform condition less restrictive. In practice κ is taken in the range of three to five.

ROBUST COCONE first filters Delaunay tetrahedra from $\mathrm{Del}\, P$ whose union approximates $\mathrm{In}(\Sigma)$. To understand the rationale behind the method, first assume that P is noise-free. As mentioned earlier, the inner polar balls approximate $\mathrm{In}(\Sigma)$. In absence of Σ, one needs a mechanism to separate the inner poles from the outer ones in order to obtain an approximation of $\mathrm{In}(\Sigma)$. Here the following *intersection depth* property comes to the rescue. Two inner (outer) polar balls cricumscribing adjacent tetrahedra intersect deeply while an inner and an outer polar ball intersect only in a shallow manner. The depth of intersection is measured by the angle at which the boundaries of the balls intersect. A breadth-first search starting from an unbounded polar ball, which is guaranteed to be outer, can collect all of them if the walk moves from an outer polar ball to an adjacent polar ball only if they intersect deeply.

The above algorithm does not work when P is noisy. A main reason for this is that the intersections between polar balls do not follow the intersection depth property if P is noisy. Dey and Goswami [**21**] circumvent this difficulty by observing that, even under noise, where the points are not perturbed more than a small fraction of the local feature size, some of the Delaunay balls behave like polar balls as in the noise-free case. These balls are identified by their relative size compared to the nearest neighbor distances. Let B be a Delaunay ball cricumscribing four points $p_i, i = 1, .., 4$. We say B is *big* if the radius of B is more than $\rho > 0$ times bigger than the average nearest neighbor distances of any of p_i. The average is taken over three nearest neighbors and ρ is taken 1.5 in the implementation. Big Delaunay

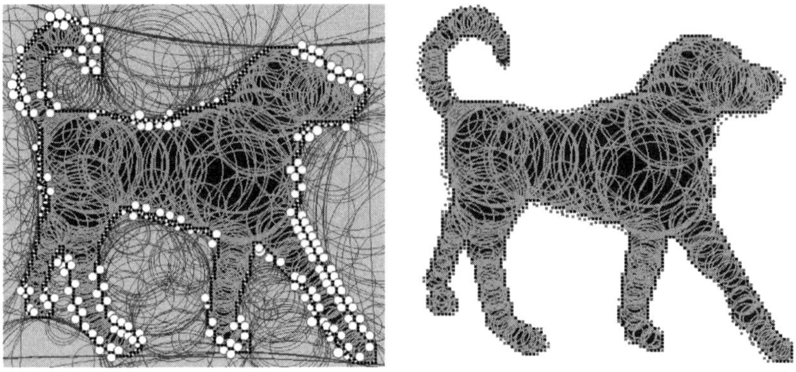

FIGURE 4. Delaunay balls where small balls are shaded white (left), union of inner big Delaunay balls (right).

balls are partitioned by the intersection depth property as in the noise-free case. A Delaunay ball is *inner* if its center lies in $\mathrm{In}(\Sigma)$ and is *outer* otherwise. It is proved that, under the specified noise model, big Delaunay balls satisfy the intersection depth property. The following theorem becomes the basis of the ROBUST COCONE algorithm.

THEOREM 3.2. *Let U be the boundary of the union of inner big Delaunay balls. The following conditions hold for U and Σ.*

(i) *U is homeomorphic to Σ,*

(ii) *Each point x in Σ is within $O(\varepsilon)f(x)$ distance from a point in U. Conversely, each point in U is within $O(\varepsilon)f(x)$ distance from a point $x \in \Sigma$.*

It follows from this theorem that the union of inner big Delaunay balls approximate $\mathrm{In}(\Sigma)$. A depth-first walk starting from an unbounded outer Delaunay ball collects all big outer Delaunay balls using the intersection depth property. The rest of the big Delaunay balls are inner. The top row of Figure 4 illustrates how the inner big Delaunay balls approximate a shape. The points $P' \subseteq P$ on the boundaries of the inner big Delaunay balls are collected. ROBUST COCONE applies the TIGHT COCONE on P' to reconstruct Σ. The results are shown in Figure 5.

4. Medial axis

Most algorithms which approximate the medial axis from samples use the Voronoi diagram of the sample [**5, 8, 7, 12, 47**]. It is known that the Voronoi vertices converge to the medial axis for a curve in 2D as the sample density approaches infinity [**13**]. Unfortunately, the same is not true in 3D [**3**]. The Voronoi vertices that are circumcenters of some flat tetrahedra called 'slivers' may come arbitrarily close to the surface no matter how dense the sample is. However, it turns out that the poles which are a subset of the Voronoi vertices approximate the medial axis [**5, 12**]. Poles provide a discrete approximation of the medial axis. For a continuous approximation they need to be connected appropriately. Amenta, Choi and Kolluri [**5**] design an algorithm that connects them with a cell complex. They consider the polar balls and then compute the medial axis of the boundary of the

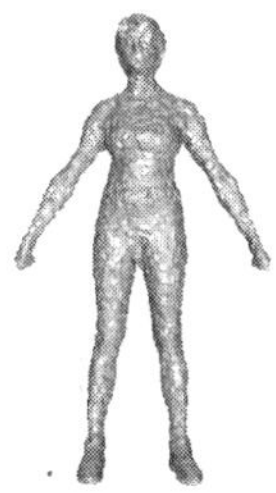

NOISY FEMALE (Robust Cocone) NOISY HORSE (Robust Cocone)

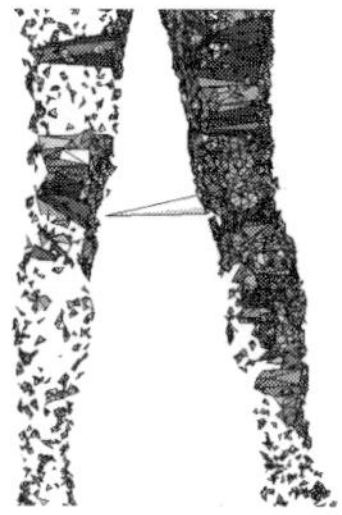

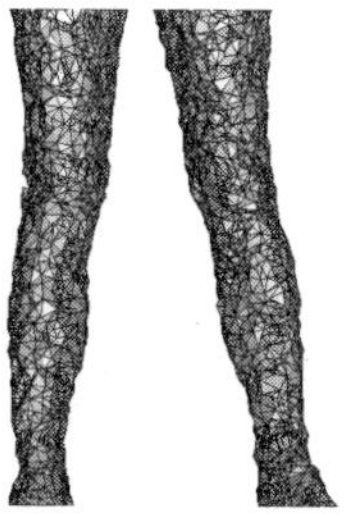

Leg (Tight Cocone) Leg (Robust Cocone)

Horse Neck (Tight Cocone) Horse Neck (Robust Cocone)

FIGURE 5. Results: Reconstruction by ROBUST COCONE and its comparison to TIGHT COCONE.

union of these balls. Since this union approximates the original object whose boundary is sampled, the computed medial axis approximates the original one. Although theoretically sound, this method requires two Voronoi diagram computations and produces noisy medial axis in some cases.

Here we describe an algorithm by Dey and Zhao [**23**] which provides a continuous approximation of the medial axis with a convergence guarantee. This algorithm approximates the medial axis directly from the Voronoi diagram. The approach does not pay any special attention to the poles, but rather computes a subcomplex from the Voronoi diagram that lies close to the medial axis and converges to it as the sampling density approaches infinity. The algorithm uses two conditions to select a subset of Voronoi facets that approximate the medial axis. These two conditions are similar to the ones used by Attali and Lachaud [**7**] though are different in details. We need the following definitions to explain these conditions.

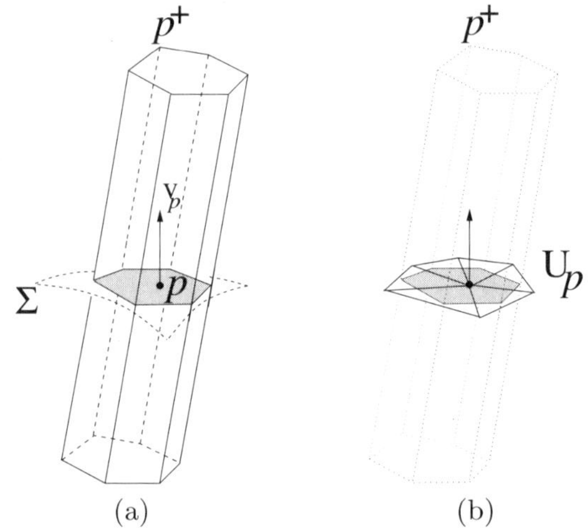

FIGURE 6. A Voronoi cell V_p. The corresponding pole, pole vector, tangent polygon (a), and the tangent umbrella (b).

DEFINITION 4.1. The *tangent polygon* for a sample point is defined as the polygon in which the plane through p with $\mathbf{v}_p$ as normal intersects V_p. See Figure 6(a) for an illustration.

Since $\mathbf{v}_p$ approximates $\mathbf{n}_p$, the tangent polygon approximates the tangent plane at p restricted within V_p. Define a dual structure to the tangent polygon from the Delaunay triangulation $\mathrm{Del}\, P$ as follows.

DEFINITION 4.2. The *tangent umbrella* U_p for a sample point p is defined as the topological disc made by the Delaunay triangles incident to p that are dual to the Voronoi edges intersected by the tangent polygon, see Figure 6(b).

4.1. Filter Conditions. The Voronoi facets approximating the medial axis M are chosen as the dual of a set of selected Delaunay edges. Two conditions, namely *Ratio* and *Angle* are used to filter these Delaunay edges from $\mathrm{Del}\, P$. To understand the rationale behind the two conditions, let us examine a medial ball B as in Figure 7. The segment pq makes an angle θ with the tangent planes at p and

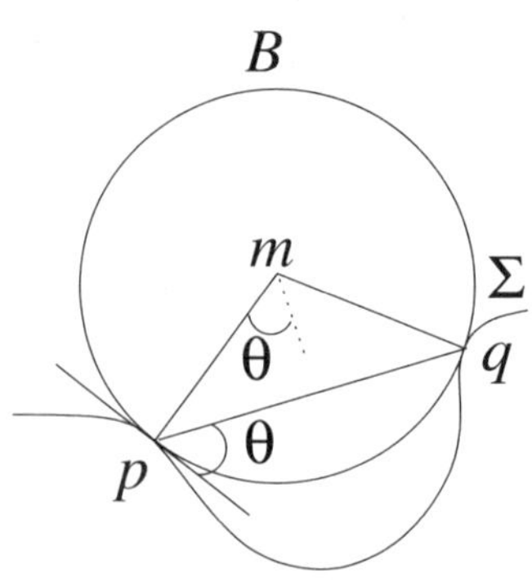

FIGURE 7. A medial axis point m, its medial angle θ and the corresponding medial ball.

q where the medial ball touches the surface Σ. If B touches Σ in more than two points, let p and q be such that the angle θ is maximum. Associate each medial axis point m with such an angle θ, which is called the *medial angle* of m. Each medial axis point with the medial angle θ greater than a threshold is approximated with the Angle condition. For the rest, the Ratio condition is applied.

4.1.1. *Angle condition.* The approximation of the medial angle θ for a medial axis point requires approximation to the tangent planes at the points where the medial ball touches the surface. The triangles in the tangent umbrellas lie close to Σ with their normals almost matching those of Σ at their vertices [**23**].

Therefore, the umbrella triangles in U_p are taken for approximating the tangent plane at a sample point p. All Delaunay edges pq that make relatively large angle with this tangent plane are determined. This is computed by measuring the angle $\angle \mathbf{n}_{ptu}, \mathbf{t}_{pq}$ between the vector $\mathbf{t}_{pq} = q - p$ and the normal $\mathbf{n}_{ptu}$ to a triangle σ in U_p. Those edges are selected that make this angle less than a threshold angle $\frac{\pi}{2} - \theta$ for all umbrella triangles in U_p for $0 < \theta < \frac{\pi}{2}$, see Figure 8. This means a selected edge pq satisfies:

$$\text{Angle condition } [\theta]\text{: } \max_{ptu \in \mathrm{U}_p} \angle \mathbf{n}_{ptu}, \mathbf{t}_{pq} < \frac{\pi}{2} - \theta.$$

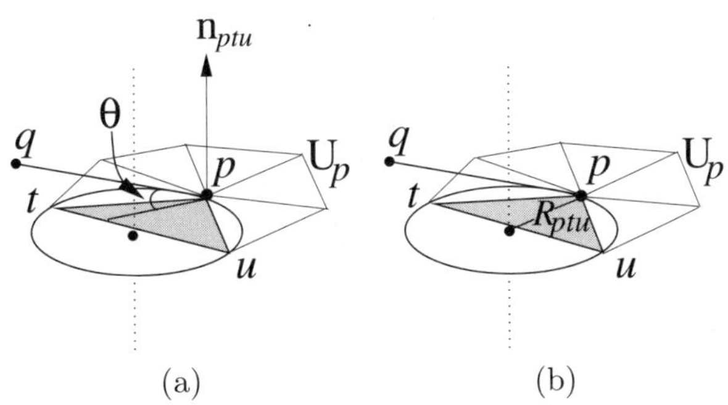

FIGURE 8. The angle θ for the Angle condition $[\theta]$ (a), radius of interest for the Ratio condition $[\rho]$ (b). The Voronoi edge shown with the dotted line is parallel to the normal of the shaded triangle.

4.1.2. *Ratio condition.* It is observed that 'only Angle condition' approximates the medial axis well only when an appropriate θ is used for each model [**23**]. If θ is fixed for all models, some of the medial axis points with medial angle below θ are not approximated. Consider the medial ball B as shown in Figure 7. From simple geometry of spheres, one obtains $\|p - q\| = 2\mu \sin\theta$ where $\mu = \|m - p\|$ is the radius of B. Therefore, if $\theta \gg \varepsilon$, where ε is the sampling density, we have $\|p - q\| \gg 2\mu \sin\varepsilon$. Also, it follows from a result in [**4**] that the radius of the umbrella triangles are only of the order of $\mu\varepsilon$. This means that the length of pq will be much larger than the circumradii of the umbrella triangles.

Therefore, if the length of the Delaunay edges is compared with the circumradii of the umbrella triangles, all medial axis points with medial angles only few times larger than ε can be approximated. Of course, the medial axis points with medial angle less than ε may not be approximated with this method, but as ε approaches zero, one gets the required convergence. Let R_{ptu} denote the circumradius of a triangle ptu, refer to Figure 8. The Ratio condition selects those Delaunay edges that satisfy:

$$\text{Ratio condition } [\rho]\text{: } \min_{ptu \in \mathrm{U}_p} \frac{\|p-q\|}{R_{ptu}} > \rho.$$

4.2. Algorithm and results. Although the theory suggests that the Ratio condition alone is sufficient to approximate the medial axis, experiments suggest that both Angle and Ratio conditions are necessary in practice [**23**]. This discrepancy can be attributed to the fact that the samples are not as dense as required by the theory in most cases. It is observed that the values of θ and ρ can remain fixed for all tested data sets when used together. With a fixed value of θ all Delaunay edges that lie away from the surface are determined by the Angle condition $[\theta]$ when θ is sufficiently large. The rest of the Delaunay edges whose duals contribute to the approximate medial axis make small angles with the umbrella triangles but are comparatively larger than their circumradii. They are captured by the Ratio condition $[\rho]$. These two-pronged strategy lets both θ and ρ fixed over different sampling densities. If the density is not high, the Angle condition captures almost all required Delaunay edges. The Ratio condition plays a more significant role when the density is high.

Experiments suggest that $\theta = \frac{\pi}{8}$ and $\rho = 8$ are appropriate for all tested data sets. With these two values the steps of the algorithm MEDIAL are enumerated below.

MEDIAL(P)
Compute Vor P and Del P;
$F = \emptyset$;
for each $p \in P$
    Compute U_p;
    for each Delaunay edge $pq \in \mathrm{U}_p$
        if pq satisfies Angle Condition $[\frac{\pi}{8}]$
            or Ratio condition [8]
            $F := F \cup \mathrm{Dual}\, pq$
        endif
    endfor
endfor
output closure(F)

Using the properties of the Voronoi diagram and ε-sampling the following result is proved [**23**].

THEOREM 4.3. *Let F_ε be the subcomplex computed by* MEDIAL *for an ε-sample of a surface $\Sigma \subseteq \mathbb{R}^3$ without boundary. We have* $\lim_{\varepsilon \to 0} F_\varepsilon = M$*, where M is the medial axis of Σ.*

Figure 9 shows the results for some example data sets. In this picture only the 'inner' medial axis lying in $\mathrm{In}(\Sigma)$ is shown. The 'inner' medial axis can be computed with the help of TIGHT COCONE as follows. Let T be the set of tetrahedra marked *in* by the marking phase of TIGHT COCONE. Only those Voronoi facets computed by MEDIAL are output that are dual to the Delaunay edges incident to a tetrahedron in T.

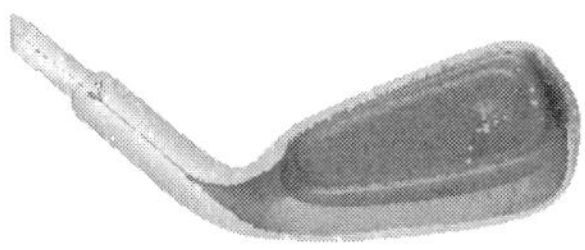

CLUB

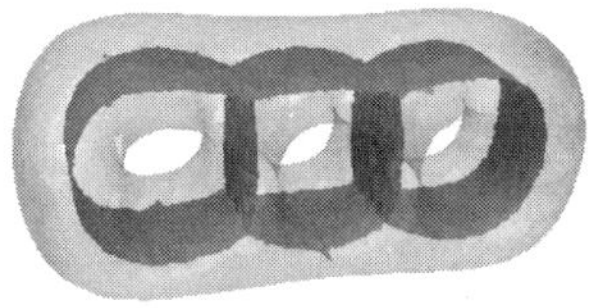

3HOLE

SCREWDRIVER

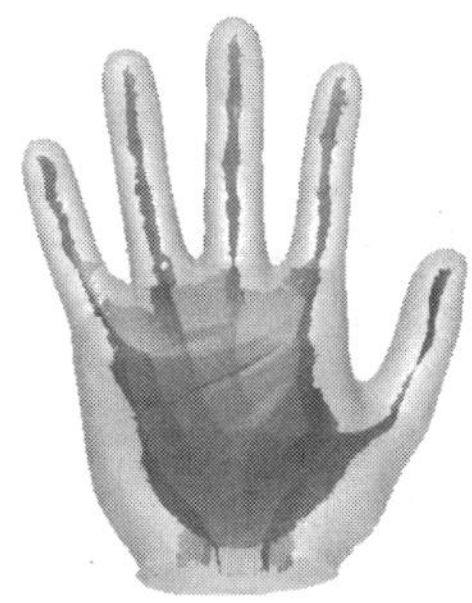

HAND

FIGURE 9. Medial axis with MEDIAL shown dark shaded.

The approximation of the medial axis is better where the data is dense. The approximation contains artifacts in the vicinity of high curvature regions, or non-smooth regions where usually undersampling happens. For example, in the CLUB data, the medial axis is well approximated in most part where the density is high. But, near the handle of the CLUB, the sampling density suffers due to high curvature and small features. As a result the approximation is poor in this region.

5. Shape segmentation and matching

In this section we consider the problems of shape segmentation and matching from a sample. Recall that the shape segmentation asks for partitioning a shape $\text{In}(\Sigma)$ into identifiable features given a sample P of Σ.

We describe an algorithm by Dey, Giesen and Goswami [**20**] which is modified to handle noisy sample [**22**]. The algorithm computes a Delaunay triangulation of the sample and figures out the tetrahedra that have to be clustered together to form segments of the shape interior. In other words, instead of building multiple data structures, such as an approximate surface mesh and then other derived structures like medial axis and Reeb graphs, the algorithm acts only on a single data structure, the Delaunay triangulation of the sample.

The segmentation algorithm has two phases. The first phase pre-processes the point set using the idea from ROBUST COCONE described in subsection 3.4. This step is specially required if the input sample is noisy. It filters Delaunay tetrahedra whose union approximates $\mathrm{In}(\Sigma)$. These tetrahedra are circumscribed by the big inner Delaunay balls. The second phase further partitions them into segments using Morse theoretic concepts. It mimicks a topological segmentation of the continuous shape $\mathrm{In}(\Sigma)$.

The Morse theoretic segmentation is exploited for matching shapes based on the principle that similar shapes have similar such segmentation. A signature of the shape is derived from its segmentation. These signatures are compared to match shapes. Only a few significant segments are selected each of which is represented with a weighted point where the weight is the volume of the segment. Then, the shape matching problem boils down to matching two small weighted point sets instead of matching large samples derived from the shape boundaries [**38**]. Furthermore, representative points of the selected segments can be precomputed and kept in a database thus reducing both space and time requirements for matching. These steps are carried out so that the entire matching process remains invariant to rotation, translation, mirroring and scaling.

5.1. Segmentation. First we describe a segmentation of the continuous shape $\mathrm{In}(\Sigma)$ and then mimic the procedure for the discrete sample P of Σ.

CONTINUOUS SHAPE. Consider the distance function $h : \mathbb{R}^3 \to \mathbb{R}$ where

$$h(x) = \min_{p\in\Sigma} \|p - x\|^2 \text{ for all } x \in \mathbb{R}^3.$$

Define a vector field $v : \mathbb{R}^3 \to \mathbb{R}^3$ that assigns to every point $x \in \mathbb{R}^3$ the direction in which h increases the most. If h is smooth at x then $v(x)$ coincides with the normalized gradient $\nabla h(x)/\|\nabla h(x)\|$. The function h is smooth everywhere except at the medial axis of Σ. However, it turns out that one can still define the direction of the steepest ascent of h at any point on the medial axis. This allows to define a vector field where $v(x)$ is a unit vector pointed in the direction of the steepest ascent of h at x. The vector field v induces a flow $\phi : [0, \infty) \times \mathbb{R}^3 \to \mathbb{R}^3$ such that the right derivative at every point $x \in \mathbb{R}^3$ matches with the vector, see Grove [**34**], Lieuter [**41**] and Giesen and John [**33**].

Given $x \in \mathbb{R}^3$ and an induced flow ϕ, the curve $\phi_x : [0, \infty) \to \mathbb{R}^3$, $t \mapsto \phi(t, x)$ is called the *orbit* of x. A point x is a *fixpoint* of ϕ if $\phi(t, x) = x$ for all $t \geq 0$. Basically, the orbit of a point is the curve it will follow if it were let move along the steepest ascent of h. The fixpoints of ϕ are the critical points of h. The critical points where h does not decrease in any direction are the minima of h. These are the points of Σ. The critical points where h decreases in all directions are the maxima of h. The rest of the critical points are *saddles*. In Figure 10, the points a and c are maxima, b is a saddle.

The stable manifold $S(x)$ of a critical point x is the set of all points that flow into x, i.e.,

$$S(x) = \{y \in \mathbb{R}^3 \; : \; \lim_{t\to\infty} \phi_y(t) = x\}.$$

The stable manifolds of all critical points partition $\mathbb{R}^3$. In particular, if we take the closure of the stable manifolds of the maxima, the stable manifolds of saddles and minima constitute its boundary. The shape $\mathrm{In}(\Sigma)$ is segmented with the closure of the stable manifolds of the maxima of h. Figure 10 shows two segments produced

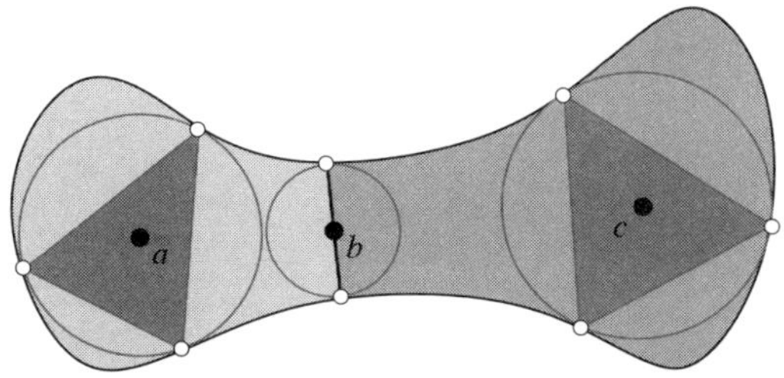

FIGURE 10. The stable manifold of the saddle b separates those of the maxima a and c producing two segments.

by this segmentation.

DISCRETIZED SHAPE. Now we mimic the above segmentation of $\text{In}(\Sigma)$ using the discrete sample P of Σ. The distance function is defined with respect to the sample as follows:

$$h(x) = \min_{p \in P} \|x - p\| \text{ for all } x \in \mathbb{R}^3.$$

It turns out that the critical points for this function are the points where the Delaunay and their dual Voronoi objects intersect. The sample points are minima. They are the intersection of their dual Voronoi cells and themselves. The Voronoi vertices that are contained in their dual Delaunay tetrahedra are the maxima. There are two types of saddles in 3D. The points where a Delaunay edge intersects its dual Voronoi facet are saddle points of type 1 and the points where a Delaunay triangle intersects its dual Voronoi edge are saddles of type 2. The segments are defined as the closure of the stable manifolds of the maxima mimicking the continuous case. Each of such segment is bounded by the stable manifolds of the saddle points of type 1 and 2 and the minima. Although the stable manifolds of the saddle points are a subset of Delaunay objects (edges) in 2D, the same is not true in 3D. Therefore, to avoid costly numerical computations, the stable manifolds of the maxima are approximated with only Delaunay objects using the Delaunay flow idea of Edelsbrunner et al. [**27**].

Let σ_1, σ_2 be two Delaunay tetrahedra sharing a triangle t. We say σ_1 *flows into* σ_2, denoted as $\sigma_1 < \sigma_2$, if σ_1 and its dual Voronoi vertex lie on the opposite sides of the plane of t. For an illustration see the triangles surrounding the big shaded triangle in the hip of the DOG in the lower left picture of Figure 11. All these three triangles flow into the shaded triangle containing a maximum. It follows from the definition that if $\sigma_1 < \sigma_2$, then the radius of the Delaunay ball of σ_1 is smaller than the radius of the Delaunay ball of σ_2. Thus, the transitive closure $<^*$ of $<$ is acyclic. For a maximum x one can collect all tetrahedra that are flowing to x by the transitive closure $<^*$. This is the method which was proposed to compute the pockets in two dimensional shapes by Edelsbrunner, Facello and Liang [**27**]. In three dimensions one faces a difficulty because a Delaunay tetrahedron may flow into two other Delaunay tetrahedra. Therefore, the approximated stable manifolds of maxima computed by the above method may not be disjoint. To overcome this problem a modified flow relation $\tilde{<}$ is introduced.

FIGURE 11. Upper row: From the Delaunay triangulation (left), only inner Delaunay triangles determined by the inner Big Delaunay balls are selected (right). Inner triangles containing maxima Voronoi vertices are shaded (right). Bottom row: segmentation by stable manifolds of the maxima (left), segments after merging (right).

Define the *strength* of a tetrahedron σ as the largest value of the distance function among all maxima that it flows into. Say $\sigma_1 \tilde{<} \sigma_2$ if (i) $\sigma_1 < \sigma_2$ and (ii) there is no other tetrahedron σ_3 with $\sigma_1 < \sigma_3$ and the strength of σ_3 is larger than σ_2. Now a segment $G(x)$ for a maximum x is defined as

$$G(x) = \bigcup_{\sigma' \tilde{<}^* \sigma} \sigma'$$

where σ is the tetrahedron containing x. One can turn this definition into an algorithm very easily. First, determine the maxima Voronoi vertices. They are the Voronoi vertices contained in their dual Delaunay tetrahedra. Then, sort these maxima in decreasing order of their h-values. This value is given by the radius of the Delaunay ball circumscribing the dual tetrahedron of the maximum Voronoi vertex. Process the maxima in this sorted order and for each maximum x collect the tetrahedra that flow into it. This can be done by a depth first search starting from the tetrahedron dual to x and then including a tetrahedron in the collection if it flows into one of the tetrahedra already in the collection. Middle picture in the lower row of Figure 11 shows the segmentation obtained by this process for a two dimensional shape.

MERGING. As one can observe from Figure 11, the segments obtained by the above algorithm may be too fine because of the sampling artifacts. Many maxima may be introduced due to discretization. To overcome this artifact, small segments are merged to coarsen the segmentation. The merging, in effect, pairs spatially adjacent saddles and maxima as opposed to the more involved pairing suggested by Edelsbrunner, Letscher and Zomorodian for topology persistence [**26**]. A segment is judged *prominent* if the h-value at the maximum is significantly larger than the h-value at the circumcenters of the triangles on the segment boundary. In other words, prominent segments are 'peaky' when seen in the graph of the distance function h. Merge all segments that are not prominent with some adjacent segments. Precisely, two segments $G(x_1)$ and $G(x_2)$ are *ρ-mergable* if they share a triangle t and the circumradii of the tetrahedra containing x_1 and x_2 are no more than $\rho > 1$ times the circumradius of t. Experiments show that merging any two 1.5-mergable segments gives good results.

Final segments of a shape are the ones obtained after the merging process. Rightmost picture in the lower row of Figure 11 illustrates the merging.

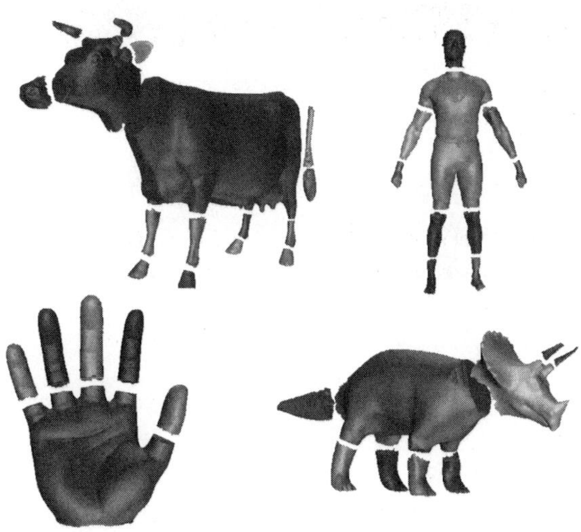

FIGURE 12. Segmentation of 3D models into so-called features.

5.2. Matching. For shape matching, a signature for a shape is generated from its segmentation. The signatures are a small set of weighted points that represent the segments.

SIGNATURE. Let $G_{P,\Sigma}$ denote the set of segments computed from a point sample P of a shape boundary Σ. To simplify notations we use G_Σ for $G_{P,\Sigma}$. By definition a segment $g \in G_\Sigma$ is a collection of Delaunay tetrahedra. For a Delaunay simplex σ let c_σ and v_σ denote the centroid and volume of σ, respectively. The *representative*

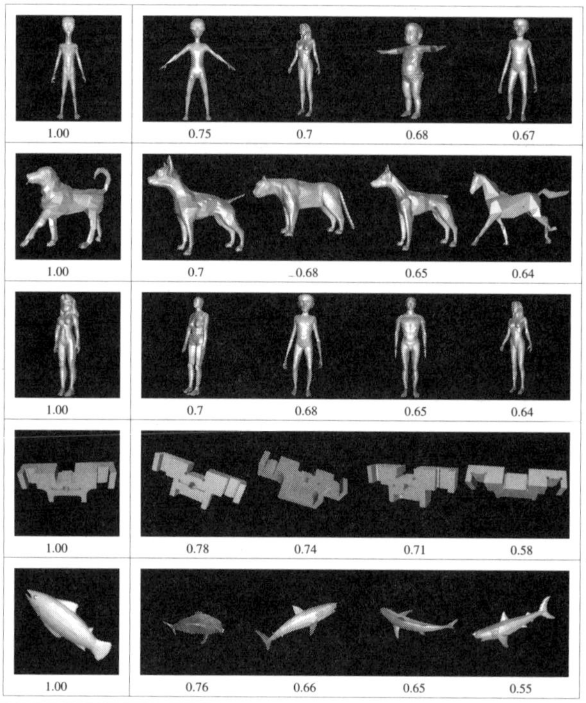

	CAD	Pig	Tiger	Dog	Boy	Alien
#pts	11646	37056	44769	55238	91775	120K
sec.	19.8	23.2	39.9	47.5	104.9	145.5

FIGURE 13. Matching result: Models with five best scores are shown for each query in a row. Timings (seconds) for segmentation are shown for some models.

point g^* of a segment g and its weight $\hat{g}$ are defined as

$$\hat{g} = \Sigma_{\sigma \in g} v_\sigma, \;\; g^* = \frac{\Sigma_{\sigma \in g}(c_\sigma \cdot v_\sigma)}{\hat{g}}.$$

That is, the weight of g is its volume and its representative point is the weighted average of the centroids of all $\sigma \in g$, weight being the volume of each simplex. Given a segmentation G_Σ of a shape Σ, the signature $\mathrm{sig}(\Sigma)$ is defined as a set of weighted points as follows:

$$\mathrm{sig}(\Sigma) \;=\; \{(g^*, \hat{g}) \mid g \in G_\Sigma\}.$$

SCORING. The amount of similarity between two shapes is measured by first scaling them with bounding boxes and then scoring the similarity between their signatures. In order to score the similarity between two signatures $\mathrm{sig}(\Sigma_1)$ and $\mathrm{sig}(\Sigma_2)$, we need to align them first.

Let g^*, h^* be the representative points in $\mathrm{sig}(\Sigma_1)$ and $\mathrm{sig}(\Sigma_2)$, respectively, with maximum weights. We first translate $\mathrm{sig}(\Sigma_2)$ so that g^*, h^* coincide. Then an alignment is obtained by rotating $\mathrm{sig}(\Sigma_2)$ so that a line segment between h^* and another point of $\mathrm{sig}(\Sigma_2)$ aligns with a line segment between g^* and another point in $\mathrm{sig}(\Sigma_1)$. Certainly, there are $\Theta(\ell n)$ alignments possible where $|\mathrm{sig}(\Sigma_1)| = \ell$ and $|\mathrm{sig}(\Sigma_2)| = n$. Since ℓ, n are typically small (less than ten), checking all alignments can be done fast.

For each alignment a score is computed based on the matching of the weighted points. Both a similarity measure (positive) and a dissimilarity measure (negative) are taken into account while computing the score. The maximum of all the scores is taken to be the amount of similarity and corresponding transformations give the best alignment.

5.3. Results. Figure 12 shows the segmentation of some 3D models. As the examples show, the segmentation mostly respect so called features. A database of signatures were created from approximately 300 point cloud data mostly collected from different web-sites and also created from 3D models. Some example matching results and segmentation timings are shown in Figure 13. The machine was a 2.8 Ghz Pentium 4 with 1 GB RAM. Matching of a query shape over the entire database took less than a second.

6. Conclusions

In this article we described algorithms for three problems in sample based geometric modeling (SBGM). The choice of the problems and the algorithms are mainly influenced by the work of the author and his co-authors. Nonetheless, the article represents a snapshot of the new developments that are taking place in the emerging area of sample based geometric modeling. The author believes that this area will outreach many application domains because of the flexibility offered by points as input.

One outcome of this SBGM paradigm is the new outlook on the sampling issues for various problems. How should one sample a surface, a volume, or a higher dimensional space to guarantee certain qualities in the approximation? The manifold learning where a manifold is to be estimated from a set of points in high dimensions is an important problem in AI and learning theory. We believe that the techniques from sample based geometric modeling will be useful for these problems.

Acknowledgements. The three software COCONE [**55**], MEDIAL [**55**] and SEGMATCH [**56**] have been implemented with the CGAL library [**54**]. These software developments would have been next to impossible without the robust, fast Delaunay triangulation code provided by CGAL.

References

[1] H. Alt and L. J. Guibas. Discrete geometric shapes: matching, interpolation, and approximation: a survey. Tech. report B 96-11, EVL-1996-142, Institute of Computer Science, Freie Universität Berlin, 1996.

[2] M. Alexa, J. Behr, D. Cohen-Or, S. Fleishman, D. Levin and C. T. Silva. Point set surfaces. *Proc. IEEE Visualization* (2001), 21–28.

[3] N. Amenta and M. Bern. Surface reconstruction by Voronoi filtering. *Discr. Comput. Geom.* **22** (1999), 481–504.

[4] N. Amenta, S. Choi, T. K. Dey and N. Leekha. A simple algorithm for homeomorphic surface reconstruction. *Internat. J. Comput. Geom. Applications*, **12** (2002), 125–141.

[5] N. Amenta, S. Choi and R. K. Kolluri. The power crust, unions of balls, and the medial axis transform. *Comput. Geom. Theory and Applications* **19** (2001), 127–153.

[6] D. Attali, J.-D. Boissonnat and A. Lieutier. Complexity of the Delaunay triangulation of points on surfaces: the smooth case. *Proc. 19th Annu. Sympos. Comput. Geom.* (2003), 201–210.

[7] D. Attali and J.-O. Lachaud. Delaunay conforming iso-surface, skeleton extraction and noise removal. *Comput. Geom. : Theory Appl.*, 2001.

[8] D. Attali and A. Montanvert. Computing and simplifying 2D and 3D continuous skeletons. *Computer Vision and Image Understanding* **67** (1997), 261–273.

[9] S. Belongie and J. Malik. Matching with shape contexts. *IEEE Trans. PAMI* **24** (2002), 509–522.

[10] S. Bouix and K. Siddiqi. Divergence-based medial surfaces. *Proc. European Conference on Computer Vision*, 2000.

[11] J. D. Boissonnat. Geometric structures for three dimensional shape representation, ACM Transact. on Graphics **3** (1984), 266–286.

[12] J. D. Boissonnat and F. Cazals. Smooth surface reconstruction via natural neighbor interpolation of distance functions. *Proc. 16th. ACM Sympos. Comput. Geom.* (2000), 223–232.

[13] J. W. Brandt. Convergence and continuity criteria for discrete approximation of the continuous planar skeletons. *CVGIP: Image Understanding* **59** (1994), 116–124.

[14] J. W. Brandt and V. R. Algazi. Continuous skeleton computation by Voronoi diagram. *Comput. Vision, Graphics, Image Process.* **55** (1992), 329–338.

[15] H. I. Choi, S. W. Choi and H. P. Moon. Mathematical theory of medial axis transform. *Pacific J. Mathematics* **181** (1997), 57–88.

[16] B. Curless and M. Levoy. A volumetric method for building complex models from range images. *SIGGRAPH '96* (1996), 303-312.

[17] T. Culver, J. Keyser and D. Manocha. Accurate computation of the medial axis of a polyhedron. *Solid Modeling '99* (1999), 179–190.

[18] T. K. Dey and J. Giesen. Detecting undersampling in surface reconstruction. *Proc. 17th Annu. Sympos. Comput. Geom.* (2001), 257–263.

[19] T. K. Dey and S. Goswami. Tight Cocone : A water-tight surface reconstructor. *J. Computing Inform. Sci. Engin.* **3** (2003), 302–307.

[20] T. K. Dey, J. Giesen and S. Goswami. Shape segmentation and matching with flow discretization. *Proc. Workshop Algorithms Data Structures* (2003), LNCS 2748, 25–36.

[21] T. K. Dey and S. Goswami. Provable surface reconstruction from noisy samples. *Proc. 20th Annu. Sympos. Comput. Geom.* (2004), 330–339.

[22] T. K. Dey, J. Giesen and S. Goswami. Shape segmentation and matching from noisy point clouds. *Proc. Eurographics Sympos. Point-Based Graphics*, M. Alexa and S. Rusinkiewicz (eds) (2004), 193–199.

[23] T. K. Dey and W. Zhao. Approximating medial axis from the Voronoi diagram with a convergence guarantee. *Algorithmica* **38** (2004), 179–200.

[24] H. Edelsbrunner and E. P. Mücke. Three-dimensional alpha shapes. *ACM Trans. Graphics* **13** (1994), 43–72.

[25] H. Edelsbrunner and N. Shah. Triangulating topological spaces. *Internat. J. Comput. Geom. Appl.* **7** (1997), 365–378.

[26] H. Edelsbrunner, D. Letscher and A. Zomorodian. Topological persistence and simplification. *Discrete Comput. Geom.* **28** (2002), 511–533.

[27] H. Edelsbrunner, M. A. Facello and J. Liang. On the definition and the construction of pockets in macromolecules. *Discrete Appl. Math.* **88** (1998), 83–102.

[28] M. Etzion and A. Rappoport. Computing Voronoi skeletons of a 3D polyhedron by space subdivision. *Tech. Report*, Hebrew University, 1999.

[29] S. Funke and E. Ramos. Smooth-surface reconstruction in near-linear time. *Proc. ACM-SIAM Sympos. Discrete Algorithms* (2002), 781–790.

[30] P. J. Giblin and B. B. Kimia. A formal classification of 3D medial axis points and their local geometry. *Proc. Computer Vision and Pattern Recognition (CVPR)*, 2000.

[31] L. Guibas, R. Holleman and L. E. Kavraki. A probabilistic roadmap planner for flexible objects with a workspace medial axis based sampling approach. *Proc. IEEE/RSJ Intl. Conf. Intelligent Robots and Systems*, 1999.

[32] H. N. Gursoy and N. M. Patrikalakis. Automated interrogation and adaptive subdivision of shape using medial axis transform. *Advances in Engineering Software* **13** (1991), 287–302.

[33] J. Giesen and M. John. The flow complex: a data structure for geometric modeling. *Proc. 14th Ann. ACM-SIAM Sympos. Discrete Algorithms* (2002), 285–294.

[34] K. Grove. Critical point theory for distance functions. *Proc. Sympos. Pure Math.* **54** (1993), 357–385.

[35] C. Hoffman. How to construct the skeleton of CSG objects. *The Mathematics of Surfaces, IV*, A. Bowyer and J. Davenport Eds., Oxford Univ. Press, 1990.
[36] H. Hoppe, T. DeRose, T. Duchamp, J. McDonald and W. Stuetzle. Surface reconstruction from unorganized points. *SIGGRAPH '92* (1992), 71-78.
[37] M. Hilaga, Y. Shinagawa, T. Komura and T. Kunni. Topology matching for fully automatic similarity estimation of 3D shapes. *Proc. SIGGRAPH 2001* (2001), 203-212.
[38] D.P.Huttelnocher, G.A.Klanderman and W.J.Rucklidge. Computing images using the Hausdorff distance. *IEEE Trans. Pattern Analysis and Machine Intelligence* **15** (1993), 850–863.
[39] A. E. Johnson and M. Hebert. Using spin-images for efficient multiple model recognition in cluttered 3-d scenes. *IEEE Trans. Pattern Analysis Machine Intelligence* **21** (1999), 433–449.
[40] F. Leymarie and B. Kimia. The shock scaffold for representing 3D shape. *Proc. 4th Internat. Workshop Visual Form.* (2001), LNCS 2059, Springer-Verlag, 216–229.
[41] A. Lieutier. Any open bounded subset has the same homotopy type than its medial axis. *Proc. 8th ACM Sympos. Solid Modeling and Applications* (2003), 65–75.
[42] C. Niblak, P. Gibbons and D. Capson. Generating skeletons and centerlines from the distance transform. *CVGIP: Graphical Models and Image Processing* **54** (1992), 420–437.
[43] R. L. Ogniewicz. Skeleton-space: A multiscale shape description combining region and boundary information. *Proc. Computer Vision and Pattern Recognition* (1994), 746–751.
[44] R. Osada, T. Funkhouser, B. Chazelle and D. Dobkin. Matching 3D Models with Shape Distribution. *Proc. Shape Modelling Int'l*, 2001.
[45] M. Pauly, R. Keiser, L. P. Kobbelt and M. H. Gross. Shape modeling with point-sampled geometry. *Proc. SIGGRAPH 2003* (2003), 641–650.
[46] T. B. Sebastian, P. N. Klein and B. Kimia. Recognition of shapes by editing shock graphs. *Proc. ICCV* (2001), 755-762.
[47] D. Sheehy, C. Armstrong and D. Robinson. Shape description by medial axis construction. *IEEE Trans. Visualization and Computer Graphics* **2** (1996), 62–72.
[48] D. Storti, G. Turkiyyah, M. Ganter, C. Lim and D. Stal. Skeleton-based modeling operations on solids. *Solid Modeling '97* (1997), 141–154.
[49] M. Teichman and S. Teller. Assisted articulation of closed polygonal models. *Proc. 9th Eurographics Workshop on Animation and Simulation*, 1998.
[50] G. Taubin and D. Cooper. Geometric invariance in computer vision, *Chapter Object Recognition Based on Moment (of algebraic) Invariants*, MIT press, 1992.
[51] R. C. Veltkamp and M. Hagedoorn. State-of-the-art in shape matching. Technical report UU-CS-1999-27, Utrecht University, the Netherlands, 1999.
[52] M. Tănase and R. C. Veltkamp. Polygonal decomposition based on the straight line skeleton. *Proc. 19th. Annu. Sympos. Comput. Geom.* (2003), 58–67.
[53] F.-E. Wolter. Cut locus & medial axis in global shape interrogation & representation. *MIT Design Laboratory Memorandum 92-2*, 1992.
[54] www.cgal.org.
[55] www.cse.ohio-state.edu/~tamaldey/cocone.html.
[56] www.cse.ohio-state.edu/~tamaldey/segmatch.html.

Department of Computer Science and Engineering, The Ohio State University, Columbus, Ohio 43210.

E-mail address: `tamaldey@cis.ohio-state.edu`

DIMACS Series in Discrete Mathematics
and Theoretical Computer Science
Volume **67**, 2005

Computational Topology and Swept Volumes

Denis Blackmore, Yuriy Mileyko, Ming C. Leu, William C. Regli, and Wei Sun

ABSTRACT. Some fundamental principles of computational topology are formulated and discussed. It is demonstrated that swept volumes comprise a class of geometric objects that is particularly well suited to the application and development of the advanced mathematics/computer science based foundations of computational topology. Two new results in this vein are presented: (1) every smooth swept volume is proved to have a Whitney regular stratification; and (2) it is shown how homology can be used to characterize surface intersections. Applications of the dynamical systems approach to swept volumes in virtual sculpting and modeling of heterogeneous material such as bones are also described.

1. Introduction

Computational topology (CT) is a new and fascinating field of inquiry created from an evolving combination of computer aided geometric design (CAGD) and geometric topology (GT). It can be likened to a much more topologically oriented version of computational geometry (CG). Although CT is a nascent discipline, and is as yet neither precisely defined nor circumscribed, it is clear that it involves elements that should make it rather attractive to computer scientists, geometric designers, manufacturing engineers, and mathematical scientists. Moreover, the problems that it addresses are important to both researchers and practitioners in computer aided design (CAD) and computer aided manufacturing (CAM).

One of our goals in this paper is to provide a clear, logical and reasonably accurate, albeit somewhat rudimentary, formulation of CT, and in so doing to help develop the foundations and describe the scope of the subject so that it can achieve the status of a full-fledged discipline, as are its primary constituent elements: CAGD and GT. In this regard, a natural first question for nonexperts might be the following: What are CAGD and GT? As we see it, CAGD can be briefly described as the discipline concerned with the development and implementation of methods - especially effectively computable (algorithmic) ones - for the analysis and graphical representation of geometric objects. GT, on the other hand, is an amalgam of those elements of algebraic geometry and topology, differential geometry and topology, and modern dynamical systems theory that are useful in characterizing the shape of geometric objects. As every definition tends to require additional explanations, so do

2000 *Mathematics Subject Classification.* 65D17, 68U05, 37N40 57N65, 57N80, 57R05, 57R40, 57R45, 53A07.

Key words and phrases. Shape, consistency, stability, sweeps, stratification, homology, virtual sculpting, heterogeneous material.

ours. For example, what is meant by the shape of a geometric object? This is another concept that we shall address in what follows.

The roots and development of CT can be traced and ascribed to a steadily increasing flow of ideas from GT to CAGD over the last two decades, the extraordinary growth in computational power realized over the same time period, the more exacting requirements for computer aided rendering of objects necessitated by the demands of modern technology, and a growing recognition that the computation of certain shape characteristics are indispensable for a growing number of CAGD applications. As the roots of CT took hold and spread, they engendered many new developments and insights relevant to the analysis and representation of geometric objects. Some excellent examples of this trend of employing advanced GT techniques to solve problems in CAGD may be found in [1]-[5], [7], [13]-[17], [19], [20], [22], [23], [25], [28], [31], [36], [39], [46], [49], [52], [54], [55], [63]-[66], [71], [74], [75], [78], [80] and [81]. The innovations in CT have come not only from CAGD and GT, they been influenced by advances in computer science, computer technology, and engineering - and they are beginning to effect a shift from the traditional combinatorial, piecewise smooth and algebraic patch decomposition representation of CG to smoother methods for rendering geometric objects.

CT has brought with it an enhancement of the theoretical foundations and algorithmic techniques of CAGD that has made it possible to more effectively address a whole new array of problems concerning the shape of computationally rendered geometric objects. The variation of shape characteristics in algorithmically generated geometric objects lies at the heart of several fundamental questions and concepts in CT. Two of the most important of these fundamental concepts are *consistency* and *stability* (*robustness*). Roughly speaking, consistency is the persistence of shape of the computed object irrespective of immanent errors found in the input data, organization of the data, mode of representation, random fluctuations, and computational and roundoff errors. Stability of a program (algorithm) for computing and representing objects means that the (approximate) shape of the rendered object is immutable and there are predictable small changes in the metric shape characteristics when there a small variations (possibly involving a random component) in the input data and parameters of the program. More precise formulations of these and other concepts from CT shall presented in the sequel. We hope that our efforts here will help stimulate the development of a cohesive, intelligible, comprehensive and widely accepted description of CT that will lead to an understanding and vision of CT comparable to those of more mature disciplines.

Swept volumes serve not only as an excellent starting point and testbed for the application of new topologically based methods for analyzing and representing geometric objects, they actually provide a paradigm for CT. A swept volume is the totality of points in a space traversed by a moving rigid or deforming object (cf. [1]-[5], [9]-[13], [41], [42], [44], [45], [48], [50], [61], [76] and [77]). Among the many areas in which swept volumes play a key role are NC machining, virtual design, manufacturing and sculpting, robotics, and the modeling and fabrication of composites. In fact, swept volumes are indispensable for the simulation and verification of machining operations and a number of other CAGD related applications. From a CT perspective, swept volumes are interesting in their own right; for example, the deforming sweeps of very simple objects can produce configurations of remarkable complexity. Moreover, the fact that swept volumes are essentially produced by flows of a lower dimensional subset of the initial object makes them particularly apt for the motivation and application of CT methodologies that embrace such concepts as homology theory, singularity/stratification theory, characteristic classes, and obstruction theory (see, e.g. [1]-[6], [13]-[17], [19], [20], [26], [34], [35], [38],[40], [43], [51], [52], [53], [68], [72],, [78], [79] and [81]).

Our exposition assumes a familiarity with both CAGD and GT (as can be found in [22], [24], [28], [32], [47], [53] and [59]), but in order to make it accessible to the widest possible audience, we shall augment it by paying more attention to the mathematical preliminaries and by providing some illuminating intuitive insights. The paper is organized as follows: In Section 2, we introduce most of the theoretical preliminaries that are employed in the sequel. Then, in Section 3, we briefly describe the class of objects of interest in CAGD applications, define shape, and discuss some fundamental shape characteristics. Precise definitions of consistency and stability are given in Section 4, along with some effectively computable shape characteristics associated with these properties. We delineate some mathematical methods for modeling, analyzing and representing swept volumes in Section 5, and we also discuss several strategies for studying consistency and stability that arise naturally from the structure of swept volumes, but have applications to a much wider class of geometric objects. The singularity/stratification theory strategy is articulated in more detail in Section 6, and it is proved that smooth sweeps of smooth objects produce swept volumes that have (Whitney) regular stratifications. Next, in Section 7, we show that transverse intersections of swept volumes can be detected using homology theory. We then show how the CT approach to swept volumes can be used to develop effective programs for virtual sculpting in Section 8. In Section 9 it is indicated how the dynamical systems formulation of swept volumes can be used to create algorithms for producing realistic models of heterogeneous structures such as those found in bones and other biological materials. Finally, in Section 10, we summarize our results, arrive at some conclusions, and describe several related avenues of future research.

2. Preliminaries

For the most part, the geometric objects of interest in CT can all be described as subsets of a Euclidean space; which is fortunate, as it will greatly simplify the mathematical nomenclature. These geometric objects are usually compact (= closed and bounded) manifolds, and almost never more complicated than compact varieties. An essentially self-contained description of these concepts is provided in this section. Our approach, which emphasizes intuition and directness of exposition, will be somewhat nonstandard (cf. [6], [26], [27], [30], [34],[37], [38], [40], [43], [53] and [62]).

Let $\mathbb{R}^n$ be a fixed Euclidean space, where n belongs to the set of natural numbers (positive integers) $\mathbb{N}$. For most applications the dimension of the ambient Euclidean space is $n = 3$. The space $\mathbb{R}^n$ is comprised of points $\mathbf{x} = (x_1, ..., x_n)$, has standard inner (dot) product $\langle \mathbf{x}, \mathbf{y} \rangle := x_1 y_1 + \cdots + x_n y_n$ inducing the norm $|\mathbf{x}| := \sqrt{\langle \mathbf{x}, \mathbf{x} \rangle}$, and has the topology generated by the open balls $B_\epsilon(\mathbf{x}) := \{\mathbf{y} : |\mathbf{y} - \mathbf{x}| < \epsilon\,(> 0)\}$.

A nonempty subset X of $\mathbb{R}^n$ is a C^0 *submanifold (sm)* of *dimension* m (*codimension* $(n-m)$), with $0 \le m \le n$, if for every $\mathbf{x} \in X$ there is a $B_\epsilon(\mathbf{x})$ such that $X \cap B_\epsilon(\mathbf{x})$ is homeomorphic with $B_\epsilon^{(m)}(\mathbf{x}) := B_\epsilon(\mathbf{x}) \cap \{\mathbf{y} : y_i = x_i, m+1 \le i \le n\}$. We recall here that two subsets X and Y of $\mathbb{R}^n$ are *homeomorphic*, denoted as $X \approx^0 Y$, if there exists a function $\Phi : X \to Y$ that is continuous and has a continuous inverse Φ^{-1}. X is a C^0 *submanifold-with-boundary (smwb)* of *dimension* m (*codimension* $(n-m)$), if for every $\mathbf{x} \in X$ there is a $B_\epsilon(\mathbf{x})$ such that $X \cap B_\epsilon(\mathbf{x})$ is homeomorphic with $B_\epsilon^{(m)}(\mathbf{x})$, for *interior* points, or with $\tilde{B}_\epsilon^{(m)}(\mathbf{x}) := B_\epsilon(\mathbf{x}) \cap \{\mathbf{y} : y_m \ge 0, y_i = x_i, m+1 \le i \le n\}$, at the *boundary* points. Observe in this case that the interior points $IntX$ comprise a C^0, m-dimensional sm, and the boundary points ∂X form a C^0, $(m-1)$-dimensional sm.

For example, the unit disk $D^n := \{\mathbf{x} \in \mathbb{R}^n : |\mathbf{x}| \leq 1\}$ is an n-dimensional smwb, where $\partial D^n = S^{n-1}$, the $(n-1)$-sphere. It is important to distinguish here between the interior and boundary points for submanifolds, and interior and boundary points for the Euclidean topology of $\mathbb{R}^n$. To illustrate this, consider the example of the $(n-1)$-sphere . As a submanifold, $IntS^{n-1} = S^{n-1}$ and $\partial S^{n-1} = \emptyset$, while in terms of the standard topology on $\mathbb{R}^n$, the interior, which we denote by $\mathrm{int}(S^{n-1})$, is empty, and the boundary or *frontier* (consisting of all points $\mathbf{x}$ such that every $B_\epsilon(\mathbf{x})$ intersect both the set and its complement), which we denote by $\mathrm{Fr}(S^{n-1})$, is the entire sphere.

$X \subseteq \mathbb{R}^n$ is a C^0 *s-subvariety (ssv)* of *dimension* m (*codimension* $(n-m)$), if for every $\mathbf{x} \in X$ there is a $B_\epsilon(\mathbf{x})$ such that $X \cap B_\epsilon(\mathbf{x})$ is comprised of a finite union of sets, each of which is homeomorphic with $B_\epsilon^{(m)}(\mathbf{x})$. The subset X is a C^0 *s-subvariety-with-boundary (ssvwb)* of *dimension* m (*codimension* $(n-m)$) if each $\mathbf{x} \in X$ is either an *interior* point with the property given above for an ssv, or a *boundary* point where $X \cap B_\epsilon(\mathbf{x})$ is comprised of a finite union of sets, each of which is homeomorphic with $\tilde{B}_\epsilon^{(m)}(\mathbf{x})$. Observe that the interior points $IntX$ form a C^0 ssv of dimension m, and the boundary ∂X is a C^0 ssv of dimension $m-1$. For example, the solid cone $\bar{K}^n$ of points in $\mathbb{R}^n$ satisfying $x_1^2 + \cdots + x_{n-1}^2 - x_n^2 \leq 0$ is an ssvwb of dimension n, with an interior that is a sm of codimension 0, and a boundary that is an ssv of codimension 1, which is the locus of the equation $x_1^2 + \cdots + x_{n-1}^2 - x_n^2 = 0$.

For smoother versions of submanifolds and subvarieties, we need some additional functional notation. Let U be an open subset of $\mathbb{R}^n$. A function (map) $f = (f_1, ..., f_m) : U \to \mathbb{R}^m$ is C^k $(k \in \mathbb{N})$ if all of the coordinate functions f_i are continuous and have continuous partial derivatives of all orders $\leq k$ at all points of U; it is *smooth*, or C^∞, if it is C^k for all $k \in \mathbb{N}$. If f is C^k on U, at every point $\mathbf{x} \in U$ the function has a derivative $f'(x) : \mathbb{R}^n \to \mathbb{R}^m$, which is a linear map that may be identified with the $m \times n$ matrix $(\partial_{x_j} f_i)$, $1 \leq i \leq m$, $1 \leq j \leq n$. The points where $f'(\mathbf{x})$ has maximal rank are called *regular points* of the function. If $\mathrm{rank} f'(\mathbf{x}) < m$, $\mathbf{x}$ is called a *singular point*. Two subsets X and Y of $\mathbb{R}^n$ are C^k*-diffeomorphic*, denoted $X \approx^k Y$, if there is a homeomorphism $\Phi : X \to Y$ such that Φ and Φ^{-1} are C^k. The map f is C_p^k if it is continuous on U and C^k on $U \setminus S$, where S is a subset of a countable union of non-accumulating sets, each of which is C^k-diffeomorphic with a q-flat $(0 \leq q < n)$. By a q-flat we mean a set of the form $\{\mathbf{x} \in \mathbb{R}^n : x_i = 0, q < i\}$. We say that X and Y are C_p^k*-isomorphic,* denoted $X \approx_p^k Y$, if there is a homeomorphism $\Phi : X \to Y$ such that Φ and Φ^{-1} are C_p^k.

Now X is a C^k *sm* (C^k *smwb*) of dimension m (codimension $(n-m)$) if homeomorphism is replaced by C^k-diffeomorphism wherever it appears in the above definition of a C^0 sm (C^0 smwb). We observe that it follows from the implicit function theorem that the submanifold property is equivalent to the statement that for each $\mathbf{x} \in X$ there exists a map $f : B_\epsilon(\mathbf{x}) \to \mathbb{R}^{n-m}$ having only regular points. This provides some insight into the following definition. To define a C^k *ssv* of dimension m, we add the property that $X \cap B_\epsilon(\mathbf{x}) = f^{-1}(0) := \{\mathbf{y} : f(\mathbf{y}) = 0\}$ for some C^k function $f : B_\epsilon(\mathbf{x}) \to \mathbb{R}^{n-m}$ with $f(\mathbf{x}) = 0$, possibly having some singular points, to the definition of a C^0 s-subvariety. Similarly, X is a C^k *ssvwb* of dimension m if it is comprised of interior points satisfying the preceding C^k *ssv* condition, and boundary points defined to satisfy the same property as boundary points of a C^0 ssv plus the condition that there exist C^k maps $f : B_\epsilon(\mathbf{x}) \to \mathbb{R}^{n-m}$ and $g : B_\epsilon(\mathbf{x}) \to \mathbb{R}^m$ with $f(\mathbf{x}) = g(\mathbf{x}) = 0$ such that $X \cap B_\epsilon(\mathbf{x}) = f^{-1}(0) \cap \{\mathbf{y} : g_i(\mathbf{y}) \geq 0, 1 \leq i \leq m\}$. With these definitions, we see that the unit disk D^n is a C^∞ smwb of dimension n, and its boundary $\partial D^n = S^{n-1}$ is a

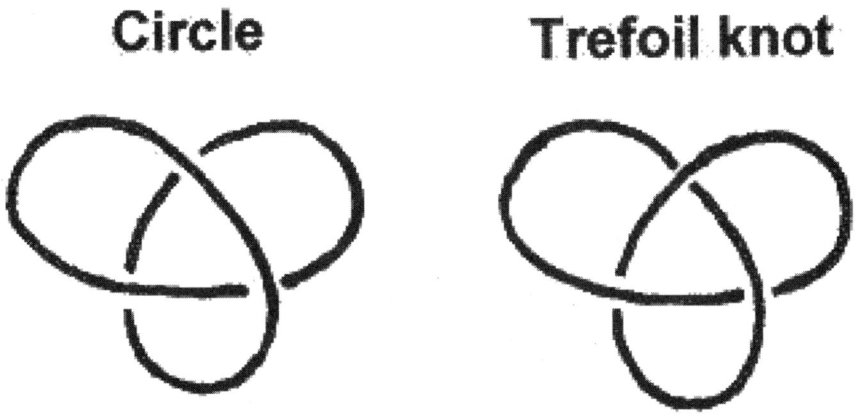

FIGURE 1. The circle and trefoil knot in three-dimensional space.

smooth sm of codimension 1, and the solid cone $\bar{K}^n$ is a smooth ssvwb of codimension 0 in $\mathbb{R}^n$, where $\partial\bar{K}^n := K^{n-1}$ is a smooth ssv of dimension $n-1$.

The piecewise-smooth versions of submanifolds and s-subvarieties can now be defined in a very direct way using local C_p^k-isomorphisms. For example, X is a C_p^k *ssvwb* of dimension m if for every $\mathbf{x} \in X$ there is a C^k ssvwb of dimension m, Y, such that $X \cap B_\epsilon(\mathbf{x}) \approx_p^k Y \cap B_\epsilon(\mathbf{y})$ for some $\mathbf{y} \in Y$ and $\epsilon > 0$. Analogous definitions of a C_p^k sm, C_p^k smwb, and C_p^k ssv are readily obtained mutatis mutandis. As an example, consider the subset of $\mathbb{R}^n$, with $n > 3$, defined as

$$M := \{\mathbf{x} : 0 \leq x_1, x_2, x_3 \leq 1 \text{ or } -1 \leq x_1, x_2, -x_3 \leq 0,\ x_i = 0,\ i > 3\}.$$

It is easy to see that M is a C_p^∞ ssvwb of dimension 3, and that ∂M, which is just the union of the faces of two 3-cubes having a common edge, is a C_p^∞ ssv of dimension 2.

3. Shape

What is the shape of a geometric object? This is a fundamental question in CT, but one that seems to have no universally accepted answer (cf. [7], [15], [17], [36], [39] and [74]). We shall address this question at some length in this section, in hope of illuminating and clarifying the concept of shape.

If all objects in of interest in CT in a Euclidean space were made of silly putty, so that donuts and coffee cups are indistinguishable soon after they are fabricated, one might be led to say that M and N in $\mathbb{R}^n$ have the same shape if they are homeomorphic. But even in such a Dali-esque world, this definition would be unsatisfactory for designers because it ignores the limitations on construction operations imposed by the ambient space $\mathbb{R}^n$. To see this, consider the circle and trefoil knot in $\mathbb{R}^3$ shown above in Fig. 1. These objects are homeomorphic, but a tearing and reconnection operation is required to deform one into the other in $\mathbb{R}^3$, because one is knotted (linked) and the other is not. Such is not the case in $\mathbb{R}^4$, however, where there is more than enough room to continuously deform the circle into the trefoil knot. At the other end of the spectrum of possible definitions for shape, it is not difficult to see that one can have objects M and N that can be continuously deformed in $\mathbb{R}^3$ into a circle and trefoil knot, respectively, such that there is a smooth length-preserving difffeomorphism from M to N. Nevertheless, it would still be imprudent to say these objects have the same shape. The problem is the same as for the unsuitable weak definition of shape: one object is knotted and the other is not. These observations illustrate that the definition of shape must somehow depend on the ambient space of the objects, which suggests the embedding categories (see [6], [30], [[51], [62] and [68]) as likely candidates for viable definitions of shape.

We first discuss some embedding analogs of the function spaces introduced in the preceding section. If M and N are two C^k submanifolds (or submanifolds-with-boundary) of $\mathbb{R}^n$, we write $M \simeq^k N$ if there is a homeomorhism $\Phi : \mathbb{R}^n \to \mathbb{R}^n$ such that $\Phi(M) = N$ and the restriction Φ_M of Φ to M is a C^k-diffeomorphism of M onto N. The isomorphism relations $M \simeq^k N$ for s-subvarieties and $M \simeq_p^k N$ for C_p^k objects are defined similarly. Yet none of these conform to our intuitive notion of shape, which demands that objects of the same shape look exactly the same, as expressed in the following:

DEFINITION 3.1. Two C^k (or C_p^k) sms (smwbs, ssvs, or ssvwbs) M and N of $\mathbb{R}^n$ have the same *C^k-shape* if $M \simeq^k N$ and Φ_M satisfies $\langle \Phi_M(\mathbf{x}), \Phi_M(\mathbf{y}) \rangle = \lambda \langle \mathbf{x}, \mathbf{y} \rangle$ for some positive constant λ and all $\mathbf{x}, \mathbf{y} \in M$ (i.e. Φ_M is a *C^k-similarity*). This shall be denoted by $M \equiv^k N$. For the case $k = 2$, where the standard differential geometric properties such as curvature of the C^2 portions of the objects can be computed by classical means, or by the Regge calculus (see [56]) for C_p^2 objects at points where they are not C^2, we simply write $M \equiv N$ and say that they have the same *shape*.

In CT one would like to determine, for the objects of interest, a *complete set of shape invariants* $\{\mathcal{I}_m(M)\}$ that is effectively computable (ec); i.e. quantities, which are usually algebraic, such that $M \equiv N$ if and only if $\mathcal{I}_m(M) = \mathcal{I}_m(N)$ for all m, all of which can be computed algorithmically. This may not always be possible even for reasonably simple objects. To see this, consider the unit circle S^1 and a knotted smooth, simple closed curve C, both embedded in $\mathbb{R}^3$. Thicken each of these curves, but not so much as to create self intersections, and let M and N be the resulting smooth smwbs with center curves S^1 and C, respectively. M and N are completely determined by their respective boundaries ∂M and ∂N, both of which are C^∞-diffeomorphic with the torus $S^1 \times S^1$, and such surfaces have an ec, complete set of diffeomorphism type invariants comprised of just one element - the Euler-Poincaré characteristic χ - and $\chi(\partial M) = \chi(\partial N) = 0$. This single invariant, however, cannot determine if M and N have the same embedding type (shape); additional invariants associated to linking are required. Moreover, if the knot C is sufficiently complicated, it follows from results of Markov, Novikov and Thue (see [68]) that N has shape invariants that are not ec (even though it can be shown that there is an ec unknotting procedure). Accordingly if one is to have any hope of finding a complete, ec set of shape invariants for a reasonably broad array of objects, the definition of shape will have to be weakened.

There is a weakened version of shape that has significant utility in CT. Most of the objects studied in CT are solid and have finite extent; more precisely, they are n-dimensional, compact C_p^k ssvwbs. Typically, one wishes to compare an algorithmically generated approximation N of a possibly scaled version M of a prototype M_0 of such an object in $\mathbb{R}^n$. In particular, M_0 is in a fixed position in $\mathbb{R}^n$ and $M = \Theta_0(M_0)$, where Θ_0 is a member of the analytic Lie group $\mathcal{S}(n)$ of maps $\mathbb{R}^n \to \mathbb{R}^n$ of the form $\Theta_0(\mathbf{x}) = \mathbf{x}_0 + \lambda A\mathbf{x}$, where λ is a positive constant (scale) and $A \in \mathrm{SO}(n)$, the Lie group of real orthogonal matrices of determinant one. An approximate algorithmic rendering of M may be viewed as a C_p^k map $\Theta : M_0 \to \mathbb{R}^n$, where $N = \Theta(M_0)$, that is ϵ close to Θ_0 on M_0 in a suitable topology such as the following: Define $\|\Theta - \Theta_0\|^{(0)}$ to be $\max\{| \Theta(\mathbf{x}) - \Theta_0(\mathbf{x}) | : \mathbf{x} \in M_0\}$ and for all $1 \leq k$, set $||\Theta - \Theta_0||^{(k)}$ to the maximum over all partial derivatives ∂ of order k of the numbers $\max\{| \partial (\Theta(\mathbf{x}) - \Theta_0(\mathbf{x})) |: \mathbf{x} \in M_0, \partial (\Theta(\mathbf{x}) - \Theta_0(\mathbf{x})) \text{ exists}\}$. Now we define a norm for these maps as (cf. [12], [26], [30] and [43])

$$||\Theta||_p^k := \sum_{j=0}^{j=k} (j+1)^{-2} \, ||\Theta||^{(j)}. \tag{3.1}$$

As M and N are solid and we can choose $||\Theta - \Theta_0||_p^k < \epsilon$, with ϵ essentially arbitrarily small, any difference in shape between N and M as ϵ increases is first manifested by a change in homeomorphism type. This suggests the following weaker characterization of shape.

DEFINITION 3.2. Let M and N be compact C_p^k ssvwbs of $\mathbb{R}^n$. Then M and N have the same *C_p^k-pseudoshape*, denoted as $M \cong_p^k N$ if, in the notation above, Θ is a homeomorphism satisfying $||\Theta - \Theta_0||_p^k < \epsilon$ for any sufficiently small ϵ. When $k = 2$, we simply say that M and N have the same *pseudoshape*.

From a CT perspective, Definition 3.2 is much more manageable than Definition 3.1. One only need show that the algorithm for approximating M can be made ϵ close in the $|| \cdot ||_p^k$ topology, and that there is an ec procedure for showing that M and N are homeomorphic. For example, consider a surface S embedded in $\mathbb{R}^3$. It is well known that the pseudoshape of S is completely determined by the Euler-Poincaré characteristic, which can be algorithmically calculated using any convenient triangulation. On the other hand, the shape (embedding type) of S may be extraordinarily difficult to calculate if the surface is knotted in a complicated way. Further illustrations of the computational advantages of dealing with pseudoshape rather than shape shall be presented in the sequel.

4. Consistency and Stability of Algorithms

From now on we shall assume that the objects of interest in CT are all compact, connected, C_p^k ssvs or ssvwbs of $\mathbb{R}^n$. The connectedness assumption involves no loss of generality, for we can consider each component of a disconnected object separately. We shall also continue to use the notation of the preceding section, often in a somewhat more detailed form.

For example, the algorithm and associated computation, embodied in the map Θ, that generates an approximation of M involves a particular algorithm $\mathfrak{A}$ and choice of data D. The approximating map should rightly be denoted as $\Theta_{(\mathfrak{A},D)}$, and it is subject to variations manifested by such factors as computational truncation errors, changes in the program used and the mode of representation (viewed as changes in the algorithm), errors in the input data, changes in the way the data is organized (viewed as a changes in the data), random fluctuations in the data, and (random) roundoff errors. An algorithm typically depends on a number of real parameters that may be represented as a vector $\mu \in \mathbb{R}^s$, so it should actually be denoted as $\mathfrak{A} = \mathfrak{A}\,(\mu)$, and the data D is selected from a class $\mathcal{D}$ of countable sets of points consistent with a preassigned numerical error bound for the algorithm. With this notation, we are able to give precise definitions of consistency and stability.

DEFINITION 4.1. Let $\mathfrak{A}$ be an algorithm for rendering approximations of geometric objects, and let $\mathcal{M}$ be a subcollection of all compact, connected, C_p^k ssvs or ssvwbs of $\mathbb{R}^n$. Then $\mathfrak{A}$ is *$\mathcal{M}$-consistent* (*$\mathcal{M}$-pseudoconsistent*) if $N = \Theta_{(\mathfrak{A},D)}\,(M_0) \equiv_p^k (\cong_p^k) M = \Theta_0\,(M_0)$ for all $M \in \mathcal{M}$ for all choices of data $\mathcal{D}$ that produce a sufficiently accurate approximation.

DEFINITION 4.2. The algorithm $\mathfrak{A}$ is *$\mathcal{M}$-stable* (*$\mathcal{M}$-pseudostable*) if for each $\epsilon > 0$ there exists a $\delta > 0$ such that, $\|\Theta_{(\mathfrak{A}(\mu),D)}(M_0) - \Theta_{(\mathfrak{A}(\mu_0),D_0)}(M_0)\|_p^k < \epsilon$ and $\Theta_{(\mathfrak{A}(\mu),D)}(M_0)$ and $\Theta_{(\mathfrak{A}(\mu_0),D_0)}(M_0)$ have the same C_p^k-shape (C_p^k-pseudoshape) whenever $| \mu - \mu_0 | < \delta$ and $D_0, D \in \mathcal{D}$, where $\mathcal{D}$ is a set of input data compatible with error tolerance of the algorithm.

A central goal in CT is to integrate subroutines for computing a complete, or nearly complete, set of shape or pseudoshape invariants into the algorithm $\mathfrak{A}$ in order to produce an enhanced algorithm $\mathfrak{A}_*$ capable of verifying consistency (pseudoconsistency) and stability (pseudostability). This is quite easy to do when the algorithm generates only solids M, where ∂M is a closed surface in $\mathbb{R}^3$. In this case $\{\chi\}$, where χ is the Euler characteristic, is a complete, ec set of pseudoshape invariants, because χ determines the homeomorphism type of M. On the other hand, if $\mathcal{M}$ is enlarged to include simple ssvwbs, the necessary invariants are less obvious. For example, suppose M is the solid cube $\{\mathbf{x} = (x_1, x_2, x_3) : 0 \leq x_1, x_2, x_3 \leq 1\}$ in $\mathbb{R}^3$, and $\mathfrak{A}$ produces an $N = M \cup \{(-\alpha, -\alpha, x_3) : 0 \leq \alpha \leq \epsilon,\ 0 \leq x_3 \leq 1\}$. Then $\chi(M\ ) = \chi(N\ ) = 1$, even though M and N are not homeomorphic. However, the integral homology H_* (see [35], [51] and [68]), which is ec in this case (see [19], [20], [35] and [81]), does provide a complete set of pseudoshape invariants. In fact, for those with only a passing familiarity with algebraic topology it is easy to show that $H_0(\partial M) = H_0(\partial N) = \mathbb{Z}$, $H_1(\partial M) = H_1(\partial N) = 0$, and $H_2(\partial M) = \mathbb{Z} \neq H_2(\partial N) = \mathbb{Z} \oplus \mathbb{Z}$. In fact, there is even a simpler single ec invariant that does the trick, namely the Euler characteristic χ_2 of the boundary of the object comprised of the vertices, edges and faces of a triangulation, which yields $\chi_2(M) = 2 \neq \chi_2(N) = 3$. These example indicate that sets of shape or pseudoshape invariants can often be found in homology groups, but they also may lie in the (dual) cohomology groups, or various homotopy and cohomotopy groups, among others. Especially useful as shape and pseudoshape invariants are the characteristic (cohomology) classes such as the Euler class, Stiefel-Whitney classes and Pontrjagin classes, all of which are ec via simplicial computations. The Euler class can be identified with the Euler-Poincaré characteristic, which determines pseudoshape for surfaces embedded in $\mathbb{R}^3$, but characteristic classes do not, in general, provide a complete set of invariants. We shall explore in the next section some of these types of invariants that suggest themselves quite naturally for the class of swept volumes.

5. Shape Invariants for Swept Volumes

When an object is moved through space, while possibly also being deformed, it traverses a set of points called its *swept volume*. The motion (possibly including deformation) that generates the swept volume is called a *sweep*. These concepts can be put on a rigorous footing with the introduction of some additional notation (cf. [1]-[5], [11], [76] and [77]).

Let M be a compact object in $\mathbb{R}^n$ that is acted upon by a *sweep* σ, defined as a continuous function

$$\sigma : I := [0, 1] \to \mathrm{Diff(c)}_p^k\left(\mathbb{R}^n\right), \tag{5.1}$$

taking values in the space of C_p^k-diffeomorphisms of $\mathbb{R}^n$ that differ from an affine map (a linear map plus a translation) at most on a compact subset, and satisfies $\sigma(0) := \sigma_0 = id$, where id denotes the identity map. We shall assume in the sequel that σ is continuously differentiable, which involves no real loss of generality since the sweeps of interest in CT are almost always at least piecewise continuously differentiable. Associated to the object

M and sweep σ_t are a *sweep map* $\Sigma : M \times I \to \mathbb{R}^n$ and *swept volume* of M generated by σ, defined as

$$(5.2) \qquad S_\sigma(M) := \operatorname{im}\Sigma = \Sigma(M \times I) = \{\sigma_t(\mathbf{x}) : (\mathbf{x}, t) \in M \times I\} \subset \mathbb{R}^n,$$

and *extended sweep map* $\Sigma^* : M \times I \to \mathbb{R}^{n+1}$, which takes values in spacetime and is defined as $\Sigma^*(\mathbf{x}, t) := (\sigma_t(\mathbf{x}), t)$, and the corresponding *extended swept volume*

$$(5.3) \qquad S_\sigma^*(M) := \operatorname{im}\Sigma^* = \Sigma^*(M \times I) = \{(\sigma_t(\mathbf{x}), t) : (\mathbf{x}, t) \in M \times I\} \subset \mathbb{R}^n \times I.$$

The trajectories of the sweep and extended sweep are generated, respectively, by the *sweep differential equation* (*SDE*), which has the general form

$$(5.4) \qquad \dot{\mathbf{x}} = X_\sigma(\mathbf{x}, t),$$

and the *extended sweep differential equation*, having the form

$$(5.5) \qquad \mathbf{z}' = (\mathbf{x}', t') = X_\sigma^*(\mathbf{x}, t) := (X_\sigma(\mathbf{x}, t), 1),$$

where the dot over a variable denotes, as usual, differential with respect to t, and the prime represents differentiation with respect to a dummy parameter s. The vector fields X_σ and X_σ^* are called the *sweep vector field* (*SVF*) and *extended sweep vector field* (*ESVF*), respectively. For example, in the case of a *rigid sweep*, where σ maps into the Euclidean group $\operatorname{Euc}(n)$ of rigid motions, the SDE has the specific form

$$(5.6) \qquad \dot{\mathbf{x}} = X_\sigma(\mathbf{x}, t) = \dot{\xi}(t) + \dot{A}(t)A^{\mathsf{T}}(t)(\mathbf{x} - \xi(t)),$$

where $A(t) \in \mathrm{SO}(n)$ for all t, and the superscript T denotes the unary transpose operator on matrices. We observe that

$$(5.7) \qquad P(S_\sigma^*(M)) = S_\sigma(M),$$

where P is the natural projection $\mathbb{R}^n \times \mathbb{R} \to \mathbb{R}^n$ defined as $P(\mathbf{x}, t) := \mathbf{x}$. This formula is more useful than it might appear at first glance, for it expresses the swept volume - which may exhibit self-intersections and singular subsets - as the image under a smooth, surjective (onto) map of a set that is regular and without self-intersections, in virtue of standard results on flows concerning the properties of autonomous systems such as (5.5) compared to nonautonomous equations such as (5.4) (see [27], [29], [37] and [67]).

The SDE leads to a useful decomposition of the boundary of a swept volume of a codimension zero object, called the *sweep flow formula* (*SFF*), namely

$$(5.8) \qquad \partial S_\sigma(M) = \partial_- M(0) \cup \partial_+ M(1) \cup \mathcal{G}_\sigma(M) \setminus \mathcal{T}_\sigma(M),$$

where $\partial_- M(0)$ is the set of initial *ingress points* where the SVF points into the interior of $M = M(0) := \sigma_0(M)$, $\partial_+ M(1)$ is comprised of the terminal ingress points where the SVF points out of the interior of $M(1) := \sigma_1(M)$, $\mathcal{G}_\sigma(M)$ is the subset of *grazing points* where the SVF points neither into nor out of the interior of $M(t) := \sigma_t(M)$, $0 \le t \le 1$, and $\mathcal{T}_\sigma(M)$ is the *trim set* consisting of all self-intersection points that lie in the interior of $S_\sigma(M)$. The SFF is particularly useful in the case when the object M is compact and of codimension 0, because the swept volume $S_\sigma(M)$ is completely determined by its boundary. Using the SDE via (5.8) to algorithmically obtain approximations of swept volume boundaries requires computing the grazing points at each incremental stage of the process in the parameter t, which tends to be computationally expensive. Fortunately, there is a variant of the SDE, called the *sweep envelope differential equation* (*SEDE*), of the form

$$(5.9) \qquad \dot{\mathbf{x}} = Y_{(\sigma, M)}(\mathbf{x}, t),$$

depending on both the sweep and the object, which has the property that its trajectories starting on the grazing set of $\partial M(0)$ generate all of $\mathcal{G}_\sigma(M)$, thereby providing the basis for very efficient swept volume algorithms.

The SFF suggests the possibility of decomposing the swept volume into simpler smoother pieces via singularity/stratification theory (see [1]-[6], [13], [14], [26], [43], [66], [72] and [79]) as a strategy in resolving the questions of consistency and stability. There is a natural way of obtaining this decomposition in terms of the sweep map $\Sigma : M \times I \to \mathbb{R}^n$, whereby the image $\Sigma(M \times I) = S_\sigma(M)$ can be expressed in the form

$$S_\sigma(M) = V_1 \cup V_2 \cup \cdots \cup V_m, \tag{5.10}$$

where the *stratification* $\mathfrak{S} = \{V_i : 1 \leq i \leq m\}$ is comprised of a finite disjoint collection of *strata* V_k that are open or closed submanifolds of $\mathbb{R}^n$ with dimensions ranging from 0 to $n_* := \min\{n, \dim(M) + 1\}$. If both the sweep map and the object are of class C^k, with $k \geq 1$, the stratification can be chosen to be of the Thom-Boardman type, where the strata of dimension less than n_* correspond to singularities of Σ, i.e. points where the derivative Σ' has less than maximal rank. We shall prove in the next section that the stratification can be chosen to be *(Whitney) regular*, meaning roughly that all the points in each stratum V_i are "equally singular", and each pair of abutting strata V_i, V_j are joined at well defined angles.

A one-dimensional reduction in the singularity/stratification characterization of swept volumes may be realized by using the flow of the SEDE (5.9), which we represent as

$$\Theta : \partial_0 M(0) \times I \to \mathbb{R}^n. \tag{5.11}$$

Application of the Thom-Boardman approach to (5.11) yields a stratification, which may also be chosen to be regular, of the form

$$\mathcal{G}_\sigma(M) = W_1 \cup W_2 \cup \cdots \cup W_q, \tag{5.12}$$

for which $\max\{\dim W_i\} + 1 = \max\{\dim V_i\}$. Here the trade-off is that Θ can be significantly more complicated than Σ.

When the objects and sweeps are fairly intricate, a complete algorithmic (approximate) characterization of a stratifications such as (5.10) or (5.12) tends to be computationally expensive, but there are useful local forms for the singularities that provide at least partial answers to the questions of consistency and stability (cf. [13], [26] and [43]). In particular, normal forms facilitate a qualitative verification of pseudostability and pseudoconsistency for swept volumes with regular stratifications using Thom-Mather theory. To briefly explain how Thom-Mather theory may be used to resolve consistency and stability, it must first be noted that a regular stratification, besides consisting of submanifolds, also has a relational structure that we denote by $V_i \prec V_j$, which indicates that the closure of V_j contains V_i. Then if $F : \mathfrak{S} \to \tilde{\mathfrak{S}}$ is a relation preserving bijection between two regular stratifications of sets such that the corresponding strata are homeomorphic, it follows from Thom-Mather theory that the two sets are homeomorphic.

One of the reasons that categorical (pseudo) shape invariants are rather accessible for swept volumes is that they are essentially isomorphic to $M \times I$ modulo self-intersection (trimming). Accordingly the characteristic cohomology classes such as the Euler class, Pontrjagin classes, and Stiefel-Whitney classes, all of which are ec for most sweeps and objects of interest, can be used with considerable effect to check for consistency and stability of swept volume rendering algorithms. They do not always provide a complete set of invariants as does the Euler class for embedded surfaces, but they are quite useful for determining inconsistency and instability of algorithms. Moreover, local versions of these

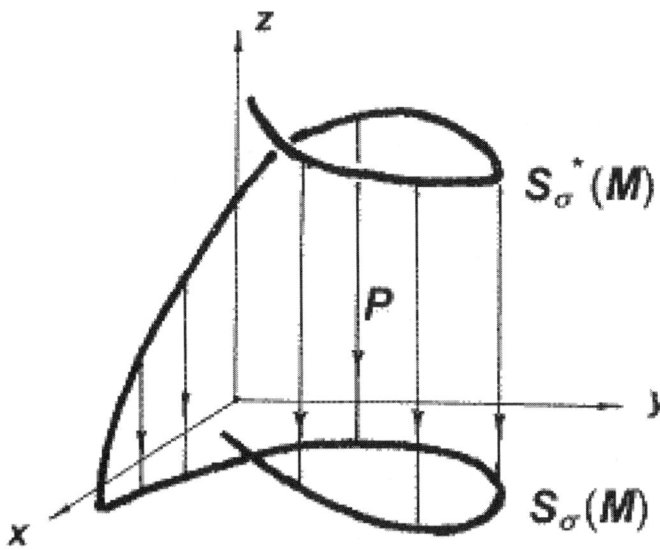

FIGURE 2. Fibers of the projection of the extended swept volume onto the swept volume.

invariants can also be helpful in detecting singular behavior such as self-intersection, as we shall show in the sequel.

There are other related, possibly ec, approaches to analyzing stability and consistency that appear quite promising owing to the unique structure of swept volumes. For example, obstruction theory, which has already been applied to some problems in CAD (see e.g. Mitchel [52]) seems to be particularly well suited to characterizing the pseudoshape of swept volumes in view of the fiber structure illustrated in Fig. 2

Is a lifting of $S_\sigma(M)$ to $S_\sigma^*(M)$ along the fibers of the projection map P obstructed by self-intersections corresponding to certain attached cells in a CW-complex structure? Obstruction theory (in particular Moore-Postnikov factorization) is a natural approach to answering this question (see [34] and [68]). The computability of the obstructions, which are typically subsets of cohomology groups with coefficients in homotopy groups, will fail in general, but for swept volumes of the kind considered in CT it appears that there are ec equivalents available that can be obtained from the simplicial or CW-complex structure of the sets.

6. Stratification of Swept Volumes

We shall show in this section that a C^1 sweep of a compact, connected, codimension-0, C^1 submanifold-with-boundary in $\mathbb{R}^n$ generates a swept volume that is an s-subvariety having a Whitney regular stratification. This result can be generalized to piecewise C^1 sweeps and objects, and objects of higher codimension in a straightforward but rather tedious manner.

Some additional notation is needed to simplify our exposition in the sequel (cf. [6], [26], [40], [43], [72] and [79]). First, we define tangent and normal spaces of a C^1 submanifold V of $\mathbb{R}^n$, as they play key roles in the definition of a (Whitney) regular stratification. Each $\mathbf{x} \in S$ has an open neighborhood U such that there is a C^1 map

$f : U \to \mathbb{R}^m$ satisfying $V \cap U = f^{-1}(0)$. The *tangent space* to V at $\mathbf{x}$, denoted by $T_{\mathbf{x}}(V)$, is defined to be the kernel of the derivative of f at $\mathbf{x}$, which is defined as $\ker f'(\mathbf{x}) := \{\mathbf{y} \in \mathbb{R}^n : f'(\mathbf{x})\mathbf{y} = 0\}$. This tangent space is an $(n-m)$-dimensional (linear) subspace of the Euclidean space $\mathbb{R}^n$. We note that it is sometimes convenient to think of the tangent space at $\mathbf{x}$ as actually passing through this point, which we can do by identifying $T_{\mathbf{x}}(V)$ with the affine subspace $\mathbf{x} + \ker f'(\mathbf{x}) := \{\mathbf{x}+\mathbf{y} : \mathbf{y} \in \ker f'(\mathbf{x})\}$. The *normal space* to V at $\mathbf{x}$ is the orthogonal complement of the tangent space defined as $N_{\mathbf{x}}(V) := T_{\mathbf{x}}(V)^{\perp} := \{\mathbf{z} \in \mathbb{R}^n : \mathbf{y} \in T_{\mathbf{x}}(V) \Rightarrow \langle \mathbf{y}, \mathbf{z}\rangle = 0\}$. Observe that the normal space is an m-dimensional (linear) subspace that yields the orthogonal direct sum decomposition $\mathbb{R}^n = T_{\mathbf{x}}(V) \oplus N_{\mathbf{x}}(V)$ for all $\mathbf{x} \in S$.

We start with a *stratification* $\mathfrak{S} = \{V_i : 1 \le i \le q\}$ of a geometric object M in $\mathbb{R}^n$. By this we mean that M can be decomposed as

$$M = V_1 \cup V_2 \cup \cdots \cup V_q,$$

where $\mathfrak{S}$ is a pairwise disjoint collection of open or closed submanifolds of $\mathbb{R}^n$, with dimensions ranging from 0 to $\dim M$, called the *strata*. In a typical stratification there exist certain strata having the property that $V_i \cap \bar{V}_j \neq \emptyset$, where the bar over the stratum denotes the (topological) closure of the set. For the purpose of defining regularity, when there is an $\mathbf{x} \in V_i \cap \bar{V}_j$, it is convenient to introduce certain tangential and normal projections; namely, we use the projection maps

$$\begin{aligned} \tau_{(V_i,\mathbf{x})} &: \mathbb{R}^n \to T_{\mathbf{x}}(V_i), \\ \eta_{(V_i,\mathbf{x})} &: \mathbb{R}^n \to N_{\mathbf{x}}(V_i), \end{aligned}$$

which are defined naturally in terms of the orthogonal direct sum decompositions mentioned above. It is also convenient to denote the unit vector in the direction of a nonzero point $\mathbf{x}$ as $\hat{\mathbf{x}}$ and the vector from the point $\mathbf{x}$ to the point $\mathbf{y}$ as $\mathbf{xy}$. We have now assembled the notation necessary to define regularity.

DEFINITION 6.1. The stratification $\mathfrak{S} = \{V_i : 1 \le i \le q\}$ of M is *Whitney regular* if it satisfies the following two properties for all $\mathbf{x} \in V_i \cap \bar{V}_j$ and $\mathbf{y} \in V_j$:

(R_a) The limiting condition

$$\lim_{\mathbf{y}\to\mathbf{x}} \eta_{(V_j,\mathbf{y})}(\mathbf{z}) = 0$$

applies for every $\mathbf{z} \in T_{\mathbf{x}}(V_i)$, which means that the distance between $\mathbf{z}$ and $T_{\mathbf{y}}(V_j)$ tends to zero as $\mathbf{y} \to \mathbf{x}$.

(R_b) The strata satisfy the limiting condition

$$\lim_{\mathbf{y}\to\mathbf{x}} \eta_{(V_j,\mathbf{y})} \left(\frac{\mathbf{yx} - \tau_{(V_i,\mathbf{x})}(\mathbf{yx})}{\left|\mathbf{yx} - \tau_{(V_i,\mathbf{x})}(\mathbf{yx})\right|} \right) = \lim_{\mathbf{y}\to\mathbf{x}} \eta_{(V_j,\mathbf{y})} \left(\widehat{\eta_{(V_i,\mathbf{x})}(\mathbf{yx})} \right) = 0,$$

which means that the distance between the unit vector in the direction of $\mathbf{yx} - \tau_{(V_i,\mathbf{x})}(\mathbf{yx})$ and $T_{\mathbf{y}}(V_j)$ tends to zero as $\mathbf{y} \to \mathbf{x}$.

To illustrate the notion of a regular stratification, consider the algebraic subvariety of $\mathbb{R}^2$ defined as $M := \{(x,y) : y(y - x^3) = 0\}$. It is easy to see from the definition that the stratification consisting of the three strata $\{(x,0) : x \in \mathbb{R}\}$, $\{(x,x^3) : x > 0\}$ and $\{(x,x^3) : x < 0\}$ is not regular, while the stratification with the five strata $\{0\}$, $\{(x,0) : x > 0\}$, $\{(x,0) : x < 0\}$, $\{(x,x^3) : x > 0\}$ and $\{(x,x^3) : x < 0\}$ is regular. Many additional illuminating examples of this type can be found in [43].

If $\mathfrak{S} = \{V_i : 1 \leq i \leq q\}$ is a regular stratification of the geometric object M, it is straightforward to show that $V_i \cap \bar{V}_j \neq \emptyset$ (and $V_i \neq V_j$) implies that $V_i \subset \bar{V}_j$ and $\dim V_i < \dim V_j$. Accordingly a regular stratification comes equipped with a partial ordering defined as $V_i \prec V_j$ if and only if $V_i \cap \bar{V}_j \neq \emptyset$ or equivalently $V_i \subset \bar{V}_j$. Thom-Mather theory (see e.g. [6], [26], [43] and [72]) therefore enables one to reduce pseudoshape comparisons of geometric objects to a finite series of shape computations with simpler objects. In particular, we have the following fundamental result.

THEOREM 6.2. *(Thom-Mather). Let M and M' be subsets of $\mathbb{R}^n$ possessing regular stratifications $\mathfrak{S}$ and $\mathfrak{S}'$, respectively. Suppose that $\Phi : \mathfrak{S} \to \mathfrak{S}'$ is a partial order preserving bijection such that V_i and $\Phi(V_i)$ are homeomorphic for all strata in $\mathfrak{S}$. Then M and M' are homeomorphic.*

With Theorem 6.2, we see an obvious advantage to analyzing regularly stratified sets. In fact, it provides a strong incentive for restricting the geometric objects of interest in CT as follows.

Recommendation: The sets of interest in CT can be reasonably defined to be just those subsets of a Euclidean space $\mathbb{R}^n$ that possess a regular stratification.

For example, the solid cube in $\mathbb{R}^3$ has a regular stratification comprised of the eight vertices, twelve edges (minus the vertices), six faces (minus the edges), and the open interior, which is the cube minus its boundary. In fact, most reasonable piecewise smooth objects (e.g. any such set having a finite triangulation) have regular stratifications. We next show that this is also the case for any swept volume generated by a smooth sweep and a smooth initial object.

THEOREM 6.3. *Let M be a compact, connected, codimension-m, C^1 submanifold-with-boundary of $\mathbb{R}^n$ and let σ be a C^1 sweep. Then the swept volume $S_\sigma(M)$ is a compact, connected, codimension-k $(0 \leq k \leq m)$ s-subvariety-with-boundary of $\mathbb{R}^n$ that has a Whitney regular stratification.*

PROOF. We note that if both the sweep and manifold are real analytic, the result follows from Whitney [79]. For concreteness, we present the proof only for the case of C^1 sweeps and codimension-0 objects in $\mathbb{R}^3$. The extension to more general sweeps in $\mathbb{R}^n$ and objects of higher codimensions is tedious but fairly routine. We could fashion our argument along the lines of a Thom-Boardman approach based on the singularities of the sweep map Σ or the sweep-envelope map Θ defined in Section 5, but we prefer to take a more geometrically intuitive tack using the definition of the swept volume as the natural projection of the extended swept volume. Recall that $P : \mathbb{R}^4 \to \mathbb{R}^3$ is defined as $P(x,t) = P(x,y,z,t) := (x,y,z)$, and $P(S_\sigma^*(M)) = S_\sigma(M)$.

First, suppose that each fiber $F_{\mathbf{x}} := \{(\mathbf{x},t) : t \in \mathbb{R}\}$ through a point in $S_\sigma(M)$ intersects $S_\sigma^*(M)$ in one and only one point, which corresponds to the case when the swept volume has no self-intersections. As $S_\sigma^*(M)$ is a piecewise C^1, codimension-0 submanifold -with-boundary of $\mathbb{R}^4$ (with all of the points at which it fails to be C^1 in the hyperplanes $t = 0$ and $t = 1$) in virtue of basic properties of C^1 flows (see e.g. [27], [30], [37], [40] and [67]), $S_\sigma(M) = P(S_\sigma^*(M))$ is a C^1, codimension-0 submanifold -with-boundary of $\mathbb{R}^3$. Accordingly $S_\sigma(M)$ has a trivial stratification comprised of $IntS_\sigma(M)$, which is an open codimension-0 submanifold of $\mathbb{R}^3$, and $\partial S_\sigma(M)$, which is a closed codimension-1 submanifold of $\mathbb{R}^3$.

It is easy to see that $\partial S_\sigma(M)$ consists of precisely those points $\mathbf{x}$ such that $F_{\mathbf{x}}$ intersects $\partial S_\sigma^*(M)$ but not $IntS_\sigma^*(M)$ - i.e. it just *grazes* $S_\sigma^*(M)$. Moreover, the boundary points of $S_\sigma(M)$, excluding the initial and terminal positions of the object during the course of the sweep, are readily seen to be projections of points where the restriction of P to $\partial S_\sigma^*(M)$, which we denote by π, has a derivative π' with rank less than three. Observe that the trivial stratification in the case with no self-intersection, given by $S_\sigma(M) = \partial S_\sigma(M) \cup IntS_\sigma(M)$, obviously satisfies the two regularity conditions (R_a) and (R_b).

We shall use the same geometric elements to construct a regular stratification of the swept volume in the case where self-intersections occur. In these cases, the 0-dimensional and 1-dimensional strata are comprised of projections of points for $\pi : S_\sigma^*(M) \to \mathbb{R}^3$ where $\text{rank}\pi' = 0$ and $\text{rank}\pi' = 1$, respectively. But once again these points shall be identified using the more geometric approach employed for the case of swept volumes having no points of self-intersection. Construction of the strata shall be done in the order of decreasing dimension.

Obviously, the three-dimensional stratum is just $IntS_\sigma(M)$, which is comprised of the points $\mathbf{x} \in S_\sigma(M)$ where the fiber $F_\mathbf{x}$ intersects points interior to $S_\sigma^*(M)$. The two-dimensional strata are the components of the set that results from subtracting the zero- and one-dimensional strata, the union of which we denote by $\mathcal{S}_{01}$, from $\partial S_\sigma(M)$. More precisely, the two-dimensional strata $V_1^2, ..., V_{k_2}^2$ are the components of $\partial S_\sigma(M) \setminus \mathcal{S}_{01}$.

Continuing in this vein, the one-dimensional strata $V_1^1, ..., V_{k_1}^1$ are then defined to be the components of $\mathcal{S}_{01} \setminus \mathcal{S}_0$, where $\mathcal{S}_0$ is the set of all 0-dimensional strata. Accordingly it remains to construct the sets $\mathcal{S}_{01}$ and $\mathcal{S}_0$ in order to obtain the desired stratification.

We now define $\mathcal{S}_{01}$ to be the totality of points $\mathbf{x}$ of $\partial S_\sigma(M)$ such that $F_\mathbf{x}$ intersects ∂S_σ^* in more than one point. Such points may be assumed to belong to one of the following two categories: There is an open neighborhood U of $\mathbf{x}$ in $\mathbb{R}^3$ such that for every $\mathbf{y} \in (U \setminus \{x\}) \cap \partial S_\sigma(M)$, $F_\mathbf{y}$ intersects $\partial S_\sigma^*(M)$ in only one point, in which case we say the point is of *type 0*, or $\pi\left(\pi^{-1}(U \cap \partial S_\sigma(M))\right)$ consists of one or more curves that are C^1 except possibly at $\mathbf{x}$, in which case we say the point is of *type 1*. The set $\mathcal{S}_0$, consisting of all 0-dimensional strata, is now defined to be the set of all type 0 points, and all type 1 points $\mathbf{x}$ for which one of the curves described above is merely piecewise C^1 at $\mathbf{x}$.

At this stage, we have now defined both $\mathcal{S}_{01}$ and $\mathcal{S}_0$. Owing to the compactness of $M \times I$ and the continuous differentiability of the sweep, we infer that there is a finite unique minimal (in terms of number of sets) decomposition of $\mathcal{S}_{01} \setminus \mathcal{S}_0$ into one-dimensional submanifolds, and these are precisely the one-dimensional strata $V_1^1, ..., V_{k_1}^1$ we are seeking. Analogously, there is a finite unique minimal decomposition of $\mathcal{S}_0$ into 0-dimensional submanifolds containing no accumulation points (some of which could actually have infinitely many points), which yields the 0-dimensional strata $V_1^0, ..., V_{k_0}^0$.

It is straightforward to show that the stratification $V_1^0, ..., V_{k_0}^0, V_1^1, ..., V_{k_1}^1, V_1^2, ..., V_{k_2}^2$, $IntS_\sigma$ satisfies the regularity conditions (R_a) and (R_b). However the computations necessary to verify regularity are complicated and tedious, so we leave them for the interested reader in order to keep the length of our argument from becoming too unwieldy. In summary then, it follows from our construction that this stratification is regular, and that the union of the strata - which comprises $S_\sigma(M)$ - is an s-subvariety-with-boundary of $\mathbb{R}^3$, so the proof is complete. □

When the object M and the sweep σ are defined in terms of relatively simple smooth functions, it is sometimes possible to find exact formulas for the strata described in the proof of Theorem 6.3 (cf. [1]-[5]). However, when the objects and sweeps are a bit more complicated, exact formulas for the strata are usually difficult or impossible to find, and approximate numerical descriptions even tend to be computationally expensive. In such cases, other approximation techniques employing certain normal forms for singularities are useful (see e.g. [12]). Even with such available methods, it still may not be practicable to compute an exact regular stratification of a swept volume. However, if one is wiling to settle for an approximate stratification of a triangulation of a swept volume, it is not difficult to envisage how this can be accomplished an effectively computable way based upon the geometric approach in the proof of the above theorem. We shall provide details of this procedure in a forthcoming paper.

7. Homological Intersection Detection

Determining when and how two objects intersect is a fundamental problem in CT that has been addressed, and partially solved, in many papers such as [12], [45], [54] and [55]. We shall show in this section how homology can be used to detect transverse intersections of submanifolds of $\mathbb{R}^n$, and also discuss how this method can be implemented in terms of available data structures.

The intersecting objects M and N (which may be identical for self-intersection) we shall study here are C^k submanifolds, with $k \geq 1$, but the results will be seen to also apply, with the necessary modifications, to topological submanifolds. A *transverse intersection point* $\mathbf{x} \in M \cap N$ is one such that the tangent spaces of the two submanifolds at $\mathbf{x}$ span $\mathbb{R}^n$, i.e. $T_{\mathbf{x}}(M) + T_{\mathbf{x}}(N) = \mathbb{R}^n$. If the submanifolds are transverse at all their intersection points - denoted as $M \pitchfork N$ - it is easy to show that $M \pitchfork N$ is a C^k submanifold and $\dim(M \pitchfork N) = \dim M + \dim N - n$ (cf. [6], [26], [30], [34], [40] and [43]). We also note that if f is a diffeomorphism, then $M \pitchfork N$ implies that $f(M) \pitchfork f(N)$. A local topological characterization of transverse intersections is provided by the following intuitively obvious (and easily proved) result.

LEMMA 7.1. *Let M and N be C^k submanifolds of $\mathbb{R}^n$ of dimensions r and s, respectively. Then $M \pitchfork N$ implies that for each $\mathbf{x} \in M \cap N$ there is an open neighborhood U of $\mathbf{x}$ such that $(M \cup N) \cap U$ is homeomorphic with $(\mathbb{R}^r \cup \mathbb{R}^s) \cap V$, where $\mathbb{R}^r$ and $\mathbb{R}^s$ are regarded as linear subspaces of $\mathbb{R}^n$, and V is some open neighborhood of the origin.*

We shall now apply some basic homology theory to describe transverse intersections of the submanifolds M and N. The reader acquainted with homology should have no difficulty following our reasoning. For those not familiar with homology, we suggest the references [35] and [68], but we also offer the following intuitive description that may be helpful. The integral homology of an m-dimensional object X of the type considered in CT is a collection of abelian groups $H_*(X) = \{H_0(X), ..., H_m(X)\}$. Each of the $H_q(X)$ is, roughly speaking, the abelian group over the integers generated by the closed submanifolds of dimension q (made up of q-dimensional faces of a triangulation of X) that are boundaries of $(q+1)$-dimensional holes of X. As illustrative examples we observe that if X is a circle in the plane, $H_0(X) = H_1(X) = \mathbb{Z}$, and if Y is the unit disk in the plane, then $H_0(Y) = \mathbb{Z}$ and $H_1(Y) = H_2(Y) = 0$. As these examples indicate, $H_0(Y)$ essentially just counts the number of components of the space.

THEOREM 7.2. *Let $M \pitchfork N$ in $\mathbb{R}^n$ with $\dim M = \dim N := r = n - 1$. Then if $\mathbf{x} \in M \cap N$, the local (relative) homology satisfies*

$$H_q(M \cup N, (M \cup N) \setminus \{\mathbf{x}\}) = \begin{cases} \mathbb{Z}^3, & \text{if } q = r \\ 0, & \text{otherwise} \end{cases}, \tag{7.1}$$

and if $\mathbf{x} \in (M \cup N) \setminus (M \cap N)$,

$$H_q(M \cup N, (M \cup N) \setminus \{\mathbf{x}\}) = \begin{cases} \mathbb{Z}, & \text{if } q = r \\ 0, & \text{otherwise} \end{cases}. \tag{7.2}$$

PROOF. We prove the second case first. If $\mathbf{x} \in (M \cup N) \setminus (M \cap N)$, there is a closed neighborhood K of $\mathbf{x}$ in $M \cup N$ that is homeomorphic to a closed r-dimensional ball D^r at the origin in $\mathbb{R}^r$. As homology is a homeomorphism type invariant, it suffices to consider the local homology of the ball. It follows from the long exact sequence for the pair $(D^r, D^r \setminus \{0\})$ that $H_q(D^r, D^r \setminus \{0\}) \approx \tilde{H}_q(D^r \setminus \{0\})$, where $\approx$ denotes isomorphism of the abelian groups. Accordingly we obtain the desired result from the fact that $D^r \setminus \{0\}$ is homotopic to the $(r-1)$-sphere S^{r-1}.

If $\mathbf{x} \in M \cap N$, it follows from Lemma 7.1 and the local nature of our computations that it suffices to consider a transverse intersection of two r-dimensional linear subspaces of $\mathbb{R}^n$ at the origin, which we denote by W. Define $X := W \cap \{x \in R^n : |x| \leq 1\}$. Again, the long exact sequence for the pair $(X, X \setminus \{0\})$ implies that $H_q(X, X \setminus \{0\}) \approx \tilde{H}_q(X \setminus \{0\})$, and it is easy to see that $X \setminus \{0\}$ is homotopic to the union of two $(r-1)$-spheres intersect along an $(r-2)$-sphere. Let $Y = Y^{r-1} := X \setminus \{0\}$ be given the cell structure of a CW-complex (see [30], [34] and [68]).

If $r > 2$, Y is comprised of one 0-cell, one $(r-2)$-cell attached to the 0-cell to form a sphere, and four $(r-1)$-cells attached to that sphere. It is straightforward to compute that the corresponding cellular chain complex in homology is

$$0 \longrightarrow H_{r-1}\left(Y^{r-1}, Y^{r-2}\right) \xrightarrow{\partial} H_{r-2}\left(Y^{r-2}, Y^{r-3}\right) \xrightarrow{\partial} \mathbb{Z} \longrightarrow 0,$$

when $r = 3$, and

$$0 \longrightarrow H_{r-1}\left(Y^{r-1}, Y^{r-2}\right) \xrightarrow{\partial} H_{r-2}\left(Y^{r-2}, Y^{r-3}\right) \longrightarrow 0,$$

when $r > 3$. Here Y^s denotes the s-skeleton of the CW-complex structure of Y consisting of all cells of dimension not exceeding s. In both cases, the kernel of ∂ is $H_{r-1}\left(Y^{r-1}\right)$, and it is not difficult to show that $H_{r-1}\left(Y^{r-1}, Y^{r-2}\right) = \mathbb{Z}^4$ and $H_{r-2}\left(Y^{r-2}, Y^{r-3}\right) = \mathbb{Z}$. Consequently, $\partial : \mathbb{Z}^4 \to \mathbb{Z}$. The actual form of ∂ is determined by the attaching maps of the $(r-1)$-cells, and it is readily verified that $\partial(i,j,k,l) = i+j+k+l$, which obviously has $\mathbb{Z}^3$ as its kernel. Therefore, $H_{r-1}\left(Y^{r-1}\right) = \mathbb{Z}^3$, the surjectivity of ∂ implies that $H_{r-2}\left(Y^{r-1}\right) = 0$, and the triviality of the other homology groups is obvious.

As the result is obvious for the case $r = 1$, it remains only to dispose of $r = 2$. In this case, Y^1 has two 0-cells and four 1-cells, and the corresponding cellular chain complex is

$$0 \longrightarrow H_1\left(Y^1, Y^0\right) \xrightarrow{\partial} H_0\left(Y^0\right) \longrightarrow 0,$$

which is actually

$$0 \longrightarrow \mathbb{Z}^4 \xrightarrow{\partial} \mathbb{Z}^2 \longrightarrow 0.$$

From the attaching map we compute that $\partial(i,j,k,l) = (i+j+k+l, -i-j-k-l)$, hence $\ker \partial = H_1\left(Y^1\right) = \mathbb{Z}^3$. Moreover, the image of ∂ is the diagonal in $\mathbb{Z}^2$. Whence $H_0\left(Y^1\right) = \mathbb{Z}$, from which we infer that $\tilde{H}_0\left(Y^1\right) = 0$. As the other homology groups are obviously trivial, the proof is complete. □

Observe that the proof of the above theorem relied only on local homeomorphisms. Hence it can be generalized to topologically transverse intersections in the following sense. Two topological submanifolds M and N of, of dimensions r and s, respectively, of $\mathbb{R}^n$ *intersect transversally* if they intersect, and for each $\mathbf{x} \in M \cap N$ there exist open neighborhoods U and V of $\mathbf{x}$ and the origin, respectively, such that $(M \cup N) \cap U$ is homeomorphic to $(\mathbb{R}^r \cup \mathbb{R}^s) \cap V$, where $\mathbb{R}^r$ and $\mathbb{R}^s$ intersect transversally in $\mathbb{R}^n$.

Theorem 7.2 gives an effectively computable method for detecting transverse intersections, whereby the necessary homological invariants can be calculated using triangulations or CW-structures for the manifolds. The homological computations are (at least in theory) much more efficient than say a brute-force direct check of faces when M and N are triangulated. This will be demonstrated in a number of effectively computable characterizations of intersections in the remainder of this section.

There are a number of interesting and useful corollaries of Theorem 7.2. For example, we have the following characterization of self intersecting hypersurfaces that is an immediate consequence of the theorem.

COROLLARY 7.3. *If a C^1 compact, connected, codimension-1 submanifold M of $\mathbb{R}^n$ has points of transverse self-intersection, then there exists an $\mathbf{x} \in M$ such that $H_{n-1}(M, M \setminus \{\mathbf{x}\}) = \mathbb{Z}^3$.*

Another closely related result, which follows directly from this result, gives an effectively computable procedure for interrogating transverse intersections.

COROLLARY 7.4. *If two C^1, codimension-1 submanifolds M and N of $\mathbb{R}^n$ intersect transversally at $\mathbf{x} \in M \cap N$ then there is an arbitrarily small closed neighborhood K of $\mathbf{x}$ such that the Euler characteristic of the $(n-2)$-skeleton, Y^{n-2}, of the CW-complex or polyhedral structure of $(M \cup N) \cap K$ satisfies $\chi\left(Y^{n-2}\right) = -3$ (5) when n is odd (even).*

The last corollary suggests an ec method of detecting transverse intersections of triangulated representations of objects that can be incorporated into the data structure of the algorithm used to render the objects. If the representation of the objects is in terms of triangulated approximations, the following result, which follows directly from Lemma 7.1 and is related to Corollary 7.4, is useful.

LEMMA 7.5. *Let M and N be as in Corollary 7.4, and suppose $\mathcal{T}$ is a triangulation of $M \cup N$. Then $M \pitchfork N$ implies that there is an $(n-2)$-simplex of $\mathcal{T}$ that is the face of exactly four $(n-1)$-simplices.*

Alternatively, we have Theorem 7.2 (with formula (7.1)) and Corollary 7.4 as means of characterizing transverse intersections using a triangulation of $M \cup N$.

The construction of effective and efficient triangulations is a nontrivial problem (see e.g. [22] and [23]). Given triangulated objects M and N in $\mathbb{R}^n$, is there an efficient method for determining if they intersect? Most of the data structures currently used to represent objects - especially those that are not manifolds - include only local means for analyzing the space. Hence, no global information is explicitly incorporated. Accordingly intersection detection for these structures requires an extensive search procedure. In particular, one has to interrogate and group all faces of the triangulation and check the conditions of Theorem 7.2, Corollary 7.4 or Lemma 7.5. This is easy to do for most of the data structures in use, but it is obviously not an efficient approach. It appears that intersection detection could be accelerated for these data structures if they are modified to compile the global data necessary to implement the criteria developed in this section.

Now that we have concluded our treatment of three of the major themes of this paper - shape, stratification, and homological intersection detection - it behooves us to examine some of the links among them. We showed that by introducing a weaker form of shape (pseudoshape), one is more apt to find a complete set of effectively computable topological shape invariants. Then by employing a regular stratification, the problem of shape characterization can be reduced to a number of simpler problems involving the strata, thereby increasing the likelihood of an effectively computable solution. Finally, self-intersection plays a key role in the shape of swept volumes, so the results of this section provide an additional algorithmic tool for shape interrogation.

8. Virtual Sculpting Applications

Maiteh, Leu and their collaborators have developed a very efficient prototype virtual sculpting system based upon an innovative combination of SDE, ray-casting and Delaunay triangulation techniques (see e.g. [41], [42], [44], [45] and [60]). We shall briefly outline in this section how our investigation of swept volumes from the CT viewpoint can be used to realize an order of magnitude improvement of the Maiteh-Leu prototype virtual sculpting system. It should be noted that there are several other approaches to virtual sculpting such as those in [18], [21] and [73].

The components of the Maiteh-Leu system are a standard fingertip mounted virtual reality (VR) tracking device to represent the sculpting motion applied to a virtual workpiece, VR shutter glasses, a transmitter, ascension unit, and a desktop computer and screen as shown in Fig. 3. The program for the system is constructed primarily upon a foundation comprised of SDE swept volume algorithm integrated into a Boolean operations algorithm that incorporates some innovative modifications of scan-rendering techniques. The sculpted object is rendered as a triangulated object, illustrated in Fig. 4, produced by updating the triangulation of the virtual workpiece via Delaunay techniques as it is sculpted.

The program for this prototype virtual sculpting environment enables the designer to initialize the three-dimensional workpiece, select and grip virtual objects that simulate standard sculpting tools, and to sweep the tools over the workpiece and view - in real-time - rather accurate graphical representations of the results of the sculpting operations. Virtual sculpting trajectories and the resulting sculpted design models are produced in the following steps: (a) The geometry modeler is used to create the virtual environment, which is then displayed on the computer screen. (b) A virtual tool is created by the geometry modeler and inserted into the VR environment. (c) The user grips the virtual sculpting tool and moves it along the workpiece to simulate the sculpting operation. (d) A tracking device records the position of the tool during its motion. (e) Virtual toolpaths and geometric models are used as inputs for the design creation procedure. (f) The geometry modeler generates the swept volumes of the sculpting tool, and performs the Boolean operations associated with the cutting of the workpiece to obtain the changes in geometry. (g) Shutter glasses provide a realistic view of the whole VR sculpting process.

We can summarize the complete Boolean operations algorithm that lies at the heart of our VR sculpting as being comprised of the following stages: (1) The initial workpiece is triangulated and the SDE is used to create an evolving swept volume of the sculpting tool as it is moved by the user. (2) Intersection points of the tool swept volume and the workpiece are computed using scan-rendering (ray-casting) techniques. Rather than the usual scan-rendering method, the computations are modified to follow the motion of the tool and made more efficient by employing bounding boxes moving with the evolving surface of the tool swept volume. (3) The modified scan-rendering approach is then used to determine which points of the tool swept volume lie inside the workpiece. (4) Boolean subtraction is then used to remove the portions of the tool swept volume from the workpiece, and then a Delaunay type algorithm is used to re-triangulate the evolving workpiece. (5) The resulting modified triangulated workpiece is rendered in piecewise linear form and the process is continued and constantly updated during the entire virtual sculpting process.

The innovative combination of an SDE produced tool swept volume approximation, scan-rendering and bounding boxes adapted to the evolving tool swept volume, and a robust Delaunay re-triangulation scheme produces a rather a natural and realistic virtual sculpting environment even with the simple equipment used to create it. A straightforward

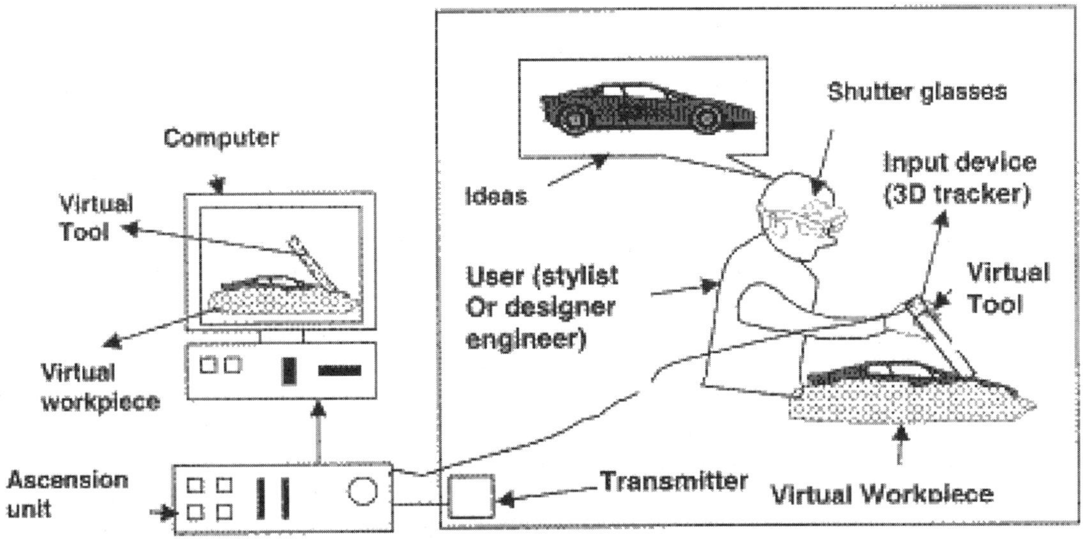

FIGURE 3. Prototype VR sculpting system.

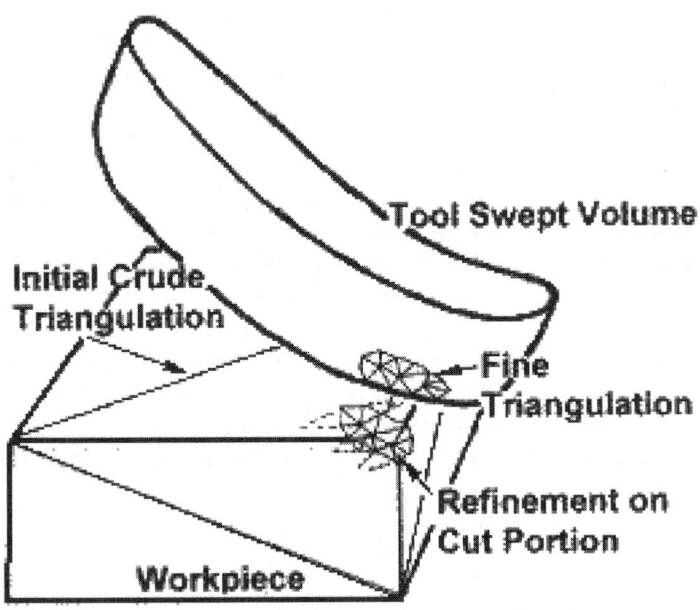

FIGURE 4. Triangulation of the sculpted workpiece.

calculation shows that the computational cost of the algorithm upon which the system is built is rather low - consistent with the real-time computations observed in several experiments. In particular, it can be readily shown that the computational complexity of the entire system algorithm is not greater that $O\left(N^{10/3}\log N\right)$, where $N^{1/3}$ is the number of vertices in the triangular approximation of the boundary of the swept volume of the tool (see [44] and [45]). We note here that smoother representations of the swept volume of the tool and the sculpted surface could be obtained using NURBS, with no appreciable increase in the computational cost of the whole Boolean operation algorithm. This would be most relevant if one wants to convert the VR sculpting environment to a manufacturing design system, where additional smoothness and accuracy are usually required.

If one substitutes an SEDE algorithm for the SDE method in representing the swept volumes of the sculpting tools in the Maiteh-Leu prototype, an order of magnitude reduction in the computational complexity of the overall Boolean operations algorithm will be obtained, and perhaps even slightly more in light of the fact that trimming is unnecessary when computing Boolean subtractions of swept volumes. This will enable us to make the sculpting environment even more realistic and flexible - by the introduction of a wider variety of tools, and the capability of performing shaping operations with the fingers and hands - while still satisfying real-time computational requirements. We can do even better by using the SEDE in an indirect but easier to implement way that employs the natural fiber structure of the extended swept volume over the swept volume described and exploited in some of the preceding sections.

To describe this alternative SEDE based sculpting approach, we assume that the object (sculpting tool) $M \subset \mathbb{R}^3$ is a smooth, compact, connected, three-dimensional submanifold-with-boundary that is identified with $M \times 0$ in spacetime $\mathbb{R}^4$. A smooth sweep σ in $\mathbb{R}^3$ produces an extended swept volume of $S_\sigma^*(M)$, which is the four-dimensional smwb generated by the flow of $M \times 0$ under the ESDE. $S_\sigma^*(M)$ can be triangulated along the trajectories of the ESDE as it evolves. As in Section 6, let π be the restriction of the projection $P : \mathbb{R}^4 \to \mathbb{R}^3$ to the boundary of $S_\sigma^*(M)$. The singular points of $\pi : \partial S_\sigma^*(M) \to \mathbb{R}^3$ are points where the rank of the derivative π' is less than three, and these points are projected onto boundary points of the swept volume $S_\sigma(M)$. Thus we may take the (small) 3-simplices (tetrahedra) on $\partial S_\sigma^*(M)$ that contain the singular points and project them onto a triangulated approximation of $\partial S_\sigma(M)$. Note that the projected singular 3-simplices are represented as 2-simplices (triangles). As these triangles are projections of faces of solid 4-simplices forming the triangulation of $S_\sigma^*(M)$, we can directly determine whether or not the projected boundary triangles intersect ∂W or are in the interior of the workpiece W, thereby eliminating the need for scan-rendering in the sculpting program. Moreover, the triangulation of any portion of $\partial S_\sigma(M)$ intersecting W can be impressed on $\partial\left(W \setminus S_\sigma(M)\right)$, which in turn can be embedded in an efficient localized re-triangulation of the surface of the (evolving) workpiece. It is not difficult to show that the resulting virtual sculpting algorithm has a computational complexity of $O\left(N^{5/3} \log N^{1/2}\right)$, so it has the potential for serving as the basis for very realistic and flexible virtual sculpting systems.

9. Modeling Heterogeneous Material

Analysis, modeling, representation and fabrication of heterogeneous materials play prominent roles in a wide variety of modern applications ranging from the design and construction of composite material structures to tissue engineering. Here we shall confine our attention to certain entities, which are important foci of interest in biomedical engineering, such as fibrous materials, bone matter, connective tissue, and growth matrices.

In a recent series of papers, Regli and Sun and their collaborators have developed several innovative methods for representing complex object properties of heterogeneous biomaterials, along with effective programs for cataloging the models that are created (see e.g. [8], [57], [58], [69]and [70]). They have crafted rather efficient algorithms for modeling and analyzing such materials, which incorporate such features as density, porosity and permeability while also capturing random aspects of the objects by integrating elements of

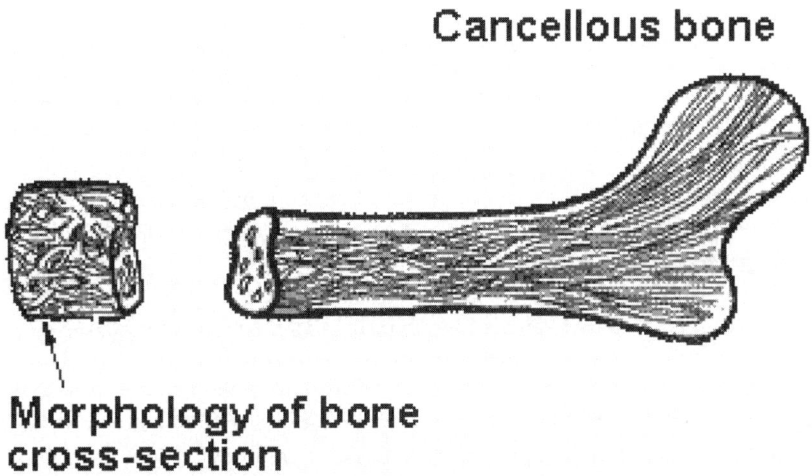

FIGURE 5. Simplified picture of internal bone morphology.

stochastic geometry with CAD and solid modeling. Furthermore, they have - in collaboration with a biomaterials company - employed their algorithms to develop manufacturing processes for creating heterogeneous objects that can be used in such applications as bone tissue replacement.

A close inspection of the structure of bones and related tissue configurations, for example, as illustrated in Fig. 5, reveals that these entities can be viewed as being comprised of biological material threads, as it were, woven or swept through the length of the object. This perspective suggests that swept volumes and sweeping techniques may provide a viable basis for the development of potentially effective algorithms for analyzing and representing complicated heterogeneous tissue structures. In what follows, we shall briefly describe a couple of ways in which sweeps can be used to address some current problems in tissue engineering related to the composition of bones.

We begin with a method of using sweep differential equations to model the tissue strands (channels) that can simulate the density of these components of the structure, but is incapable of producing the branching structure that is often encountered in bone morphology. Our description will be for the most natural space for such application, $\mathbb{R}^3$, but the approach can easily be extended to any finite dimension. Assume that the axis along the bone of length L is the z-axis, and at $z = 0$ the channels intersect the cross-section in circles $\partial D_1, ..., \partial D_m$ that bound disks $D_1, ..., D_m$, respectively. We want to flow these disks along the length of bone (to create channels) so that although they may twist around one another, the density of the disks is either constant or multiplied by some scale factor at each cross-section $z =$ constant.

To do this we use a differential equation of the following form, which may be obtained from a smooth planar SDE or an even more general construction

$$\dot{\mathbf{u}} = U(\mathbf{u}, z), \quad \dot{z} = 1, \tag{9.1}$$

where $\mathbf{u} := (x, y)$. For this equation to have the desired density preserving or scaling property on cross-sections, all we have to do is make it subject to the following restriction (satisfied by rigid or similarity SDEs):

$$\partial_{\mathbf{u}} U(\mathbf{u}, z) = \kappa(z) A(\mathbf{u}, z), \tag{9.2}$$

with $A \in \mathrm{SO}(2)$. Here the positive quantity $\kappa(z)$ is the scale factor, so the density of disks bounded by the channels walls at any cross-section $z = z_*$, with $0 \leq z_* \leq L$, is just $\kappa(z_*) \times$(density at $z = 0$). If φ_t is the flow generated by (9.1), the solid channels obtained will have the form $\{\varphi_t (D_k \times 0) : 0 \leq t \leq L\}$, $1 \leq k \leq m$, and can be constructed in a wide variety of very intricate patterns that are capable of producing fairly realistic models of bone morphology.

The above process of using a sweep type differential equation to construct nonbranching channel patterns for bones can also be accomplished using a combination of empirical data and numerics as follows: Let the length of the bone be partitioned into equally spaced cross-sections $z = 0 < z_1 < z_2 < \cdots < z_q = L$, where the increment is small. Measure the positions of the channel disks at each corresponding cross-section. We can join what appear to be corresponding points on disks belonging to the same channels on cross-sections z_i and z_{i+1} by vectors that point in the positive z-direction and have z-component equal to one. If we choose a fairly fine subdivision of points to represent the disks along the cross-sections, we have enough data to use a numerical differential equation solver, such as a Runge-Kutta scheme, to create a rather accurate representation of the flow of a differential equation that produces the channel structure of the bone. This method may be particularly useful for bone replacement applications, where it is usually important to match the actual bone structure as closely as possible.

To create patterns in which the channels branch at various points along the length of the bone, it is necessary to modify the above sweeping procedure. All that it takes to accomplish this is to multiply the system of equations by a factor that vanishes at points in a way that forces the flows of the disks to split into two, three or more parts at selected locations along the length of the bone. To this end, we consider the flow of the system of differential equation of the form

$$\dot{\mathbf{u}} = \rho(\mathbf{u}, z) U(\mathbf{u}, z), \quad \dot{z} = \rho(\mathbf{u}, z), \tag{9.3}$$

where ρ is a nonnegative function that vanishes only along certain curve segments lying in cross-sections of the bone, and otherwise this differential equation satisfies the same (cross-sectional area preserving) condition as (9.1). More specifically, ρ is chosen to vanish only on finitely many curves that cut across the cross-sections of the flows of the disks in ways that partition them into two, three or more parts (as if they were pieces of a pie). In addition, ρ is constructed to be 1 outside of a very small neighborhood of each of these cross-sectional branching curves. In this way, one can create rather complicated branched channel patterns that are defined by the solid channels $\{\psi_t (D_k \times 0) : 0 \leq t\} \cap \{(\mathbf{u}, z) : 0 \leq z \leq L\}$, $1 \leq k \leq m$, where ψ_t is the flow generated by (9.3).

We note finally that the experimental/numerical procedure described for the nonbranching construction can readily be modified to fit our branched sweeping approach. All we need to do is approximate the positions where and the manner in which the channels branch by studying changes in successive cross-sections (with bifurcations in successive cross-sections signaling the branching of channels). Then we define an approximate factor $\rho(\mathbf{u}, z)$ to fit the subsets where the channels bifurcate, and multiply the discrete vector field obtained just as in the nonbranching case by this factor. This will produce a model that is capable of simulating even the most complicated branched channel bone morphology if the mesh is chosen to be fine enough.

10. Epilogue

In this paper we have endeavored to identify and carefully delineate the categories of objects, some of the principal concepts such as shape, consistency and stability, several fundamental questions, and a variety of approaches that lie at the very heart of computational topology. We have used the notion of a swept volume as a paradigm to apprehend and illuminate an array of techniques from dynamical systems, geometry and topology that can be used to great advantage in the analysis and representation of geometric objects - specifically those that can be implemented algorithmically to obtain categorical classifications of the entities being studied. It remains to be seen whether or not, or to what extent, our efforts have the desired effect of aiding in the establishment of a focussed perspective for and hastening the codification of the exciting new discipline of computational topology.

In addition to attempting to formulate some of the basic principles of computational topology, we have obtained two new results in this field. We have proved that every smooth swept volume is a subvariety of its ambient space possessing a Whitney regular stratification, and we have obtained local homology criteria for characterizing transverse intersections of submanifolds. In each of these new results we have what appears to be an effectively computable means for analyzing and representing the shape of computer generated objects that reduces the computational complexity of the necessary calculations. In the first case, the problem is simplified by comparing the strata, which are simpler pieces of the entire object, and in the second case the problem is reduced to the computation of elementary groups, or just integers. Naturally, it will take considerable additional research to determine just how useful these results may be in practical applications.

Finally, we have shown how some mathematical approaches to analyzing and representing swept volumes - inspired by the computational topology perspective - can be used in applications. In particular, we have indicated how the natural fiber structure over a swept volume associated to the extended swept volume can be employed to develop what appears to be a very efficient program for virtual sculpting and design, and we have sketched some new methods for modeling heterogeneous bone structures that are based on flows of dynamical systems that resemble those that generate swept volumes. Admittedly, we have only adumbrated these applications to virtual sculpting and heterogeneous material modeling here. Accordingly, one of our current research priorities is to work out the details of these new approaches and develop programs for their effective and efficient implementation.

Acknowledgments

The research in this paper was supported in part by NSF/DARPA CARGO Grant CCR-0310619. Thanks are due to a reviewer whose insightful remarks led to the current improvement of the original manuscript.

References

[1] K. Abdel-Malek and H.J. Yeh, Geometric representation of the swept volume using Jacobian rank-deficiency conditions, CAD 29 (1997), 457-468.

[2] K. Abdel-Malek, H.J. Yeh and S. Othman, Swept volumes: void and boundary identification, CAD 30 (1998), 1009-1018.

[3] K. Abdel-Malek, W. Seaman and H.J. Yeh, NC verification of up to 5-axis machining processes using manifold stratification, ASME J. Manufacturing Sci. and Engineering 122 (2000), 121-135.

[4] K. Abdel-Malek, J. Yang and D. Blackmore, On swept volume formulations: implicit surfaces, CAD 33 (2001), 113-121.

[5] K. Abdel-Malek, D. Blackmore and K. Joy, Swept volumes: foundations, perspectives and applications, Int. J. Shape Modeling (submitted).

[6] V. Arnold, S. Gusein-Zade and A. Varchenko, Singularities of Differentiable Maps, Vols. I & II, Birkhäuser, Boston, 1985.

[7] M. Bern et al., Emerging Challenges in Computational Topology, Report of NSF Workshop on Computational Topology, June 11-12, 1999, Miami Beach, Florida, 1999, arXivcs.CG/9909001 v1 1 Sep 1999.

[8] D. Bespalov, A. Shoukoufandeh, W. Regli and W. Sun, Scale-space representation and classification of 3D models, ASME J. Computers and Information Science in Engineering 3 (2003), 315-324.

[9] D. Blackmore, M.C. Leu and F. Shih, Analysis and modelling of deformed swept volumes, CAD 26 (1994), 315-326.

[10] D. Blackmore, M.C. Leu and L.P. Wang, The sweep-envelope differential equation algorithm and its application to NC machining verification, CAD (1997), 629-637.

[11] D. Blackmore, M.C. Leu, L.P. Wang and H. Jiang, Swept volumes: a retrospective and prospective view, Neural, Parallel and Scientific Computations 5 (1997), 81-102.

[12] D. Blackmore, R. Samulyak and M.C. Leu, Trimming swept volumes, CAD 31 (1999), 215-223.

[13] D. Blackmore, R. Samulyak and M.C. Leu, A singularity theory approach to swept volumes, Int. J. Shape Modeling 6 (2000), 105-129.

[14] B. Chazelle, H. Edelsbrunner, L. Guibas and M. Sharir, A singly exponential stratification scheme for real semi-algebraic varieties and its application, Proc. 16th Int. Coll. Automata, Languages and Programming, Lect. Notes in Computer Science, No. 372, Springer-Verlag, Berlin. 1989, pp. 179-192.

[15] H.L. Cheng, H. Edelsbrunner and P. Fu, Shape space from deformation, Proc. 6th Pacific Conf. Comput. Graphics Appl., Singapore, 1998, pp. 104-113.

[16] Y. Cointepas, I. Bloch and L. Garnero, A cellular model for multi-objects multi-dimensional homotopic deformations, Pattern Recognition 34 (2001), 1785-1798.

[17] N. Cornish and J. Weeks, Measuring the shape of the universe, Not. AMS 45 (1998), 1463-1471.

[18] F. Dachille IX, H. Qin and A. Kaufman, A novel haptics-based interface and sculpting system for physics based geometric design, CAD 33 (2001), 403-420.

[19] C. Delfinado and H. Edelsbrunner, An incremental algorithm for Betti numbers of simplicial complexes on the 3-sphere, CAGD 12 (1995), 771-784.

[20] T. Dey and S. Guha, Computing homology groups of simplicial complexes in $\mathbb{R}^3$, J. ACM 45 (1998), 266-287.

[21] J. Dorman and A. Rockwood, Surface design using hand motion with smoothing, CAD 33 (2001), 389-402.

[22] H. Edelsbrunner, Geometry and Topology for Mesh Generation, Cambridge Univ. Press, Cambridge, 2001.

[23] H. Edelsbrunner and N. Shah, Triangulating topological spaces, Int. J. Computational Geometry and Applications 7 (1997), 365-378.

[24] G. Farin, Curves and Surfaces for Computer Aided Design: A Practical Guide, Academic, Boston, 1988.

[25] R. Farouki, M. al-Kandari and T.Sakkalis, Structural invariance of spatial pythagorean hodographs, CAGD 19 (2002), 895-907.

[26] M. Golubitsky and V. Guillemin, Stable Mappings and Their Singularities, Springer-Verlag, New York, 1973.

[27] J. Guckenheimer and P. Holmes, Nonlinear Oscillations, Dynamical Systems, and Bifurcations of Vector Fields, Springer-Verlag, New York, 1983.

[28] L. Guibas, Computational geometry and visualization: problems at the interface, in Scientific Visualization of Physical Phenomena, N. Patrakalikas (ed.), Springer-Verlag, New York, 1991, pp.45-59.

[29] P. Hartman, Ordinary Differential Equations, Birkhäuser, Boston, 1982.

[30] M. Hirsch, Differential Topology, Springer-Verlag, New York, 1976.

[31] C. Hoffmann, The problems of accuracy and robustness in geometric computations, Computer 22 (1989), 31-41.

[32] C. Hoffmann, Geometric and Solid Modeling - An Introduction, Morgan-Kaufman, San Mateo, 1989.

[33] X. Hu, J. Pang, M. Atwood, W. Sun and W. Regli, A survey of design rationale and intent: representation, capture, and retrieval, Engineering with Computers 16 (2000), 209-235.

[34] D. Husemoller, Fibre Bundles, Springer-Verlag, New York, 1975.

[35] T. Kacyzynski, K. Mischaikow and M. Mrozek, Computational Homology, Springer-Verlag, New York, 2004.

[36] K. Kase, A. Makinouchi, T. Nakagawa, H. Suzuki and F. Kimura, Shape error evaluation method of free-form surfaces, CAD 31 (1999), 495-505.

[37] A. Katok and B. Hasselblatt, Introduction to the Modern Theory of Dynamical Systems, Cambridge University Press, Cambridge, 1995.

[38] Kobayashi and K. Nomizu, Foundations of Differential Geometry, Vols. I & II, Interscience, New York, 1963 & 1969.
[39] B. Kvasov, Algorithms for shape preserving local approximation with automatic selection of tension parameters, CAGD 17 (2000), 17-37.
[40] H. Lawson, The Qualitative Theory of Foliations, Regional Conf. Series in Math., No. 27, AMS, Providence, 1975.
[41] M.C. Leu, D. Blackmore and B. Maiteh, Deformed swept volume analysis of NC machining with cutter deflection, in Machining Impossible Shapes, B.K. Choi and R. Jerard (eds.), Kluwer, Boston, 1999, pp. 158-166.
[42] M.C. Leu, B. Maiteh, D. Blackmore and L. Fu, Creation of freeform solid models in virtual reality, Annals of CIRP 50 (2000), 73-76.
[43] Y.C. Lu, Singularity Theory and an Introduction to Catastrophe Theory, Springer-Verlag, New York, 1976.
[44] B. Maiteh, M.C. Leu, D. Blackmore and L. Abdel-Malek, Swept volume computation for machining simulation and virtual reality application, J. Materials Processing & Manufacturing Science 7 (1999), 380-390.
[45] B. Maiteh, D. Blackmore and M.C. Leu, A virtual reality system for freeform conceptual design (preprint).
[46] R. Malgouyres and A. Lenoir, Topology preservation within digital surfaces, Graphical Models 62 (2000), 71-84.
[47] A. Mäntylä, An Introduction to Solid Modeling, Computer Science Press, Rockville, 1988.
[48] R. Martin and P. Stephenson, Sweeping of three-dimensional objects, CAD 22 (1990), 223-234.
[49] D. Meek and D. Walton, On the surface normal and Gaussian curvature approximations given data sampled from a smooth surface, CAGD 17 (2000), 521-543.
[50] J. Menon, R. Marisa and J. Zagajac, More powerful solid modeling through ray representation, IEEE Computer Graphics & Appl. 14 (1994), 22-35.
[51] J. Milnor and R. Stasheff, Characteristic Classes, Princeton University Press, Princeton, 1974.
[52] J. Mitchel, Topological obstructions to blending algorithms, CAGD 17 (2000), 673-694.
[53] E. Moise, Geometric Topology in Dimensions 2 and 3, Springer-Verlag, New York, 1977.
[54] C. Mow, T. Peters and N. Stewart, Specifying useful error bounds for geometry tools: an intersector exemplar, CAGD 20 (2003), 247-251.
[55] N. Patrikalakis, T. Sakkalis and G. Shen, Boundary representation models: Validity and rectification, Proc. 9th Conf. on Mathematics of Surfaces, Cambridge University, Sept. 2000,R. Cipolla and R. Martin (eds.), Springer-Verlag, New York, 2000, pp. 389-409.
[56] T. Regge, General relativity without coordinates, Nuovo Cimento 19 (1961), 558-571.
[57] W. Regli and V. Cicirello, Digital libraries for engineering, design and manufacturing, CAM 32 (2000), 119-132.
[58] W. Regli, J.H. Han and M. Pratt, Algorithms for feature recognition from solid models: a status report, IEEE Trans. Robotics and Automation 16 (2000), 982-996.
[59] A. Requicha, Representations of solid objects - theory, methods and systems, ACM Computing Surveys 12 (1980), 437-464.
[60] M. Rivara, N. Hitschfeld and B. Simpson, Terminal-edges Delaunay (small-angle based) algorithms for the quality triangulation problem, CAD 33 (2001), 263-267.
[61] J. Rossignac and J. Kim, Computing and visualizing pose interpolating 3-D motions, CAD 33 (2001), 279-291.
[62] T. Rushing, Topological Embeddings, Academic Press, New York, 1973.
[63] P. Saha and A. Rosenfeld, The digital topology of sets of convex voxels, Graphical Models 62 (2000), 343-352.
[64] T. Sakkalis, G. Shen and N. Patrikalakis, Representational validity of boundary representation models, CAD 32 (2000), 7-9-726.
[65] T. Sakkalis, G. Shen and N. Patrikalakis, Topological and geometric properties of interval solid models, Graphical Models 63 (2001), 163-175..
[66] V. Shapiro, Maintenance of geometric representations through space decompositions, Int. J. Computational Geometry and Appl. 7 (1997), 383-393.
[67] S. Smale, The Mathematics of Time, Springer-Verlag, New York, 1980.
[68] E. Spanier, Algebraic Topology, McGraw Hill, New York, 1966.
[69] W. Sun and P. Lal, Recent developments in computer-aided tissue engineering - a review, J. Computer Methods and Programs in Biomedicine 67 (2002), 85-103.
[70] W. Sun, A. Darling, B. Starly and J. Nam, Computer-aided tissue engineering: overview, scope and challenges, Biotechnology & Applied Biochemistry 39 (2004), 29-47.

[71] T. Tautges and S. Mitchel, Whisker weaving: invalid connectivity resolution and primal construction algorithm, Proc. 4th Int. Meshing Roundtable, 1995, pp. 115-128.
[72] R. Thom, Ensemble et morphisms stratifiés, Bull. AMS (1969), 240-284.
[73] T. Thompson, D. Johnson and E. Cohen, Direct haptic rendering of sculpted models, Computer Graphics (1997 Sympos. on Interactive 3D Graphics) (1997), 167-176.
[74] G. Vegter, Computational topology, in Handbook of Discrete and Computational Geometry, J. Goodman and J. O'Rourke (eds.), CRC Press, Boca Raton, 1997, pp. 517-536.
[75] J. Wallner, R. Krasauskas and H. Pottmann, Error propagation in geometric constructions, CAD 32 (2000), 631-634.
[76] W.P. Wang and K.K. Wang, Geometric modeling for swept volume of moving solids, IEEE CG/A 12 (1986), 8-17.
[77] G. Wang, J. Sun and X. Hua, The sweep-envelope differential equation algorithm for general deformed swept volumes, CAGD 17 (2000), 399-418.
[78] J. Weeks, Reconstructing the global topology of the universe from the cosmic microwave background, Classical Quantum Gravity 15 (1998), 2599-2604.
[79] H. Whitney, Tangents to an analytic variety, Annals of Math. 81 (1965), 496-540.
[80] C. Wollmann, Estimation of the principal curvatures of approximated surfaces, CAGD 17 (2000), 621-630.
[81] A. Yao, Decision tree complexity and Betti numbers, Proc. 25th Sympos. on Theory of Computing, ACM, 1994, pp. 615-624.

DEPARTMENT OF MATHEMATICAL SCIENCES AND CENTER FOR APPLIED MATHEMATICS AND STATISTICS NEW JERSEY INSTITUTE OF TECHNOLOGY NEWARK, NJ 07102-1982

DEPARTMENT OF MATHEMATICAL SCIENCES AND CENTER FOR APPLIED MATHEMATICS AND STATISTICS NEW JERSEY INSTITUTE OF TECHNOLOGY NEWARK, NJ 07102-1982

DEPARTMENT OF MECHANICAL AND AEROSPACE ENGINEERING AND ENGINEERING MECHANICS UNIVERSITY OF MISSOURI-ROLLA ROLLA, MO 65409-0050

DEPARTMENT OF COMPUTER SCIENCE DEPARTMENT OF MECHANICAL ENGINEERING AND MECHANICS DREXEL UNIVERSITY PHILADELPHIA, PA 19104

DEPARTMENT OF COMPUTER SCIENCE DEPARTMENT OF MECHANICAL ENGINEERING AND MECHANICS DREXEL UNIVERSITY PHILADELPHIA, PA 19104

DIMACS Series in Discrete Mathematics
and Theoretical Computer Science
Volume 67, 2005

Elements of Computational Metrology

Vijay Srinivasan

ABSTRACT. Metrology deals with the art and science of measurements. It has been an integral part of manufacturing industry for centuries. In recent times, computational issues brought about by the use of sophisticated inspection instruments, such as Coordinate Measuring Machines and scanning probes, have pushed to the forefront the problems associated with processing and interpreting such measurement data. Computational problems in this domain can be broadly described as fitting and filtering discrete geometric data. This paper examines the industrial need that motivates these problems, and their classification and presentation as optimization and convolution problems.

Key words and phrases. Metrology, fitting, filtering, computation, optimization, convolution.

1. Introduction

We start with a very old problem that caught the attention of ancient Greeks. Marveling at the pyramids of Egypt, they asked how tall was the biggest (which they called the pyramid of Cheops) of them all. This simple question had no simple answer, till around 600 B.C. when Thales, one of the seven wise men of the ancient, famously replied "measure the length of the pyramid's shadow when your own shadow exactly equals your height." Since then this anecdote has been used countless times as a clever example of an early application of properties of similar triangles. We also would have left it there and wouldn't have had any need to mention it here but for the fact that the same problem was revisited by Joseph Fourier around 1800 A.D.

When Napoleon invaded Egypt in the late 18th century, he quite unwittingly turned it into one of the most fruitful scientific expeditions. A small band of gifted French scientists who accompanied him to Egypt ventured to measure the height of the tallest pyramid. This time Fourier recorded a different means of computing it. Adding the measured heights of each of the 203 steps would give the pyramid's height, he reported rather plainly, but added a crucial observation that its uncertainty was 14 times the uncertainty in measuring a single step. Where did the mysterious number 14 come from? It is the nearest integer to the square root of 203. So Fourier explicitly acknowledged the presence of uncertainty in such dimensional measurements, and provided a simple statistical method for estimating the uncertainty associated with a result of computation on these measured data. It is worth pointing out that statistical theories of errors were just emerging around that time.

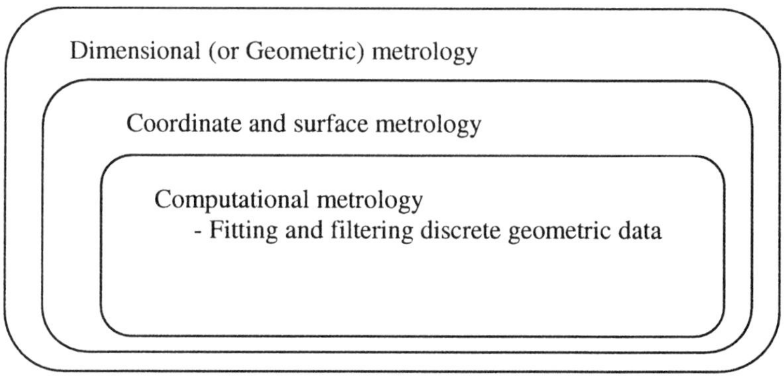

FIGURE 1. A hierarchical view of dimensional, coordinate, and computational metrology.

Thus computation on measured geometric data has a long and colorful history, and such computation will be our prime field of study. Before we proceed too far, we should provide some basic definitions of computational metrology. Metrology is the art and science of measurements. Measurement is the association of one or more numerical values or attributes to physical objects and characteristics. So a measurement of the diameter of a shaft using a vernier caliper might produce a numerical value in millimeters, and an inspector examining the same shaft diameter using a functional gage might come up with an attribute that the diameter is too large, too small, or just right. Both are deemed measurements of the shaft diameter.

Our focus is on geometric measurements and computations on them. Specifically, our focus is on *fitting* and *filtering* of discrete geometric data. Pictorially, we can classify the subject of our study using the Venn diagram shown in Figure 1. Dimensional (or, more precisely, geometric) metrology deals with all geometric measurements. A subset of this is the field of coordinate and surface metrology, where coordinates of points on an object's surface are measured. Computational metrology, which deals mainly with fitting and filtering geometric data, is a further subset of this. In our context, we will see that fitting implies optimization, which involves both continuous optimization (e.g., least squares fitting) and combinatorial optimization (e.g., minimax fitting). We will also see that filtering implies convolution, which involves both convolutions of functions (e.g., Gaussian filters) and convolutions of sets (e.g., envelope filters using Minkowski sums).

2. Industrial Setting

Dimensional (or geometric) metrology plays a critical role in manufacturing industry. There are at least two major reasons for this. The first is the issue of *product conformance*. Here we ask whether a manufactured object is within designer specified tolerances. The second reason deals with manufacturing *process characterization*. The concern here is the quantification of the manufacturing process capability and ensuring that it remains under control over the duration of the production run. We will revisit specification tolerances and process capabilities in later sections. In each case, measurements are made on the surface of a manufactured part and the measured data are reduced to a few numbers or attributes by increasingly sophisticated computational techniques – and this is the domain of computational metrology.

As we enter the realm of manufacturing industry, it is useful to establish some basic facts about manufacturing and measurement. These are best captured in two axioms that can be called (1) axiom of manufacturing imprecision, and (2) axiom of measurement uncertainty. Let's look at each in some detail.

2.1. Axiom of Manufacturing Imprecision

If manufacturing science were to be formulated as a logical system, one of its axioms would be that of manufacturing imprecision. A formal definition of this axiom can be stated as follows.

> *Axiom of manufacturing imprecision*: All manufacturing processes are inherently imprecise and produce parts that vary.

In fact, this axiom of manufacturing imprecision contains a hidden primal fact that no man-made artifact has ideal geometric form. That is, no manufactured object can be perfectly cylindrical or perfectly spherical and so on. The engineering practice of tolerancing is rooted in this "axiom of imprecision." Because it is stated as an axiom, we don't have to prove it. It is, however, reasonable to ask how old and established this axiom is. A celebrated passage from Plato provides an early answer.

In what has come to be known as the seventh letter, written around 350 B.C., Plato explained his theory of ideas using a geometric example. "There is an object called 'circle,'" he wrote. "Its name is the word that I have just uttered. Next comes its definition ... 'the thing whose extremities in every direction are equidistance from its center.' Third, there is the representation, which can be drawn and rubbed out or turned on a lathe and later destroyed; none of these things can happen to a real circle. Fourth, there is knowledge and understanding and true belief about these things ... the actual entity which is the object of knowledge and truly exists must be regarded as the fifth."

Plato went on to argue that no concrete representation of a circle could be absolutely circular because any tangent drawn to a concrete circle would coincide with it over certain distance. Had Plato owned a factory, he might have described this phenomenon as the geometric form deviation and handled it using circularity tolerancing. Instead, he started an academy and encouraged the pursuit of ideal form geometry. Almost all of our geometric knowledge comes from his school of ideal forms. According to legend, the inscription at the entrance to his school forbade entry to anyone who did not know geometry. The Platonic notion of ideal form and the difficulties associated with achieving it in practice have since been discussed quite extensively by philosophers.

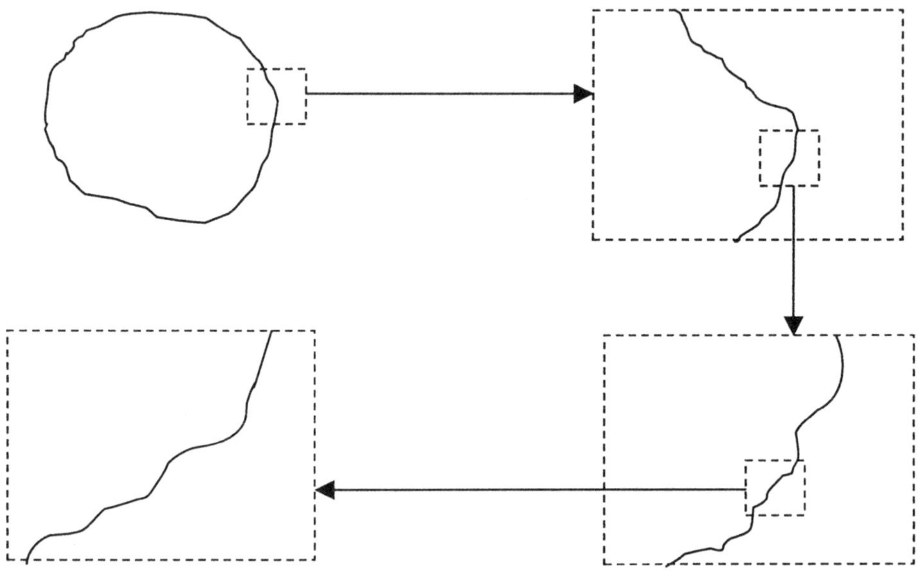

FIGURE 2. Illustration of details at four scales in a hypothetical "circular" object.

There is increasing experimental evidence that the geometry of manufactured surface behaves more like a *fractal* over the range of engineering scales – from a nanometer to a kilometer (a dynamic range of 10^{12} in decimal scale, or 2^{40} in dyadic scale). The illustration in Figure 2 is typical of the geometry of manufactured parts; as we look closer, more details are revealed. In contrast, a surface of ideal geometric form will appear flatter as we get closer, as seen in Figure 3. We don't want to claim that manufactured surfaces are *mathematical* fractals, which have details at *all* scales, any more than these surfaces are mathematically perfect planes, cylinders etc. As we reach the atomic scale, the notion of continuum inherent in our geometric models breaks down. We only observe that, over the rather wide range of engineering scales just quoted, manufactured surfaces show details, and they are better modeled by fractals than smooth surfaces.

This observation has a profound practical implication – it mean that the only information we can obtain about a manufactured surface comes from a discrete set of points sampled on that surface and the scale of measurement has a strong influence on the model of the surface we build from these measurements for further application. We can never gather information about all points on a manufactured surface and have to settle for a discrete set of point data gathered at a pre-determined scale. This scale is determined by what we want to do with the data. So issues of product functionality and process characterization have strong influence on the scale at which we want to examine a manufactured part.

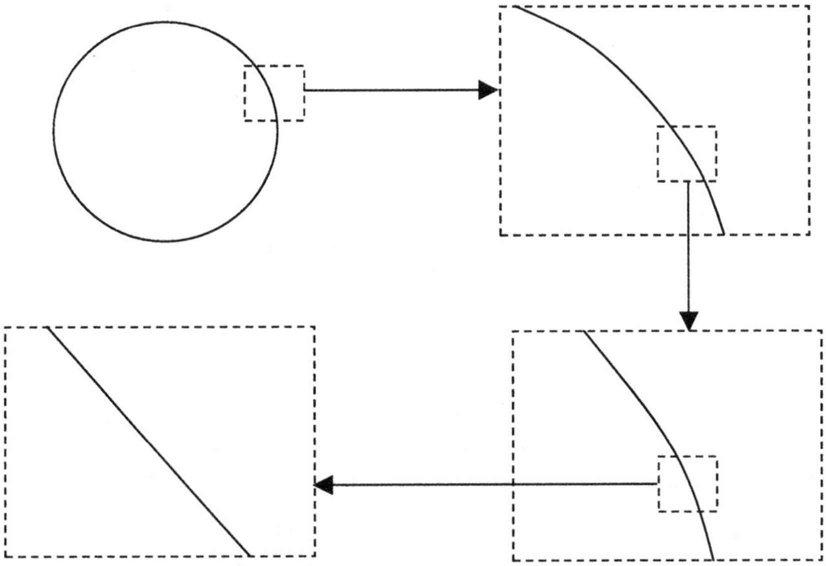

FIGURE 3. Illustration of details at four scales in an ideal "circular" object.

Thus far we have argued rather convincingly that no manufactured part can have ideal geometric form. Let's call this *intra*-part variability; in fact it is an *intra*-feature variability. An extension of this observation – and this is a major leap – is that no two manufactured parts can be geometrically identical. Why? Because we assert that all manufacturing processes are inherently imprecise and they produce parts that vary. Let's call this *inter*-part variability. This is the axiom of manufacturing imprecision.

So we suffer from double deficiency – no manufactured part can have ideal geometric form and no two parts we make can be exactly alike. That is, we have both intra- and inter-part variability. The axiom of manufacturing imprecision explicitly addresses the inter-part variability. But it should be pointed out that this is strongly influenced by intra-part variability. We will address both intra- and inter-part variability in our discussion of computational metrology.

The axiom of manufacturing imprecision is not a novel concept. Engineers have implicitly acknowledged the presence of variability in their products. A skilled engineer is one who has learned the art of controlling this variability. In support of this observation, we now recount several notable examples of how engineers have progressively tamed the problem of imprecision in manufacturing.

By boring the cylinders for his engines in the 18th century, James Watt was proud to have attained uniformity of diameter "within the thickness of a worn shilling." Realizing the need for making measurements to control the geometric variations of his engines, he even invented the first screw micrometer. This is an early example of metrology being recognized as a necessary technology in controlling product variability.

In 1785 Thomas Jefferson, who was then minister to France, wrote to John Jay, "An improvement is made here in the construction of the musket which it may be interesting to Congress to know. ... Supposing it might be useful to the U.S. I went to the workman. He presented me with the parts of 50 locks taken to pieces and arranged in compartments. I put several together myself, taking pieces at hazard as they came to hand, and they fitted in the most perfect manner. The advantages of this, when arms need repair, are evident."

The improvement Jefferson observed was originated by a French General, Jean-Baptiste de Gribeauval. It was based on the notion of interchangeability of manufactured parts that go into the assembled product. This idea became a reality in large-scale manufacture when the U.S. War Department aggressively championed the cause and the American workmen perfected the production techniques. And it took almost a century of work and large sums of money to realize. Why? Because interchangeable manufacture had to await the invention and refinement of two "enabling technologies" – gaging and special purpose tooling – and their acceptance by industry.

In the early 19th century America, armament manufacturers understood the power of the Platonic models. These early American armsmakers viewed a model weapon as a Platonic model possessing ideal form. They fully realized that all production arms were but imperfections of this ideal (but real) model. Jigs and fixtures were designed on some common, rational basis and the model provided this basis. Where dimensions and fits were critical, gages were made on this model. With such gages and fixtures, parts produced in machine tools approximated comparable parts of the model. This enabled arms manufacturers to achieve interchangeability on a large scale. The success of the armament industry later influenced other manufacturers to adopt this system of fixtures and gages in their factories.

In a 1906 advertisement, Henry Ford claimed, "We are making 40,000 cylinders, 10,000 engines, 40,000 wheels, 20,000 axels, 10,000 bodies, 10,000 of every part that goes into the car ... all exactly alike." It was, of course, an exaggeration. It wasn't until the mid-1920s when the Model A was under development that Ford was forced to pay close attention to precision. Then he bought the Swedish gage-making operation of Carl Johansson. The Johansson gage blocks provided Ford tool makers with the best dimensional standards in the world at that time.

While manufacturing his gages, Johansson had to set tolerances for them because even these high precision gages were subject to the tyranny of the axiom of manufacturing imprecision. He established what he called a progressive tolerance, which was proportional to the size of the gage. He was granted a Swedish patent in 1909 for "measurement with progressive tolerances." His gage systems and progressive tolerances were extensively used throughout the world. By 1915, the U.S. War Department had established the Johansson system as the standard in all workshops manufacturing war material.

The Second World War accelerated the acceptance and codification of the notion of product variability. Ordnance manuals evolved into military standards for dimensioning and tolerancing. Later the military and industry standards merged into national standards. Now we are attempting to unify national standards under an international standard called ISO standard. These standards start with the assumption that product geometry variation is inevitable and provide means for accommodating and controlling this variability.

Of late, the axiom of manufacturing imprecision has been challenged by some prominent precision engineers. They argue that there is nothing inherent about imprecision in automated manufacturing processes. They state emphatically that, in their experience, observed non-repeatabilities had all been traceable to systematic sources that were correctable. All one needs, they argue, are time, money, and the right attitude. The basic premise of this deterministic school is that the only exception to deterministic laws of nature occur at atomic scale and, therefore, have no practical significance in the field of manufacturing. The late John Loxham, the patron saint of this school, observed from his close study of automatic manufacturing processes that they operate under the influence of deterministic natural laws. The Loxham Principle states (in paraphrase) "An automated manufacturing process always operates perfectly. It may not do what is required, but that is because it has not been set up suitably."

Advances made in the past 50 years in precision science and engineering have indeed been impressive. The drift of the structural loop between the tool and the workpiece in a diamond-turning machine has been brought under 25 nanometers and held at that level for weeks. X-ray microscopes have been diamond-bored to a diameter within 50 nanometers. Using float polishing technology, we can produce magnetic recording surfaces to a flatness of 10 nanometers. (Soon we will define what we mean by flatness more precisely.) Such levels of precision have been achieved by long and painful studies to understand the causes that affect precision. The list of causes includes human errors, thermal effects, vibrations, friction and lubrication, dirt, and variations in supply of power and compressed air. A determinist believes that the list is short enough to understand and control precision to a very high level.

One can detect a Platonic idealism in the determinists' position. It encourages a positive attitude toward precision. As such, its value should not be underestimated. It drives the desire to improve manufacturing process capabilities and decrease the cost of precision. Precision engineers may rightly take exception to what they perceive as the negative message contained in the axiom of imprecision. They may argue that blind acceptance of the axiom will stifle progress in precision science and engineering. For their sake, should the axiom of manufacturing imprecision be abandoned?

The current defense of the manufacturing imprecision axiom is based on some recent understanding of the fractal nature of engineered surfaces quoted earlier, and economics. Even if technically feasible, tracking of all the assignable causes of product variations does reach a point of diminishing economic returns. For purely business reasons, industry is not willing to pay an unlimited price for precision. Nor is it required by the product's function. The tradeoff between profit and precision will continue, and this alone will ensure the survival of the axiom of manufacturing imprecision well into the future.

2.2. Axiom of Measurement Uncertainty

The Greek roots for the word geometry (*geo* – earth, *metron* – to measure) strongly suggest the origin of geometry in land measurements. It was clear from the very beginning that for measurements to have any value, they must be subject to comparison. If values of distances and angles measured at different places and at different times are to be compared, then there must be standards for units of such measures.

The Mesopotamians set the standard for units of angle measurement. One full turn (say, around a circle) is equally divided into 360 degrees, each degree is equally divided into 60 minutes, and each minute is equally divided into 60 seconds. The Mesopotamian sexagesimal (base-60) system may seem odd, but attempts to change the units to a decimal system (for example, during the French Revolution) have failed. So we live with this standard. An important aspect of the unit for angle measure is that it can be defined without any external reference – once we agree that a full turn is to be divided into 360 equal parts to define a degree, nothing more needs to be said. The situation is quite different in standardizing units of distance measures because it requires an external reference.

The earliest civilization that seems to have standardized length measurement is the Harappan civilization that flourished along the Indus river about 5000 years ago. Bricks found in different sites are remarkably uniform in size. Sets of standard weights in cubical shapes have been unearthed. But in most of the early world, units such as foot, cubit, and yard were based on anatomical features and lacked any rigorous standardization, which made the comparison of lengths and distances difficult. Surprisingly this predicament continued for most of human history. It is only in the last two centuries we have progressively succeeded in establishing a universal length standard of meter (and yard in terms of meter.)

Standardization of units of measure goes a long way towards our ability to compare the measurements. But this does not eliminate uncertainties in these measurements. The ANSI B4.4M-1987 standard on inspection of workpieces furnishes a rationale for the presence of measurement uncertainties. "No measurement can be given an absolute value," it states flatly, "because each measuring system and measuring instrument contains inherent sources of error." One can justify this causality reasoning on the basis of the axiom of manufacturing imprecision just described. In addition, it can be argued that there is a deeper scientific basis because quantization of space at very small scales, Heisenberg's uncertainty principle, and thermal noise place some theoretical lower limit to attainable uncertainty in dimensional measurements.

More formally, we can define the axiom of measurement uncertainty as follows.

> *Axiom of measurement uncertainty*: No measurement can be absolutely accurate and with every measurement there is some finite uncertainty about the measured attribute or measured value.

How can we express the uncertainty in a particular measurement? Here is an amusing quote from C.H.Meyer reporting on his group's measurement of the heat capacity of ammonia (ca 1970): "We think our reported values is good to 1 part in 10,000: we are willing to bet our own money at even odds that it is correct to 2 parts in 10,000. Furthermore, if by chance our value is shown to be in error by more that 1 part in 1000, we are prepared to eat the apparatus and drink the ammonia." This is a colorful way of expressing measurement uncertainty, but one would like a more formal language to document it.

ISO (International Organization for Standardization) took a major step towards this goal by issuing a "Guideline for Expression of Uncertainty in Measurement" in 1993. Following this, the National Institute of Standards and Technology (NIST), the ultimate authority on measurement standards in the U.S.A., issued a policy codifying this principle of measurement uncertainty. This policy requires that NIST measurement results be accompanied by quantitative uncertainty statements and that a uniform approach to expressing measurement uncertainty be followed. So if we send a gage block to NIST for calibration, they would issue a certificate for its dimensional value and an estimation of uncertainty in this value.

Ideally we would like to apply the NIST policy to all industrial measurements. Quality control inspectors can satisfactorily carry out conformance tests only if they have quantitative information about uncertainties in the measurements made. But this adds a sobering realization to our computational endeavor. If we take the discrete set of point measurements as input to our computations, it is good to remember that these input values can never be taken as absolutely accurate. It is equally important to remember that results of our computations should be accompanied by statements about their uncertainty, which is partly inherited from the input uncertainty and partly attributable to the computational scheme itself. This is a major challenge faced by computational metrology.

Finally, we observe that the axioms of manufacturing imprecision and measurement uncertainty can be treated as independent. It is possible to build a modeling framework where manufacturing is imprecise but measurements have negligible uncertainty, and vice versa. But to model reality more closely, both axioms should be considered operative.

3. Fitting

We have already seen that no manufactured object can have ideal geometric form, and the best primary information we can get from a manufactured surface is a discrete set of points. But there are several practical reasons, which we will describe shortly, that demand that we *fit* Platonic ideal forms to these measured set of points.

More formally, fitting is the act of associating ideal geometric objects (straight lines, circles, planes, cylinders, spheres, etc) to a discrete set of points sampled on a manufactured surface. Engineers are interested in fitting at least for the following two reasons:

- *Deviation assessment*: It is often important to determine how far a manufactured surface has deviated from its intended ideal geometric form. This deviation can be quantified by fitting.
- *Datum establishment*: Datum is a reference geometric object of ideal form established on one or more non-ideal geometric forms on a manufactured part. Datums are used for relative positioning of geometric objects in parts and assemblies of parts.

Historically such fitting was accomplished by the use of surface plates, collates, mandrels, and specialized measurement fixtures. More recently, manufacturing industry has started using modern measurement devices such as Coordinate Measuring Machines (CMM). This has accelerated the use of fitting by computation. Its initial success is placing increasingly complex demands on our ability to compute. So let's look at what we mean by deviation assessment and datum establishment in some detail.

3.1. Deviation Assessment

Recall how Plato was certain that no concrete representation of a circle, either drawn by hand or produced on a lathe, can be perfectly circular, and James Watt was pleased that he could attain uniformity of diameter of cylinders in his engines "within the thickness of a worn shilling." Since then engineers have come up with a more precise graphical language, using stylized drawings and symbolic annotations, to quantify how much variation from the ideal geometric form is acceptable. This is the domain of what has come to be known as geometric dimensioning and tolerancing (GD&T).

Figure 4 shows a simple example of form tolerancing using the GD&T language. It shows projected views of a rectangular plate of some finite thickness with a cylindrical hole. In its ideal form, its boundary (that is, the interface between the part and its surrounding environment) is made up of several flat patches and one cylindrical patch. Each patch is also called a surface feature – so we have several planar features and one cylindrical feature. One of the flat features and the lone cylindrical feature have been selected in Figure 4 to indicate form tolerances on them, using a standardized syntax (that is, symbolic annotations). Figure 5 illustrates the semantics (that is, the meaning behind the symbols) of this specification. This semantics takes the form of a "theoretical inspection procedure", where we assume that a manufactured part with non-ideal surface features has been given to us and we have to prescribe the procedure to inspect it to see if it conforms to the tolerance specifications.

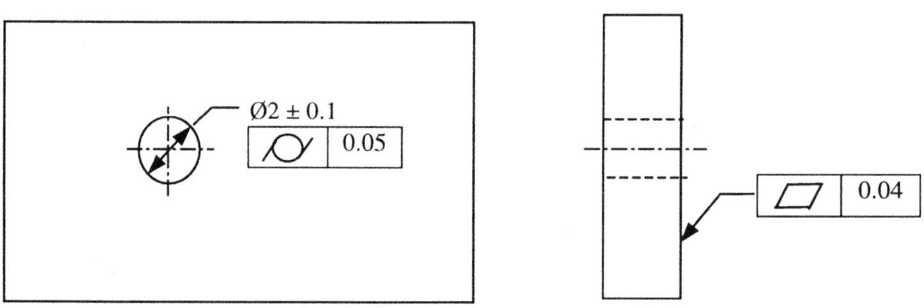

FIGURE 4. An example of form tolerancing.

It is customary in standards to illustrate both the syntax and the semantics in cartoon-like style: an annotated picture under the title "this on the drawing" or "drawing indication" will illustrate the syntax as shown in Figure 4, and another annotated picture under the title "means this" or "drawing interpretation" will define the semantics as shown in Figure 5. So the cylindricity tolerance specified on a nominally cylindrical feature in Figure 4 means that the corresponding actual cylindrical feature on a manufactured part, illustrated in Figure 5, should be contained within an annular tolerance zone bounded by two coaxial cylinders whose radii differ by 0.05 units – the cylindricity tolerance specified in Figure 4 – for conformance. Note that the actual radius value of inner or outer cylinder in this tolerance zone is not important. Nor is the orientation or location of this tolerance zone important. Thus we have a theoretical inspection procedure to check for conformance to the form tolerance specification of a cylinder. And this is more precise and less ambiguous than James Watt's notions of "uniformity of diameter within the thickness of a worn shilling" for his cylinders.

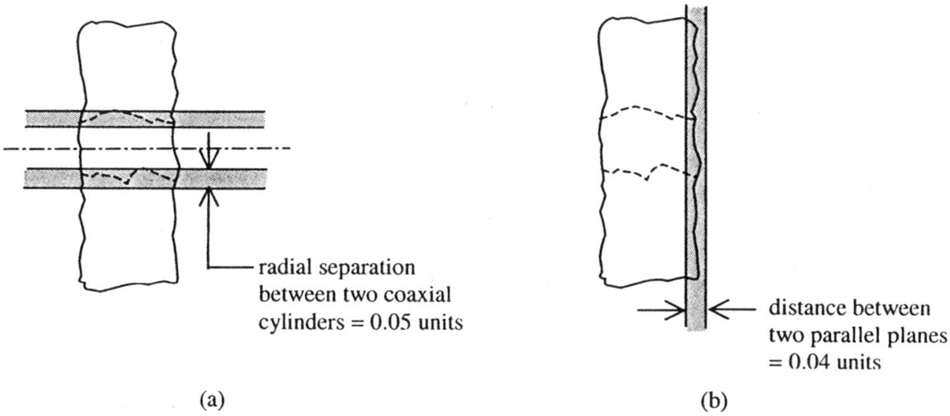

FIGURE 5. Checking form tolerances specified in Figure 4 for conformance. (a) Cylindricity specifies a tolerance zone bounded by two coaxial cylinders. (b) Flatness specifies a tolerance zone bounded by two parallel planes.

Another form tolerance illustrated in Figure 4 is for flatness of a nominally flat feature. Its semantics, as shown in Figure 5, is that the corresponding feature on a manufactured part should be contained within a tolerance zone bounded by two parallel planes separated by 0.04 units, which is the flatness tolerance specified, for conformance. The actual orientation or location of the tolerance zone is not important here. So again we have a theoretical inspection procedure that gives the semantics of flatness tolerance specification.

Flatness and cylindricity are but two of several form tolerances defined by standards. A comprehensive list of standardized form tolerances developed thus far is shown in Table 1. The purpose of the standards is to describe the usage of this syntax and the associated semantics, both in cartoon-like fashion and in mathematical terms. All these form tolerances have the following three common notions:

1. a non-ideal feature A defined by a (potentially dense) set of points,
2. a tolerance zone B bounded by ideal geometric forms, and
3. a conformance rule in the form of an existential statement: *If B can be moved by a rigid motion such that it contains A, then A is within specified tolerance.*

TABLE 1. List of Standardized Form and Profile Tolerances.

Type	Characteristic	Symbol
Form	Straightness	—
	Flatness	⏥
	Roundness (Circularity)	○
	Cylindricity	⌭
Profile	Profile of a line	⌒
	Profile of a surface	⌓

Let's examine how these three notions apply to flatness tolerance as a means of illustrating the role of fitting in conformance assessment.

Let's start with the assumption that an actual part, such as the one illustrated in Figure 5, has been delivered to us and we need to check if it conforms to the flatness tolerance specified in Figure 4. A traditional way to accomplish this on a shop floor is to use commonly available tools such as a surface plate and a dial indicator, as shown in Figure 6. A surface plate is a specially constructed, sturdy table with a very flat top surface, which can be taken to be the best practical approximation to a mathematically ideal plane. In the method indicated in Figure 6, the feature whose flatness is being inspected is placed on the surface plate into which a dial indicated has been inserted as shown. The stylus in the dial indicator is sufficiently compressed so that the tip of the stylus is always in contact with the feature under inspection. The part is then traversed on the surface plate so that the stylus tip covers as much of the feature as possible, and if the range of the dial indicator movement is less than 0.04 units then the feature is deemed to be within the specified flatness tolerance of 0.04 units.

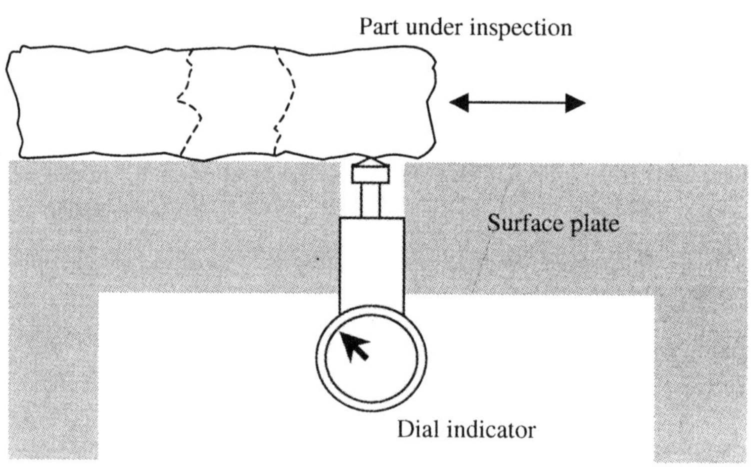

FIGURE 6. Example of inspecting for flatness using surface plate and dial indicator.

The method thus outlined for checking flatness conformance has the great advantage of simplicity. It can be carried out even in small industrial establishments. But it also has several drawbacks. Chief among them is that it may not always be possible to position the part on the top of a surface plate, if the part is big or the feature to be inspected is located in some inaccessible portion of the part. Also, it is not clear how to traverse the part on the surface plate so that "all" the points in the feature are covered in the measurement. The best we can do is to cover as much as possible. The finite radius of the stylus tip in the dial indicator cannot touch all the points of the feature; in fact, it is not quite clear how to interpret the dial indicator movement. Add to this the imperfections inherent in the surface plate itself and the uncertainty associated with the dial indicator reading, and we have a method that is less than satisfactory in many cases.

Some of these problems can be avoided if we have some means of measuring the coordinates of many points on the feature in question and then doing some computations on them. This is precisely what is attempted in modern Coordinate Measuring Systems, which consist of a Coordinate Measuring Machine (CMM) that measures coordinates of points on the surface in question and a software system that does the computations on them. So let's assume that we can measure the coordinates, say the (x, y, z) coordinates, of several points, say n points, on the feature, which is checked for flatness. We can then ask whether these n points can be contained within two parallel planes that are 0.04 units apart. This is purely a mathematical problem that can be attacked algorithmically by transforming an existential question to a procedural problem. In fact, this is one of the most celebrated problems in computational metrology. So it is worth examining it in some detail.

Given a set A of points in three-dimensional space, mathematicians define the "width" of this set as the smallest distance between two parallel planes within which A can be contained. It can be easily seen that the width of A is less than or equal to t if and only if A can be contained within two parallel planes that are t units apart. So the assessment of conformance to flatness tolerance specification can be reduced to width computation. There are several algorithms and software implementations to compute the width of a set of points in two and three dimensions.

Thus we have a computational scheme to solve a flatness tolerance conformance assessment problem. But what has this to do with fitting? It turns out that we can pose an equivalent plane fitting problem as follows.

<u>Minimax Plane</u>: *Given a set of n points in space, find a plane P that minimizes the maximum deviation of the points from this plane.*

Here the deviation of a point from a plane is the distance between the point and the plane measured as the perpendicular distance of the point from the plane. It is always non-negative (that is, zero if the point is on the plane, and positive if it is off the plane). This is a classical minimization problem – it is a minimax problem because it minimizes the maximum deviation. The plane thus defined is also called the Chebyshev best-fit plane. If we can find such a plane, then we can also find the width (which is just twice the minimized objective function) and check for conformance to flatness tolerance specification.

As we examine the minimax plane fitting scheme to assess deviation from flatness, it is natural to ask if there are other measures of out-of-flatness that can also be useful. For example, we could fit a least squares plane as in the following optimization problem.

Least Squares Plane: *Given a set of n points in space, find a plane P that minimizes the sum of the squares of the deviations of the points from this plane.*

Then the root of the mean of the squares (RMS) of the deviations, computed from this least squares plane, or some constant multiple of it, can be taken as a measure of out-of-flatness. The RMS of the deviation can be thought of as the standard deviation σ. If we construct two parallel planes, each parallel to the least squares plane at 3σ distance from it on each side, then "most" of the points will be contained between these two planes. So 6σ can be considered to be the "statistical" width, and one can use this as a measure of the out-of-flatness.

Having introduced the minimax plane and the least squares plane, it is worth noting their relationship using some simple statistical ideas. In elementary, univariate statistics we learn that given a set of numbers, their spread (or dispersion) can be characterized by the range R or the standard deviation σ. Extending this idea to a set of points in space, we first consider a plane P and the distances of the points from P. These distances are the set of numbers whose spread can be characterized by their range or standard deviation. These statistics strongly depend on the plane P we choose. By design, the minimax plane gives the smallest value for the range and the least squares plane gives the smallest standard deviation. Both are useful measures of spread. If there is a strong reason to favor range, then the minimax plane will do the job. If the standard deviation is what we choose, then the least squares plane is the best choice and it serves as the "mean" plane. On other words, the minimax plane leads to an unbiased estimation of the range and the least squares plane leads to an unbiased estimation of the standard deviation.

There is considerable theoretical justification to use the 6σ measure to characterize out-of-flatness, especially to estimate the uncertainty associated with this "statistical" width. As we will see soon, it is also easier to compute the least squares plane than the minimax plane. But the least squares plane is not the one used for the standardized definition for flatness. One can only speculate why this is so. While it is easier to compute the least squares plane from the input coordinates of n point, there is no simple way to come up with the RMS value for the deviation using surface plates and dial indicators as was done for flatness in Figure 6. So one may argue that the lack of computational resources in the past, combined with the ready availability of surface plate and dial indicator, had led to the definition of flatness tolerance as shown in Figure 5. But this issue may be revisited in light of wide spread availability of inexpensive computational resources these days.

Even if the 6σ value computed from the least squares plane is not a standardized measure for out-of-flatness, it can be used for assessing the capability of manufacturing processes that are employed to produce the planar feature. To see this, we need to consider how to characterize the precision of manufacturing processes, such as planing, shaping, milling, grinding etc. Note that this is not the same as the tolerance specification problem; we would like to choose a manufacturing process that is capable of producing parts that are within specified tolerances, but the capability of the manufacturing process can be assessed quite independently. Let's take a planar surface as an example and examine some of its production processes.

A flat surface can be manufactured using several manufacturing processes, some of which are listed in Table 2 along with some information about typical "deviation from flatness" values associated with them. These numbers should not be taken literally. In an actual case, other factors such as speed of cut, feed rate, and coolant can affect the deviation from flatness significantly. However, it is important to have some quantitative characterization of the capability of a flat-surface producing process. How are these deviation measures computed? The deviation from flatness values can be the width itself. Then these values can be directly related to a flatness tolerance specified on a planar feature. But the 6σ value computed with fitting the least squares plane can also serve as a measure of deviation from flatness because these numbers are used merely as a guide to choose a manufacturing process and not for checking conformance of a part to specified tolerances.

In summary, there are two types of applications that call for deviation assessment. One is the conformance assessment to specified tolerances. The other is the capability assessment of manufacturing processes. In both cases, fitting can be used as a basic tool to obtain some useful measures of deviation. For planar surface features, a minimax fit yields a plane from which the width of the input set of points can be easily computed. Statistically, this corresponds to finding a plane from which the point distances have the smallest range of variation. A least squares fit yields a different plane from which a 6σ measure for the deviation from flatness can be computed. Statistically, this corresponds to the plane from which the point distances have the smallest standard deviation. These ideas can be extended to non-planar features. Minimax and least squares criteria can be applied to fit cylinders, spheres, cones etc for tolerance conformance and process capability assessment.

TABLE 2. Typical Deviation from Flatness Values.

Process	**Deviation from flatness**
Sand casting	4.0 mm
Die casting	1.0 mm
Injection molding	1.6 mm
Planing and shaping	0.5 mm
Milling	0.5 mm
Grinding	0.1 mm

Observed over an area of $1\text{m} \times 1\text{m}$ in some manufacturing processes that produce planar features.

3.2. Datum Establishment

In the last section on deviation assessment, we dealt mainly with individual features. In manufactured parts, one also has to deal with relationships among various features and assess their deviation from the ideal. For example, there are three pairs of planar features in Figure 4 that are supposed to be parallel. But, as we see in Figure 5, such parallelism between features is not achieved in actual parts. Thus it is important to deal with deviations in such relationships and control them. We start with some general remarks about relative positioning ideal geometric objects.

There are four basic geometric constraints often invoked in relative positioning problems. They are incidence, parallelism, perpendicularity, and chirality.

- Incidence is a constraint that states that a geometric element lies within or on another geometric element. These constraints can be enumerated as point-on-point, point-on-line, point-on-plane, line-on-line, line-on-plane, and plane-on-plane. Some of these constraints can also be referred to as coincidence (for example, point-on-point), concentric, collinear (line-on-line), coaxial, or coplanar (plane-on-plane) constraints. Tangency is a special type of incidence constraint. Note that more than two elements can be involved in an incidence constraint.
- Parallelism applies to lines and planes. There can be parallelism constraint between line and line, line and plane, and plane and plane. More than two elements can be involved in a parallelism constraint.
- Perpendicularity also applies to lines and planes. There can be perpendicularity constraint between two lines in a plane, between a line and a plane, and between two planes.
- Chirality refers to the left- or right-handedness of an object. It is an important constraint because left- and right-handed objects are usually not interchangeable. This comes up, for example, in the relative positioning of two skew lines.

There is a fundamental result in geometry that states that it is possible to position arbitrary geometric objects (of both ideal and non-ideal forms) relative to each other using only points, lines, planes, and helices as reference elements. It follows from the fact that any geometric object belongs to one and only of the seven classes of symmetry listed in Table 3. The second column in that table gives us a set of points, lines, planes, and helices that can serve as reference elements for the corresponding row object. These reference element(s) are important because the associated object can be positioned using the reference element, as we will see in what follows.

TABLE 3. Classes of Symmetry and the Associated Reference Element(s).

CLASS	REFERENCE ELEMENT(S)
Spherical	Center (point)
Cylindrical	Axis (line)
Planar	Plane
Helical	Helix
Revolute	Axis (line) and a point on the axis
Prismatic	Plane and a line on the plane
General	Plane, line, and point

1. For a sphere, its center is the reference element. This holds true even for a solid sphere or a set of concentric spheres.
2. For a cylinder, its axis is the reference element. This is also true for a solid cylinder, a cylindrical hole, or a set of coaxial cylinders.
3. For a plane, the reference element is the plane itself. If we have several parallel planes, then any plane parallel to them can serve as a reference element. For a slot or a slab, which is bounded by two parallel planes, it is customary to take their middle (median) plane and treat it as the reference element.
4. For a helical object, such as a screw thread, a helix having the same chirality, axis and pitch will serve as a reference element. In practice, we just take the axis of the helix and use it to position the helical object.
5. Any surface of revolution is a revolute object. A cone, for example, belongs to the revolute class. Its reference elements consist of the axis of the cone and the vertex of the cone (which lies on the axis of the cone).
6. An oval shaped slot belongs to a prismatic class. Its reference elements consist of a plane, which could be the plane of symmetry of the slot, and a line in that plane along the generator of the prism.
7. Finally, if an object does not belong to any of the six described so far, it belongs to the general class. We then need to find a plane, a line in the plane, and a point on the line and treat them as the reference elements for this general object.

As observed earlier, the advantage of reference element(s) is that they can be used to dimension the relative positioning of one object with respect to another. For example, the relative position of a sphere and a cylinder can be dimensioned by dimensioning the distance between the center of the sphere and the axis of the cylinder. Similarly, the relative position between two planes, if they are not coplanar, can be dimensioned by the distance between them if they are parallel and by the angle between them if they intersect. So the relative positioning of objects of ideal geometric form can be accomplished using the reference elements described thus far. But we don't have ideal geometric forms in manufactured parts. How can we deal with variations in their relative positions? There are two solutions to this problem.

The first solution is to fit ideal geometric objects to each of the non-ideal features involved and see how these ideal forms are related to each other – and how far this relationship has deviated from the nominal specification. Let's call this solution the *parametric tolerancing* approach. In this approach, for example, one may be interested in the parallelism between the two planar features indicated as *C1* and *C2* in Figure 7(a). So we may independently fit a plane to each of the two actual features, as shown in Figure 7(b), and compute the angle between these two fitted planes. Ideally, this should be 0° and any deviation from this value can be considered as a measure of out-of-parallelism of these two features. Here we are using the concept of fitting to both features in question. While this may appear to be a reasonable solution, this is not the standardized solution to the problem.

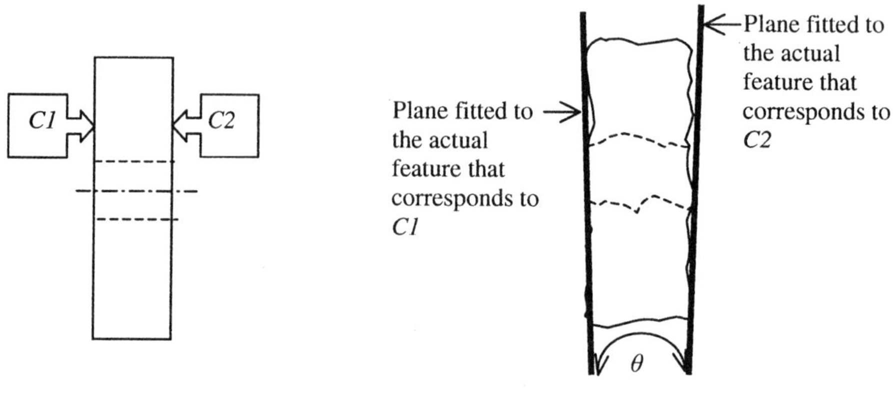

(a) (b)

FIGURE 7. (a) Two planar features whose parallelism is of concern. (b) One method to check for deviation from parallelism. Here θ is the angle between the two fitted planes. Its deviation from 0° is a measure of out-of-parallelism.

The second solution, which is the one supported by standards, is to pick one of the two planes as the datum and use it as a reference object. Let's call this approach the *zone tolerancing* approach. In Figure 8(a) the planar feature on the right has been chosen as the datum feature C and the parallelism of the planar feature on the left is toleranced with respect to C. To interpret this syntax, a datum plane is first fitted to the actual feature on the right as shown in Figure 8(b). Then we check if the actual feature on the left is within a tolerance zone bounded by two parallel planes separated by 0.025 units, with the constraint that the tolerance zone is parallel to the datum plane just established. This type of tolerance specification can be checked using surface plate and dial indicator, as shown in Figure 9. By moving the part under inspection on the surface plate, while maintaining the contact between the upper feature and the dial indicator, one can check if the full indicator movement is within the specification tolerance. This type of inspection can be carried out in any shop floor. This may explain the widespread adoption of parallelism standard just mentioned and illustrate in Figure 8.

While the "surface plate and dial indicator" technique can work for parallelism in some cases, it suffers from the same kinds of drawback outlined for checking flatness in the last section. If one were to use CMMs to collect point coordinate data on both actual planar features, it is possible to develop computational techniques to check for parallelism between these features, as we describe now.

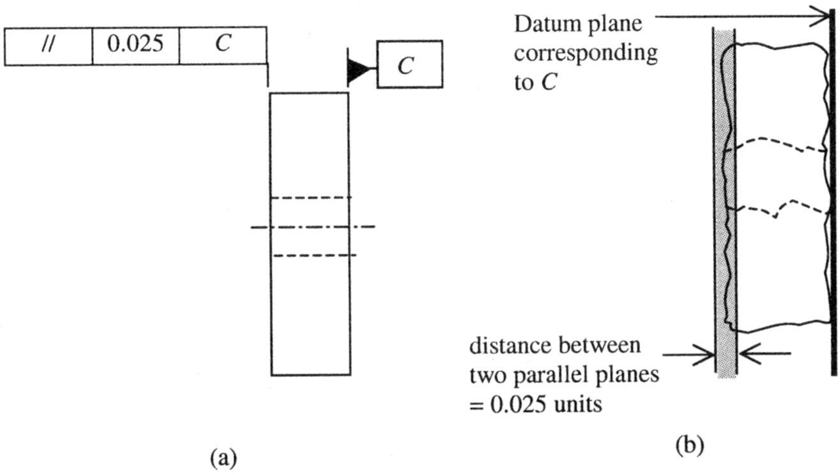

FIGURE 8. (a) Standardized specification of parallelism tolerance, and (b) its interpretation. The shaded tolerance zone is parallel to the established datum plane in (b).

In the parametric tolerancing approach, a plane will be fitted to each of the measured sets of points in Figure 7b. It is not clear at the outset whether we should fit a least squares plane or a minimax plane – we may adopt either scheme provided it is agreed to, in some standardized way, by all parties involved. In the zone tolerancing approach, only one plane will be fitted in the example of Figure 8b. And one may want this plane to be as close an approximation to the surface plate illustrated in Figure 9. This brings up some interesting computational issues. For example, we may demand that the fitted plane should contact at least three of the measured points from the outside of the part. We will examine these issues in the next section, where the fitting problem is formally posed as an optimization problem.

Once a datum plane is fitted in the zone tolerancing approach, the distances of points measured on the other planar feature to this datum plane can be easily computed and their range can be considered as a measure of out-of-parallelism to conform to the current standards. Or, one may want to consider 6 times the standard deviation of these distances as a measure of out-of-parallelism and use this to characterize manufacturing processes that produce parallel surfaces. This brings us back to the same set of issues discussed in the last section on flatness, including the estimation of uncertainties in these computations.

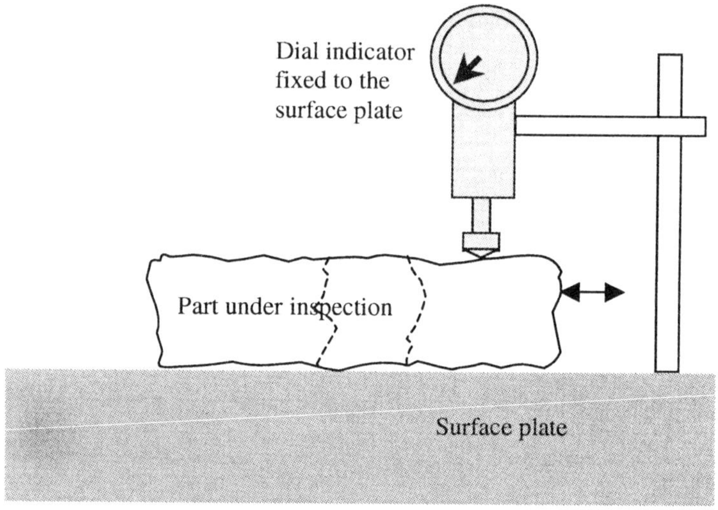

FIGURE 9. Example of inspecting for parallelism using surface plate and dial indicator.

Parallelism is just one of many relational characteristics that can be toleranced. Table 4 shows the types of zone tolerances that control variability in the relationship between features, which are supported by national and international standards. In each case, one or more datums are established and these datums are of ideal geometric forms. Then tolerance zones are defined and constrained with respect to these datums, and actual features are required to be contained within these tolerance zones for conformance to specified tolerances.

TABLE 4. Tolerancing Relational Characteristics.

Type	Characteristic	Symbol
Orientation	Parallelism	//
	Perpendicularity	⊥
	Angularity	∠
Location	Position	⌖
	Concentricity	◎
	Symmetry	⌯
Runout	Circular runout	↗
	Total runout	⌰
Profile	Profile of a line	⌒
	Profile of a surface	⌓

In summary, datums are important reference objects used to specify, characterize, and control variability in the relationship among geometric objects. Datum establishment involves fitting, which yields a point, a line, a plane, or a combination of these to form reference objects.

3.3. Fitting as Optimization

In the last two sections, we examined the critical role played by fitting in manufacturing industry. We will now pose fitting formally as an optimization problem. In general, fitting can be posed as follows.

Given a set X of points, find ideal geometric object(s) Y that minimize an objective function involving distances between X and Y, subject to certain constraints.

Since the objective function to be optimized involves distances between sets of points, we should define such distances carefully. Let p_i and p_j denote two points in the traditional Euclidean space with coordinates (x_i, y_i, z_i) and (x_j, y_j, z_j), respectively. The distance between them is the Euclidean metric $d(p_i, p_j) = \{(x_i\text{-}x_j)^2 + (y_i\text{-}y_j)^2 + (z_i\text{-}z_j)^2\}^{1/2}$. We now define the distance $d(p_i, S)$ between a point p_i and a set S as the shortest (or infimum) of the distances between p_i and points in S. So the distance between a point and a plane is the perpendicular distance of the point from the plane, and so on. Using these notions, we are now ready to define a distance between two sets X and Y.

If there are n points in the set X, and we denote the distance between $p_i \in X$ and an ideal geometric object Y by d_i, then we define an l_p-norm for the distance between X and Y as $\left\{\sum_{i=1}^{n} |d_i|^p\right\}^{1/p}$. Most of the fitting problems can be posed as optimization problems involving this distance measure. Let's illustrate this by looking at the plane fitting problems in some detail. There are at least four types of plane fitting problems, as described below.

<u>Fitting Planes</u>:

Given a set X of n points in space, find a plane that

1. keeps X to one side of the plane and minimizes the sum of the distances of the points from the plane.
   - This problem minimizes the l_1–norm.
2. minimizes the sum of the squares of the distances of the points from the plane.
   - This problem minimizes the l_2–norm. It is the least squares plane described in an earlier section.
   - We may add a constraint that the set X should lie on one side of the plane. This will result in a plane that is more suitable for datum establishment.
3. minimizes the maximum distance of the points from the plane.
   - This problem minimizes the l_∞–norm. It is the minimax plane described in an earlier section.

- This is equivalent to finding the width of the given set of points. That is, it is equivalent to finding the thinnest slab, bounded by two parallel planes, that contains the given points.
- Here again we may want to add a constraint that the set X should lie on one side of the plane, to make it more suitable as a datum plane.

In addition, a fourth plane fitting problem arises when we have two sets of points that can be separated by a plane; such sets are called linearly separable sets. The optimization problem can be posed as follows:

4. Given a set X of points in space that can be partitioned into two linearly separable sets X_1 and X_2, find a plane that separates X_1 and X_2, and maximizes the minimum distance of the points in X from the plane.
   - This is a maximin plane – a companion to the minimax plane defined earlier.
   - This is equivalent to finding the thickest separating slab.

This collection of plane fitting problems is summarized in Table 5 for easy reference. The last column in Table 5 gives a designation for each of the planes fitted under these four criteria.

TABLE 5. Plane Fitting as Optimization Problems.

Find	Input	Objective function to be optimized	Constraints	Comments	Designation
A plane Y	A set X of n points	Minimize l_1-norm	X lies to one side of Y	Minimum 3-point touching plane	$Plane_1C$
		Minimize l_2-norm	None	Least squares plane	$Plane_2$
			X lies to one side of Y	Constrained least squares plane	$Plane_2C$
		Minimize l_∞-norm	None	Minimax plane	$Plane_\infty$
			X lies to one side of Y	Constrained minimax plane	$Plane_\infty C$
	Linearly separable sets X_1 and X_2	Maximize the minimum distance of points in $X_1 \cup X_2$ from Y	Y linearly separates X_1 and X_2	Maximin plane	$PlaneSep$

Of the four plane fitting problems, fitting $Plane_2$ is well studied as a *total least squares* (also as an *orthogonal regression*) problem by numerical analysts and statisticians; in fact, solution to this problem is so well known that we can give a surprisingly compact algorithm to compute it as follows:

Algorithm $Plane_2$
Input: Coordinates (x_i, y_i, z_i), $i = 1,\ldots, n$.
Output: A point p_0 on $Plane_2$ and a vector v normal to $Plane_2$

1. Find p_0 as the mean of the n input points. Denote $p_0 = (x_0, y_0, z_0) = \frac{1}{n}\left(\sum_{i=1}^{n} x_i, \sum_{i=1}^{n} y_i, \sum_{i=1}^{n} z_i\right)$.
2. Form the central coordinate matrix M of size $n\times 3$, whose i^{th} row is $[(x_i-x_0)\ , (y_i-y_0)\ , (z_i-z_0)]$.
3. Find v as the singular vector corresponding to the smallest singular value of M.
4. Return p_0 and v.

End

A line-by-line MATLAB implementation of this algorithm is straightforward and is as compact as the pseudo-code shown here. Such an implementation has been used in a reference software to test commercial software. It is difficult to find a pithier yet efficient code! At the expense of slight numerical stability, one can dispense with the singular value decomposition of M in Step 3 of the algorithm by computing v as the eigenvector corresponding to the smallest eigenvalue of the 3×3 matrix M^TM, which makes the entire computation of $Plane_2$ a simple classroom exercise. This may explain why the least squares plane is very popular with CMM users. It is a poster child for a non-trivial, continuous optimization problem.

By contrast, computing $Plane_1C$, $Plane_\infty$, $Plane_\infty C$, or *PlaneSep* is subtler and comes under the class of combinatorial optimization problems. These particular plane-fitting problems are closely related to the topic of convex hull, whose theory and computation have been studied quite extensively in the computational geometry literature over the past 25 years. As we saw in earlier sections, $Plane_1C$ is more appropriate for planar datum establishment, whereas $Plane_\infty$ is more appropriate for conformance assessment of standardized flatness tolerance. *PlaneSep* arises, for example, when a machined slot is inspected for conformance to tolerance specifications.

Let's briefly summarize other fitting problems posed as optimization problems. We will not dwell on their solutions and applications here. Suffice it to say that part of their applications is dictated by the need to find reference elements listed in the last column of Table 3 to establish datums. And some of these problems can be solved using Voronoi diagrams and Delaunay triangulations.

<u>Fitting Lines in 2D</u>:
This is identical to the plane fitting problems discussed thus far; here all the input points lie in a plane, that is, in two-dimensional space. The goal is to fit a straight line in the plane.

<u>Fitting Lines in 3D</u>:
Given a set of n points in space, find a line that

1. minimizes the sum of the squares of the distances of the points from the line.
   - This problem minimizes the l_2–norm.
2. minimizes the maximum distance of the points from the line.
   - This problem minimizes the l_∞–norm.
   - This is equivalent to finding the smallest circumscribing cylinder for the given set of points.

<u>Fitting Circles in 2D</u>:
Given a set of n points in a plane, find a circle in the plane that

1. minimizes the sum of the squares of the distances of the points from the circle.
   - This problem minimizes the l_2–norm.
2. minimizes the maximum distance of the points from the circle.
   - This problem minimizes the l_∞–norm.
   - This is equivalent to finding the thinnest annulus that contains the points. So it is called the Minimum Radial Separation Circle.
3. has the minimum radius while enclosing all the points.
   - This is called the Minimum Circumscribing Circle. It is also called the minimum spanning circle.
   - This equivalent to finding a point (center) that minimizes the maximum distance of the points from the center. It then becomes a "point fitting" problem that minimizes the l_∞–norm.
4. has the maximum radius, while the circle has its center within the convex hull of the set of points, and does not include any of the points in its interior.
   - This is called the Maximum Inscribing Circle. It is also called the largest empty circle.
   - This is equivalent to finding a point (center) that lies within the convex hull of the set of points, and maximizes the minimum distance of the points from the center. So this is a maximin problem.

In addition, here is a constrained minimax circle problem: Given a set of n points in a plane, find a circle that encloses all the points and minimizes the maximum distance of the points from the circle. It minimizes the l_∞–norm under constraints, and it is not necessarily the same as the minimum circumscribing circle.

<u>Fitting Circles in 3D</u>:
Given a set of n points in space, find a circle that

1. minimizes the sum of the squares of the distances of the points from the circle.
   - This problem minimizes the l_2–norm.
2. minimizes the maximum distance of the points from the circle.
   - This problem minimizes the l_∞–norm.
   - This equivalent to finding the minimum circumscribing torus.
3. lies within the "doughnut" of the points and maximizes the minimum distance of the points from the circle.
   - This is equivalent to finding the maximum inscribing torus.

Fitting Spheres:
This is identical to the problems of fitting circles in 2D; here the points lie in three-dimensional space and we want to fit a sphere.

Fitting Cylinders:
This is also identical to the problems of fitting circles in 2D; here the points lie in three-dimensional space and we want to fit cylinders.

Fitting Cones:
Given a set of n points in space, find a cone that

1. minimizes the sum of the squares of the distances of the points from the cone.
   - This problem minimizes the l_2–norm.
2. minimizes the maximum distance of the points from the cone.
   - This problem minimizes the l_∞–norm. This equivalent to finding the minimum zone cone.

Fitting Torus:
Given a set of n points in space, find a torus that

1. minimizes the sum of the squares of the distances of the points from the torus.
   - This problem minimizes the l_2–norm.
2. minimizes the maximum distance of the points from the torus.
   - This problem minimizes the l_∞–norm. This is equivalent to finding the minimum zone torus.

For minimum circumscribing torus and maximum inscribing torus, see fitting circles in 3D.

3.4. Soft Gaging

Some mechanical assembly fit requirements can be specified directly using tolerance specification, as in the example shown in Figure 10(a). Its interpretation is shown in Figure 10(b), where a hard gage, also known as a functional gage, of certain dimensions is built and an actual part is checked to see if it would fit within the hard gage. The hard gage itself is subjected to manufacturing imprecision, but we could assume that it is manufactured much more precisely than the set of parts it is supposed to check.

If we want to replace the hard gage by, say, a soft gage in the foregoing example, we could proceed as follows. Assume that on an actual part a set of points X_1 is measured on the larger (diameter) cylindrical surface feature and a set of points X_2 is measured on the smaller (diameter) cylindrical surface feature. Now consider two coaxial cylinders C_1 and C_2 of diameters 25.6 and 15.1 units, respectively. We can then pose a soft gaging problem as follows: Is there a rigid motion r such that rC_1 contains X_1 and rC_2 contains X_2, still subject to the constraint that C_1 and C_2 are coaxial?

More generally, we can pose the soft gaging problem as,

Deterministic version: *Given two sets A and B, is there a rigid motion r such that A is contained in rB?*

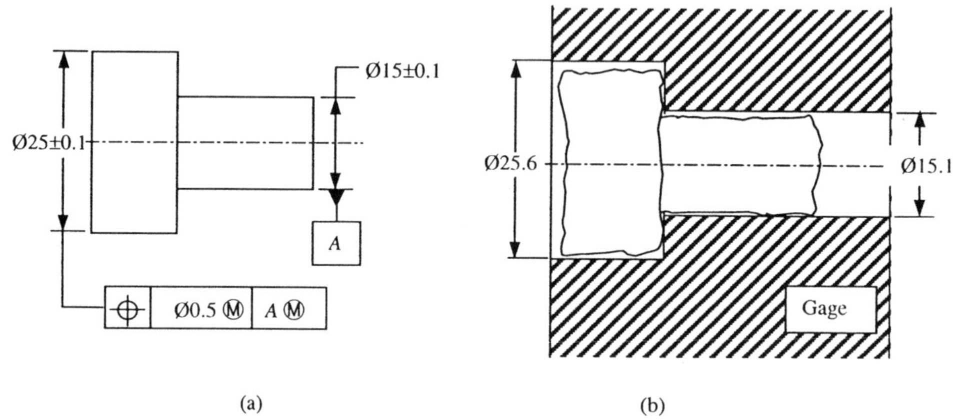

FIGURE 10. Example of use of hard gage to check a tolerance specification. (a) Tolerance syntax. (b) Its semantics using a gage. An actual manufactured part should fit within the gage.

It is not obvious how this problem belongs to the type of fitting problems described thus far. We only need a yes or no answer to the above question. Yet, the soft gaging problem seems to be related to the fitting problem. Also, it offers us a simple way to accommodate uncertainty by posing the problem as follows.

Probabilistic version: *If A is given with some uncertainty, what is the probability of finding a rigid motion r such that A is contained in rB?*

Such problems are being considered only recently.

4. Filtering

Filtering is the task of obtaining scale-dependent information from measured data. Recall that due to the fractal nature, geometry of manufactured surfaces reveals details over the range of engineering scales. Filtering allows us to focus on a particular scale and obtain information in that scale. At a more mundane level, filtering can be used to remove noise and other unwanted information from the measured data. In the context of engineering metrology, engineers are interested in filtering mainly for the following two reasons.

- *Surface finish*: Many engineering functions depend on how rough or smooth a piece of surface is. Designers define bounds on certain roughness parameters obtained by observation on rather small scale to ensure functionality of parts. These small-scale variations are subtracted from the surface measurement data before form and other deviations, which were discussed in the last section on fitting, are assessed.
- *Manufacturing process diagnosis*: Manufacturing processes leave tool marks on surfaces. By measuring surfaces at a fine scale, it is possible to track the tool erosion and its effect on the surface quality.

Historically, filtering techniques were pioneered by communication theorists. Developments in analog and digital signal processing strongly influenced how filtering was carried out in surface metrology. More recently, developments in digital image processing have been influencing computational surface metrology.

The general computational method used for filtering is one of convolution. Engineers use the following two types of convolutions.

- *Convolution of functions*: Filtering is often implemented as discrete convolution of functions. Such filters are linear filters. In the most popular version, called the Gaussian filter, the measured data is convolved with the Gaussian function. It has a smoothing effect on the surface data.
- *Convolution of sets*: Morphological filters are implemented using Minkowski sums. These can be regarded as convolutions where the input set is convolved with a round or flat structuring element.

Recent developments in wavelets seem to indicate that the scale-dependent information can be processed more effectively using wavelet filters.

Let's take a brief look at surface metrology first, and then focus on linear and morphological filters.

4.1. Surface Metrology

Motivated by classical Fourier analysis, surface metrologists tend to decompose variations in surfaces into small, medium, and long wavelength components. Filtering is the means to achieve such decomposition. In fact, it is becoming apparent that this decomposition should be done on the basis of small, medium, and long scales. The corresponding variation characteristics are often called roughness, waviness, and form deviations, respectively, in surface metrology literature. In addition, there are variations smaller than the roughness scale; these are tiny, microscopic tears and cracks. And there are variations larger than the form deviation scale; these correspond to variations in size and relative positions.

Different measurement techniques and instruments are needed to observe variations at these different scales. Visual assessment using microscopy was used, and is still used, to observe and classify very small scale variations such as micro burr or torn material. Stylus based instruments are capable of measuring small, medium, and moderately large scale variations. For truly large scale variations, one will have to resort to the likes of CMMs.

It has been well established that surface variations at different scales are caused by different aspects of the processes used to manufacture the surface. Manufacturing processes such as flame cutting and sawing produce very rough surfaces that have details over longer scales. Milling and broaching produce smooth surfaces. Still smoother surfaces are produced in polishing and lapping, leading to lesser variations even at smaller scales. This has made it possible to produce correlation tables that relate manufacturing processes to surface roughness, waviness, and form deviations to aid engineers in manufacturing process planning.

It has also been widely accepted that surface variations at different scales affect the functionality of the produced parts differently. But – and this is a crucial distinction – this type of correlation is far from being well established. In specific instances we know the relationship; for example, variations at different scales affect the performance of crankshafts and bearings differently. But a more comprehensive understanding is yet to be developed. Suffice it to say that such correlation studies strongly depend on clearly defined filtering techniques and their careful implementations on actual measurement data. So it is appropriate that current focus is on defining and standardizing a suite of computational techniques for filters, which broadly fall into linear filters and morphological filters, to which we now turn.

4.2. Linear Filters

Intuitively, linear filters map each point on a profile or a surface to a weighted average of points in its neighborhood. Mathematically, a linear profile filter can be defined as a linear operator on functions, as in

$$y(x) = \int_{-\infty}^{+\infty} K(x,s)z(s)ds$$

where $z(x)$ is the unfiltered input profile, $y(x)$ is the filtered profile, and $K(x,s)$ is the kernel function with finite or infinite support. In this integral transform, we require the kernel $K(x,s)$ to be real, symmetric and space invariant (that is, shift invariant). The kernel is the weighting applied to "average" the input function. Different choices of the kernel lead to different linear filters. If $K(x,s) = K(x-s)$ the filtering is a convolution given by

$$y(x) = \int_{-\infty}^{+\infty} K(x-s)z(s)ds.$$

Metrology standards for linear filters consider only those kernels that lead to the convolution form as shown above.

The constraints imposed on the kernel function have some important consequences. First, such filters are not causal filters because the kernel is symmetric. Also, the symmetry implies that there is no preferred direction of scanning. As $K(x,s) = K(x-s)$, the filter is shift invariant.

For practical applications in metrology, the input profile or surface comes from sampled data. In these cases, it is advantageous to employ discrete linear filters. An extracted profile can be represented by a finite length vector. The length n of the vector is equal to the number of data points. The sampling is assumed to be uniform, i.e., the sampling interval is constant. The ith data point of the profile is the ith component of the column vector $Z = (z_1, z_2, \ldots, z_i, \ldots, z_n)^T$. A filtered profile is also represented discretely by a column vector Y of length n.

A linear filter is given by an $n \times n$ square matrix A so that the filtered profile is obtained by a simple matrix multiplication as $Y = AZ$, which is often called the filter equation. (Sometimes, instead of A its inverse is more easily defined and this is then used to form the filter equation. This method is used to define spline filters.) Elements of the matrix A are determined by the kernel K. If the filter is phase correct, then A is symmetric, i.e., $a_{ij}=a_{ji}$. The sum of the matrix elements in each row equals unity, i.e.,

$$\sum_j a_{ij} = 1.$$

If the kernel is of the form $K(x-s)$, each row of the matrix is a mere shift of the one above so that the matrix elements may be represented by only one row, that is, $a_{ij} = s_k$, $k = i-j$. The values s_k form a vector S of length n. This vector is a discrete representation of the weighting function (kernel). Usually, S contains lots of contiguous zeros at both ends and the length of the non-zero part of S is much smaller than n.

The most popular linear filter is the Gaussian filter, which corresponds to a continuous weighting function

$$s(x) = \frac{1}{\alpha\lambda_c} \exp\left[-\pi\left(\frac{x}{\alpha\lambda_c}\right)^2\right]$$

of infinite support, where x is the distance from the center (maximum) of the weighting function, λ_c is the so called cut-off wavelength, and α is a constant given by $\sqrt{(\log 2)/\pi}$ = 0.4697. Figure 11 provides an illustration of this weighting function. This continuous function is sampled at equal intervals of Δx to yield

$$s_k = \frac{1}{C} \exp\left[-\pi\left(\frac{\Delta x}{\alpha\lambda_c}\right)^2 k^2\right]$$

where the normalization constant is

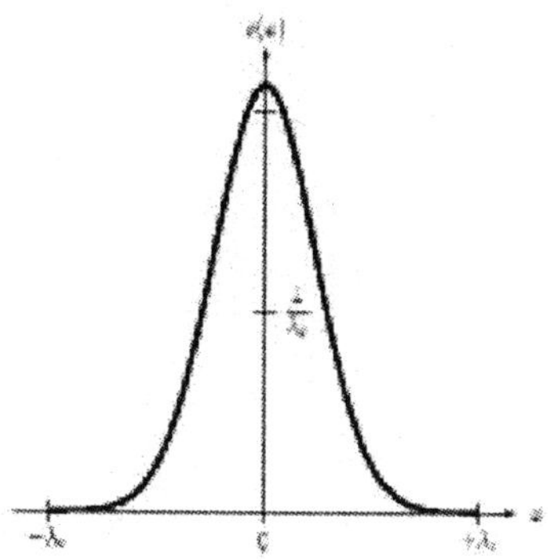

FIGURE 11. Gaussian weighting function.

$$C = \sum_k \exp\left[-\pi\left(\frac{\Delta x}{\alpha\lambda_c}\right)^2 k^2\right]$$

The matrix A for Gaussian filters contains rows that are merely one-column shifts of the row vector defined by s_k. The result is that each component y_i in the filtered output vector is the weighted average of all components of the unfiltered input vector Z. However, due to the rapid decline to zero of the weighting function $s(x)$ beyond the cut-off wavelength in either direction from y_i, the practical result is that the averaging of Y is more local than global. Figure 12 shows the result of applying Gaussian filter superposed on an unfiltered input profile. It can be seen from this figure why linear filters, such as the Gaussian filter, are also known as mean line filters because the filtered profile can be considered as the "mean" of the input profile.

More recently, another linear filter called spline filter has been proposed as an alternative to the familiar Gaussian filter. As the name implies, spline filters are characterized by the use of spline functions to define the filtered output. Splines are low degree polynomials that can be pieced together to obtain very smooth functions that can be changed locally to suit the need. Recent experiments with spline filters have proven their flexibility and usefulness for industrial applications. Spline filters are linear filters – however, their weighting functions cannot be given in a simple closed form. Instead the filter equations are defined for the spline filters, and where necessary, a numerical calculation of the weighting function is provided. For example, the filter equation for non-periodic open profiles using cubic spline is given by

$$(1 + \alpha^4 Q)Y = Z$$

where Q is the $n \times n$ square matrix given by

$$Q = \begin{bmatrix} 1 & -2 & 1 & & & & & \\ -2 & 5 & -4 & 1 & & & & \\ 1 & -4 & 6 & -4 & 1 & & & \\ & & \cdot & \cdot & \cdot & \cdot & & \\ & & & \cdot & \cdot & \cdot & \cdot & \\ & & & & 1 & -4 & 6 & -4 & 1 \\ & & & & & 1 & -4 & 5 & -2 \\ & & & & & & 1 & -2 & 1 \end{bmatrix}$$

and the parameter $\alpha = 1/(2\sin(\pi\Delta x/\lambda_c))$. If the sampling interval Δx is small enough, the weighting function $s(x)$ can be approximated by the continuous function given by

$$s(x) = \frac{\pi}{\lambda_c}\sin(\sqrt{2}\frac{\pi}{\lambda_c}|x| + \frac{\pi}{4})\exp(\sqrt{2}\frac{\pi}{\lambda_c}|x|)$$

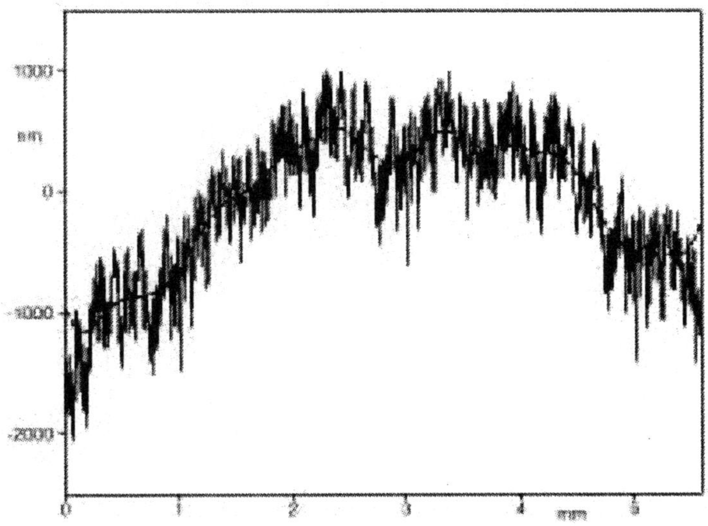

FIGURE 12. Output of the Gaussian filter superposed on an unfiltered input profile, with $\lambda_c = 0.8$ mm.

which is illustrated in Figure 13. It is instructive to compare this with the weighting function defined for the Gaussian filter and shown in Figure 11. Figure 14 shows the result of applying the cubic spline filter superposed on an unfiltered input profile.

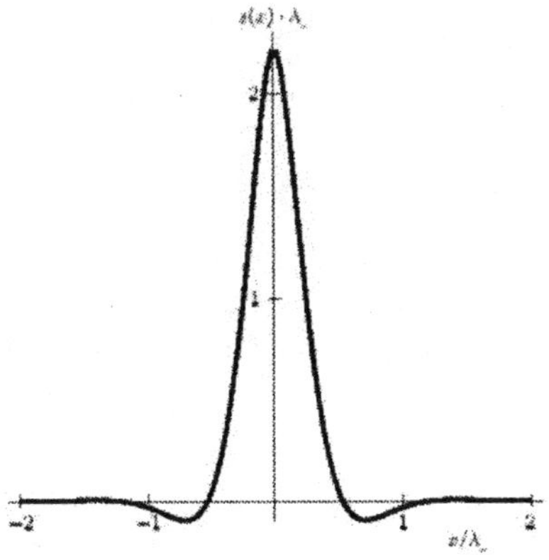

FIGURE 13. Weighting function for cubic spline filter.

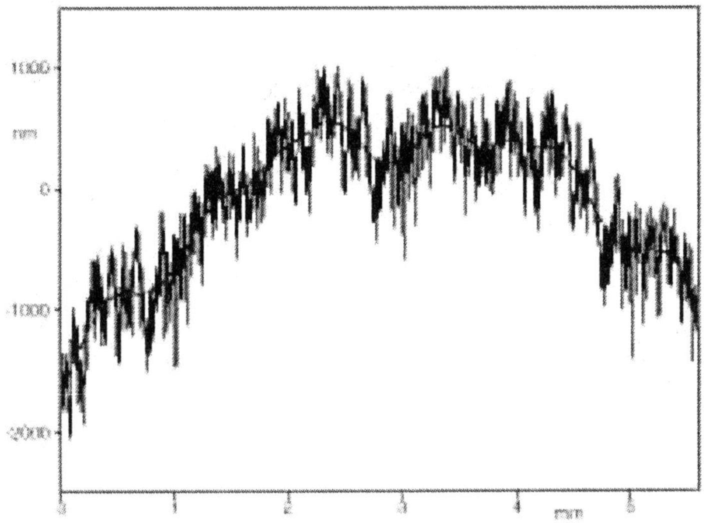

FIGURE 14. Output of a cubic spline filter superposed on an unfiltered input profile, with $\lambda_c = 0.8$ mm.

Finally, we should mention the potential for wavelet filters in surface metrology. Wavelets provide an attractive alternative to traditional Fourier analysis. In the field of metrology, their use has received recent attention because they can be used to perform multi-resolution analysis and they have good diagnostic capabilities. Wavelet analysis consists of decomposing a profile into linear combination of wavelets $g_{a,b}(x)$, all generated from a single mother wavelet. A mother wavelet is a function of one or more variables, which forms the basic building block for wavelet analysis. See Figure 15 for an example mother wavelet. Usually a mother wavelet integrates to zero and is localized in space with finite support.

If $g(x)$ is a mother wavelet, then the wavelet family is generated as $g_{a,b}(x) = a^{-1/2} g((x-b)/a)$ where a is the dilation (scaling) parameter and b is the translation parameter. This is similar to Fourier analysis, which decomposes a profile into a linear combination of sine waves, but unlike Fourier analysis wavelets can identify the location as well as the scale of a feature in a profile. As a result they can decompose profiles where the small-scale structure in one portion of the profile is unrelated to the structure in a different portion, such as localized changes (e.g., scratches). Wavelets are ideally suited for non-stationary profiles. Basically, wavelets decompose a profile into building blocks of constant shape but of different scales. The discrete wavelet filter is defined as

$$y_i(a) = \Delta x \sum_j z_j g_{a,j\Delta x}(j\Delta x)$$

with mother wavelet $g(x)$. The dilation parameter a is also restricted to discrete values. Typically consecutive values of a have a fixed ratio of 2.

Of particular interest are spline wavelets, which are families of wavelets whose corresponding weighting functions are B-splines. Figure 16 shows a milled profile measured with a 5μm tip stylus with successively smoothed profiles using the so-called (4-6) cubic spline wavelet. The separation of details at different scales is clearly seen in this figure. Future developments in linear filters for metrology might well come from wavelet filters.

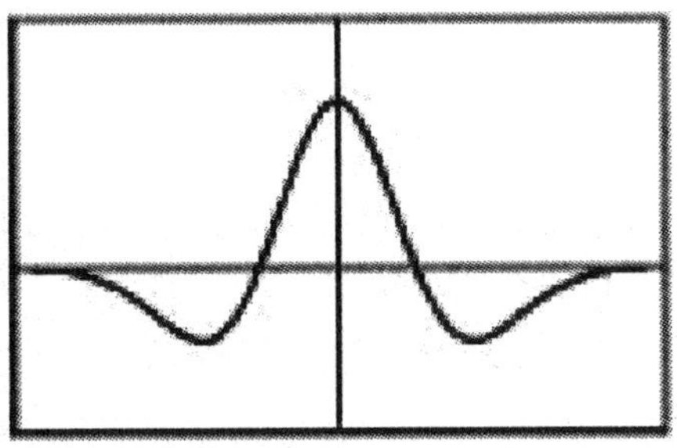

FIGURE 15. A mother wavelet.

4.3. Morphological Filters

Morphological filters are mathematically well-defined operations that are rooted in Minkowski sums; hence their association with convolutions of sets. Such filters can be defined for both profiles and surfaces. Morphological filters are particularly attractive for metrology applications where surfaces are scanned using a tactile probe having a specially shaped tip. Intuitively, morphological filters operate by sweeping the input geometric object using a second geometric (structuring) element. Let's start with a brief definition of Minkowski sums and give a quick description of several morphological filters.

If A and B are two sets in $\mathbb{R}^n$ then their Minkowski addition is denoted and defined as $A \oplus B = \{a + b : a \in A, b \in B\}$. Minkowski addition is also known as the vector addition of sets. It can be viewed geometrically as the translational sweep of one of the sets by the other. It can be seen that the Minkowski addition is commutative, that is, $A \oplus B = B \oplus A$. Minkowski subtraction is defined and denoted as $A \Theta B = \overline{\overline{A} \oplus B}$ where overbar denotes complementation. Next come several morphological operations, which are defined in terms of Minkowski sums. Here it is customary to call one of the sets, say A, as the input set and the other, B, as the structuring element. A symmetric version of B is obtained by a reflection of B through the origin and is denoted $\check{B} = \{-b : b \in B\}$.

With these terms and notations, we have the following two primary morphological operations:

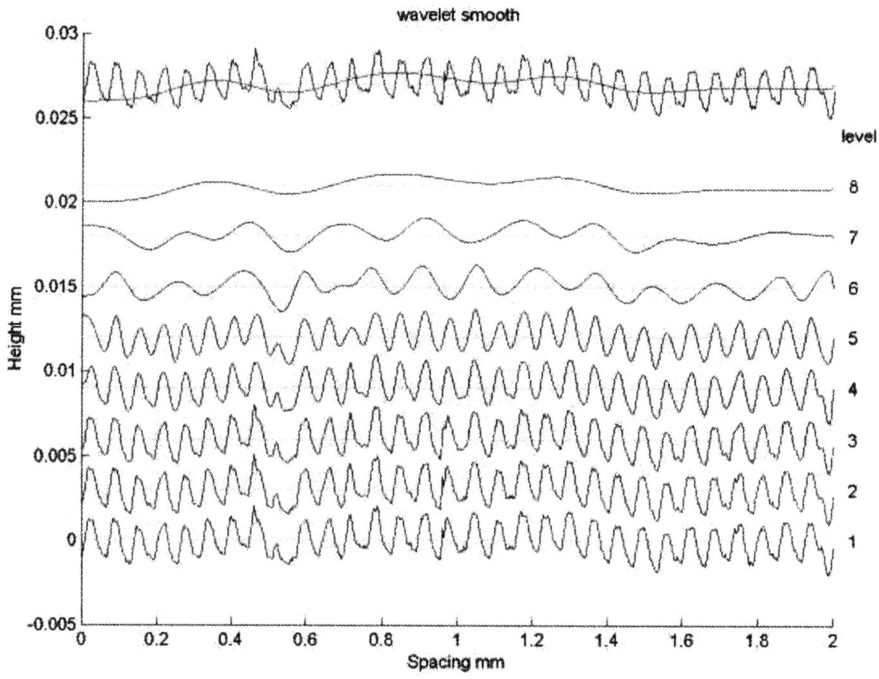

FIGURE 16. Successively smoothed profiles of a milled profile using (4-6) cubic spline wavelet filter.

- *Dilation*: Dilation of *A* by *B* is $D(A,B)=A \oplus \check{B}$.
- *Erosion*: Erosion of A by *B* is $E(A,B)=A \ominus \check{B}$. Dilation expands the input set by the structuring element, while erosion shrinks it.

Applying dilation and erosion consecutively, two secondary morphological operations are obtained as:

- *Opening*: Opening of *A* by *B* is $O(A,B)=D(E(A,B), \check{B})$.
- *Closing*: Closing of *A* by *B* is $C(A,B)=E(D(A,B), \check{B})$.

It is possible to apply these operations further in sequence to obtain other complex morphological operations.

In image processing literature morphological operations defined above are referred to as morphological filters. In binary morphology the input image *A* and the structuring element B are treated as sets in $\mathbb{R}^2$. In gray-scale morphology the filtering operations are invoked on functions defined over a domain in $\mathbb{R}^2$. A mathematical link between morphological operations on sets and functions can be established through umbra transformations. Informally, the umbra of a function is the set of points under the graph of the function. Without elaborating the details, we simply observe that morphological filtering of functions of n variables can be shown to be equivalent to the corresponding morphological operation on sets in $\mathbb{R}^{n+1}$. As a general rule, morphological operations on sets should be taken as the fundamental mathematical definitions because they are more general, and morphological filtering of functions should be derived and justified from the set definitions.

For metrological filtering applications we follow the convention established in signal and image processing. This means that we will speak of morphological filters for functions of one and two variables. For simplicity of exposition, morphological filters for metrology are first developed for functions of one variable (also called signals) so that they can be applied and tested for profiles. These can be extended conceptually to functions of two variables that cover the case of surfaces. In general, signal processing deals with both analog (continuous) and digital (discrete) signals. In computational metrology, we focus on discrete morphological filters. It has been recently shown that these discrete filers can be computed efficiently and robustly.

It is well known that a morphological operation the *erosion* can be used to define a mechanical surface when information about the surface is gathered by collecting data on the locus of centers of a spherical ball that is rolled on the real surface. To illustrate this idea, Figure 17 shows an upper curve that is the locus of a 50 μm radius disk rolled over a profile. The mechanical surface is the lower curve obtained mathematically from the upper curve by the morphological operation of erosion.

After the introduction of morphological filters, it was recognized that several envelope filters used in metrology data processing are special cases of morphological filters. Figure 18 shows how an envelope filter using a 50 μm radius disk operates on an input data. This envelope filter was implemented as a morphological filter the *closing filter*. Note how this fills valleys while preserving the peaks . A similar morphological filter the *opening filter* knocks out the peaks and preserves the valleys. When applied in alternating sequence, these two filters can be used to selectively eliminate features of any given "size" from the input data.

5. Summary

Over the last decade, computational metrology has evolved as a discipline in its own right. This is largely due to the rapid migration of industry to the information age. Increasing use of general purpose measurement tools, such as CMMs and scanning probes, in industry has accelerated the use of fitting and filtering techniques as routine practices in day-to-day industrial operations. This, in turn, has placed urgent demands on clear understanding, definition, and standardization of computational aspects of such techniques.

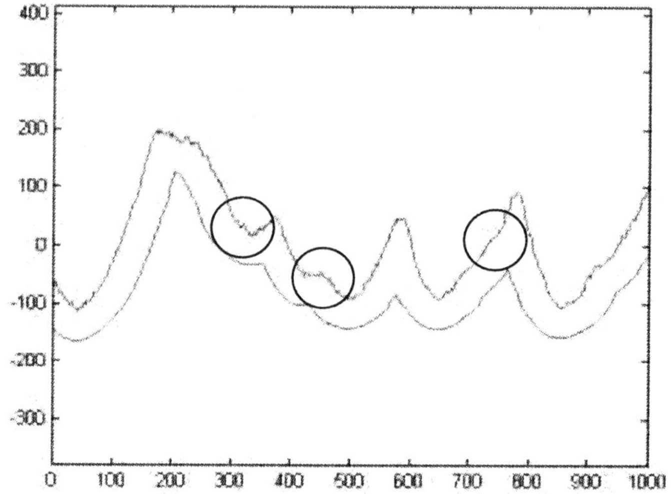

FIGURE 17. Output of a morphological operation (erosion) shown as a lower curve. The input is the upper curve. The lower curve is the mechanical surface corresponding to a 50 μm radius disk, shown in three distinct places.

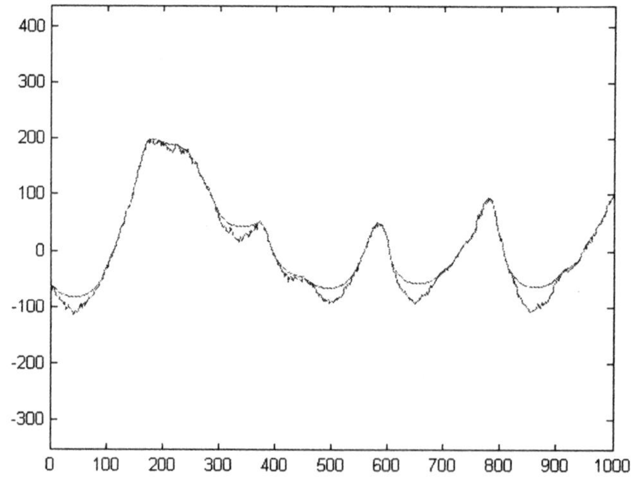

FIGURE 18. Output of an envelope filter, implemented as a morphological filter called closing filter, is shown as the upper curve. The input to the filter is the lower curve.

This paper shows how seemingly different practices that have evolved in industrial applications are now being consolidated under optimization (fitting) and convolution (filtering). We can now provide better scientific basis for these applications. Several problems still remain open, especially involving measurement uncertainty. Industrial need will continue to drive further innovations and developments in this field, ensuring productive and profitable work in this area for quite a while more.

6. Notes and References

Hounshell (1985) describes the early history of American manufacturing industry. Most of the historical accounts mentioned in this paper on the evolution of manufacturing precision are drawn from his book. Hume (1980) provides a delightful account of the history of engineering metrology. He traces the contribution of measurement instruments, such as Watt s micrometer, initially to the industrial revolution and later to mass production. Klein (1974) gives a general historical survey of the science of measurements. His account of the evolution of length measurements is quite exhaustive and insightful. Stigler (1986) is a standard reference for the history of statistics as it relates to measurement of uncertainty before 1900. He draws attention to Fourier s attempt to assess the uncertainty of the great pyramid s height. His description of the historical development of statistics as the measurement of uncertainty (viz. uncertainty of measurements) is very relevant to today s problems.

Doiron and Stoup (1997) give a technical account of how measurement uncertainty is handled in dimensional calibration services of a national laboratory. Srinivasan (2003) provides a mathematical theory of dimensioning. ASME Y14.5M-1994 and ASME Y14.5.1M-1994 are the American national standards on tolerance specifications and their mathematical interpretations. There are international standards (ISO) corresponding to these ASME dimensioning and tolerancing standards. These national and international standards govern bulk of the technical documentation in industry.

Forbes (1991) and Shakarji (1998) report how national laboratories developed least-squares fitting reference algorithms for testing commercial fitting software. Anthony et al. (1993) report how the same task was accomplished for Chebyshev fitting software at U.K. ASME B89.4.10-2000 describes standardized performance evaluation methods for fitting software for CMMs. These are valuable sources for practical solutions to the fitting problems described in this paper. The national laboratories have been at the forefront in providing reference software to test and compare commercial software for fitting.

Computational geometry literature on various fitting problems are too numerous to mention here. Houle and Toussaint (1988) describe the first systematic account of the minimax line and plane fitting problems from the computational geometric viewpoint, which emphasizes combinatorial optimization. A historic note: this was presented as the very first paper in the First ACM Computational Geometry Symposium held at Baltimore in 1985. Smid and Janardan (1999) expand upon this to include the case of minimax fitting of circles. A steady stream of papers continues to pour on other challenging fitting problems. A separate survey paper, from a computational geometry viewpoint, is needed to fully describe the attempts, successes, and failures in finding theoretical and practical solutions to various fitting problems described in this paper.

Whitehouse (2002) provides a readable account of surface metrology, especially for non-specialists. Latest developments in filtering techniques are surveyed by Raja et al (2002). More references on filtering can be found in their paper. Some details of morphological filters can be found in Srinivasan (1998). Scott (2000) describes the scale-space techniques pursued in filtering problems.

7. Bibliography

ASME Y14.5M-1994. *Dimensioning and Tolerancing*, The American Society of Mechanical Engineers, New York, 1995.

ASME Y14.5.1M-1994. *Mathematical Definition of Dimensioning and Tolerancing Principles*, The American Society of Mechanical Engineers, New York, 1995.

ASME B89.4.10-2000. *Methods for Performance Evaluation of Coordinate Measuring System Software*, The American Society of Mechanical Engineers. New York, 2002.

Anthony, G.T. et al. *Chebyshev Best-fit Geometric Elements*, NPL Report DITC 221/93, National Physical Laboratory, U.K., 1993.

Doiron, T. and Stoup, J. Uncertainty and Dimensional Calibrations, *Journal of Research of the National Institute of Standards and Technology*, Vol. 102, 1997, pp. 647-676.

Forbes, A. *Least-squares Best-fit Geometric Elements*, NPL Report DITC 140/89, National Physical Laboratory, U.K., 1991.

Houle, M.E. and Toussaint, G.T. Computing the width of a set, *IEEE Transactions on Pattern Analysis and Machine Intelligence*, Vol. 10, 1988, pp. 761-765.

Hounshell, D. *From the American System to Mass Production, 1800-1932*, Johns Hopkins University Press, Baltimore, 1985.

Hume, K.J. *A History of Engineering Metrology*, Mechanical Engineering Publications, London, 1980.

Klein, H.A. *The Science of Measurement*, Dover Publications, New York, 1974.

Raja, J., Muralikrishnan, B., and Fu, S. Recent advances in separation of roughness, waviness and form, *Precision Engineering*, 2002, pp. 1-14.

Scott, P.J. Scale-space techniques. In *Proceedings of the Xth International Colloquium on Surfaces*, Chemnitz University of Technology, Germany, 2000.

Shakarji, C.M. Least-squares fitting algorithms of the NIST algorithm testing system, *Journal of Research of the National Institute of Standards and Technology*, Vol. 103, 1998, pp. 633-641.

Smid, M. and Janardan, R. On the width and roundness of a set of points in the plane, *International Journal of Computational Geometry & Applications*, Vol 9, 1999, pp. 97-108.

Srinivasan, V. Discrete morphological filters for metrology. In *Proceedings of the 6th ISMQC Symposium on Metrology for Quality Control in Production*, TU Wien, Austria, 1998.

Srinivasan, V. *Theory of Dimensioning*, Marcel Dekker, New York, 2003.

Stigler, S. M. *The History of Statistics*, Harvard University Press, Cambridge, 1986.

Whitehouse, D. *Surfaces and their Measurements*, Taylor & Francis, New York, 2002.

Vijay Srinivasan, IBM Corporation and Columbia University, New York.
E-mail address: vasan@us.ibm.com

DIMACS Series in Discrete Mathematics
and Theoretical Computer Science
Volume **67**, 2005

Combinatorial approaches to Geometric Constraint Solving: Problems, Progress and Directions

Meera Sitharam

ABSTRACT. We formulate the fundamental combinatorial problems in geometric constraint solving that have emerged in the past 15 years within the mechanical CAD context. We discuss recent progress on these problems and identify future directions along with promising techniques from related areas.

Organization

1. Background

Geometric constraint systems have been studied in the context of variational constraint solving in CAD for nearly 2 decades [**VL81, Nel85, Tod89, Rol91, Kra92, aAM96, aCR97, aRAaGLK93**] [**aIFaCMHaJCaRP95, Fud95, aRV95, aRJa97, aALaMS97a**] [**aALaMS98a, aALaMS01a, aALaMS01b, aALaMS99**] [**aBY00, aCSC01, aALaMS99, aSCC98a, aSCC98b, Owe91**] [**Owe93, Bru86,**

Key words and phrases. Geometric constraint solving, Algorithms, Rigidity theory, Geometric modeling, Mechanical computer aided design.
The author was supported in part by NSF Grants EIA 0096104 and CCR 9902025.

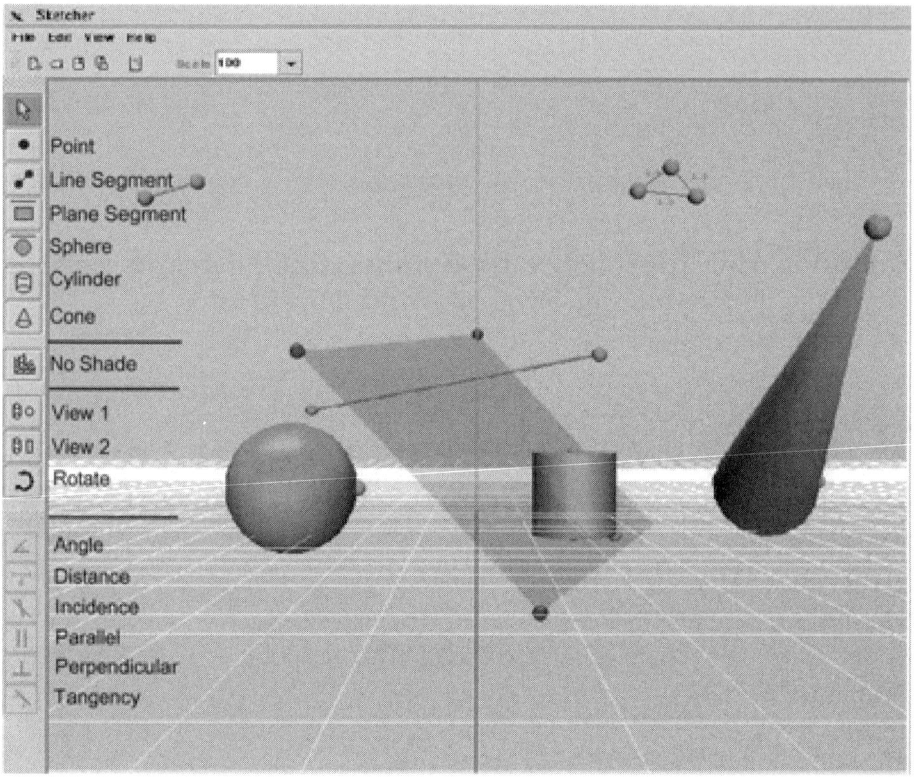

FIGURE 1. Typical 3D sketcher: icons show repertoire of object and constraint typ es

Owe, Pab93, aRJaDM93] [**aMSaBY04, aMS04, aMSaBMaAA01, Sit04b**]. For recent reviews of the extensive literature on geometric constraint solving see, e.g, [**aALaMS98a, Kra92, Fud95**].

A *geometric constraint system* consists of a finite set of geometric objects and a finite set of constraints between them. See Figure 1. The constraints can usually be written as algebraic equations and inequalities w hose variables are the coordinates of the participating geometric objects. For example, a distance constraint of d between two points (x_1, y_1) and (x_2, y_2) in 2D is written as $(x_2 - x_1)^2 + (y_2 - y_1)^2 = d^2$ Most of the constraint solvers so far deal with 2D constraint systems, although some of the newer approaches including [**aALaMS97b, aALaMS98b, aALaMS01a, aALaMS01b**] [**aALaMS99, Bru86, Owe**] [**aMSaBY04, aMS04, aMSaBMaAA01, Sit04b**], extend to 3D constraint systems. A *solution or realization* of a geometric constraint system is the (set of) real zero(es) of the corresponding algebraic system. In other words , the solution is a class of valid instantiations of (the position, orientation an d any other parameters of) the geometric elements such that all constraints are sa tisfied. Here, it is understood that such a solution is in a particular geometry, for example the Euclidean plane, the sphere, or Euclidean 3 dimensional space. A constraint system can be classified as *overconstrained*, *well-constrained*, or *underconstrained*. Well-constrained systems have a finite, albeit potentially very large number of *rigid* solutions; th eir solution space is a zero-dimensional variety.

Underconstrained systems have infinitely many solutions; their solution space is not zero-dimensional. Overconstrained systems do not have a solution unless they are *consistently overconstrained.* Well or overconstrained systems are called *rigid* systems.

1.0.1. *Scope.* Geometric constraints have been used as succinct, minimal representations of geometric composites in many other applications besides CAD, such as robotics, molecular modeling and teaching geometry. As a result, there is a vast body of literature related to geometric constraint solving. This manuscript's scope is restricted to combinatorial methods for geometric constraint solving, specifically within the CAD application context.

Technically, geometric constraint solving straddles combinatorics, algebra and geometry. Specifically, it overlaeps combinatorial rigidity theory, geometric methods for kinematics, robotics and mechanisms, automated geometry theorem proving, geometric invariant theory, and computational algebraic geometry (solving specific classes of polynomial systems arising from geometric constraints).

The question: *to what extent can geometric constraint problems be approached combinatorially?* is highly non-trivial and important. It provides a framework for a systematic discussion of progress in the area and leads to a natural organization of the problems to be discussed in this paper.

1.1. Constraint Graphs and Degrees of Freedom. A geometric constraint graph $G = (V, E, w)$ corresponding to geometric constraint system is a weighted graph with vertex set (representing geometric objects) V and edge set (representing constraints) E; $w(v)$ is the weight of vertex v and $w(e)$ is the weight of edge e, corresponding to the number of degrees of freedom available to an object represented by v and number of degrees of freedom (dofs) removed by a constraint represented by e respectively.

For example Figures 2, 3, 4 show 2D and 3D constraint systems and their respective dof constraint graphs. Several more 3D constraint systems whose graphs have vertices of weight 3, 5 and edges of weight 1,3 can be found in Figures 9, 8, 5, 6, 7.

Note that the constraint graph could be a *hypergraph*, each hyperedge involving any number of vertices. A subgraph $A \subseteq G$ that satisfies

$$\sum_{e\in A} w(e) + D \geq \sum_{v\in A} w(v) \tag{1.1}$$

is called *dense*, where D is a dimension-dependent constant, to be described below. Function $d(A) = \sum_{e\in A} w(e) - \sum_{v\in A} w(v)$ is called *density* of a graph A.

The constant D is typically $\binom{d+1}{2}$ where d is the dimension. The constant D captures the degrees of freedom of a rigid body in d dimensions. For 2D contexts and Euclidean geometry, we expect $D = 3$ and for spatial contexts $D = 6$, in general. If we expect the rigid body to be fixed with respect to a global coordinate system, then $D = 0$.

Next, we give some purely combinatorial properties of constraint graphs based on density. These will be later shown to be related to properties of the corresponding constraint systems.

A dense graph with density strictly greater than $-D$ is called *overconstrained.* A graph that is dense and all of whose subgraphs (including itself) have density at

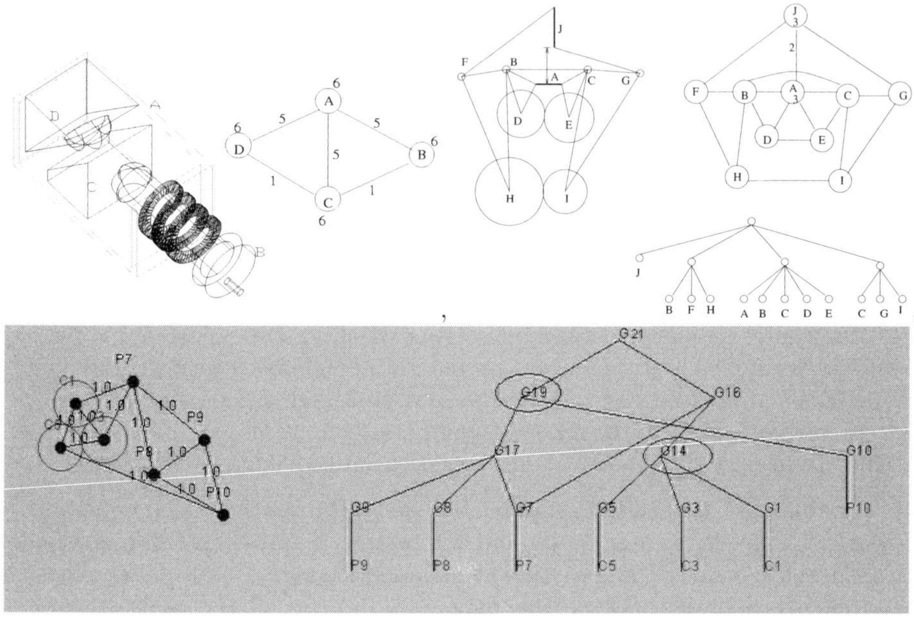

FIGURE 2. *Left:* An underconstrained 3D example with 1 extra dof and its dof constraint graph: each of 4 rigid objects has 6 dofs, B, C, D are constrained by incidence with the housing A, which removes 5 dofs in each case, so that each of the pairs AB, AC, AD is underconstrained and has one extra degree of freedom. The remaining 2 pairs BC and CD are connected by distance constraints, which remove 1 dof each. *Right:* A 2D constraint system with points or fixed radius circles and distance constraints, except for A and J which are line segments with a perpendicularity and distance constraint between them. In the constraint graph, all unnumbered vertices have weight 2 and unnumbered edges have weight 1. *Below:* A 2D example, constraint graph, DR-plan.

most $-D$ is called *wellconstrained.* A graph G is called *well-overconstrained* if it satisfies the following: G is dense, G has atleast one overconstrained subgraph, and has the property that on replacing all overconstrained subgraphs by wellconstrained subgraphs (in any manner), G remains dense. A graph that is wellconstrained or well-overconstrained is called *cluster.* A dense graph is *minimal* if it has no dense proper subgraph. A graph that is not a cluster is said to be *underconstrained.* If a dense graph is not minimal, it could in fact be an underconstrained graph: the density of the graph could be the result of embedding a subgraph of density greater than $-D$.

1.2. Inadequacy of a pure dof analysis. Note that all minimal dense subgraphs are clusters but the converse is not the case.

To illustrate the challenge involved, we need to discuss how the graph theoretic properties described above based on *degree of freedom (dof) analysis* relate to corresponding properties of the corresponding constraint *system.* For this, we need to introduce the notion of *genericity.* Informally, a constraint system is generically

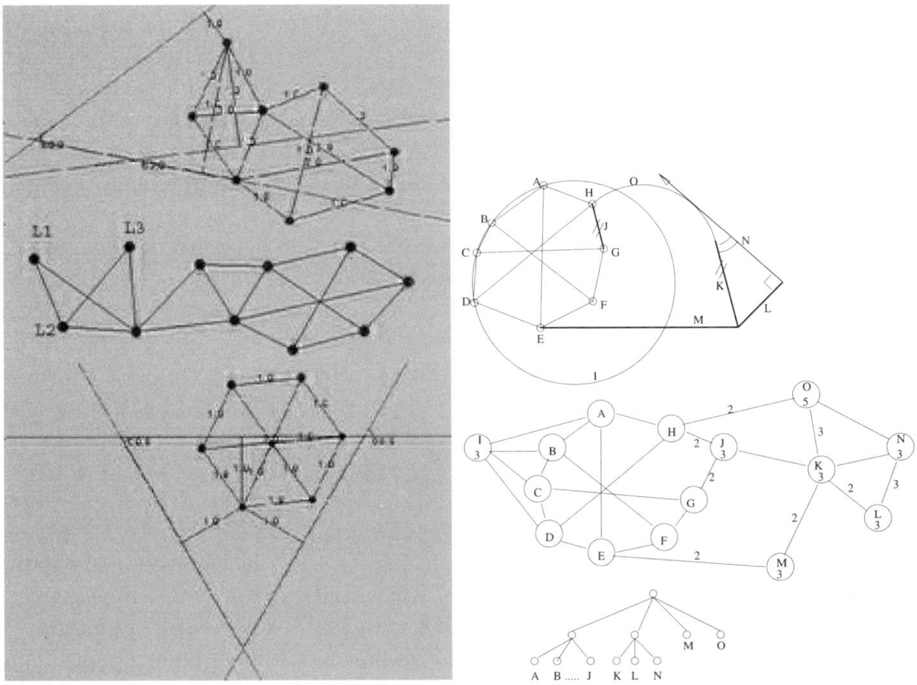

FIGURE 3. Two more 2D constraint systems examples; constraint graphs; DR-plans; all unnumbered vertices and edges have dof weights 2 and 1. *Left:* vertices L1, L2, L3 in constraint graph represent the line objects; solution shown at bottom. *Right:* Example of Variable radius Circle, Arc and Ray objects, Angle, Paralellism and Tangency constraints

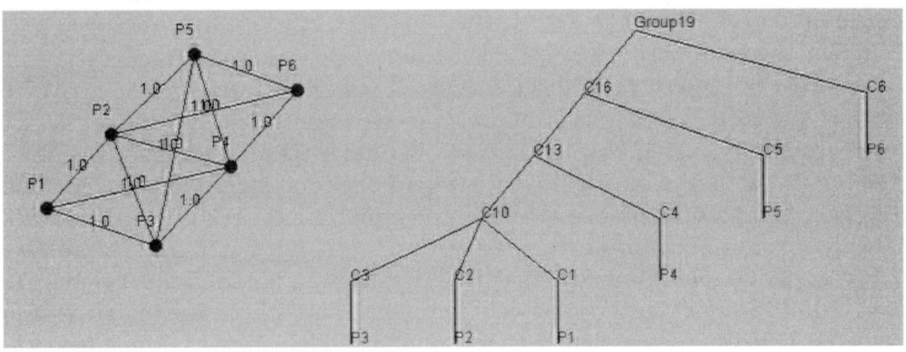

FIGURE 4. Simple 3D constraint system drawn on 2D canvas with points and distances and DR-plan

rigid if it is rigid (does not flex or has only finitely many non-congruent, isolated solutions) for most of choices of coefficients of the system. More formally we use the notion of genericity of e.g, [**aJLaDO98**]. A property is said to hold *generically* for polynomials $f_1, \ldots, f_n$ if there is a nonzero polynomial P in the coefficients of the f_i such that this property holds for all $f_1, \ldots, f_n$ for which P does not vanish.

Thus the constraint system E is generically rigid if there is a nonzero polynomial P in the coefficients of the equations of E - or the parameters of the constraint system - such that E is solvable when P does not vanish. For example, if E consists of distance constraints, the parameters are the distances. Even if E has no overt parameters, i.e, if E is made up of constraints such as incidences or tangencies or perpendicularity or parallelism, E in fact has hidden parameters capturing the extent of incidence, tangency, etc., which we consider to be the parameters of E.

A generically rigid system always gives a cluster, but the converse is not always the case. In fact, there are well-constrained, dense clusters whose corresponding systems are not generically rigid and are in fact generically not rigid due to the presence of generic *constraint dependences*. Consider for example Figures 5, 6, 7 that illustrate the "bananas" problem of [**aBSaHS93**] caused by generic constraint dependence. This problem is also present in Figure 9, and can be detected as the root cause beneath large class of combinatorial misclassifications, although this detection is nontrivial.

A combinatorial dof analysis of the 3D constraint system in Figure 6(top) would correctly report the left and right subsystems ($P1, P2, P3, P4, P5$ and P_1, P_6, P_7, P_8, P_5 respectively) and the whole system to be wellconstrained clusters. Figure 7 (bottom) has the same number of constraints, but a dof analysis

would correctly report both that the left subgraph as overconstrained and the whole as underconstrained. Figure 5 (left) also has the same number of constraints and appears to be a wellconstrained cluster according to a dof analysis. However, while the left and right subsystems are (in fact) wellconstrained clust ers, the whole system is generically overconstrained. In a well-defined sense, this is an overconstraint that is not *explicit*, but is caused by a constraint *dependence* that is not detectable by a dof or density count. However, when *restricted to consistently overconstrained situations* (those choices of distances - such as in this example - that are guaranteed to admit a solution), the system in Figure 5 (left) is generically underconstrained, although the system on Figure 5(right) is generically wellconstrained.

In fact, a constraint system has a constraint dependence if the common overlap of *any* subset of its wellconstrained clusters is underconstrained. The above "bananas" is a special case of this. However, among these situations, the dof analysis is inaccurate only in the "bananas" case. To date, there is no known, tractable characterization of generic rigidity of systems for 3 or higher dimensions, based purely on combinatorial properties of the constraint graph [**Whi97**], [**aBSaHS93**], although several conjectures exist.

However, it should be noted that in 2 dimensions, according to Laman's theorem [**Lam70**], if all geometric objects are points and all constraints are distance constraints between these points then any minimal dense cluster represents a generically rigid system. But there are no known combinatorial characterizations of 2D rigidity, when other constraints besides distances are involved. For methods that address 2D systems with angle and incidence constraints, see Section 8 and [**JJO04**].

Note. In this manuscript, we will nevertheless rely on the above dof analysis, augmented, however, to check for many types of constraint dependence (including the bananas situation and others such as the so-called "hinge" situation [**Cra79, Cra82**]). Since a complete description of this augmented property of a

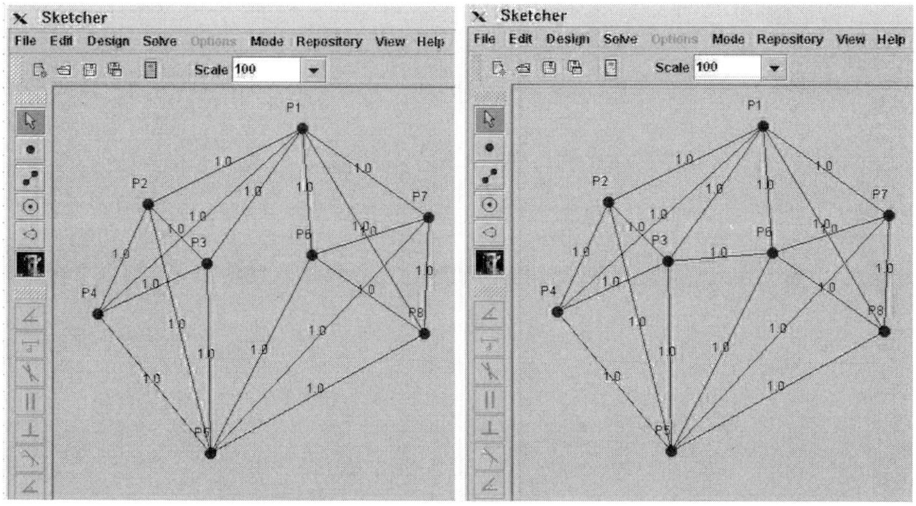

FIGURE 5. Constraint dependences in 3D: constraint system drawn on a 2D canvas; corresponding constraint graphs have vertices of weight 3 and edges of weight 1

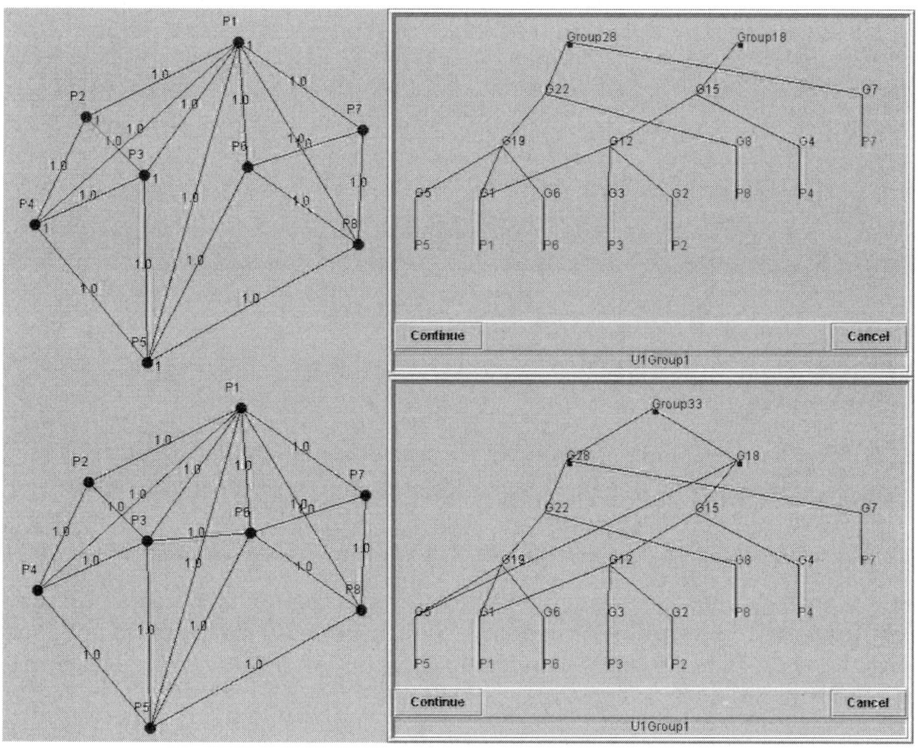

FIGURE 6. Modifications to 3D system in Figure 5: Combinatorially well con strained (DR-plan has single source, top) and underconstrained (DR-plan has many sources, bottom)

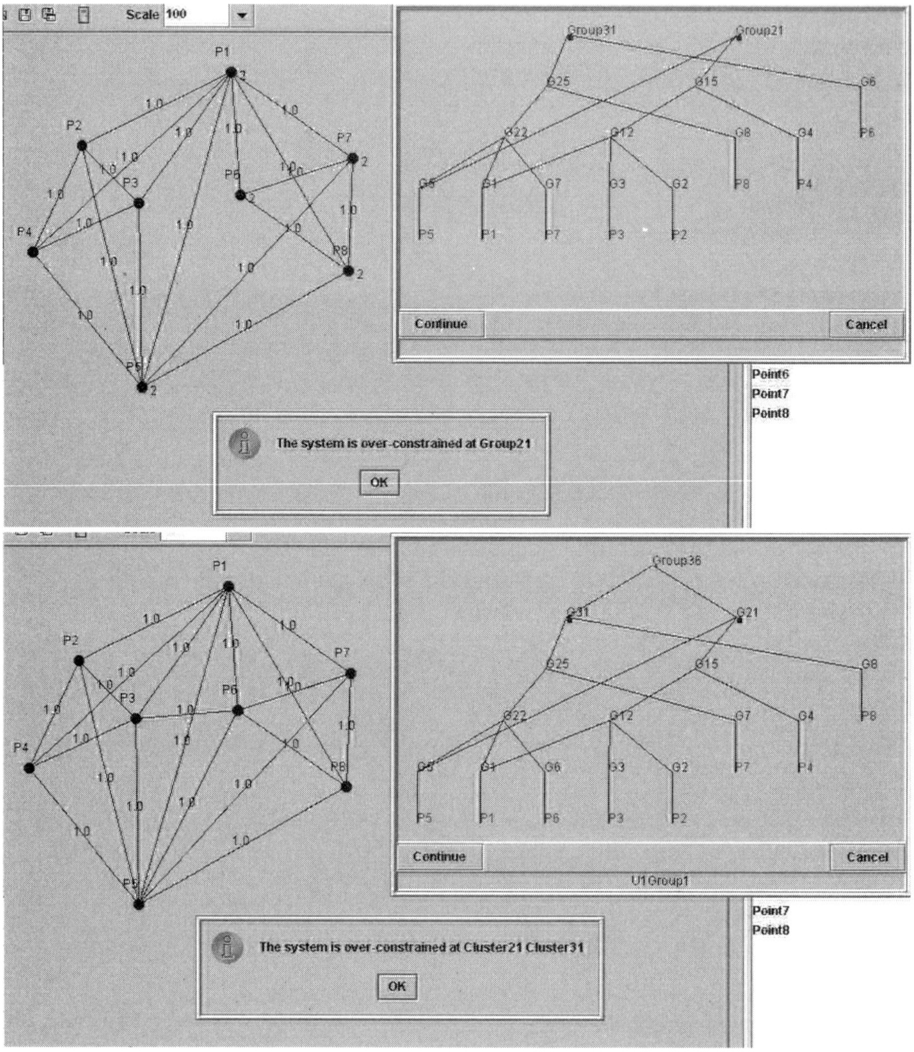

FIGURE 7. Combinatorially or explicitly overconstrained clusters in well (single source in DR-plan) and underconstrained (multiple sources in DR-plan) graphs

cluster (called *module-rigidity*) [**aYZ04b**] is involved and outside the scope of this paper, from now on we will use the terms *rigid system* and the definition of *cluster* interchangeably. In Section 4 however, we will sketch an intuitive, simple algorithm that approximates a check for module-rigidity.

2. Basic Input-Output requirements of a geometric constraint solver

The generic input to an effective geometric constraint solver for CAD applications consists of the following.

(1) A 2D or 3D geometric constraint system (see Section 1), S, i.e, a set of primitive geometric objects, each of a specified type, and constraint types relating pairs (or

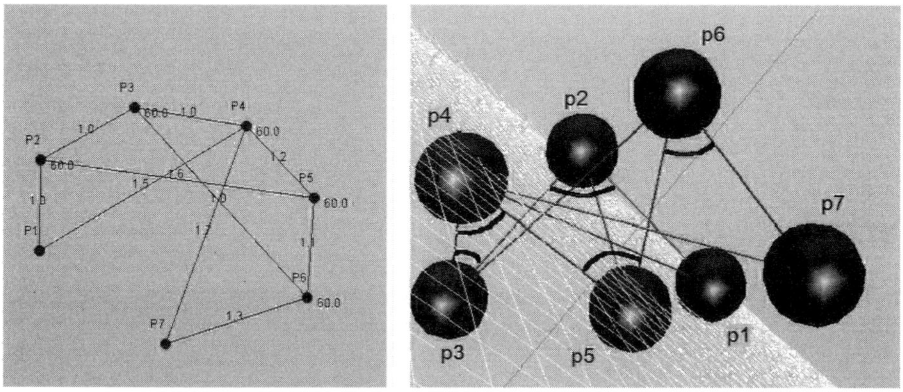

FIGURE 8. 3D constraint system drawn on 2D canvas with line and point objects and distance and angle constraints (graph has vertices of weight 3(points and lines); edges of weight 3 (incidences) and 1 (distances and angles). Solution (right)

larger subsets) of these objects. See Figure 1. Some constraint types have a metric value specification, for example, a distance constraint between 2 points and a value for the distance; or a torsion angle constraint between a pair of line segments in 3D and a value for the angle. The values associated with the constraints could be constants or variables related to other such values in the case of so-called *relational or engineering constraints or navigation constraints*, see Section 5. In addition, *intervals* of values are permitted for some of the incidence or intersection constraints which are effectively chirality or oriented matroid constraints, declaring (non) intersections of convex hulls of subsets of geometric objects, and used primarily for navigation of the solution space, for example, the constraint that 2 line segments should or should not interesect. The constraint system, also in 3D, is often sketched on a 2D canvas. See Figures 9 and 8.
(2) An input partial decomposition of the constraint system P obtained from one or more views of a feature, part or subassembly design hierarchy. The features or subassemblies are nodes of directed acyclic graph (*dag*) and are intended to be manipulated independently within their own local coordinate systems, as well as moved around relative to each other in the coordinate system of a higher level feature or subassembly. There could be a *priority order* given to the resolution of the features or subassemblies: a linear order degenerates to so-called parametric constraint solving. See Figure 10.
(3) Some of the features or subassemblies in the input hierarchy may have been already solved and may be chosen from a repertoire of commonly occuring such features, parts or subassemblies. In this case, they may not be input as constraint systems; they should be flagged and treated en-bloc as already solved rigid bodies.
(4) Additions, changes, updates to all of the above. See Figure 28.
(5) Other interactive user input requested by the geometric constraint solver during solution space navigation.

A geometric constraint solver's desired outputs are the following.

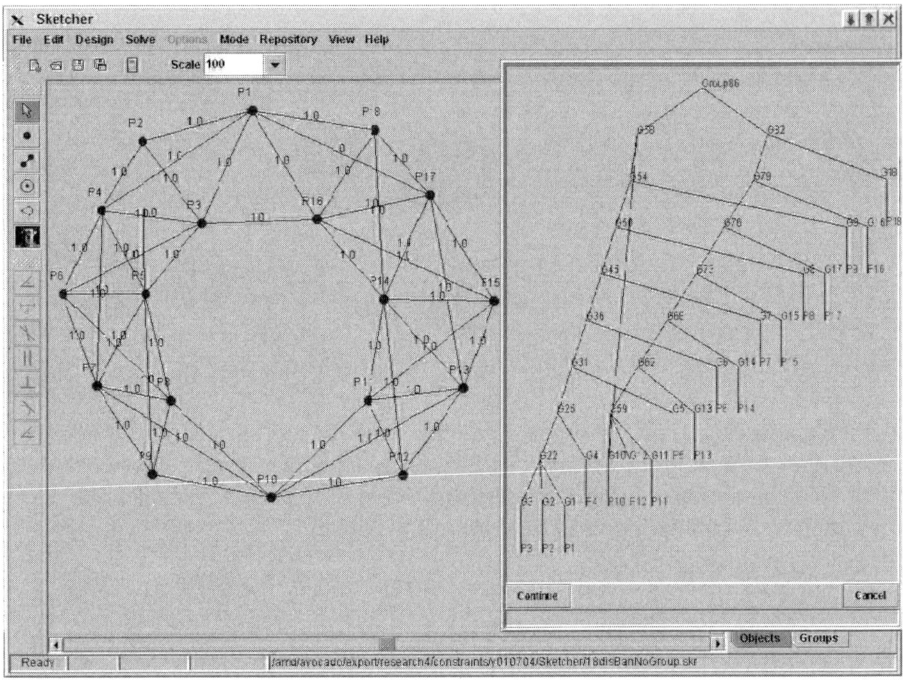

FIGURE 9. 3D constraint system (with generic constraint dependence not detectable by a simple dof count), sketch on a 2D canvas; it is wellconstrained for the given set of distances; corresponding DR-plan

(1) *Classification* of the input constraint system S as (generically) wellconstrained, overconstrained or underconstrained. Classification of each feature or subassembly in the hierarchy P as wellconstrained, overconstrained or underconstrained. See Figures 6, 7, 5.
(2) In the case the system is classified as wellconstrained, a method for systematically *navigating* the solution space with or without user intervention, specifically, steering through the various solutions, realizations or conformations of each of the wellconstrained features, parts and subassemblies of the associated input partial decomposition P, in priority order (if one is given) See Figures 20, 21, 22, 23, 24.
(3) In case combinatorially overconstrained, potentially inconsistent, features, parts, subassemblies or subsystems are present, see Figures 7, for any requested collection of them, give a complete, navigable representation of the wellconstrained *reductions*; for example an exhaustive list of existing constraints any one of which can be removed without generically making the entire system or any of the subsystems - i in the collection - underconstrained. For each such removal, a second exhaustive list of constraints that can be removed, and so on, resulting in a sequence of lists.
(4) In case of those (sub)systems classified as underconstrained (*ambiguous*), a *complete* decomposition into maximal clusters (i.e., where all the maximal clusters are present). In addition, a complete, tractable, navigable representation of its wellconstrained *completions*. This question again has a well-defined interpretation using rigidity matroids which we omit here. One such navigable representation

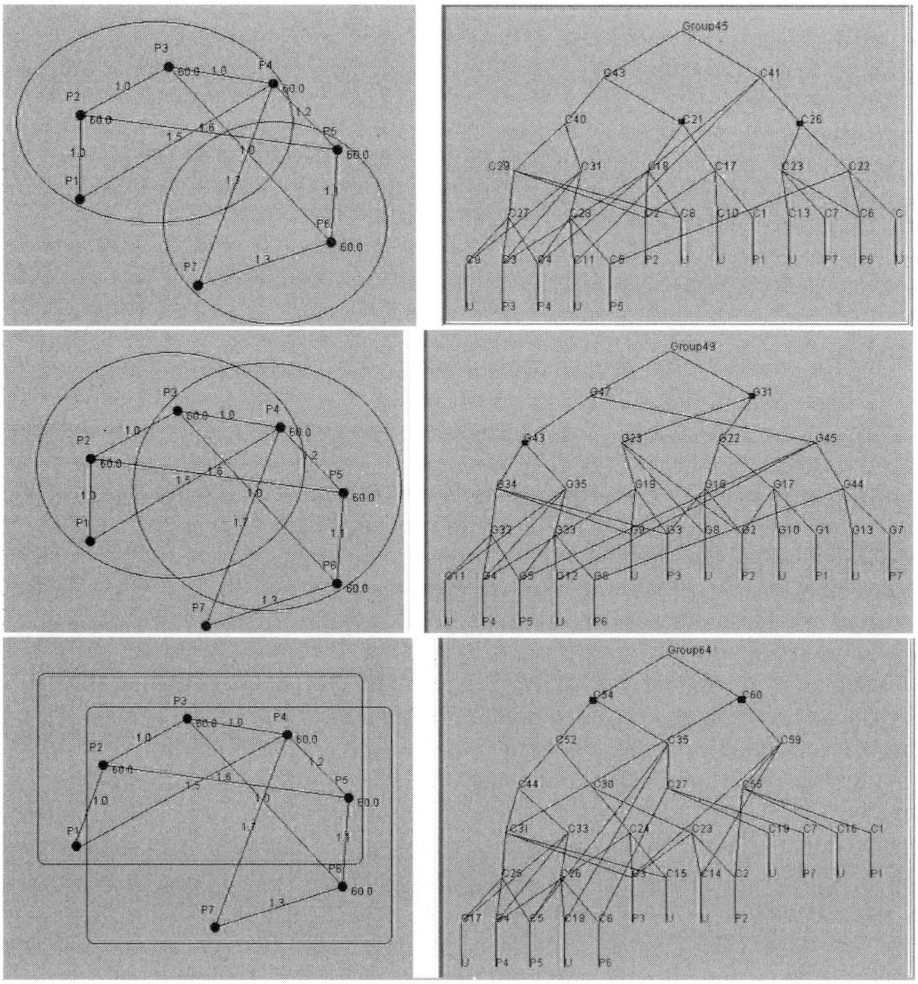

FIGURE 10. Left: input partial decompositions for 3D constraint system shown in Figure 8; groups are features or subassemblies. Right: 3 different DR plans incorporating corresponding input decomposition. Features appear as clusters or, if underconstrained, their complete set of maximal clusters. Features may (not) intersect on (non) trivial subgraphs.

could be an exhaustive list of constraints (participating objects and type of constraint, but not value) that can be added without generically making the system overconstrained. For each such addition, a second exhaustive list of constraints that can be added, and so on, resulting in a sequence of lists.

(5) A method of efficiently *updating* the output when incremental changes are made to the input. For example: a constraint value is changed, a constraint is added or removed, an object is added or removed, a feature is added etc. These could be offline or online updates (see settings of a typical constraint solver below).

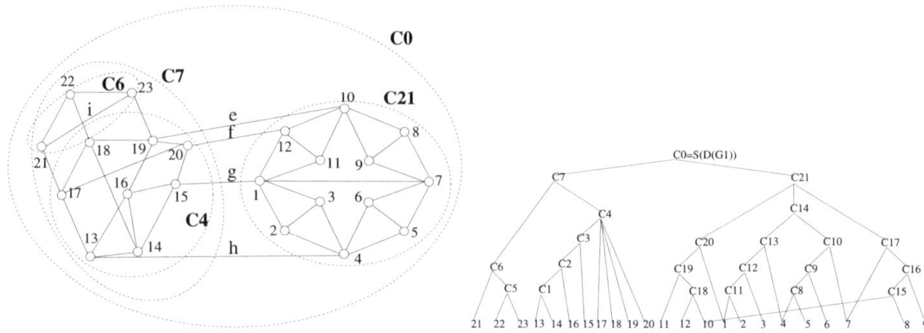

FIGURE 11. 2D constraint graph G1 and DR-plan; all vertices have weight 2 and edges weight 1

2.1. The need for decomposition: DR-plans and their properties. The overwhelming cost of solving a geometric constraint system is the size of the largest subsystem that is solved using a direct algebraic/numeric solver. This size dictates the practical utility of the overall constraint solver, since the time complexity of the constraint solver is at least *exponential* in the size of the largest such subsystem.

Therefore, an effective constraint solver should *combinatorially* develop a *plan* for (recursively) *decomposing* the constraint system into small subsystems, whose solutions (obtained from the algebraic/numeric solver) can be (recursively) *recombined* by solving other small subsystems. Such a recombination is straightforward, provided all the subsystems are generically rigid (have only finitely many solutions). The DR-planner is a graph algorithm that outputs a *decomposition-recombination plan (DR-plan)* of the constraint *graph*. In the process of combinatorially constructing the DR-plan in a bottom up manner, at stage i, it locates a wellconstrained subgraph or cluster S_i in the current constraint graph G_i, and uses an abstract *simplification* of S_i to to create a transformed constraint graph G_{i+1}.

Although recursive decompositions were used for geometric solving from the very beginning, DR-plans and their properties were formally defined for the first time in [**aALaMS01a**]. See Figures 3, 2, 4. Formally, a DR-plan of a constraint graph G is a directed acyclic graph (dag) whose nodes represent clusters in G, and edges represent containment. The leaves or sinks of the dag are all the vertices (primitive clusters) of G. The roots or sources are all the maximal clusters of G. There could be many DR-plans for G. See Figures 10 and 11. An *optimal* DR-plan is one that minimizes the maximum fan-in. The *size* of a cluster in a DR-plan is its fan-in (it represents the size of the corresponding subsystem, once its children are solved).

Besides solving efficiency purposes, it is straightforward that a DR-plan is required for the classification problem (the output requirement (1) above) and for the underconstrained detection and completion problem (output requirement (4) above). In addition, it will be clear from Sections 8 and Section 7, that it is indispensable also for navigation, dealing with overconstraints and for efficient updates (output requirements (2), (3) (5) above).

A few other properties of DR-plans are of interest. We would like the *width* i.e, *number* of clusters in the DR-plan to be small, preferably linear in the size of

G: this reflects the complexity of the planning process and affects the complexity of the solving process that is based on the DR-plan. In addition, it is desirable for DR-plans to have the *cluster minimality* property. I.e., the child clusters of any given cluster are guaranteed to be maximal proper sub-clusters that form a minimal covering set for the cluster – this property is used both in determining rigidity even using pure dof analysis [**aMS04**], and for building stable algebraic systems of low complexity corresponding to the clusters [**aAAaYZaNK04, aJPaYZ04**].

A byproduct of the augmented definition of cluster of [**aYZ04b**], described in Section 1 and discussed in Section 4 is that a somewhat modified definition of a DR-plan is used that incorporates another partial order called the *solving priority order*, which is consistent with the DR-plan's dag order, but could be more refined. The intent is that clusters that appear later in the order are not self-contained, but *dependent* and need to be solved after the clusters that appear earlier. These clusters become wellconstrained clusters only in the transformed constraint system after earlier clusters in the solving order are already solved. This type of priority order can be leveraged for assembly constraint systems, or to express procedural histories where later features are dependent upon geometric objects that are not initially present, but are the results of operations applied to earlier features.

Another important, related property is that the DR-plan *incorporate an input partial decomposition.* I.e., given an input dag P whose nodes are subgraphs of G and whose edges represent containment, a DR-plan of every node in P should be embedded in the output DR-plan for G. This is discussed in [**aYZ04a**].

All properties defined above for DR-plans transfer as performance measures of the *DR-planners* or DR-planning algorithms. It is shown in [**aMS04**], that the problem of finding the optimal DR-plan of a constraint graph is NP-hard, and approximability results are shown only in special cases. Nonapproximability results are not known. See Section 9. However, most DR-planners make adhoc choices during computation (say the order in which vertices are considered) and we can ask of how well (close to optimal) the *best* computation path of such a DR-planner would perform (on the worst case input). We call this the *best-choice approximation factor* of the DR-planner.

2.2. A geometric constraint solver's phases, solving options, modes and settings. As discussed in Section 1, an effective constraint solver has a *planning* phase, which is purely combinatorial, mainly graph theoretic resulting in a DR-plan; and a *solving* phase where the polynomial systems corresponding to the nodes of the DR-plan are actually solved using an algebraic-numeric solver. In the latter phase, two solving options should be available. The *autosolve* option in which the constraint solver searches and displays a solution or realization of the entire input system (in case it is wellconstrained) There could be exponentially many such solutions, the process is not steered by the user and each solution is displayed as it is found. See for example the full solution displayed in 8. In the *navigate* option, the user should be permitted to guide the process closely by picking one of the available solutions (displayed on a separate window) for each subsystem appearing as a node in the DR-plan, which should include all the maximal well or well-overconstrained subsystems of the user's input conceptual feature, part or subassembly hierarchy. Backtracking starting at any subsystem of the DR-plan should be allowed, and partially solved systems should be displayed. See Figures 10, 20, 21, 22, 23, 24. A constraint solver's *generate* mode is used when the input is new; the *update* mode is used when incremental changes are being made to any of the part of an earlier input; the underlying algorithms of the planning and solving phases permit efficient updates for the changed input. See Figure 28. One important advantage of modularizing the phases is that in the update mode, the DR-planning phase is entirely omitted when, for example, only value changes are made to existing constraints – these do not generically change the DR-plan. A constraint solver's *offline* setting would be used if the entire input is constructed (sketched using the GUI) before the planning and solving phases are begun. The *online* setting is used when an instant resolution of each constraint is required even as it is being input, typically by sketching on the GUI screen. See Figure 29.

Note. This manuscript is illustrated by screenshots of the FRONTIER constraint solver's sketcher. FRONTIER is an opensource 3D geometric constraint solving software suite developed by the author's group at the University of Florida.

3. The Decomposition-Recombination (DR) planning problem

Although decomposition algorithms based on constraint graphs have been proposed since the early 90's, [**aCMH97, Owe91, Owe93, Owe**] [**aAM96, aCR97**], [**Pab93, aRJaDM93**]. prior to [**aALaMS01a**], the DR-planning problem and appropriate performance measures for the planners were not formally defined.

In this section, we give a table comparing 3 main types of DR-planners, with respect to the input-output requirements of Section 2 and expand on the latest of the 3 DR-planners, [**aALaMS01b, aALaMS99, aMS04, aMSaBMaAA01, aAAaYZaNK04, aYZ04b, Sit04b, Sit04a, aYZ04a, aJPaYZ04**] the Frontier vertex Algorithm (FA) which which was designed specifically to excel simultaneously in these strongly competing requirements.

3.1. Two early DR-planners. We concentrate on two primary types of early algorithms for constructing DR-plans using constraint graphs and geometric degrees of freedom.

Note. We leave out constraint solving methods such as [**aTH88, Hav91, Hsu96**] not only because they do not fit the graph based DR-planner description but also since they are nondeterministic and rely on symbolic computation, or they are randomized or generally exponential and based on exhaustive search.

The first type of algorithms, which we call SR for *constraint Shape Recognition* (e.g.[**aCMH96b, Owe91, Owe93, aIFaCMHaJCaRP95**], [**aPJV94, aRV95, aCMH96b, aCMH97**]), concentrates on recognizing specific wellconstrained subgraphs of known shape, most commonly, patterns such as triangles. Such graphs are called *triangle decomposable.* The second type, which we call MM for *generalized Maximum Matching* (e.g. [**aRJaDM93, Pab93, aAM96, Kra92**], [**aWW85, Hen92**]) is based on first isolating certain wellconstrained subgraphs by transforming the constraint graph into a bipartite graph and finding a maximum generalized matching, followed by a connectivity analysis to obtain the DR-plan. This is related to rigidity matroid based methods (generally only for 2D constraint systems with points and distances) as discussed briefly in 1, see [**aBSaHS93**]. measures defined in the previous section.

Informally, one inherent drawback of the SR algorithms is their inability to perform a generalized degree of freedom analysis. For example, SR would require an infinite repertoire of patterns (common ones are listed in [**aPJV94, aRV95**], [**aCMH96a**]). Similarly, a decomposition of underconstrained constraint graphs into wellconstrained components is possible for SR algorithms but only subject to the pattern limitations. In many cases, MM algorithms do not satisfy cluster minimality, i.e., will output DR-plans with larger, nonminimal subgraphs S_i that may contain smaller wellconstrained subgraphs. This affects the best-choice approximation factors adversely. Most significantly, neither is able to incorporate input partial decompositions, deal with algebraically overconstrained subgraphs that often arise in 3D (or in 2D in the presence of angle constraints), or give a *complete* decomposition of underconstrained graphs into maximal clusters.

Below is a formal performance table of SR, MM and the FA (Frontier vertex) algorithms – choosing a representative algorithm (typically the best performer) in each class – using the performance measures given above. The FA DR-planner is sketched later.

3.2. Comparing performance.

Performance measure	SR	MM	FA
Generality	No	Yes†	Yes
Underconstrained decomposition	No(Yes*)	No	Yes
Incorporate Design decomposition	No(Yes$^{*,\circ}$)	No	Yes
Cluster minimality	No(No*)	No	Yes
Best-choice approximation factor	0 $(O(\frac{1}{n})^{*})$	$O(\frac{1}{\sqrt{n}})$	$O(\frac{1}{2})$
Complexity(all have $O(n)$ width)	$O(s^2)^{*}$	$O(n^{D+1}s)$	$O(n^3 s)$

Note. The variable s in the complexity expressions denotes the number of vertices, n, plus the number of edges, m, of the constraint graph. Recall that the constant D refers to the number of degrees of freedom of a rigid object in the input geometry. ◇

The superscript '*' refers to narrow classes of DR-plans: those that require the clusters to be based on triangles or a fixed repertoire of patterns. The superscript '†' refers to general 2D systems, but not 3D. The superscript '∘' refers to a strong restriction on the design decompositions that can be incorporated into DR-plans by SR.

3.3. The Frontier Vertex DR-plan (FA DR-plan). Intuitively, an FA DR-plan is built by following two steps repeatedly:

(1) *Isolate* a cluster C in the current graph G_i (which is also called the *cluster graph* or *flow graph* for reasons that will be clear below). Check and ensure a complete, maximal, dof rigid decomposition of C.
(2) *Simplify* C into $T(C)$, transforming G_i into the next cluster graph $G_{i+1} = T(G_i)$ (the recombination step).

3.3.1. *Isolating Clusters.* The isolation algorithm, first given in [**aALaMS97a, aALaMS98a**] is a modified incremental network maximum flow algorithm. The key routine is the *distribution* of an edge in the constraint graph G. For each edge, we try to *distribute* the weight $w(e) + D + 1$ to one or both of its endpoints as *flow* without exceeding their weights, referred to as "distributing the edge e." This is best illustrated on a corresponding bipartite graph G^*: vertices in one of its parts represent edges in G and vertices in the second part represent vertices in G; edges in G^* represent incidence in G. As illustrated by Figure 12, we may need to redistribute (find an augmenting path).

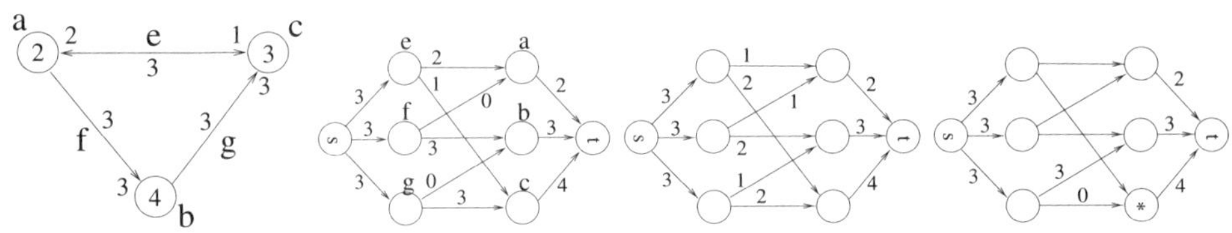

FIGURE 12. From Left. Constraint graph G with edge weight distribution. D is assumed to be 0 (system fixed in coordinate system); A corresponding flow in bipartite G^*. Another possible flow. Initial flow assignment that requires redistribution

If we are able to distribute all edges, then the graph is not dense. If no dense subgraph exists, then the flow based algorithm will terminate in $O(n(m+n))$ steps and announce this fact. If there is a dense subgraph, then there is an edge whose weight plus $D+1$ cannot be distributed (edges are distributed in some order, for example by considering vertices in some order and distributing all edges connecting a new vertex to all the vertices considered so far). It can be shown that the search for the augmenting path while distributing this edge marks the required dense graph. It can also be shown that if the found subgraph is not overconstrained, then it is in fact minimal. If it is overconstrained, [**aALaMS97a, aALaMS98a**] give an efficient algorithm to find a minimal (non-trivial, if one exists) dof cluster inside it. Then [**aMS04**] gives a method to ensure a complete, maximal, dof rigid decomposition of C.

The found cluster then needs to be checked for possible dof misclassification due to the presence of constraint dependences and rotational symmetry as described in Section 1 and [**aYZ04b**]. An intuitive and simplified version of these checks are described in Section4.

3.3.2. *Cluster Simplification.* This simplification was given in [**aALaMS01b, aALaMS99**]. The found cluster C interacts with the rest of the constraint graph through its *frontier vertices*; i.e., the vertices of the cluster that are adjacent to vertices not in the cluster. The vertices of C that are not frontier, called the *internal vertices*, are contracted into a single *core* vertex. This core is connected to each frontier vertex v of the simplified cluster $T(C)$ by an edge whose weight is the the sum of the weights of the original edges connecting internal vertices to v. Here, the weights of the frontier vertices and of the edges connecting them remain unchanged. The weight of the core vertex is chosen so that the density of the simplified cluster is $-D$, where D is the geometry-dependent constant. This is important for proving many properties of the FA DR-plan: even if C is overconstrained, $T(C)$'s overall weight is that of a wellconstrained graph, (unless C is rotationally symmetric and trivial, in which case, it retains its dof or weight). Technically, $T(C)$ may not be wellconstrained in the precise sense: it may contain an overconstrained subgraph consisting only of frontier vertices and edges, but its overall dof count is that of a wellconstrained graph.

Figure 13 illustrates this iterative simplification process ending in the final DR-plan of Figure 11.

The graph transformation performed by FA is described formally in [**aALaMS01b, aALaMS99**] using *simplifier maps* that provide the vocabulary for proving certain

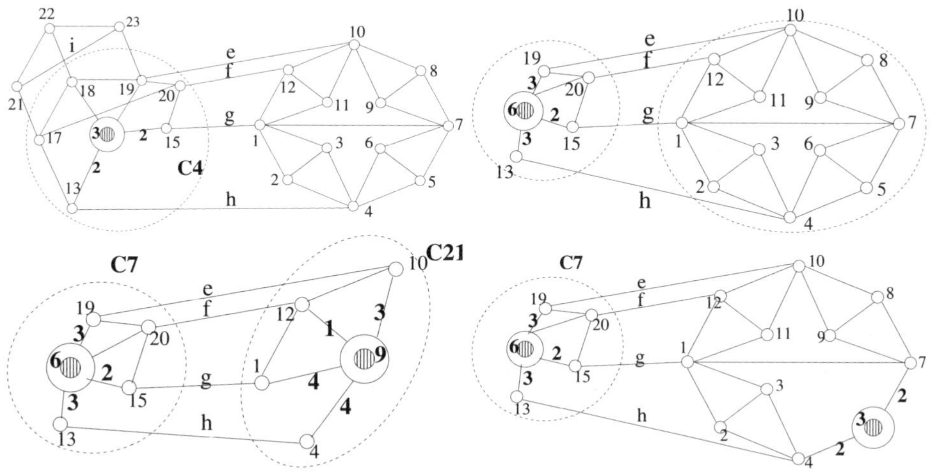

FIGURE 13. From left: FA's simplification of graph givin DR-plan in Figure 11; clusters are simplified in their numbered order: C4 is simplified before C7 etc.

properties of FA that follow directly from this simplification mechanism. However, other properties of FA require details of the actual DR-planner that ensures them. The FA algorithm was developed in stages; it and the data structures for its implementation can be found in [**Sit04b, aMSaBMaAA01, aMS04, Sit04a, aYZ04a, aYZ04b**]. The challenge met by FA is that it provably meets the several competing requirements of the table above. Some of these requirements are discussed in the next section. Others, such as incorporating an input feature hierarchy [**aYZ04a**], are omitted.

3.4. The Frontier Vertex DR-planner (FA DR-planner). Note: a detailed pseudocode of the FA DR-planner (the existing version, as well as incorporating the module-rigidty algorithm of this paper) can be found in [**Sit04b**], [**Sit04a**]. The pseudocode has been implemented as part of the downloadable, opensource FRONTIER geometric constraint solver [**aMSaBMaAA01**], [**Sit04b**], [**Sit04a**], [**Sit04c**].

The basic FA algorithm is based on an extension of the *distribute* routine for edges (explained above) to vertices and clusters in order for the isolation algorithm to work at an arbitrary stage of the planning process, i.e, in the cluster or flow graph G_i.

First, we briefly describe this basic algorithm. Next, we sketch the parts of the algorithm that ensure 3 crucial, inter-related properties of the output DR-plan:
(a) ensuring cluster minimality - this is needed to ensure true clusters even just according to the dof-based definition of Section 1;
(b) for underconstrained graphs: outputing a complete set of maximal clusters as sources of the DR-plan;
(c) controlling width of the DR-plan to ensure a polynomial time algorithm.

Three datastructures are maintained. The current *flow or cluster graph*, G_i the current *DR-plan* (this information is stored entirely in the hierarchical structure of clusters at the top level of the DR-plan), and a *cluster queue*, which is the top-level clusters of the DR-plan that have not been distributed so far, in the order that they were found (see below for an explanation of how clusters are distributed). We start with the original graph (which serves as the cluster or flow graph initially, where the clusters are singleton vertices). The DR-plan consists of the leaf or sink nodes which are all the vertices. The cluster queue consists of all the vertices in an arbitrary order.

The method *DistributeVertex* distributes all edges (calls DistributeEdge) connecting the current vertex to all the vertices considered so far. When one of the edges cannot be distributed and a minimal dense cluster C is discovered, its simplification $T(C)$ (described above) transforms the flow graph. The flows on the internal edges and the core vertex are inherited from the old flows on the internal edges and internal vertices. Notice that undistributed weights on the internal edges simply disappear. The undistributed weights on the frontier edges are distributed (within the cluster) as well as possible. However, undistributed weights on the frontier edges (edges between frontier vertices) may still remain if the frontier portion of the cluster is severely overconstrained. These have to be dealt with carefully. (See discussion on dealing with the problems caused by overconstraints in Section 7.) The new cluster is introduced into the DR-plan and the cluster queue.

Now we describe the method *DistributeCluster* Assume all the vertices in the cluster queue have been distributed (either they were included in a higher level cluster in the DR-plan, or they failed to cause the formation of a cluster and continue to be a top level node of the DR-plan, but have disappeared from the cluster queue). Assume further that the DR-plan is not complete, i.e., its top level clusters are not maximal. The next level of clusters are found by distributing the clusters curently in the cluster queue. This is done by filling up the "holes" or the available degrees of freedom of a cluster C being distributed by D units of flow. The *PushOutSide* method successively considers each edge incident on the cluster with 1 endpoint outside the cluster. It distributes any undistributed weight on these edges + 1 extra weight unit on each of these edges. It can be shown that if C is contained inside a larger cluster, then atleast one such cluster will be found by this method once all the clusters currently in the cluster queue have been distributed. The new cluster found is simplified to give a new flow graph, and gets added in the cluster queue, and the DR-plan as described above.

Eventually, when the cluster queue is empty, i.e, all found clusters have been distributed, the DR-plan's top level clusters are guaranteed to be the complete set of maximal dof rigid subgraphs of the input constraint graph. See [**aMS04**] for formal proofs.

Note: Throughout, in the interest of formal clarity, we leave out ad hoc, but highly effective heuristics that find simple clusters by avoiding full-fledged flow. One such example is called "sequential extensions" which automatically creates a larger cluster containing a cluster C and a vertex v provided there are atleast D edges between C and v. These can easily be incorporated into the flow based algorithm, provided certain basic invariants about distributed edges is maintained (see below).

This completes the description of the backbone of the basic FA DR-planner. Next we consider some details ensuring the properties (a) – (c) above.

3.4.1. *Ensuring Cluster Minimality.* First we intuitively explain why cluster minimality is a crucial property. In Figure 14, after C_1 and C_2 are found, when C_1

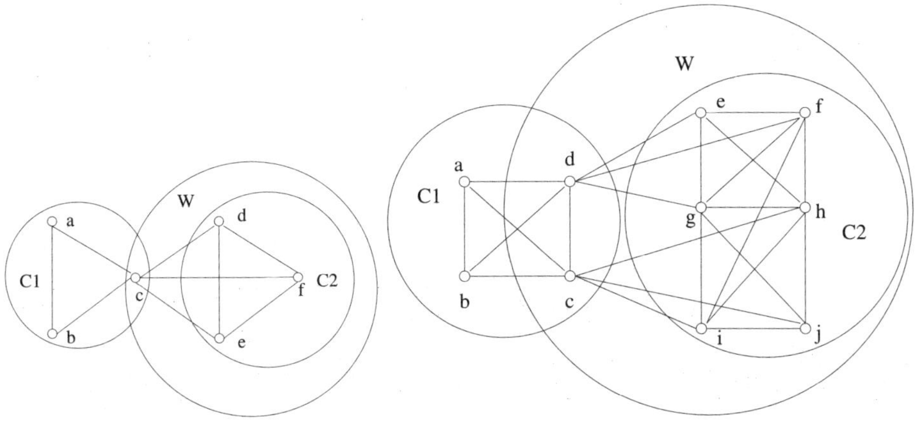

FIGURE 14. Finding W first will prevent dof misclassification: Left 2D example, Right 3D example.

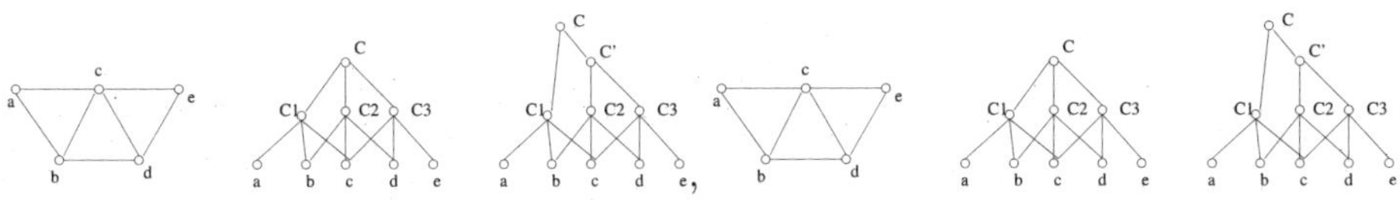

FIGURE 15. Ensuring Cluster Minimality: E is a set of essential clusters that must be present in any subset of the children of C that form a cluster. In this case, E itself forms a cluster. C' is a cluster made up of a proper subset of at least 2 of C's children

is distributed, C_1 and C_2 would be picked up as a cluster, although they do not form a cluster. The problem is that the overconstrained subgraph W intersects C_1 on a trivial cluster, and W itself has not been found. Had W been found before C_1 was distributed, W would have been simplified into a wellconstrained subgraph and this misclassification would not have occurred.

It has been shown in [**aMS04**] that this type of misclassification can be avoided (W can be forced to be found after C_2 is found), by maintaining three invariants. The first two are described here. The third is highly related to property (b) and is described in the next subsection.

The first is the following invariant: always distribute all undistributed edges connecting a new found cluster C (or the last distributed vertex that caused C to be found), to all the vertices distributed so far that are outside the cluster C. Undistributed weight on edges inside C are less crucial: if they become internal edges of the cluster, then this undistributed weight "disappears" when C is simplified into a wellconstrained cluster; there is also a simple method of treating undistributed weight on frontier edges so that they also do not cause problems - the method and proof can be found in [**aMS04**]).

The second invariant that is useful for ensuring cluster minimality is that for any cluster in the DR-plan, no proper subset of atleast 2 of its child clusters forms a cluster. FA ensures this using a generalization of the method *Minimal* of [**aALaMS97a, aALaMS98a**] which finds a minimal dense subgraph inside a dense subgraph located by *DistributeVertex* and *DistributeEdge*.

See Figure 15. Once a cluster C is located and has children $C_1, \ldots, C_k$, for $k \geq 2$, a recursive method *clusMin* removes one cluster C_i at a time (replacing earlier removals) from C and redoes the flow inside the flow graph restricted to C, before C's simplification. If a proper subset of atleast 2 C_j's forms a cluster C', then the clusMin algorithm is repeated inside C' and thereafter in C again, replacing the set of child clusters of C that are inside C' by a single child cluster C'. If instead no such cluster is found, then the removed cluster C_i the *essential.* I.e., it belongs to every subset of C's children that forms a cluster. When the set of clusters itself forms a cluster E (using a dof count), then *clusMin* is called on C again with a new child cluster E replacing all of C's children inside E.

3.4.2. *(b) Finding a complete set of maximal clusters in underconstrained graphs.* While the DR-planner described so far guarantees that at termination, top level clusters of the DR-plan are maximal. It also guarantees that the original graph is dof underconstrained only if there is more than one top level cluster in the DR-plan. However, in order to guarantee that *all* the maximal clusters of an underconstrained graph appear as top level clusters of the DR-plan, we use the observation that any pair of such clusters intersect on a subgraph that reduces (once incidence constraints are resolved) into a trivial subgraph (a single point in 2D or a single edge in 3D). This bounds the total number of such clusters and gives a simple method for finding all of them. Once the DR-planner terminates with a set of maximal clusters, other maximal clusters are found by simply performing a *Pushoutside* of 2 units on every vertex (in 2D) or every vertex and edge (in 3D), and continuing with the original DR-planning process until it terminates with a larger set of maximal clusters. This is performed for each vertex in 2D and each edge in 3D which guarantees that all maximal clusters will be found. See [**aMS04**] for proofs.

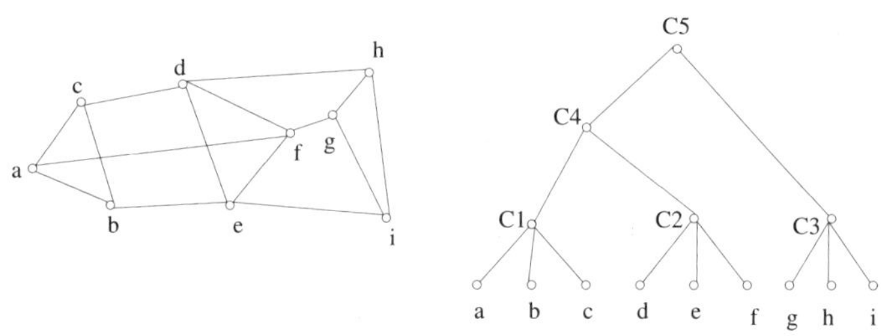

FIGURE 16. Prevent accumulation of clusters

3.4.3. *(c) Controlling width of the DR-plan.* FA achieves a linear bound on DR-plan width by maintaining the following invariant of the cluster or flow graph: *every pair of clusters in the flow graph (top level of the DR-plan) at any stage intersect on at most a trivial subgraph.* FA does this by repeatedly performing 2 operations each time a new potential cluster is isolated.

The first is an *enlargement* of the found cluster. In general, a new found cluster N is enlarged by any cluster D_1 currently in the flow graph, if their nonempty intersection is *not* a rotationally symmetric or trivial subgraph. In this case, N neither enters the cluster graph nor the DR-Plan. Only $N \cup D_1$ enters the DR-plan, as a parent of both D_1 and the other children of N. It is easy to see that the *sizes* of the subsystems corresponding to both $N \cup D_1$ and N are the same, since D_1 would already be solved.

For the example in Figure 16, when the DR-plan finds the cluster C_2 after C_1, the DR-planner will find that C_1 can be enlarged by C_2 The DR-planner forms a new cluster C_4 based on C_1 and C_2 and puts C_4 into the cluster queue, instead of putting C_2 to cluster queue.

The second operation is to iteratively *combine* $N \cup D_1$ with any clusters $D_2, D_3, \ldots$ based on a nonempty overlap that is not rotationally symmetric or trivial. In this case, $N \cup D_1 \cup D_2$, $N \cup D_1 \cup D_2 \cup D_3$ etc. enter the DR-plan as a staircase, or chain, but only the single cluster $N \cup D_1 \cup D_2 \cup D_3 \cup \ldots$. enters the cluster graph after removing D_1, D_2, D_3 ....

Ofcourse, both of these processes are distinct from the original flow distribution process that *locates* clusters.

4. Correcting the 3D dof misclassifications

As pointed out in Section 1 (see also Section 9), it is not known despite several conjectures and heuristics in the combinatorial rigidity literature [**aBSaHS93**], whether a tractable, purely combinatorial analysis exists to correctly classify a 3D constraint system (or 2D constraint graph with other constraints besides distance constraints) as generically rigid. The few constraint solvers that deal with 3D systems for example the proprietary, commercial D-cubed solver [**Owe**], which handles certain 3D constraint systems such as the "pipe routing problem" and FRONTIER [**aMSaBMaAA01, Sit04b, Sit04a, aMS04, aYZ04b**], which deals with fairly general 3D systems, must grapple with this problem in some systematic fashion.

In this section, we discuss an intuitive simplification of a method used by the FRONTIER constraint solver [**aYZ04b**] that uses a concept called module-rigidity to rectify dof misclassification of a large class of 3D constraint graphs. Rotationally symmetric, trivial subgraphs and clusters underlie these misclassifications. Below we discuss both their *direct* and *indirect* involvement in such misclassifications and how to rectify them - the latter is especially nontrivial (see Figures in Section 1 concerning the "bananas" problem) and both are complicated by the presence of incidence constraints.

Methods for handling other common misclassifications discussed in 1 caused by 2D angle constraints are discussed only briefly in Section 8, under the context of online solving.

4.1. Rotationally symmetric, trivial subgraphs. Some common 3D examples of rotationally symmetric, trivial subgraphs are: a single point in 3D or 2D (does not cause any misclassifications); 2 points in 3D incident on a line; 2 points in 3D and a distance constraint between them, along with other incident points, line segments and lines, reducing to a trivial fixed-length line segment cluster. Here are some examples of dof analysis problems *directly* involving the latter two types of subgraphs and how they are rectified by the FRONTIER constraint solver's FA-based DR-planner [**Sit04b, Sit04a, aYZ04b**].

4.1.1. *Detecting rotational symmetric and underconstrained trivial subgraphs.* For the 3D example in Figure 17, the points a and b are incident to the line l. The whole object has 6 degrees of freedom. It would be deemed a cluster in the usual dof analysis. However, it is underconstrained and should not be picked up as a cluster. Intuitively, the object loses one degree of freedom because of its rotational symmetry - a *geometric* property, which compensates for its underconstrainedness, thereby "fooling" the simple dof count into classifying it as wellconstrained.

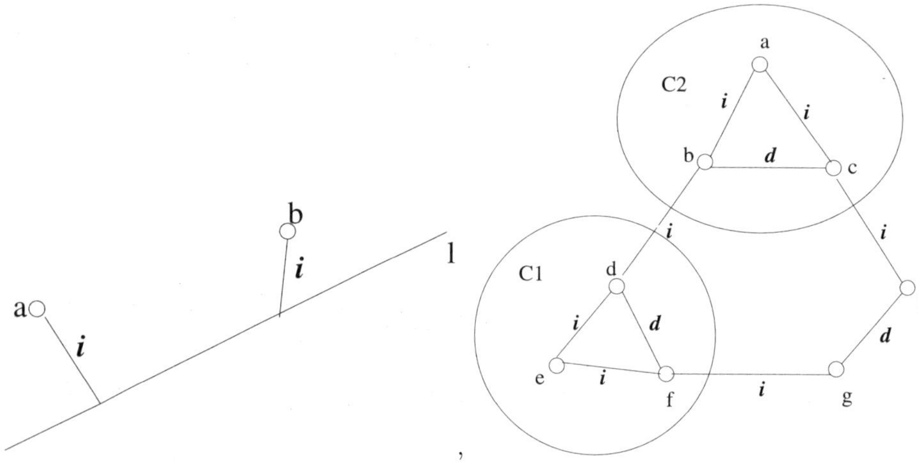

FIGURE 17. Rotationally symmetric object: detecting by organizing incidences (left); special case of FA simplification (right); d denotes distance and i incidence constraint

To solve this problem, in 3D, after a new subgraph is found to be a cluster by dof analysis by the DR-planner, it checks whether it is a trivial, rotationally symmetric subgraph. Detecting this involves organizing potentially numerous incidence constraints: these are divided into 3 groups, those objects that are incident on the left endpoint, the right endpoint and on both. Once the rotational symmetry is detected, if this subgraph has 6 degrees of freedom, then it is not a cluster and should be discarded. If it has 5 or fewer degrees of freedom (reducing to a fixed-length line segment in 3D), it is kept as a cluster and marked as rotationally symmetric, as a crucial help during navigation of its solution space by as discussed in Section 5.

4.1.2. *Enlarging and combining clusters based on nontrivial overlap.* One of the important features of any general, tractable DR-planner that avoids a combinatorial explosion is that it restrict the width or number of clusters in the DR-plan. For this purpose, we now show that the check for rotational symmetry - by organizing incidences - given in subsection 4.1.1 is crucial.

As described in Section 3, FA achieves a linear bound on DR-plan width by maintaining the following invariant of the cluster or flow graph: every pair of clusters in the cluster graph at any stage intersect or overlap on at most a rotationally symmetric or trivial subgraph. This is done by enlarging and combining clusters based on nontrivial overlap.

For the example in Figure 18, the overlapped part of $D_1 = C_1$ and $N = C_2$ has a DR-plan which has 2 source clusters, ab and cd. Once the incidences are organized, the overlap is seen to be a trivial rotationally symmetric subgraph (albeit not a cluster) and hence N would *not* be enlarged by D_1.

In one special case, clusters should be enlarged even when their overlap is a trivial rotationally symmetric cluster. For example, as in Figure 18, either N or D_1 is a rotationally symmetric cluster and they overlap on a rotationally symmetric cluster, then N can in fact be enlarged by D_1 (even without any other constraint between them). Here, $N = C_2$ is a rotationally symmetric cluster and can be enlarged by $D_1 = C_1$.

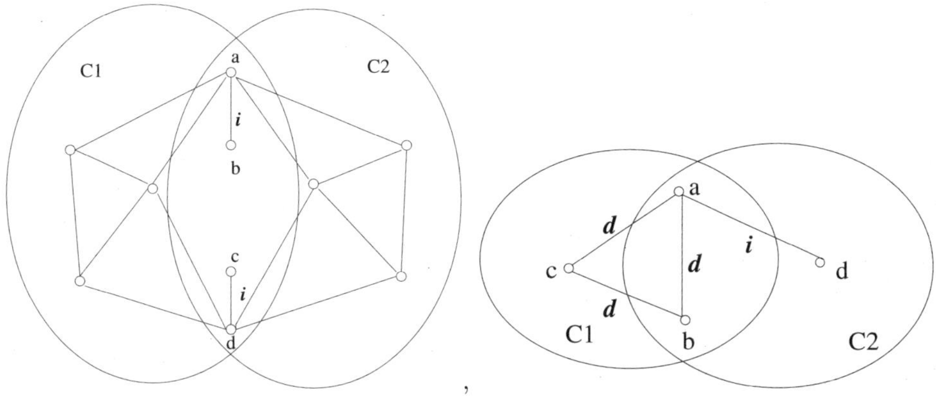

FIGURE 18. Enlarging clusters: organizing incidences to (left) detect rotationally symmetric subgraphs in overlap and (right) detect a rotationally symmetric participating cluster

4.1.3. *Rotational symmetry and overconstraints.* For the 3D example in Figure 17 (right) the line segment C_1, with 5 degrees of freedom, is an overconstrained cluster based on dof. To retain many of their properties, FA and other DR-planners would have to simplify such clusters (see Section 3) into a wellconstrained cluster with 6 dof's. In this case, however, such a simplification would cause FA or any standard dof analysis to fail to find this simple triangle as a cluster! Since linesegments and triangles are basic building blocks of constraint systems, this problem needs to be corrected.

To solve this problem. when the DR-planner detects a rotationally symmetric cluster with 5 degrees of freedom, it not only marks it as rotationally symmetric, but simplifies it into cluster that maintains its 5 degrees of freedom. Intuitively, we view the loss of 1 dof due to rotational symmetry as yet another type of overconstraint a *geometric* overconstraint (distinct from the combinatorial and constraint dependences discussed in Section 5). These clusters behave in many ways like wellconstrained even though they possess 1 less degree of freedom.

4.2. The "Bananas" Problem. We now discuss a classic case where dof analysis fails to correctly classify a large class of constraint graphs due to the constraint dependence caused by the *indirect* involvement of rotationally symmetric, trivial clusters or subgraphs. Some example figures illustrating this were given in Section 1. These errors must be checked for as soon as a potential cluster is isolated by flow during the FA DR-planning process. We briefly describe an intuitive simplification of how the FRONTIER constraint solver ([**Sit04b, aYZ04b**]) handles these errors, thereby correctly classifying a large class of graphs for which a pure dof analysis is inadequate. As mentioned in Section 1, complete process involves ensuring an augmented notion of rigidity called module-rigidity and can be found in [**aYZ04b**].

4.2.1. *Problem Description.* For the 3D example in Figure 5, if the DR-planner operates purely by dof analysis, it will locate this graph as a cluster. In fact, the graph is generically overconstrained since the distance between the 2 overlap points in the center is determined independently both by the left and right clusters. The graph (right) however is combinatorially underconstrained: *If* these distances were consistent (equal), then the graph would be underconstrained. It has a dof of 6 (appears to be wellconstrained) because its underconstrainedness is compensated by its overconstrainedness.

The example in Figures 7, 6 is very similar to the above example, except that it has a distance constraint between the two overlap points. In this case, the left and right clusters are combinatorially overconstrained and will be simplified into well-constrained clusters. The combined cluster will not be picked up by dof analysis unless there is one more constraint between the two clusters, say a distance constraint as in Figures 7, 6 (left). In both of these cases, the dof classification is in fact correct.

In fact, as pointed out in Section 1, the constraint system has constraint dependences, if the common overlap O of *any* subset of its wellconstrained clusters is underconstrained; the undetermined edges in the overlap are independently determined by each of the clusters, which is the source of the overconstraint - i.e, many equations with dependent or even identical left hand sides, but generically independent right hand sides. The above "bananas" is a special case of this, where the overlap is a rotationally symmetric trivial subgraph that is not a cluster. This type of constraint dependence causes the dof analysis to be inaccurate only in this specific "bananas" case. See Claim 1 below.

Different responses are possible when an constraint dependence is detected.
(a) The user could be asked to correct the offending underconstrained overlap O by adding explicit constraints to the overlap. This would now cause explicit combinatorial overconstraints, which can be detected and corrected (see Section 7, [**aMSaBY04**]). This could also cause new subgraphs to become self-contained clusters, where previously they would have been directly made part of larger clusters.
(b) We assume that the source of constraint dependence is intentional and known to the user who has otherwise ensured that the overconstraint is consistent (the right hand sides have the same dependence as the left hand sides of the corresponding equations). In this case, the user, after being informed of an constraint dependence, may choose not to correct it. I.e, it is desirable that all constraint dependences (atleast those within clusters located by the flow algorithm or other dof analysis) be detected, since they could cause it is sufficient to *treat* only those cases of constraint dependence that actually cause a misclassification by a dof analysis.

4.2.2. *A recursive solution.* When a new cluster C is found in 3D, we describe how it is checked for dof misclassification and how it is treated –in the case when it consists of exactly 2 child clusters which have been correctly classified. Addressing this case is sufficient because of the following 2 claims.

Claim 1: Let C be a cluster in 3D detected by flow or dof analysis. Assume C consists of child clusters that have all been correctly classified as clusters (i.e., they have passed the dof misclassification check being described here). Furthermore, assume that C satisfies the cluster minimality property described in Section 3. I.e, for every cluster in the DR-plan, no proper subset of its child clusters form a cluster. Finally, assume that C has been misclassified due to an constraint dependence caused by more than one cluster independently fixing a single distance between a pair of points not connected by an explicit distance constraint. Then C must have consisted exactly of a *pair* of child clusters forming the "bananas" structure, i.e., whose overlap is a trivial (rotationally symmetric) cluster having 6 dofs.

To prove the claim, first consider the case where C consists of a pair of child clusters – in this case, while their overlap could be an underconstrained graph with arbitrarily many clusters, it is not hard to see that a dof misclassification takes place exactly when the pair forms the "bananas structure."

Next, consider the case where cluster-minimal C consists of 3 or more clusters. Also assume some subset of its child clusters have a common overlap that is an underconstrained system causing the source of an constraint dependence of the type described above. It is not hard to see that a dof misclassification would result exactly when some pair of clusters form the "bananas" structure. But that would imply that pair would have been picked up as cluster (according to dof analysis) contradicting the cluster-minimality of C.

In fact, the proof of the claim is further simplified in the case of the FA DR-planner: due to the enlarging and combining operations it uses to control width of the DR-plan, (see Section 3) - if the parent cluster C has more than 2 child cluster s, then for every pair of them, their overlap is guaranteed to be equivalent (via incidence constraints) to a trivial subgraph.

Claim 2: The FA DR-plan has the *cluster minimality* property. I.e, for every cluster in the DR-plan, no proper subset of its child clusters form a cluster.

Claim 2 is proven true (by [**aMS04**] and Section 3) only for the FA DR-plan built by a purely flow based dof analysis. In particular, that FA DR-planner first finds (by dof flow analysis) a potentially nonminimal cluster S consisting of clusters $D_1, \ldots, D_k$ and then runs a minimality-ensuring algorithm (that recursively performs a flow based dof analysis again) and constructs a sub-DR-plan of S, i.e., rooted at S, whose leaves are $D_1, \ldots, D_k$ (see [**aMS04**], Section 3). However, the FA DR-planner based on purely dof flow analysis may misclassify the minimal "clusters," thus found and thereby the entire sub-DR-plan may contain misclassified clusters, including the original nonminimal "cluster" S.

Thus, given Claim 1, the most efficient way to modify the FA DR-planner to avoid misclassifications due to this type of constraint dependence, is to perform the base case of the "bananas" correction (described below) *during* the minimality-ensuring portion of the FA algorithm. I.e., the base case is performed during the construction of the sub-DR-plan for S, assuming that the clusters $D_1, \ldots D_k$ have already been checked and corrected. I.e, it is performed each time a cluster C consisting of a *pair* of (already checked and corrected) child clusters C_1 and C_2 is located during and is being considered as an element in the sub-DR-plan for the cluster S. Once the "bananas" correction described below is completed: in the first 2 cases below, C is inserted into the sub-DR-plan for S and in the last case, C is *not inserted*. And the minimality-ensuring method continues.

At the end of the process, a sub-DR-plan for it is found, with cluster-minimal clusters corrected for dof misclassification, and either S is found to be a true cluster, in which case the sub-DR-plan has a single root or sink, namely S itself, or S is not a cluster, in which case the sub-DR-plan's roots or sinks correspond to the complete set of maximal clusters within S.

4.2.3. *The base case of bananas correction.* When a cluster is found consisting of 2 child clusters C_1 and C_2, during the the minimality-ensuring method, first check whether the entire overlap between C_1 and C_2 reduces to a rotationally symmetric subgraph with 5 or 6 dofs. (after incidences are organized as described in the previous subsections). If it is not a rotationally symmetric subgraph, we put $C = C_3$ in the DR-plan (after checking for potential enlargement described in previous sections, and checking for and announcing any constraint dependences, i.e, if the overlap is not wellconstrained).

If the overlap is a rotationally symmetric *cluster*, i.e., it reduces either to a single point (after resolving incidences) or there is a distance constraint in the overlap, and there are explicit constraints as in Figure 6, 7, then C_3 is retained as a normal wellconstrained cluster.

If the overlap is a rotationally symmetric subgraph which is not a cluster, *and* there are other constraints between C_1 and C_2, retain C_3, but mark it as a cluster containing a constraint dependence (and announce it); it is then simplified into a cluster with 6 dofs, retained but marked to be "handled with care" during solving and navigation (Section 5).

If there are no other constraints between C_1 and C_2, then the constraint dependence must be announced. there is a dof misclassification and in fact, the found cluster $C = C_3$ should be thrown away. However we need to prevent it from being "found" again, and moreover, we need to prevent it from causing a dof misclassification of a larger cluster that includes both C_1 and C_2, and is found at a later stage.

To do so, we remove the constraint dependence by adding an edge - representing a "virtual" distance constraint - to the overlap. The following corrections need to be done:

a) all the existing clusters that contain this edge must be resimplified so that they continue to have a dof of 6. b) the edge must be appropriately flagged as "belonging" to C_1 or C_2 (its distance value was fixed by C_1 or C_2), so that a cluster D containing this edge that will be found in the future by the DR-planner (and passes the "bananas" check), have to be handled as follows: (i) if D remains a cluster on removal of this edge, then the source of a potential constraint dependence has been detected – this need not be announced until the point when it causes an underconstrained overlap between D and a partner cluster which is picked up as a future cluster. (ii) if D requires the edge to be picked up as a cluster by dof analysis, then D should inherit the flag of the edge so that during the solving phase D is solved only *after* atleast one of the original clusters C_1 or C_2 is solved (and thereby the distance associated with the edge has been fixed).

c) both the clusters C_1 and C_2 in the pair are distributed again (see description of flow method in Section 3)

Note. It is important to notice that the above method utilizes the second partial ordering or the *solving priority order* of the clusters of the DR-plan given in Section 1. Recall that this ordering is consistent with the DR-plan's inherent partial order. More significantly, it uses the modified definition of DR-DAG given in Section 1, where not every node of the DR-plan represents an independent cluster in the original graph or constraint system. A node represents a generically wellconstrained system in the *transformed system or transformed graph after* all the clusters before it in the solving order have been solved or simplified.

Figure 19 shows examples of the bananas situation that are correctly classified by the method described above.

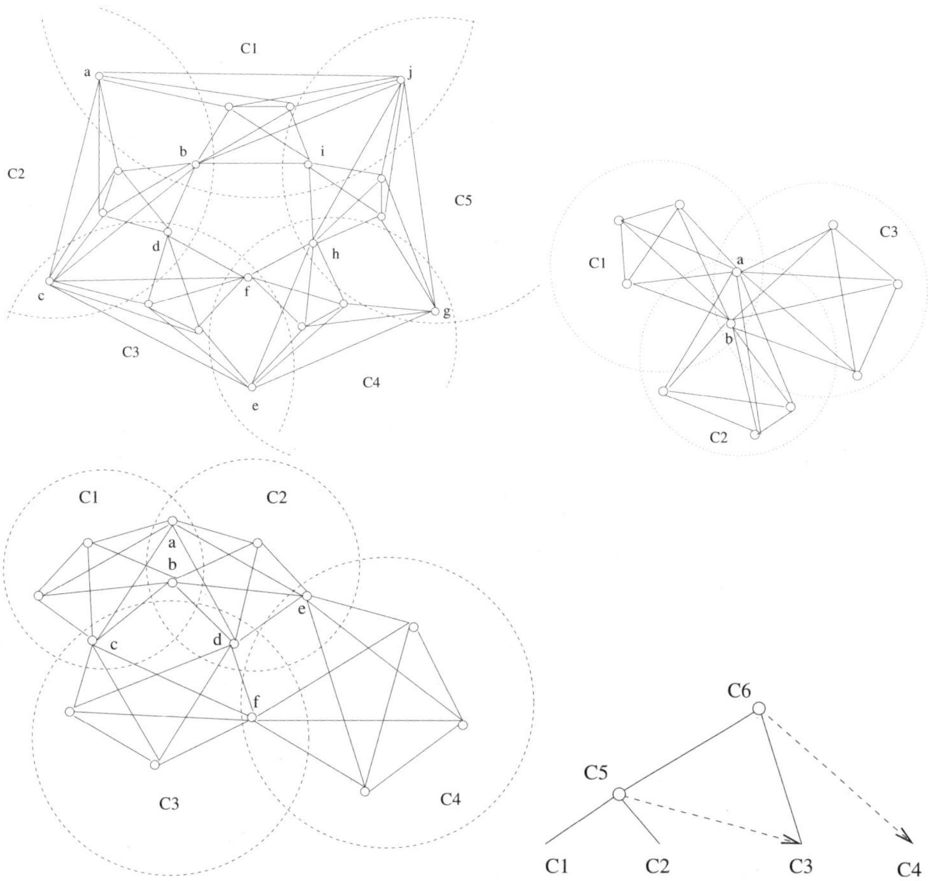

FIGURE 19. Examples where bananas correction works. See text for explanation.

In Figure 19 Top Left: the graph is a cluster according to dof analysis, but not really rigid, as detected by the bananas correction and as seen by the decomposition shown. Top Right: rigid, but no pair of self-contained clusters shown forms a truly rigid subgraph, as detected by the bananas correction although they do form clusters according to dof analysis. Bottom: not rigid as detected by the bananas correction, but a cluster according to dof analysis; complete maximal module decomposition and solving priority orders found by the bananas correction: the pair C_1, C_2 forms a rigid C_5 but that C_5 can be solved only after C_3 is solved; I.e., before the virtual edge (c, d) is added, C_1 and C_2 would not be picked up together as a cluster candidate. Similarly, it will also determine that C_5, C_3 form a rigid C_6, but only according to solving priority order shown. Bottom Right: DR-plan for left constraint system with 2 sources or roots: C_6 and C_4.

5. Solving and Navigating the solution space

In this section we discuss combinatorial approaches to solving and navigating the solution space of a cluster, during the solving phase of a constraint solver's operation, once the planning phase has already output a DR-plan. For underconstrained systems the situation is more complex, see Section 7, and Section 9.

A wellconstrained system may have exponentially many solutions in the number of geometric primitives. For example, even a simple ruler and compass construction in 2D permits 2 solutions for each constructed point (reflections are considered distinct since they cannot be obtained by a physical euclidean transformation such as rotation or reflection) Existing approaches to controlling this combinatorial explosion fall under three broad categories.

One standard approach [**Fud95, aCMH96b, Owe**] attempts to automatically pick a solution that is as close as possible to the appearance of the input sketch. Typically, chirality (or relative orientation) of the geometric primitives in the sketch is used as the guide. A related approach uses explicitly defined "navigation" constraints or "declarative solution selectors" provided by the user. The formal basis for this began with [**Hen92**], and some of the more recent methods are surveyed in [**aAS03**]. The use of genetic algorithms for dealing with these extra constraints are given in [**aMLaAS03**]. In general, these constraints could include chirality or relative orientation constraints, and other constraints related to chirality (by oriented matroid theory [**aMLVaBSaNWaGZ93**]) such as intersection or nonintersection of convex hulls or subsets of the points. They reduce to specific polynomial inequalities obtained from determinants of matrices of indeterminates. Another type of navigational constraint are the so-called relational or engineering constraints [**aCMH97**].

A second approach gives the user an indexing [**aIFaCMHaJCaRP95**], of the solution space using a Decomposition Recombination (DR) plan of the constraint system as a guide. These DR-plans are generated by most constraint solvers to efficiently guide the algebraic-numeric solvers. Specifically, they break down the large polynomial system that arises from the entire geometric constraint system into small subsystems that can be handled by algebraic-numeric solvers. This is crucial since the latter typically have exponential time complexity.

The methods of [**aIFaCMHaJCaRP95**] provide tractable steering for simple, 2D constraint systems that are triangle decomposable, or have linear DR-plans, similar to ruler and compass constructible systems, that permit solving for one geometric primitive at a time.

In this case, each such addition provides two "bifurcation" choices that are typically two reflections of the newly solved-for primitive. However, such DR-plans exist only for special constraint systems and in overconstrained cases they may not incorporate a user's conceptual feature decomposition, potentially denying the user's prefered navigation path. This problem is explained in detail in [**aYZ04a**].

A *third* mixed approach (used by the FRONTIER constraint solver [**aMSaBMaAA01, Sit04b, Sit04a, aAAaYZaNK04**]), is a hybrid method that both uses navigation constraints and offers two options: a *visual walkthrough* with backtracking or *automatic search* for a solution. See Figures 20, 21, 22, 23, 24.

Furthermore, the walkthrough is based on *general* DR-plans (for 2D or 3D systems), such as the FA DR-plan (output of the Frontier vertex Algorithm) of

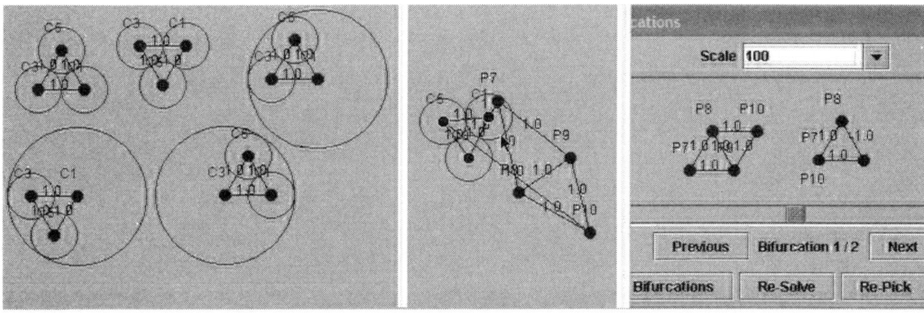

FIGURE 20. Left: display of all solutions of the cluster formed by the 3 circles in constraint system of Figure 2. Middle: partially solved sketch after choosing one of them; Right 2 solutions of the diamond of points and distances

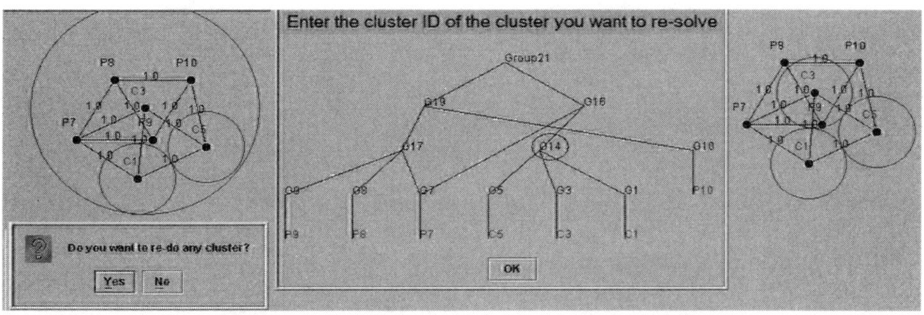

FIGURE 21. Left: complete solution of root cluster of Figure 2 after choosing one of the solutions for each of the child clusters (the 3 circles and the diamond of Figure 20). Right: new solution of root cluster after backtracking to repick a different solution of the 3 circle cluster

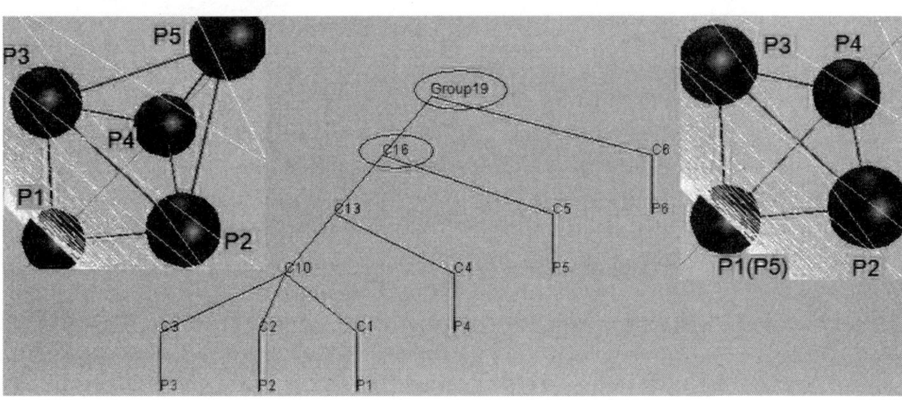

FIGURE 22. Solution navigation for constraint system in Figure 4: Two solutions for the child cluster of the root marked in the DR-plan.

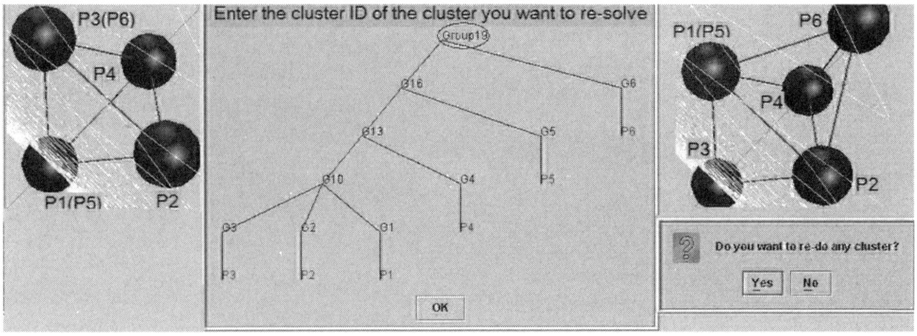

FIGURE 23. After picking the solution on right in Figure 22, 2 final solutions for root cluster with backtracking option.

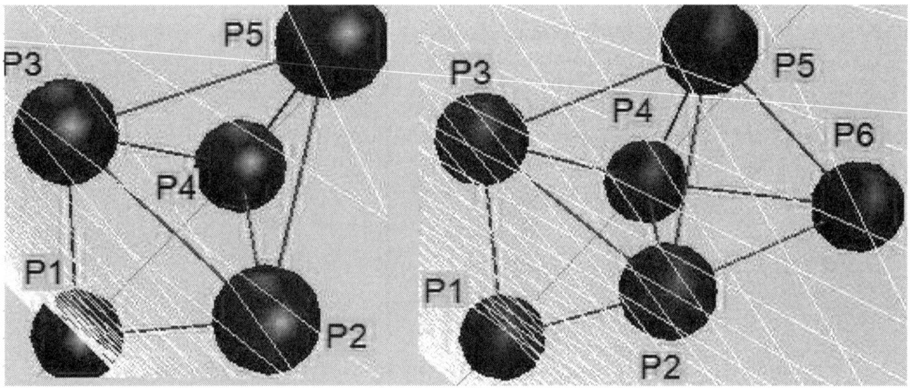

FIGURE 24. After backtracking and repicking the left solution of child cluster in Figure 22, two final solutions for root cluster (complete system of Figure 4).

Section 3, that incorporates a user's desired conceptual feature or subassembly decomposition, which is crucial for a user-directed search for the solution space.

The method described completely in [**aAAaYZaNK04**] has the advantage that it can be a stand-alone module: it takes as input the constraint graph, the DR-plan (output of *any* DR-planner) and navigation constraints, and outputs to the GUI (resp. stores in the DR-plan structure) the realizations or solutions - of the subsystems encountered during navigation. Other interactive input includes the user's choice of solutions to the clusters. The method provides fully flexible *backtracking* for redoing such choices starting from *any* cluster. See Figure 28. In the FRONTIER constraint solver, this method is called the *Equation and Solution Manager (ESM).* [**aMSaBMaAA01, Sit04b, Sit04a, aAAaYZaNK04**]. The module in addition efficiently deals with updates or changes to the constraint graph and the navigation constraints; these are discussed in Section 8.

There are 4 crucial parts of the ESM. The first is the overall control flow based on a standard recursive traversal of the DR-plan and an efficient datastructure for maintaining chosen realizations, which maintains clarity and controls complexity during navigation, and provides fully flexible backtracking or *steering*.

See [**aAAaYZaNK04, aMSaBMaAA01, Sit04b, Sit04a**] for descriptions of this part of the ESM. The second part is a method for building independent algebraic systems corresponding to the clusters of the DR-plan. The third part is a purely combinatorial method [**aJPaYZ04**] for optimizing the algebraic complexity of these systems - this method interestingly uses minimum spanning trees on a graph obtained from the cluster structure. The weights are determined based on a cluster solving method that uses rigid body mechanics and quaternions. The fourth part leverages the modularity of the ESM as well as the above characteristics to incorporate navigation constraints.

Note. We restrict our description here to the second and the fourth parts described above.

6. Building algebraic systems - the stability issue

The polynomial system corresponding to a cluster C typically has 9 variables (entries of the rotation/translation matrix in homogeneous coordinates) corresponding to *each* child cluster of C. Three of these variables are dependent, via equations of the type $s^2 + c^2 = 1$, where s and c represent the sine and cosine of one of the rotation angles. These could be reduced using a quaternion or other forms for optimization of algebraic complexity [**aJPaYZ04**] which is a separate problem that does not influence our discussion here. In addition there are polynomial equations for the constraints that relate primitive geometric elements of the child clusters. These are the constraints associated with C and need to be resolved to obtain a solution to C given the solutions of its children. The solutions of the children determine the relative positions of the geometric elements within the child clusters: these appear on the left hand sides of the polynomial system for C. But in the cases where C appears after one of its children (or some other cluster) in the solving priority order (see Section 2) one or more of C's distance constraints (values) could be determined by the solution of the child clusters (or other clusters). These appear on the right hand sides of the polynomial system corresponding to C. (These equations could result in an undesirable compounding of numerical error, which is outside the scope of our discussion here).

In order to avoid even more basic instabilities, it is usually necessary to remove overconstraints and dependent constraints: a stable, wellconstrained system generically has as many variables as equations. The number of equations could exceed the number of variables, because of explicit overconstraints that appear as edges in the constraint graph or because of overconstraints as in the "bananas" case, where some of the constraints are dependent. The former can be removed using the method of [**aMSaBY04**], which detects and removes so-called *reducible* edges whose removal does not make the cluster underconstrained. The latter are explicitly marked by a good DR-planner such as [**aYZ04b**] as discussed in Section 2: the method of dealing with these is discussed later in this section.

The bulk of this section, however, deals with wellconstrained subsystems with no constraint dependences. Even for these, picking an independent set of algebraic equations is a nontrivial task. Some of these equations are incidence or overlap equations (equating 2 variables) caused either by explicit incidence constraints or by geometric elements that are shared by two or more overlapping child clusters.

Figure 25 illustrates the problem for an example in 3D, the entire cluster C has solved child clusters C_1, C_2, C_3. To solve C, i.e, to solve for the rotation/translation

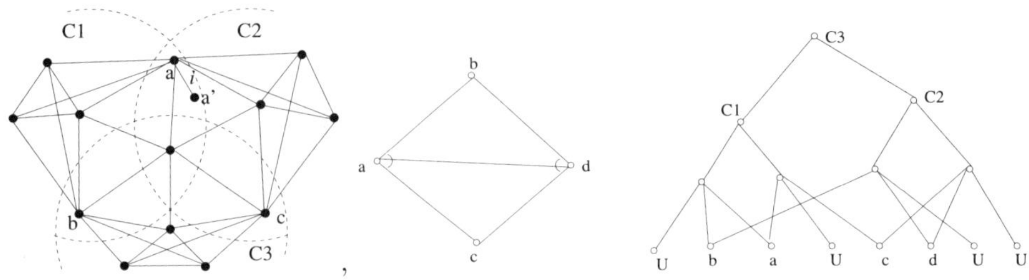

FIGURE 25. Left:Assigning overlap equations. Right:Dealing with constraint dependences during solving.

of C_2 and C_3, say, into the coordinate system of C_1 (chosen to be the coordinate system for C) several overlap equations, that require that, for instance, point A in C_1 fall in the same location as point A in C_2. We treat these as distinct from explicit incidence constraints, which directly enter the dof analysis.

For generic stability (independence), when m child clusters are present in the parent cluster, the number of equations to solve the parent cluster be at most $D(m-1)$ (D is 3 in 2D and 6 in 3D). This includes the number of equations caused by any other constraints between C_1 and C_2 (in this example, there are none) and the overlap constraints.

Even in wellconstrained parent clusters C although the dof analysis of the cluster C (appropriate inclusion-exclusion count of the dofs of the child clusters and their overlaps) would show the the cluster to be wellconstrained, the number of potential overlap equations could exceed the $D(m-1)$ as in the above example: each of the points A, A', B, and C generates 3 overlap equations and the point D generates 3*2 = 6, totaling 18 overlap equations which exceeds $D(m-1) = 6 * 2 = 12$.

The ESM uses a method (see the *getEquation* and *getOverlapEquation* methods of the ESM pseudocode in [**Sit04a, Sit04b**]) that partitions overlaps and other constraints between the child clusters of C in a way that they constitute a stable and and algebraically independent set of $D(m-1)$ constraints for solving C.

We decide on which overlap equations to discard as follows. Let n be the number of other (nonoverlap) constraints relating the child clusters of the parent cluster C being solved. If C is wellconstrained, then $D(m-1) \geq n$. To choose and partition the overlap equations, we stipulate the following.

- Any cluster (including trivial clusters) in the overlapped region should participate in atmost as many overlap constraints as the number of its degrees of freedom, although the total number of position variables of its geometric primitives could be higher; For example, in 3D, any line segment (trivial cluster), even one that is of variable length should participate in at most 5 overlap equations although its two end points constitute 6 position variables, since they belong to a cluster and hence the distance between them has been fixed by that cluster (this is the constraint dependence or overconstraint caused by the classic bananas structure of Section 1). any underconstrained overlap is a source of constraint dependences and overconstraints which must be eliminated in building a stable system, regardless whether it causes a misclassification during dof analysis.

- Any coordinate position of a geometric element in a cluster (an expression in the rotation/translation variables of the cluster) should participate either in an explicit incidence constraint relating child clusters of C or an overlap constraint, but not both; otherwise when we set everything else except that variable to constants, we get either 2 identical equations, or 2 inconsistent equations with the variable occuring with the same coefficient, i.e, 1;
- If two geometric elements participate in an explicit incidence constraint within a cluster C_i, then only one of them can participate in an overlap equation;
- Any non-rotationally symmetric child cluster should participate in at least D (3 in 2D; 6 in 3D) equations (overlap and nonoverlap combined); for points in 2D, this bound is 2, and for fixed length line segments in 3D, this number is 5.
- Any pair of clusters should be related by at most D overlap or nonoverlap equations (this number is less if $m >= 2$, or if one of the clusters is rotationally symmetric).
- Total of $D(m-1)-n$ overlap equations should be chosen.

It can be shown that for generically wellconstrained cluster C, *any* assignment or partitioning algorithm for choosing the overlap equations results in a generically stable and independent system, provided it satisfies these criteria.

As mentioned in Section 3 a significant advantage of using an FA DR-planner in this context is the following. In case C consists of a just a pair of child clusters, then their overlap could be an underconstrained graph with arbitrarily many clusters, but ensuring the first requirement above is relatively easy, by simply obtaining the DR-plan of the overlapped part. If, on the other hand, the parent cluster C has more than 2 child clusters, then because of the cluster-minimality property of Section 3, the overlap of every pair of them is guaranteed to be equivalent (via incidence constraints) to a rotationally symmetric, trivial subgraph. See below for how rotationally symmetric trivial clusters and subgraphs involving incidence constraints are detected and treated. In this case as well, ensuring the first criterion becomes easy. To do so, for each subset of child clusters of C with a nonempty intersection (overlap) subsystem, a DR-sub-plan for the overlap is created. Although there are exponentially many such (necessarily trivial) overlaps in the number of child clusters, the number of distinct overlaps and their total size is bounded by the size of the cluster C, i.e, most of the subsets generate overlaps that are empty or identical to others.

An advantage of using FA to create these DR-plans as well is that they would be created in one pass (as part of a common DR-plan of the entire cluster C) simply by providing an input decomposition whose nodes correspond to these nonempty overlap subsystems. For each cluster in these overlap DR-plans, the assignment of overlap equations should ensure that the first criterion above should hold.

Once these DR-plans are created, since there are atmost $D(m-1)$ overlap equations to be chosen, a standard assignment algorithm - ensuring the 5 above criteria - will take only $O(m)$ time to choose and partition the set of overlap equations.

In the example above, this method will assign 3*2 overlap equations for the point shared by all the clusters, which we denote d: $X_{d,C_1} = X_{d,C_2} = X_{d,C_3}$; $Y_{d,C_1} = Y_{d,C_2} = Y_{d,C_3}$; and $Z_{d,C_1} = Z_{d,C_2} = Z_{d,C_3}$. And 2 overlap equations

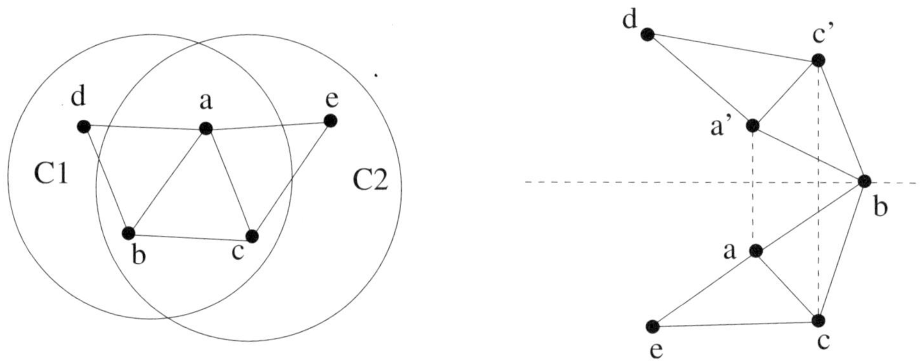

FIGURE 26. A 2D example with extraneous solutions

each for the points a, b and c: $X_{a,C_1} = X_{a,C_2}$; $Y_{a,C_1} = Y_{a,C_2}$; $X_{b,C_2} = X_{b,C_3}$; $Y_{b,C_2} = Y_{b,C_3}$; $X_{c,C_1} = X_{c,C_3}$; and $Y_{c,C_1} = Y_{c,C_3}$.

6.0.4. *Extraneous solutions.* While the set of equations thus obtained for solving the cluster C is guaranteed to be stable and algebraically independent, and their solution set will be finite contain all the valid realizations of C, some of the solutions obtained may not be valid. Consider the following example:

For the 2D example in Figure 26, the overlapped part of C_1 and C_2 consists a, b, c. After ESM performs the above assignment method of overlap equations for C_3, one possible choice of overlap equations could be: $X_{a,C_1} = X_{a,C_2}$; $Y_{b,C_1} = Y_{b,C_2}$; $X_{c,C_1} = X_{c,C_2}$.

Retaining C_1's coordinate system for C_3, the rotation/translation of C_1 is fixed to be the identity hence the left hand sides of the above equations are constants. The right hand sides are expressed in terms of the 4 nonzero rotation/translation variables (2 translations and the sine s and cosine c of the rotation angle) of C_2, which need to be solved for. Setting the fourth equation to be $s^2 + c^2 = 1$, the algebraicSolver is called on the 4 equations in 4 variables. There is only one valid solution for C_3 for any given choice of solutions for C_1 and C_2. However, Two solutions are returned, but in one of them, the locations of a and c in C_1 do not match the locations of a and c in C_2 (although their X coordinate values match).

Hence the various solutions of the current cluster C (assuming fixed solutions for the child clusters) need to be inspected and extraneous solutions discarded.

6.0.5. *Stable algebraic systems in the presence of overconstraints and constraint dependences.* Overconstrained clusters C fall into two types that are handled differently.

The first type of cluster is overconstrained according to the definition in Part I, Section 2 and can be detected using a traditional dof analysis. For such *combinatorially* overconstrained clusters C, overconstraints are first reduced using the method given [**aMSaBY04**]. Inductively, solved child clusters are assumed to be wellconstrained. The above method of partitioning overlap constraints for wellconstrained clusters is then applied. Once the solutions are returned by the algebraic-numeric solver, these discarded equations are checked and only those solutions that satisfy them are retained.

The second type of overconstrained cluster is also generically overconstrained but the source of the overconstaint is constraint dependence. See Figures in Section 1 and description in Section 4, explaining the "bananas" problem, and how a good DR-planner marks these as constraint dependences to be "handled with care" during solving. In some of these cases, the left hand sides of these equations are dependent and the right hand sides do not reflect the same dependence. We illustrate with an example how stable systems of equations are built in this case.

For the 3D example in Figure 25(middle, with DR-plan shown on right), the overlapped part of C_1 and C_2 consists of b, c. When ESM builds the algebraic equations for solving C_3, as discussed above, it would obtain 6 equations for b and c in the overlapped part and 1 equation for the distance constraint between a and d: 7 equations but only 6 variables corresponding to the rotation/translation of C_2 into the coordinate system of C_1, chosen as the home cluster, to be the coordinate system of C_3. In fact, we obtain 10 equations including 3 of the form $s^2 + c^2 = 1$ and 9 variables including the s's and the c's.

To solve the problem, the following modification is made to the method used by ESM. For the vertices in the overlapped part of two child clusters, ESM first checks whether the distances in the local coordinate of each child cluster are consistent (equal within a tolerance). If not, the cluster is returned as inconsistent or unsolvable. If they are, ESM assumes there is one distance constraint between the two vertices in the overlapped part and continues in the usual way, i.e. it only obtains 5 equations for the overlapped part, since it is a rotationally symmetric trivial cluster.

6.1. Incorporating Navigation constraints. Navigation constraints can be used to convert the above ESM methods of user-directed navigation and backtracking into a (tractable) automatic search for desired solutions for *any* cluster of the DR-plan. These constraints can either be given apriori or interactively during a user-directed navigation, to prune off the realizations of the DR-plan's clusters encountered thus far.

Leveraging the modularity of the ESM method discussed above - i.e., its separation from the DR-planner, and its applicability to *any* input DR-plan (generated by any method) - permits methods developed along with other DR-planners for incorporating navigation constraints - to be transferred *practically unchanged* to the FA DR-planner and the ESM methods discussed above. We describe the two common types of navigation constraints below.

One type of constraint used for navigation or picking out desired solutions for any cluster of the DR-plan are Relational or Engineering constraints which typically relate metric parameters of one constraint with that of another without fixing any of them. For example, the sum of the squares of 3 distances (between 3 pairs of points) could be forced to a constant; or the tangent of an angle (between two lines) could be forced to the ratio of some pair of distances (between 2 pairs of points), etc. A method for incorporating these types of constraints into a Shape Recognition (SR) DR-planner (see Section 3) is given in [**aCMH97**] for triangle decomposable systems).

The other type of common navigation constraints are chirality [**Fud95, aCMH96b**], also called order type, and related *oriented matroid* constraints. Chirality constraints can be inferred from the input sketch. They describe relative orientations of the objects in the input sketch. For example, for point objects, they assign signs to the $D+1$ by $D+1$ determinants of the homogeneous coordinate matrices obtained from each set of $D+1$ point objects. I.e, these constraints are $D+1$ degree polynomial inequalities. These correspond to a complete oriented matroid specification. Alternatively, a *partial* oriented matroid specification could be specified by the user: these assert intersection or separation of certain pairs of convex hulls of the point objects (called respectively circuits and co-circuits in oriented matroid terminology). For example, two line segments could be constrained to intersect, or a point could be forced to lie in the convex hull of some other set of points. These, too could be written as polynomial inequalities. Depending on their degree, only adhoc approaches are known for dealing with these constraints and these work for any DR-plan. See Section 9.

7. Dealing with inconsistencies and ambiguities

Here we discuss methods for dealing with combinatorially overconstrained and underconstrained systems. Overconstraints caused by constraint dependences such as the "bananas" problem are discussed in Section 4.

7.1. Underconstrained systems. Underconstrained sketches arise often in practice. In fact, as explained in Section 3, the complete set of maximal wellconstrained clusters of underconstrained systems can be read off as the "sources" or "roots" of a FRONTIER's DR-plan [**Sit04b**], [**aMS04**]. For underconstrained systems, FRONTIER also provides a sequence of exhaustive lists of constraints(types, not values) that can be generically added (a so-called *completion*) without making the system overconstrained. This was implemented in [**Sit04b**] (2001 version for general 2D constraint systems, 2002 version for a small class of 3D constraint systems, and by the 2003 version for most 3D constraint systems). A proof of why this method works is given below. As pointed out in Part I, Section 1, this should be compared with a method for obtaining completions for the class of triangle decomposable 2D constraint systems which is the subject of the paper [**aASRaSVMaJVP03**].

Claim: Let S be the set of root or source clusters of a DR-plan of a 3D or 2D constra int graph G output by FRONTIER. The new constraints that can be added without making any of the clusters (more) overconstrained are exactly those new constraints that relate clusters in S, i.e., those constraints that are not within any of the clusters of S.

Proof. As described in Section 3, and proven in [**aMS04**], S consists of a complete set of maximal clusters in G. Clearly adding a new constraint within a cluster of S will cause that cluster to become more overconstrained than before. To prove the converse, assume to the contrary that one of the new constraints causes an overconstraint between two clusters C_1 and C_2 in S. This would imply that either $C_1 \cup C_2$ was a cluster before the addition of the constraint, contradicting the fact that S consists of maximal clusters, or that $C_1 \cup C_2$ contains another wellconstrained cluster D as a subgraph, which overlaps both C_1 and C_2 (atleast one of the overlaps must be on a trivial cluster since $C_1 \cup C_2$ is not a cluster) and D is not contained in either C_1 or C_2. This contradicts the completeness of S.

Once any of these constraints is added, the FA-DR-planner can *update* the DR-plan (as described below) to potentially provide a new list of source vertices, and a new

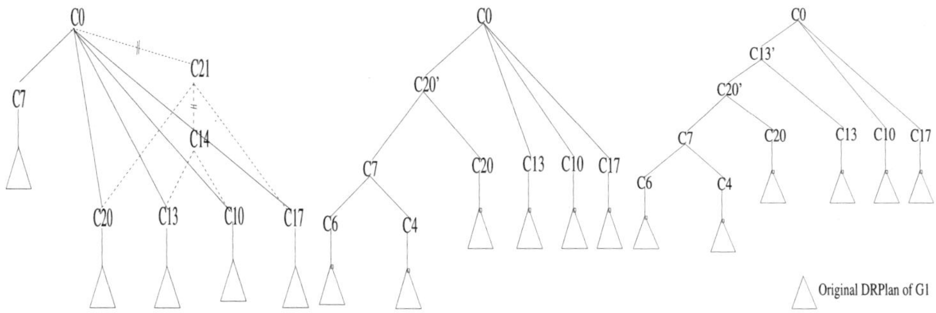

FIGURE 27. Left: unoptimized DR-plan of $G1$ in Figure 11, after reduction of edge $(1,7)$; modification path is marked; Middle and Right: two optimization steps

list of constraints between them. This iterative process ends when the DR-plan has a single source vertex, i.e, when the system is wellconstrained. The process gives the sequence of lists required by Part I, Section 3 to obtain a wellconstrained completion of the input underconstrained system.

To obtain the realizable intervals of metric values associated with these constraints is a more difficult problem and is related to the question of describing and navigating the infinite solution space of underconstrained systems (see Section9). However, for the CAD application, there are effective heuristics to infer these values [**aCMH97, aRAaGLK95, Yua99**] from the input sketch of the constraint system.

7.2. Overconstraints. Detection of combinatorially overconstrained clusters is achieved by many DR-planners, including [**aAM96**], [**aCSC01**], and a modification of [**Pab93**] described in [**aALaMS01a**]. See the message in Figure 7. However, efficiently utilizing the DR-plan to isolate the *minimal but complete* set of overconstraints that can be deleted, and efficiently updating the DR-plan after deletion have been addressed only recently in [**aMSaBY04**]. Most solvers ask the user to remove constraints in an over-constrained situation but give poor guidance which constraints to delete. The method in [**aMSaBY04**] is based on showing that there is a unique, well-defined set of *reducible* constraints that can be deleted without making the graph underconstrained. See Figure 27.

The method is further based on key properties of the FA-DR-plan and gives an efficient (generally linear time) solution that retains full generality and moreover:
(i) Traverses the given DR-plan top down to incrementally output this unique set of constraints in reverse solving order, minimizing the need to solve again previously solved portions of the DR-plan;
(ii) Selects constraints from those parts of the constraint system that the user identifies;
(iii) Isolates information that can be routinely stored and maintained as part of the DR-plan, making the above process even more efficient; and
(iv) Automatically *updates* (see Figure 27) the DR plan with optimal reorganization, once one of the reducible constraints is removed, which may cause many lower level clusters of the DR plan to become underconstrained.

This method is described in [**aMSaBY04**] and is part of the FRONTIER solver [**Sit04b**]. This method is purely combinatorial, deceptively simple but makes sophisticated use of the FA DR-plan and hence requires an intricate proof of correctness.

8. Efficient Updates and Online Solving

8.1. Updating a solved constraint system. We discuss the question of efficient updates to the solution of a geometric constraint system (the output in Section 2), given incremental updates to any portion of the input in Section 2.

As it is to be expected, a good DR-plan can be used as a datastructure to systematically realize efficient updates and also to analyze efficiency. Similarly, a good DR-planner (Section 3) and ESM method (Section 5) can be judged by their clarity and efficiency in handling updates.

Before the DR-planning problem was formally defined and the quality and performance of DR-plan structure shown to be crucial for the entire geometric constraint solving process in [**aALaMS01a**], the question of updates was mostly ignored. Moreover, since the constraint systems are often simple and triangle decomposable, with effectively linear DR-plans whose subsystems were of small size and typically involved solving a single quadratic equation, those constraint solvers such as [**Owe**] that permit updates, implement them by simply redoing the entire system from scratch. For general 3D constraint systems and more complex DR-plans, this approach is intractable.

The properties of FA DR-plans in conjunction with the modularity and clarity of the ESM method described in Section 5 [**aAAaYZaNK04**], can be used to perform efficient updates [**aMSaBMaAA01, Sit04b, Sit04a**].

We briefly describe several possible input updates: for each, we list which portions of the output need to be updated and in what possible ways, and sketch methods to do so. In FRONTIER's update mode, changes to the input are allowed at the end of the DR-planning phase, or at any point during the solving and navigation phase of the ESM. See Figure 28.

8.1.1. *Changing Constraint Values.* The first and simplest case is changing the metric value of a constraint e. In this case, no DR-planner is involved. Furthermore, let C be the active cluster where e is resolved (i.e., e is a constraint relating child clusters of C). The only relevant clusters whose realizations will change (for which the algebraicSolver needs to be called again) are those ancestors on the unique path from the active cluster C to a source or root cluster of the DR-plan. In the generic case, there will be no change in the classification of any cluster, or change in the number of realizations of any cluster, or the realization-choices of non-relevant clusters (in order to ensure a solution of the entire system). Hence in the generic case, the update simply requires the ESM to resolve and navigate the realizations of the relevant clusters, keeping all other realization choices fixed.

See Section 4 for a discussion of nongeneric cases and earlier in this section for the underconstrained case where e has no active cluster (it relates 2 source nodes of the DR-plan).

8.1.2. *Adding a Constraint.* Constraints are typically added to convert an underconstrained system into a wellconstrained one. See Section 7.1.

Adding a constraint could require the involvement of the DR-planner. since in this case, the classification of the clusters as well as the entire system could change.

I.e, these clusters could now become (combinatorially) generically overconstrained if they were wellconstrained before or the entire system could now become wellconstrained if it was underconstrained before. The former case is immediate to detect. The latter case requires the involvement of the DR-planner. This can be viewed simply as a continuation of FA's DR-planning process from before the addition of e: now e will have to be distributed in the final flow graph containing the source clusters of the DR-plan before the edge was added; and the DR-plan will have to be updated with any new clusters found. For this purpose, the final flow assignments are retained in addition to the cluster structures in the DR-plan, at the end of the DR-planning phase.

Once the DR-planning phase of the update is complete, the ESM's solving phase proceeds as described in the previous subsection on changing constraint value since again, the only clusters affected by the addition of a constraint e are its active and relevant clusters in the new DR-plan.

8.1.3. *Removing a Constraint.* The first case for constraint removal is to convert an overconstrained system into a wellconstrained one. This requires the involvement of the DR-planner and could significantly change the DR-plan and its clusters at every level, since many of the previous clusters may now no longer be wellconstrained. The method of Section 7 [**aMSaBY04**] achieves this with the minimal disruption.

Since the method already deals with previous clusters becoming underconstrained, it directly extends to the second case of constraint removal when the entire wellconstrained system becomes underconstrained.

A third case of constraint removal causes a combinatorially overconstrained or even wellconstrained system to now develop a constraint dependence as in the figures describing the bananas problem in Section 1. However, if no other changes are made, the system continues to have the same set of realizations.

Since constraints have only been removed, all the previous solutions of the the constraint system remain. This is true of any unchanged cluster in the DR-plan. However, constraint removal changes the DR-plan significantly, as described in Section 7, and hence the navigation of the solution space by the ESM would have to be redone atleast for all the new clusters in the DR-plan.

8.1.4. *Adding or Removing a new object or feature.* Addition is handled by putting the new object (and constraints relating it to the existing objects) in the flow graph as an undistributed cluster along with its undistributed incident edges e; and adding the object to the DR-plan as a new source vertex. Now the method reduces to the "Adding a constraint" method discussed above for the undistributed edges e.

Removal is handled by simply removing the constraints that relate the object to the remainder of the constraint system (using the method above).

Adding a new feature to the input decomposition relies on details of the method described in [**aYZ04a**], incorporated into the FA DR-planner and can also be found in [**Sit04b, Sit04a**]. Removal of a feature involves direct removal of those clusters - that have now become inessential - from the DR-plan. The process does not involve the DR-planner.

Note: none of these update methods attempt to preserve optimality of the DR-plan, i.e, the sizes of the clusters may not be optimal once the update is complete and is hence not sustainable for large number of updates. See open problem in Section 9.

8.2. Solving a geometric constraint system online. The problem of *online solving* is the following: as the user inputs (sketches) each geometric primitive and constraint of the input constraint system and each feature of the input decomposition, the output - corresponding to the subsystem sketched so far - is immediately available to the user for inspection. This is a difficult question since partially input constraint systems are often underconstrained and realizing those is a a difficult problem. See Sections 7 and 9. The difficulty is however ameliorated by the fact that while the the online solving and partial output process need to be fast and inexpensive, some incorrectness and incompleteness can be tolerated, since the partial output is usually only meant to guide the user and the errors will be rectified when the input is complete and the (correct and complete) offline algorithm is run.

The typical approach to online solving is to run an inexpensive *simplesolver* method that attempts to extend or modify the displayed solution - of the subsystem input so far - to include the newly added geometric element or constraint. This method must handle highly underconstrained systems in a simple, lightweight, local fashion, choosing an adhoc extended or modified solution that is as close as possible to the input sketch of the constraint system. If the simplesolver is unable to extend or modify the current solution, then either it outputs the closest approximation it can find, or it then calls the offline algorithm to perform the update and displays the extended solution thus obtained. In the online setting, finding a solution close to the appearance of the input sketch is typically important, and the offline update process must include it as a navigation constraint, see Section 8 and Section 5).

The best performing online simplesolver (for 2D and specific class of "pipe routing" 3D systems), to date is the proprietary commercial D-cubed solver of [**Owe**]. FRONTIER's simplesolver [**aMSaBMaAA01, Sit04b, Sit04a, JJO04**] works only for 2D and iteratively switches between two methods. Method (i) handles parameterless or nonmetric constraints such as incidence, tangency, etc. as well as angle constraints [**JJO04**] (ignoring distance constraints) and uses nontrivial observations leveraging dependencies of angle constraint graphs. See Figure 29. Method (ii) is a simple, repeated use of the triangle inequality that handles distance (ignoring angles and nonmetric) constraints.

9. Open Problems, Related areas and techniques

9.0.1. *Inadequacy of combinatorial classification.* This is a longstanding open problem in the *rigidity theory* (see [**aBSaHS93**]) with many conjectures and heuristic approaches but no satisfactory solution; any improvement is desirable: give a completely combinatorial characterization (and associated polynomial time algorithm) for determining if a constraint graph is generically rigid, i.e, wellconstrained, overconstrained or underconstrained. The problem is open not only for 3D systems, but also for 2D systems other than those that contain only points and distances, for example for those that include angle and incidence constraints.

9.0.2. *Underconstrained systems.* This is another longstanding open problem to obtain a satisfactory description and systematic navigation of the solution spaces of underconstrained systems. This is related to the *configuration spaces of generalized 2D and 3D mechanisms* and to *robot arm (inverse) kinematics* with wide ranging applications including newer ones such as *molecular modeling* [**aAJRaLAKaMFT01,**

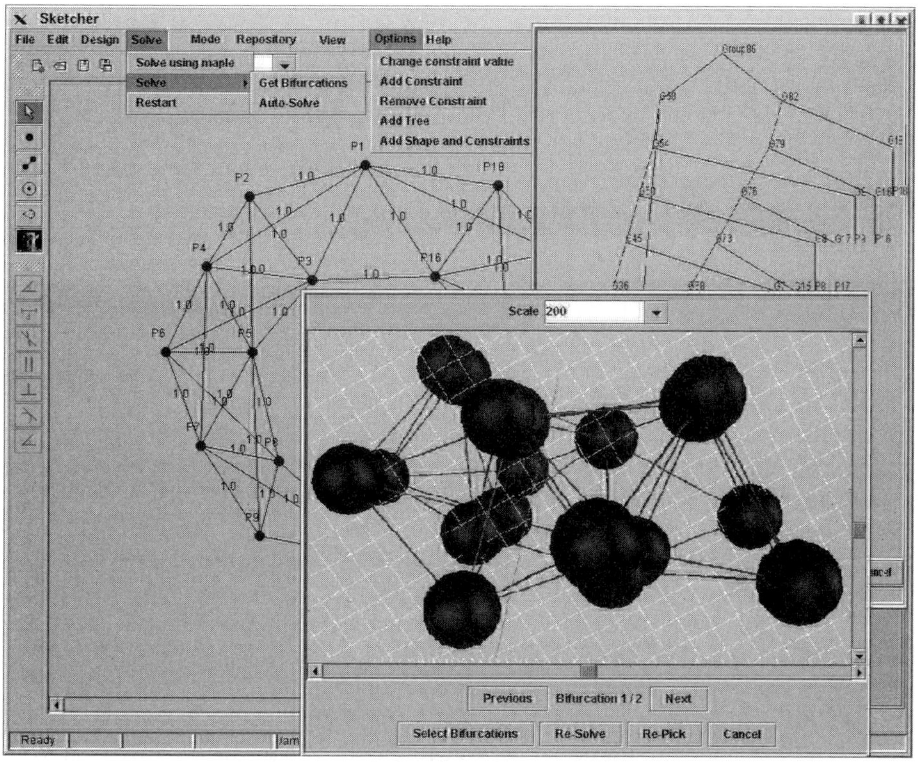

FIGURE 28. Update possibilities shown on pull down menu – updates are possible after the DR-planning stage, during the navigation and after the final solution is displayed

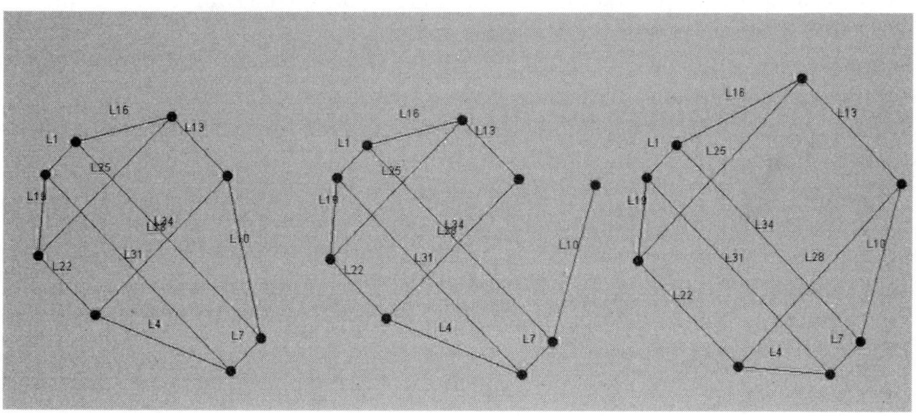

FIGURE 29. Simplesolver that handles angle and incidence constraints online

aTH88, CNS] The problem has both combinatorial and algebraic underpinnings. The algebraic part is related to roadmapping real algebraic varieties arising from geometric constraints and also to geometric groups and Lie algebra techniques.

A recent investigation could shed new light on the problem and was kickstarted by a proof of [**aEDaGR03**] of the *Carpenter's rule* conjecture that a polygonal linkage can be convexified without self-intersections. Crucial to the proof is the natural, but clever use of a standard (rigidity based) formalization the concept of *expansive motions*. A constructivization of these expansive motions was provided by [**Str00**], which introduced highly versatile structures called *pseudotriangulations* (wellconstrained completions of 2D polygonal linkage mechanisms which are highly underconstrained systems). These have been used to elucidate several questions [**aGRaBSaIS03, aDOaGRaFSaBSaHSaDSaISaWW03, aFSaIS02, aFSaIS03, Str03**] related to rigidity, topological constraints on configuration spaces of simple 2D mechanisms and promises [**aWW**] a combinatorial rigidity approach to 3D robot arm kinematics, molecular modeling etc.

9.0.3. *Inequality constraints and navigation constraints.* The above mentioned problem of describing intervals of realizable distance values - for underconstrained systems - is related to another longstanding problem of dealing with constraint graphs whose edges are *given* intervals of distance values.

More generally, how to deal with constraint systems which include fixed types of polynomial inequalities? Highly structured polynomial inequalities such as oriented matroid and chirality constraints are used as navigation constraints as mentioned in Section 5 and while that section lists adhoc methods of dealing with such constraints which are effective for limited 2D CAD applications, no formal, provably efficient methods are known: these would be necessary for more complex 3D situations.

9.0.4. *Inverse problem and global constraints.* We consider the *inverse* problem of determining a good, minimal set of local constraints to impose in order to include a desired set of solutions and exclude others. The desired set could have been described by global constraints. This is a crucial problem related to the underconstrained or even wellconstrained solution space navigation problem and is related to the overall issue of using constraints to represent geometric composites.

A typical route is to first decide on which global constraints to impose and then localize them. A classical example is *symmetry constraints*. These are natural in many physical applications as they correspond to minimum energy constraints. These often force solution uniqueness and moreover permit the reduction to a local constraint system. Moreover, symmetry often forces an otherwise underconstrained system, with infinitely many solutions, to become a well-constrained system with few solutions.

The first challenge is in describing the group of symmetries to impose upon the composite and the second is to reduce it into a system of local constraints. These are independently interesting research problems in *computational invariant theory* and *automated geometry theorem proving*, specifically concerning ideal membership and alternative sets of generators.

9.0.5. *Topological and Assembly constraints.* Assembly constraints are of two types: one type explicitly imposes a partial order on the way the constraint system is solved, i.e., an input design decomposition along with a priority order imposed upon its nodes which represent subsystems. As pointed out in Section 3, one of the main advantages of the FRONTIER constraint solver is its ability to generate a DR-plan that incorporates such input decompositions [**aYZ04a**] and priority orders. The other type of assembly constraint only implicitly imposes such an order: it concerns intersections that must be avoided during the process of assembly. These are effectively topological constraints on paths in the configuration spaces of the subassemblies. Together with natural assumptions about starting configurations, an assembly (partial) order can be computed.

As mentioned earlier, the primary use of the pseudotriangulations introduced in [**Str00**] is to guarantee expansive infinitesimal motions to deal with topological constraints on paths in the configuration space of an 2D underconstrained system. Generalization to 3D of constructions that guarantee expansive motions would be highly desirable. Initial efforts can be found in [**aWW**]

9.0.6. *Optimal DR-planning.* As pointed out in Section 3, the problem of finding an optimal DR-plan - i.e., one that minimizes the size of largest subsystem (minimizes the max fan-in of the DR-DAG) - is NP-complete. No guaranteed (deterministic or randomized) approximation algorithms (or nonapproximability results) are known except for special cases [**aMS04**].

9.0.7. *Analyzing Updates and Efficient Online operation.* Detailed methods have been presented in Section 8 to the question of efficiently updating the solution of a constraint system to reflect incremental changes in the input. The methods use the FA DR-plan as a crucial datastructure. However none of the methods attempt to preserve optimality of the DR-plan. As a result of which, the methods "tolerate" only a limited number of updates: to restore optimality, the DR-plan has to be reconstructed and its clusters resolved from scratch periodically. More sophisticated update methods and (perhaps amortized) analysis is called for.

No efficient online 3D constraint solvers currently exist. They require 2 ingredients: efficient, accurate, offline updates (discussed above); and fast, inexpensive simplesolvers that obtain an approximate solution of a (highly) underconstrained system (special case of a problem discussed below); in particular, they extend an existing solution to an underconstrained system when a constraint or geometric element is added.

9.0.8. *Repeatability and Persistent Naming.* A problem with several of the constraint solvers discussed here is that two different, *nonisomorphic* output DR-plans (and hence the solution space navigation process) could be obtained for the same input constraint system during 2 different sessions, simply because the geometric elements and constraints were input in a different order and considered by the DR-planner in a different order (or perhaps by a different sequence of updates) *Repeatability* is desirable: i.e., to ensure that the DR-planner outputs isomorphic DR-plans DR-plan for isomorphic constraint graphs. Even more desirable would be a systematic manner in which the user could choose one out of several nonisomorphic DR-plans. The ability of a DR-planner to handle an input design decomposition helps, since this restricts the possible output DR-plans to those that incorporate the input design decomposition, but it is not adequate.

One technique to handle this problem is to solve the problem of *persistent naming*: i.e., the constraint solver first renames or renumbers the vertices and edges of the input constraint graph typically based on their local neighborhood information. Regardless of the input ordering of the vertices and edges, the process results in the same renumbering.

References

[aAAaYZaNK04] M Sitharam and A Arbree and Y Zhou and N Kohareswaran, *Solution management and navigation for 3d geometric constraint systems*, submitted, available upon request (2004).

[aAJRaLAKaMFT01] D. J. Jacobs and A. J. Rader and L. A. Kuhn and M. F. Thorpe, *Protein flexibility predictions using graph theory*, Proteins: Structure, Function, and Genetics **44** (2001), 150–165.

[aALaMS97a] C. M. Hoffmann and A. Lomonosov and M. Sitharam, *Finding solvable subsets of constraint graphs*, Constraint Programming '97 Lecture Notes in Computer Science 1330, G. Smolka Ed., Springer Verlag (Linz, Austria), 1997.

[aALaMS97b] ———, *Finding solvable subsets of constraint graphs*, Springer LNCS 1330 (Smolka G., ed.), 1997, pp. 463–477.

[aALaMS98a] ———, *Geometric constraint decomposition*, Geometric Constraint Solving (Bruderlin and Roller Ed.s, ed.), Springer-Verlag, 1998.

[aALaMS98b] ———, *Geometric constraint decomposition*, Geometric Constr Solving and Appl (Bruderlin B. and Roller D., ed.), 1998, pp. 170–195.

[aALaMS99] ———, *Planning geometric constraint decompositions via graph transformations*, AGTIVE '99 (Graph Transformations with Industrial Relevance), Springer lecture notes, LNCS 1779, eds Nagl, Schurr, Munch, 1999, pp. 309–324.

[aALaMS01a] ———, *Decomposition of geometric constraints systems, part i: performance measures*, Journal of Symbolic Computation **31** (2001), no. 4.

[aALaMS01b] ———, *Decomposition of geometric constraints systems, part ii: new algorithms*, Journal of Symbolic Computation **31** (2001), no. 4.

[aAM96] R. Latham and A. Middleditch, *Connectivity analysis: a tool for processing geometric constraints*, Computer Aided Design **28** (1996), 917–928.

[aAS03] B. Bettig and A. Shah, *Solution selectors: A user oriented answer to the multiple solution problem in constraint solving*, Journal of Mechanical Design **125** (2003), no. 3, 445–451.

[aASRaSVMaJVP03] R. Joan-Arinyo and A. Soto-Riera and S. Vila-Marta and J. Vilaplana-Pastó, *Transforming an under-constrained geometric constraint problem into a well-constrained one*, ACM Symposium on Solid Modeling and Applications,Proceedings of the eighth ACM symposium on Solid modeling and applications, 2003, pp. 33–44.

[aBSaHS93] Jack E. Graver and Brigitte Servatius and Herman Servatius, *Combinatorial rigidity*, Graduate Studies in Math., AMS, 1993.

[aBY00] C. M. Hoffmann and B. Yuan, *On spatial constraint solving approaches*, Proc. ADG 2000, ETH Zurich, 2000, p. in press.

[aCMH96a] I. Fudos and C. M. Hoffmann, *Constraint-based parametric conics for CAD*, Computer Aided Design **28** (1996), 91–100.

[aCMH96b] ———, *Correctness proof of a geometric constraint solver*, J. Comp. Geometry and Applic. **6** (1996), 405–420.

[aCMH97] ———, *A graph-constructive approach to solving systems of geometric constraints*, ACM Transactions on Graphics **16** (1997), 179–216.

[aCR97] A. Middleditch and C. Reade, *A kernel for geometric features*, ACM/SIGGRAPH Symposium on Solid Modeling

Foundations and CAD/CAM Applications, ACM press, 1997.

[aCSC01] C. M. Hoffmann and C. S. Chiang, *Variable-radius circles in cluster merging, Part I: translational clusters*, CAD **33** (2001), in press.

[aDOaGRaFSaBSaHSaDSaISaWW03] Ruth Hass and David Orden and Gunter Rote and Francisco Santos and Brigitte Servatius and Herman Servatius and Diane Souvaine and Ileana Streinu and Walter Whiteley, *Planar minimally rigid graphs and pseudo-triangulations*, Proc. ACM Symp. Comp. Geometry (SoCG) San Diego, California, 2003, pp. 154–163.

[aEDaGR03] Robert Connelly and Erik Demaine and Gunter Rote, *Straightening polygonal arcs and convexifying polygonal cycles*, Discrete and Computational Geometry **30** (2003), 205–239.

[aFSaIS02] Gunter Rote and Francisco Santos and Ileana Streinu, *On the number of embeddings of minimally rigid graphs*, Proc. ACM Symp. Computational Geometry (SoCG) Barcelona, June 5-7, 2002, 2002, pp. 25–32.

[aFSaIS03] ———, *Expansive motions and the polytope of pointed pseudo-triangulations*, Discrete and Computational Geometry - The Goodman-Pollack Festschrift, 2003, pp. 699–736.

[aGRaBSaIS03] Oswin Aichholzer and Gunter Rote and Bettina Speckmann and Ileana Streinu, *The zigzag path of a pseudo-triangulation*, Proc. 8th International Workshop on Algorithms and Data Structures (WADS) , Ottawa, Canada, 2003, pp. 377–388.

[aIFaCMHaJCaRP95] W. Bouma and I. Fudos and C. M. Hoffmann and J. Cai and R. Paige, *A geometric constraint solver*, CAD **27** (1995), 487–501.

[aJLaDO98] D. Cox and J. Little and D. O'Shea, *Using algebraic geometry*, Springer-Verlag, 1998.

[aJPaYZ04] M Sitharam and J Peters and Y Zhou, *Solving minimal, wellconstrained, 3d geometric constraint systems: combinatorial optimization of algebraic complexity*, submitted to ADG 2004, available upon request (2004).

[aMLaAS03] R. Joan-Arinyo and M.V. Luzon and A. Soto, *Genetic algorithms for root multiselection in geometric constraint solving*, Computers and Graphics **27** (2003), no. 1, 51–60.

[aMLVaBSaNWaGZ93] A. Bjorner and M. Las Vergnas and B. Sturmfels and N. White and G. Ziegler, *Oriented matroids. encyclopaedia of mathematics. vol. 46. edited by g-c. rota*, Cambridge University Press, 1993.

[aMS04] A. Lomonosov and M. Sitharam, *Graph algorithms for geometric constraint solving*, submitted, 2004.

[aMSaBMaAA01] J. J. Oung and M. Sitharam and B. Moro and A. Arbree, *Frontier: fully enabling geometric constraints for feature based design and assembly*, abstract in Proceedings of the ACM Solid Modeling conference, 2001.

[aMSaBY04] C Hoffman and M Sitharam and B Yuan, *Making constraint solvers more useable: the overconstraint problem*, to appear in CAD (2004).

[aPJV94] C. M. Hoffmann and P. J. Vermeer, *Geometric constraint solving in R^2 and R^3*, Computing in Euclidean Geometry (D. Z. Du and F. Hwang, ed.), World Scientific Publishing, 1994, second edition.

[aRAaGLK93] N. Sridhar and R. Agrawal and G. L. Kinzel, *An active occurrence-matrix-based approach to design decomposition*, CAD **25** (1993), 500–512.

[aRAaGLK95] ———, *Algorithms for the structural diagnosis and decomposition of sparse, underconstrained design systems*, CAD **28** (1995), 237–249.

[aRJa97] C. M. Hoffmann and R. Joan-arinyo, *Symbolic constraints in geometric constraint solving*, J. for Symbolic Computation **23** (1997), 287–300.

[aRJaDM93] S. Ait-Aoudia and R. Jegou and D. Michelucci, *Reduction of constraint systems*, Compugraphics, 1993, pp. 83–92.

[aRV95] C. M. Hoffmann and R. Vermeer, *Geometric constraint solving in R^2 and R^3*, Computing in Euclidean Geometry (Du D. Z. and Hwang F., ed.), 1995, pp. 266–298.

[aSCC98a] X. S. Gao and S. C. Chou, *Solving geometric constraint systems. I. a global propagation approach*, CAD **30** (1998), 47–54.

[aSCC98b] ———, *Solving geometric constraint systems. II. a symbolic approach and decision of rc-constructibility*, CAD **30** (1998), 115–122.

[aTH88] G. Crippen and T. Havel, *Distance geometry and molecular conformation*, John Wiley & Sons, 1988.

[aWW] Ileana Streinu and Walter Whiteley, *Single-vertex origami and 3-dimensional expansive motion*, submitted,Dec. 2003.

[aWW85] T. Tay and W. Whiteley, *Generating isostatic frameworks*, Topologie Structurale **11** (1985), 21–69.

[aYZ04a] M Sitharam and Y Zhou, *Mixing features and variational constraints in 3d*, submitted, available upon request (2004).

[aYZ04b] ———, *A tractable, approximate, combinatorial 3d rigidity characterization*, submitted to ADG 2004, available upon request (2004).

[Bru86] B. Bruderlin, *Constructing three-dimensional geometric object defined by constraints*, ACM SIGGRAPH, Chapel Hill, 1986.

[CNS] CNS, *http://cns.csb.yale.edu/v1.0/*, NMR structure software.

[Cra79] Henry Crapo, *Structural rigidity*, Structural Topology **1** (1979), 26–45.

[Cra82] ———, *The tetrahedral-octahedral truss*, Structural Topology **7** (1982), 52–61.

[Fud95] I. Fudos, *Constraint solving for computer aided design*, Ph.D. thesis, Dept. of Computer Sciences, Purdue University, August 1995.

[Hav91] T. Havel, *Some examples of the use of distances as coordinates for Euclidean geometry*, J. of Symbolic Computation **11** (1991), 579–594.

[Hen92] B. Hendrickson, *Conditions for unique graph realizations*, SIAM J. Comput. **21** (1992), 65–84.

[Hsu96] Ching-Yao Hsu, *Graph-based approach for solving geometric constraint problems*, Ph.D. thesis, University of Utah, Dept. of Comp. Sci., 1996.

[JJO04] Y. Zhou J. J. Oung, M. Sitharam, *A fast, simple solver for constraints without parameters*, In preparation (2004).

[Kra92] G. Kramer, *Solving geometric constraint systems: a case study in kinematics*, MIT Press, 1992.

[Lam70] G. Laman, *On graphs and rigidity of plane skeletal structures*, J. Engrg. Math. **4** (1970), 331–340.

[Nel85] G. J. Nelson, *A costraint-based graphics system*, ACM SIGGRAPH, 1985, pp. 235–243.

[Owe] J. Owen, *www.d-cubed.co.uk/*, D-cubed commercial geometric constraint solving software.

[Owe91] ———, *Algebraic solution for geometry from dimensional constraints*, ACM Symp. Found. of Solid Modeling (Austin, Tex), 1991, pp. 397–407.

[Owe93] ———, *Constraints on simple geometry in two and three dimensions*, Third SIAM Conference on Geometric Design, SIAM, November 1993, To appear in Int J of Computational Geometry and Applications.

[Pab93] J.A. Pabon, *Modeling method for sorting dependencies among geometric entities*, US States Patent 5,251,290, Oct 1993.

[Rol91] D. Roller, *An approach to computer-aided parametric design*, CAD **23** (1991), 303–324.

[Sit04a] M Sitharam, *Frontier, an opensource 3d geometric constraint solver: algorithms and architecture*, monograph, in preparation (2004).

[Sit04b] M. Sitharam, *Frontier, opensource gnu geometric constraint solver: Version 1 (2001) for general 2d systems; version 2 (2002) for 2d and some 3d systems; version 3 (2003) for general 2d and 3d systems*, http://www.cise.ufl.edu/~sitharam, http://www.gnu.org, 2004.

[Sit04c] M Sitharam, *Graph based geometric constraint solving: problems, progress and directions*, AMS-DIMACS volume on Computer Aided Design (Dutta and Janardhan and Smid, ed.), 2004.

[Str00] Ileana Streinu, *A combinatorial approach to planar non-colliding robot arm motion planning*, Proc. 41st ACM Annual Symposium on Foundations of Computer Science (FOCS), 2000, pp. 443–453.

[Str03] ———, *Combinatorial roadmaps in configuration spaces of simple planar polygons*, Proceedings of the DIMACS Workshop on on Algorithmic and Quantitative Aspects of Real Algebraic Geometry in Mathematics and Computer Science, 2003, pp. 181–206.

[Tod89] P. Todd, *A k-tree generalization that characterizes consistency of dimensioned engineering drawings*, SIAM J. Discrete Mathematics **2** (1989), 255–261.

[VL81] D. C. Gossard and R. Light V. Lin, *Variational geometry in computer-aided design*, ACM SIGGRAPH, 1981, pp. 171–179.

[Whi97] W. Whiteley, *Rigidity and scene analysis*, Handbook of Discrete and Computational Geometry, CRC Press, 1997, pp. 893 –916.

[Yua99] B. Yuan, *Research and implementation of geometric constraint solving technology*, Ph.D. thesis, Dept. of Computer Science and technology, Tsinghua University, China, November 1999.

Current address: Department of Computer and Information Science, University of Florida, Gainesville, Florida 32611

E-mail address: `sitharam@cise.ufl.edu`

DIMACS Series in Discrete Mathematics
and Theoretical Computer Science
Volume **67**, 2005

Immobilization: Analysis, Existence, and Output-Sensitive Synthesis

A. Frank van der Stappen

ABSTRACT. We give an overview of research on analysis, existence, and synthesis issues in part immobilization. The fundamental *analysis* question asks to determine whether a given set of contacts along its boundary immobilizes a part. The *existence* question asks to identify classes of parts that can be immobilized with a given set of contacts. Finally, *synthesis* is concerned with the computation of immobilizations of a given part. We highlight two recent contributions on the existence of immobilizations for serial chains of hinged polygons, and on the output-sensitive synthesis of all immobilizations of a polygon.

1. Introduction

Many manufacturing operations, such as machining and assembly, require the objects, or *parts*, that are subjected to these operations to be held in such a way that not even the smallest motion is possible. Immobilization of parts, fundamental to the common tasks of grasping and fixturing is a problem that has been studied extensively, see e.g. [**CSU99, Lak78, MNP90, MSS87, Ngu88, Ohw81, Reu76, RB98a, RB98b**].

We give an overview of classical and recent results on immobilization, focussing on analysis, existence, and synthesis. The discussion on analysis deals with the question whether a given set of contacts along the boundary of a part constrains all possible motions of that part. We briefly review the almost 130 years old sufficient condition of form closure by Reuleaux [**Reu76**] as well as the very recent necessary and sufficient condition of second-order immobility by Rimon and Burdick [**RB98a, RB98b**]. The section on existence addresses how many contacts are sufficient or necessary to satisfy a certain condition for immobility. The results can be different for different classes of parts and/or contacts. Finally, we shall focus on algorithms for synthesizing immobilizations. Algorithms exist for the computation of a single immobilization, an optimal immobilization according to some quality measure, a set of immobilizations, or all immobilizations.

The sections on existence and synthesis distinguish between general and modular settings. In the general setting the placement of the contacts is unconstrained,

1991 *Mathematics Subject Classification.* Primary 65D18, 68U05; Secondary 68Q25, 68W40.
Key words and phrases. Grasping, Fixturing, Computational geometry.

as is generally the case in grasping. In the modular setting that is often encountered in commercially available fixturing systems, the placement of the contacts is in some way constrained by an underlying grid. Such systems offer advantages like reusability and rapid reconfigurability at the cost of less freedom in placing contacts. Modular systems fit well in the RISC (=Reduced Intricacy in Sensing and Control) robotics philosophy proposed by Canny and Goldberg [**CG94**].

We will highlight some of our recent results on the existence and synthesis of immobilizations. Nearly all results on sufficient and necessary numbers of contacts for immobilization focus on rigid parts. As one of the first steps towards the immobilization of non-rigid parts, we have studied the immobilization of serial chains of polygons connected by rotational joints, or hinges. We review these results in Section 3. Nearly all algorithms for synthesizing immobilizations have an asymptotic running time proportional to the maximum output size. In most cases the actual output size will be much smaller. In Section 4, we will discuss some of our recently developed algorithms for computing all immobilizations of polygonal parts by means of frictionless point contacts; the running times of these algorithms largely depend on the size of the reported output, and are therefore generally more efficient than their non-output-sensitive counterparts.

2. Analysis

An attractive theoretical model for immobility was formulated by Reuleaux in 1876 [**Reu76**]. He defined a rigid body to be in *form closure* if a set of contacts along its boundary constrains all *infinitesimal* motions of the body. This notion provides a sufficient but not necessary condition for immobility, as for instance an equilateral triangle with a point contact in the middle of each edge is immobilized, but is not in form closure (it permits an infinitesimal rotation around its center). Form closure depends only on the position of the contacts and their normals, and is invariant with respect to the curvature of body and contacts. This is not true for immobility in general: if we replace the equilateral triangle in the example by its inscribed circle, contacts and normals remain identical, but the body is no longer immobilized.

Reuleaux' graphical form closure analysis for two-dimensional parts considers instantaneous velocity centers in the space of the part itself (see also Mason's text book on robotic manipulation [**Mas01**]). It is based on the observation that every infinitesimal motion of a rigid part can be seen as either a positive or a negative infinitesimal rotation around a point in the plane; infinitesimal translations are rotations around a point at infinity. As a consequence, every infinitesimal rotation can be characterized by a point in the plane and a sign. Now let us assume that a point contact is placed at some interior point p of an edge e. Let $l(p)$ be the directed line through p perpendicular to e and directed towards the interior of the part. Negative rotations about points to the right of $l(p)$ cause the contact at p to enter the part, and are therefore impossible (see Figure 1a). Positive rotations about those points are still possible as they move the part away from the contact. Similarly, negative rotations are possible around points right of $l(p)$ and impossible around points left of $l(p)$. Reuleaux' analysis conservatively assumes that positive and negative rotations are possible around all points on $l(p)$, because any such rotation locally slides the part along the contact. (Taking the curvature of the motion of the part boundary into account, as is done by more recent immobility

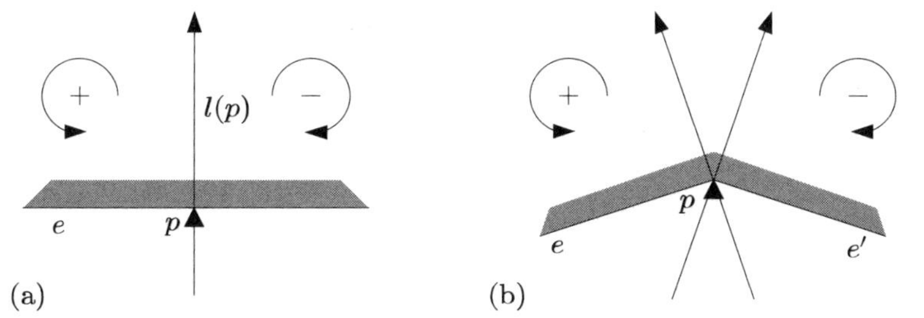

FIGURE 1. Motions allowed by (a) a point contact at an interior point p of an edge e, (b) a point contact at a concave vertex p with incident edges e and e'.

analyses such as those by Czyzowicz et al. [**CSU99**] and Rimon and Burdick [**RB98a, RB98b**], shows that rotations about points above p are prevented by the contact at p whereas rotations around points below p remain possible.) When a point contact is at a concave vertex, it induces two half-plane constraints, as such a vertex is considered to belong to both incident edges (see Figure 1b). A part is in form closure if the half-plane constraints induced by the contacts along the part boundary jointly rule out all possible centers of positive and negative rotations.

A more general form-closure analysis, applicable also to three-dimensional parts, takes place in the so-called wrench space (see e.g. Mason's textbook on robotic manipulation [**Mas01**]). The wrench space is three-dimensional for two-dimensional parts and six-dimensional for three-dimensional parts. A wrench is a force-torque pair $(f, p \times f)$, where p is a point on the boundary of a part, and f is the force applied at p. A contact at p_i induces a wrench $w_i = (f_i, p_i \times f_i)$, where f_i is the normal to the part boundary if the contact is frictionless, and can span the entire Coulomb friction cone if there is friction between the contact and the boundary. A set of frictionless contacts provides form closure if and only if these contacts can resist any external wrench w; in other words, for any wrench w there exist $\lambda_i \geq 0$ such that $-w = \sum \lambda_i w_i$. This is equivalent to saying that there must exist $\lambda_i > 0$ such that $\sum \lambda_i w_i = 0$, so the origin must be strictly contained in the convex hull of the wrenches. We use this condition in Section 4.1.2. In the case of friction between the contacts and the part boundary, the part is said to be in force closure if the condition on the wrenches is satisfied.

Czyzowicz et al [**CSU99**] studied the immobilization of polygonal parts by less than four point contacts. They identified a condition under which three frictionless points at interior points immobilize a polygonal part P. For an interior point p of an edge e, let $l(p)$ be the directed line through p perpendicular to e directed towards the interior of the P. Let $H(e)$ be the half-plane bounded by the supporting line of e and locally containing the interior of P. Three edges e, e', and e'' of a polygon P are said to form a triangular triple if $H(e) \cap H(e') \cap H(e'')$ is a bounded nonempty triangle (see Figure 2). Let p, p', and p'' be interior points of e, e', and e'' respectively. Czyzowicz et al [**CSU99**] proved that frictionless point contacts at p, p', and p'' immobilize P if and only if (i) $l(p)$, $l(p')$, and $l(p'')$ meet at a common point, and (ii) the edges e, e', and e'' form a triangular triple. We use this condition in Section 4.1.3.

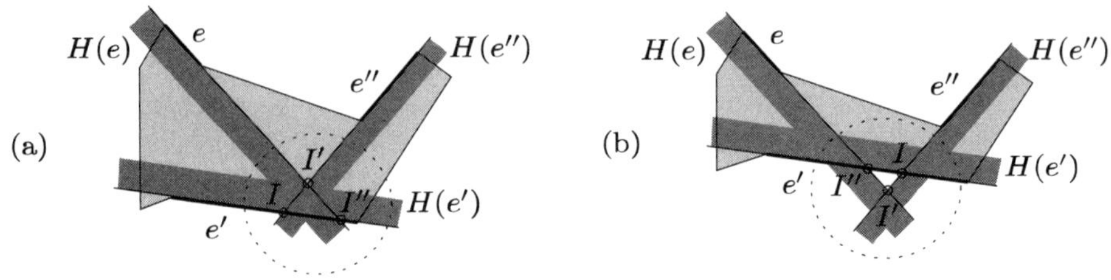

FIGURE 2. The edges e, e', e'' in (a) are a triangular triple, while those in (b) are not.

Rimon and Burdick [**RB98a, RB98b**] analyzed the mobility of a part in the configuration space of that part. Each contact is regarded as an obstacle that limits the part's ability to move. The set of all placements in which the part intersects that obstacle defines a forbidden region in the configuration space. If the forbidden regions induced by all contacts separate the current placement from any other non-forbidden point in configuration space, or, in other words, if the current placement consitutes a single-point cell of the non-forbidden portion of the configuration space, then the part is (second-order) immobilized: there exists no collision-free motion that moves the part out of its current placement. The shapes of a contact and the part determine the shape of the corresponding forbidden region in configuration space; it is therefore clear that the shapes of the part and contacts play a role in whether or not a part is immobilized according to Rimon and Burdick's notion of second-order immobility. The conditions for immobilization formulated by Rimon and Burdick [**RB98a, RB98b**] and Czyzowicz et al. [**CSU99**] are both necessary and sufficient.

Related to immobilization is the notion of caging, defined by Rimon and Blake [**RB99**]. Rather than really isolating the current placement of the part, the forbidden regions in configuration space enclose that placement, preventing the existence of a path to infinity. Informally speaking, a part is caged if it cannot escape the contacts, and move to a location outside the proximity of the contacts. Gopalakrishnan and Goldberg [**GG04**] built on second-order immobility and caging to analyze the mobility of a deformable object held by contacts. Their finite-element-based approach analyzes the mobility in the configuration space of the vertices of the triangulated object; the dimension of this space is proportional to number of vertices in the triangulation.

3. Existence

3.1. General setting.

3.1.1. *Rigid parts.* Reuleaux [**Reu76**] showed that at least four contacts are necessary for form closure in the plane. This can be readily understood from both form closure analyses outlined in the previous section: four half-plane constraints are necessary to exclude all positive and negative rotation centers, and four wrenches are necessary to put the origin strictly inside their convex hull. Markenscoff et al. [**MNP90**] showed that four contacts also suffice for any piecewise-smooth planar part, excluding circles.

Seven contacts are necessary to put any three-dimensional object in form closure [**Lak78, Som00**], since seven wrenches are necessary to put the origin of the six-dimensional wrench space in the interior of their convex hull. Markenscoff et al. [**MNP90**] and Mishra et al. [**MSS87**] established that twelve contacts are sufficient to hold any three-dimensional object in form closure, except for objects with rotational symmetries. Markenscoff et al. [**MNP90**] and Meyer [**Mey93**] proved that seven contact suffice for polyhedral objects. In the case of friction between the contacts and the object, any object allows for force closure with three contacts in 2D and four contacts in 3D. Hong et al. [**HLMT90**] showed that force-closure immobilizations with two contacts exist for smooth objects.

Wentink et al. [**WvdSO97**] (see also [**Wen98**]) studied the existence of form-closure immobilizations involving linear contacts. They showed that any polygon that has no edges parallel to edges of its convex hull can be held in form closure by one line and two points, and by two lines and one point. Moreover, two orthogonal lines and one point can put any convex polygon in form clsoure.

Czyzowicz et al. [**CSU99**] proved that any polygon without parallel edges can be immobilized by three frictionless points. This result generalizes to curved objects whose largest enclosed circle intersects the object boundary in three points of which no two are antipodal on that circle. The authors also identified a class of parts with parallel edges that require four frictionless points to be immobilized.

Rimon and Burdick [**RB95**] used their second-order immobility analysis to show that three frictionless convex fingers with sufficiently flat curvature can immobilize generic piecewise-smooth planar objects. Rimon [**Rim01**] extended these results to 3D by proving that four frictionless convex fingers with sufficiently flat curvature can immobilize generic piecewise-smooth 3D objects.

Czyzowicz et al. [**CSU99**] showed that $2d$ points are sufficient but may also be necessary to immobilize any d-dimensional polytope. If the normals to $d+1$ faces of the polytope are linearly independent, then $d+1$ points suffice.

3.1.2. *Non-rigid parts.* All aforementioned results on necessary and sufficient numbers of contacts for immobilization deal with rigid parts. As one of the first steps in the direction of immobilizing non-rigid parts, such as complex structures of interconnected rigid bodies and of deformable objects, we study the immobilization of a serial chain of n polygons connected by rotational joints, or hinges. Besides the recent work by Gopalakrishnan and Goldberg [**GG04**] on the analysis of the mobility of deformable parts, very little has been done along this line. A notable paper is that by Baraff et al. [**BMK97**] who show that the problem of determining the smallest set of contacts sufficient to stabilize a collection of contacting rigid bodies is NP-hard.

Let $(\mathcal{B}_1, ..., \mathcal{B}_n)$ be a serial chain of n hinged closed polygons. The hinge v_i connects the adjacent polygons $\mathcal{B}_i$ and $\mathcal{B}_{i+1}$ and is assumed to be a vertex of both these polygons. It allows the adjacent polygons $\mathcal{B}_i$ and $\mathcal{B}_{i+1}$ to rotate relative to each other. We will denote by $\mathcal{E}_i$ a largest enclosed circle of $\mathcal{B}_i$.

Our aim is to identify how many frictionless point contacts suffice to immobilize any serial chain of n hinged polygons in a priorly specified placement. For practical reasons it is undesirable to place point contacts at convex vertices since that may lead to damage to the part (see e.g. [**BG96**]), and the process of establishing such a contact is inherently unstable. As a result, we forbid the placement of contacts at convex vertices. Note that it is therefore impossible to have a contact at a hinge. Our intuitive analysis of possible motions of serial chains resembles that of

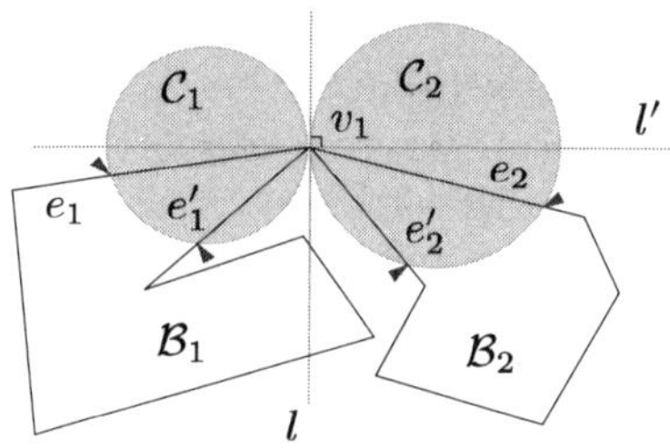

(a)

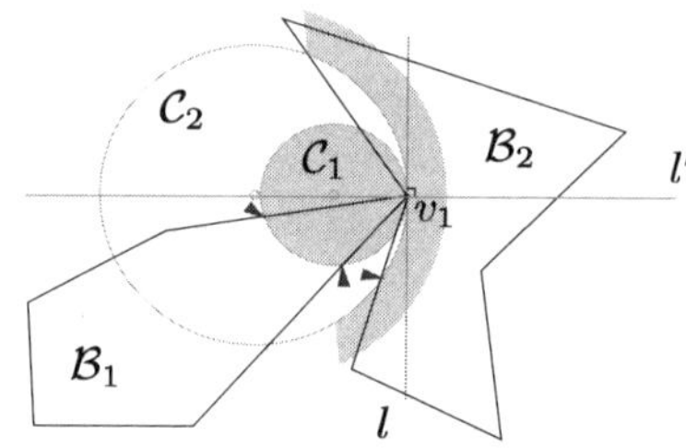

(b)

FIGURE 3. Immobilization of a two-polygon chain $(\mathcal{B}_1, \mathcal{B}_2)$ with four contacts when the hinge v_1 is (a) a convex vertex of $\mathcal{B}_2$, and (b) a concave vertex of $\mathcal{B}_2$.

Czyzowicz et al. [**CSU99**] in that it also takes place in the two-dimensional space of the object itself, and it also takes into account the curvature of the motions.

We first discuss the immobilization of serial chains of with few polygons. The results serve as building blocks for the immobilization of longer chains. We recall that three points are sufficient to immobilize any polygon without parallel edges, and that four points may be necessary but are also sufficient to immobilize polygons with parallel edges.

The construction of an immobilization of two hinged polygons relies on an observation about the possible motions of the common endpoint v of two edges of a polygon due to point contacts at interior points p and p' on these edges. If C is the circle through v, p, and p' then the point contacts prevent v from moving outside C if v is convex inside C if v is concave.

We are now ready to immobilize the two-polygon chain $(\mathcal{B}_1, \mathcal{B}_2)$. Let e_1 and e'_1, and e_2 and e'_2 be the edges incident to v_1 of $\mathcal{B}_1$ and $\mathcal{B}_2$ respectively. The hinge v_1 must be a convex vertex of $\mathcal{B}_1$ or $\mathcal{B}_2$; we assume without loss of generality that v_1 is a convex vertex of $\mathcal{B}_1$.

- If v_1 is a convex vertex of $\mathcal{B}_2$ (as in Figure 3a) then there exists a line l through v_1 that strictly separates e_1 and e'_1 from e_2 and e'_2. Let l' be the line through v_1 perpendicular to l. Let $\mathcal{C}_1$ be a circle centered on l', through v_1, and intersecting the interiors of both e_1 and e'_1. Let $\mathcal{C}_2$ be a similar circle intersecting the interiors of both e_2 and e'_2.
- If v_1 is a concave vertex of $\mathcal{B}_2$ (as in Figure 3b) then there exists a line l through v_1 that leaves e_1, e'_1, e_2, and e'_2 strictly on one side. Let l' be the line through v_1 perpendicular to l. Let $\mathcal{C}_2$ be a circle centered on l', through v_1, and intersecting the interiors of both e_2 and e'_2. Let $\mathcal{C}_1$ be a strictly smaller circle centered on l', through v_1, and intersecting the interiors of both e_2 and e'_2.

In both cases, we place point contacts at the intersections p_1 and p'_1 of $\mathcal{C}_1$ with e_1 and e'_1, and at the intersections p_2 and p'_2 of $\mathcal{C}_2$ with e_2 and e'_2. The four contacts at p_1, p'_1, p_2, and p'_2 immobilize $(\mathcal{B}_1, \mathcal{B}_2)$ because of our earlier observation.

We have not succeeded in finding a generic way of immobilizing three-polygon chains $(\mathcal{B}_1, \mathcal{B}_2, \mathcal{B}_3)$ with five frictionless contacts. Although this would mark a striking exception, we therefore conjecture that there exist chains of three hinged polygons that cannot be immobilized with five frictionless point contacts. It is

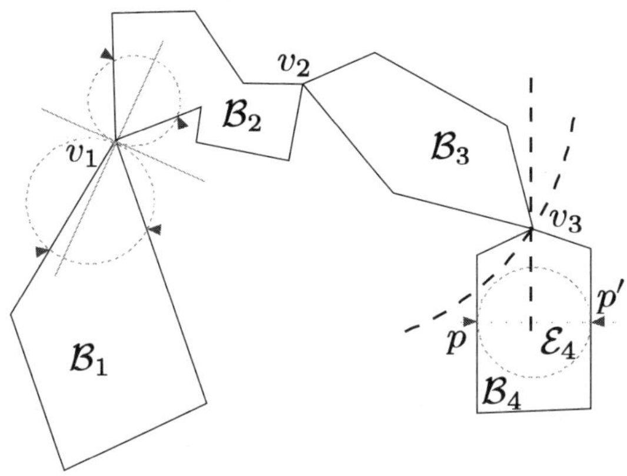

FIGURE 4. Immobilization of a four-polygon chain $(\mathcal{B}_1, \mathcal{B}_2, \mathcal{B}_3, \mathcal{B}_4)$ with six contacts when the largest enclosed circle $\mathcal{E}_4$ intersects $\mathcal{B}_4$ at two points that are antipodal on $\mathcal{B}_4$.

straightforward to immobilize any chain $(\mathcal{B}_1, \mathcal{B}_2, \mathcal{B}_3)$ with six point contacts. One simply uses four contacts to immobilize the subchain $(\mathcal{B}_1, \mathcal{B}_2)$, and then add two contacts on the edges of $\mathcal{B}_3$ incident to v_2 to immobilize $\mathcal{B}_3$ as well.

We concentrate on immobilization of the chain $(\mathcal{B}_1, \mathcal{B}_2, \mathcal{B}_3, \mathcal{B}_4)$. If the largest enclosed circles $\mathcal{E}_1$ and $\mathcal{E}_4$ intersect the boundaries of $\mathcal{B}_1$ and $\mathcal{B}_4$ respectively at three points of which no two are antipodal, then we immobilize $\mathcal{B}_1$ and $\mathcal{B}_4$ with three contacts each. Alternatively, let us assume without loss of generality that $\mathcal{E}_4$ intersects the boundary of $\mathcal{B}_4$ at two antipodal points p and p'. In that case we immobilize $(\mathcal{B}_1, \mathcal{B}_2, \mathcal{B}_3, \mathcal{B}_4)$ by immobilizing the subchain $(\mathcal{B}_1, \mathcal{B}_2)$ with four contacts and placing two additional contacts at p and p' (see Figure 4). The immobility of $(\mathcal{B}_1, \mathcal{B}_2)$ constrains v_3 to a circular arc while the contacts at p and p' simultaneously constrain it to a line segment: the chain is immobilized.

The immobilization of long chains relies on the following lemma. It follows easily from the fact that the motion of the hinge v_{i+1} is simultaneously constrained to two circular arcs that locally have only one point in common—see [**CvdSG$^+$04**] for details.

LEMMA 3.1. *A chain $(\mathcal{B}_1, ..., \mathcal{B}_n)$ with $n \geq 4$ is immobilized if, for some $1 \leq i \leq n-3$, the subchains $(\mathcal{B}_1, ..., \mathcal{B}_i)$ and $(\mathcal{B}_{i+3}, ..., \mathcal{B}_n)$ are both immobilized.*

The lemma allows to immobilize a chain $(\mathcal{B}_1, ..., \mathcal{B}_7)$ without parallel edges with nine points by immobilizing $\mathcal{B}_1$, $\mathcal{B}_4$, and $\mathcal{B}_7$, with three points each.

Now we have all the ingredients to immobilize a chain of polygons without parallel edges of arbitrary length $n \neq 3$. We proceed by repeatedly considering the trailing four polygons, until either one, two, four, or seven polygons remain. Of each such trailing quadruple we jointly immobilize the last two using the scheme for chains of two polygons. After immobilizing the final one, two, four, or seven polygons using the appropriate scheme, and repeatedly applying Lemma 3.1 we obtain the following result.

THEOREM 3.2. *A serial chain of $n \neq 3$ hinged polygons without parallel edges can be immobilized with $n+2$ frictionless point contacts. Six such contacts can immobilize three polygons without parallel edges.*

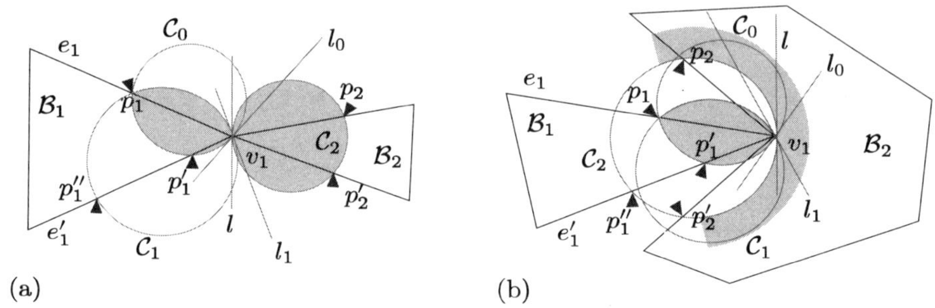

FIGURE 5. Robust immobilization of a two-polygon chain $(\mathcal{B}_1, \mathcal{B}_2)$ with five contacts when the hinge v_1 is (a) a convex vertex of $\mathcal{B}_2$, and (b) a concave vertex of $\mathcal{B}_2$.

If the polygons in the chain can have parallel edges, we can no longer be sure that a single polygon can be immobilized by three points. As a consequence, the nine-point immobilization of a chain of seven polygon also becomes invalid. To handle chains of arbitrary length n, we proceed again by considering trailing quadruples, but now continue until at most four polygons remain. If one or two remain we use four points; if three or four remain we use six points. We get the following result.

THEOREM 3.3. *A serial chain of n hinged polygons with parallel edges can be immobilized with $n+2$ frictionless point contacts if n is even, and with $n+3$ frictionless point contacts if n is odd.*

An essential difference between three contacts providing second-order immobility (or immobility according to Czyzowicz et al.) and four contacts providing form closure of a rigid part lies in their sensitivity to slight perturbations. In the former case small perturbations of the contacts—caused e.g. by misplacement of the contacts along the part boundary—are highly unlikely to retain immobility. In the latter case, immobility is definitely maintained for sufficiently small perturbations. Our immobilizations with $n+2$ point contacts behave consistently: small perturbations of the contacts destroy the immobility. This has motivated us to also investigate the number of point contacts that is sufficient to obtain a more robust form of immobilization, which has the property—like form closure—that any contact can be perturbed slightly along the edges without destroying the immobility.

The construction of a robust immobilization of a serial chain $(\mathcal{B}_1, \mathcal{B}_2)$ departs from the immobilization of that chain. Once more we assume that v_1 is a convex vertex of $\mathcal{B}_1$. Let e_1, e_1', e_2, e_2', l, $\mathcal{C}_2$, p_2, p_2' be as before, and recall that the edges e_1 and e_1' lie on the same side of l. Recall also that $\mathcal{C}_2$ is tangent to l at v_1. We rotate l by a clockwise angle around v_1 that is sufficiently small to keep e_1 and e_1' on the same side. Let l_0 be the resulting rotated copy of l. Similarly, we rotate l by a counterclockwise angle to obtain a line l_1. Let l_0' and l_1' be the lines through v_1 perpendicular to l_0 and l_1. We construct circles $\mathcal{C}_0$ and $\mathcal{C}_1$ centered on l_0' and l_1' respectively that pass through v_1, through the same point in the interior of e_1, and through (different points in) the interior of e_1'. (See Figure 5a and 5b for the circles in the case that v_1 is a convex and concave vertex of $\mathcal{B}_2$ respectively.) The intersection of $\mathcal{C}_0 \cap \mathcal{C}_1$ with the interior of $\mathcal{C}_2$—if v_1 is a convex vertex of $\mathcal{B}_2$—or the exterior of $\mathcal{C}_2$—if v_1 is a concave vertex of $\mathcal{B}_2$—equals the current location of v_1.

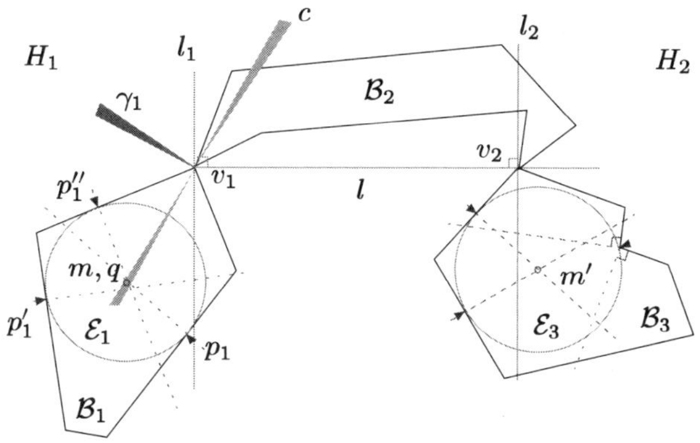

FIGURE 6. Robust immobilization of a three-polygon chain $(\mathcal{B}_1, \mathcal{B}_2, \mathcal{B}_3)$ with six contacts in the case that $\mathcal{E}_3$ intersects a vertex of $\mathcal{B}_3$.

We place point contacts at the common intersection p_1 of $\mathcal{C}_0$ and $\mathcal{C}_1$ with e_1, at the distinct intersections p_1' and p_1'' of $\mathcal{C}_0$ and $\mathcal{C}_1$ with e_1', and at the intersections p_2 and p_2' of $\mathcal{C}_2$ with e_2 and e_2'. See Figure 5. The five contacts at p_1, p_1', p_1'', p_2, and p_2' robustly immobilize $(\mathcal{B}_1, \mathcal{B}_2)$, as small perturbations maintain the topology of the intersections of $\mathcal{C}_0$, $\mathcal{C}_1$, and $\mathcal{C}_2$, and hence maintain the immobility.

We move on to consider the robust immobilization of a chain $(\mathcal{B}_1, \mathcal{B}_2, \mathcal{B}_3)$ whose polygons have no parallel edges. Let m and m' be the centers of the largest enclosed circles $\mathcal{E}_1$ and $\mathcal{E}_3$ respectively. Let l be the line through the hinges v_1 and v_2. For ease of discussion we assume without loss of generality that l is horizontal and v_1 lies to the left of v_2.

We focus on $\mathcal{B}_1$. If $\mathcal{E}_1$ intersects at least one vertex of $\mathcal{B}_1$ then we can robustly immobilize $\mathcal{B}_1$ with three contacts at the intersection points p_1, p_1', and p_1''. Alternatively, the circle $\mathcal{E}_1$ intersects the boundary of $\mathcal{B}_1$ at three points p_1, p_1', and p_1'' in the interior of edges. Now we apply the ideas from Reuleax' graphical analysis from Section 2. We can perturb p_1, p_1', and p_1'' to create a small triangle of positive rotation centers if the intersection of the orthogonals $l(p_1)$, $l(p_1')$, and $l(p_1'')$ lies below l and negative rotation centers if that intersection lies above l (see Figure 6).

We treat $\mathcal{B}_3$ similarly to $\mathcal{B}_1$. The single difference lies in the perturbation to be applied if $\mathcal{E}_3$ intersects the boundary of $\mathcal{B}_3$ at three interior points p_3, p_3', and p_3''. It must be such that the triangles consist of opposite rotation centers.

We place contacts at the resulting points p_1, p_1', p_1'', p_3, p_3', and p_3''. A careful analysis shows based on the half-plane constraints from Section 2 that any motion of $(\mathcal{B}_1, \mathcal{B}_2, \mathcal{B}_3)$ will move v_1 or v_2 outside the slab bounded by the lines l_1 and l_2 perpendicular to l and through the initial locations of v_1 and v_2 (see Figure 6). The rigidity of $\mathcal{B}_2$ prevents such a motion (see [**CvdSG$^+$04**] for the details). The immobilization is robust because the perturbation has been arbitrary to some extent.

As the two final cases we robustly immobilize a chain $(\mathcal{B}_1, \mathcal{B}_2, \mathcal{B}_3, \mathcal{B}_4)$ with eight points by robustly immobilizing $\mathcal{B}_1$ and $\mathcal{B}_4$, using four points on each, and a chain $(\mathcal{B}_1, \mathcal{B}_2, \mathcal{B}_3, \mathcal{B}_4, \mathcal{B}_5)$ with nine points by robustly immobilizing $\mathcal{B}_1$ with four points, and the subchain $(\mathcal{B}_4, \mathcal{B}_5)$ with five points.

Now we have all the ingredients to robustly immobilize arbitrary-length chains of polygons without parallel edges. We proceed by repeatedly considering the trailing five polygons. Of each such trailing sequence of five we robustly immobilize the last three using the scheme for chains of three polygons. After immobilizing the remaining polygons using the appropriate scheme, and repeatedly applying Lemma 3.1 we obtain the following result.

THEOREM 3.4. *A serial chain of n hinged polygons without parallel edges can be immobilized robustly with $\lceil 6(n+2)/5 \rceil$ contacts.*

If the polygons can have parallel edges we can no longer use the six-point immobilization of $(\mathcal{B}_1, \mathcal{B}_2, \mathcal{B}_3)$. We obtain an alternative robust immobilization of the chain $(\mathcal{B}_1, \mathcal{B}_2, \mathcal{B}_3)$ with seven points by robustly immobilizing the subchain $(\mathcal{B}_1, \mathcal{B}_2)$ with five contacts and placing two additional contacts along the interiors of the edges e_3 and e_3' of $\mathcal{B}_3$ incident to v_2.

We deal with chains of arbitrary length by repeatedly considering trailing quadruples again. Of each quadruple we robustly immobilize the last two using the scheme for two polygons. We arrive at our final result.

THEOREM 3.5. *A serial chain of n hinged polygons with parallel edges can be immobilized robustly with $\lceil 5(n+2)/4 \rceil$ contacts.*

3.2. Modular settings. Modular systems offer the advantage of reusability and rapid configurability at the price of a limited set of contacts and less freedom to place these contacts. The systems typically include a square lattice of tightly toleranced holes to which locators and clamps can be attached. Locators are rigid cylinders while clamps can extend along the directions of the principal axes of the lattice. Assuming unit distance between neighboring holes, the holes essentially limit the contact provided by a locator to $\mathbb{Z} \times \mathbb{Z}$ and by a clamp to $(\mathbb{Z} \times \mathbb{R}) \cup (\mathbb{R} \times \mathbb{Z})$.

Zhuang and Goldberg [**ZG96**] showed that for any D there exist polygons with diameter larger than D that cannot be put in form closure by three locators and a clamp (on a unit-resolution lattice). They did so by showing that there always exists a large polygon that can never simultaneously touch three lattice holes. Mishra [**Mis91**] showed that any rectilinear polygon whose edges all have length at least four can be held in form closure by six clamps. Zhuang and Goldberg [**ZG96**] proved that any rectilinear polygon with edges of at least unit length can be put in form closure by four clamps. Van der Stappen [**vdS00**], finally, showed that any polygon, regardless of its size, can be held by two pairs of points on two perpendicular lines. This implies that there always exist contact locations in $(\mathbb{Z} \times \mathbb{R}) \cup (\mathbb{R} \times \mathbb{Z})$ that provide form closure to a polygon.

4. Synthesis

4.1. General setting.

4.1.1. *Known results.* Several papers on the existence of immobilizations also give algorithms for synthesizing a single immobilization. Mishra et al. [**MSS87**] gave linear-time algorithms to compute form-closure immobilizations of planar objects using six points and of 3D objects using twelve points. Markenscoff et al. [**MNP90**] also presented linear algorithms for determining form-closure immobilizations. The output-sensitive computation of all form-closure immobilizations with at most four points is discussed in Section 4.1.2. Meyer [**Mey93**] developed an algorithm with a worst-case running time of $O(n^{3/2}\sqrt{\log n})$ for synthesizing a form-closure immobilization of a convex polyhedron with n vertices.

Czyzowicz et al. [**CSU99**] gave an algorithm for computing three locations for an n-vertex polygon without parallel edges such that point contacts at these locations immobilize the polygon according to their necessary and sufficient conditions. The algorithm finds these locations by searching for the largest enclosed circle of the polygon, which intersects the part boundary in three points with the desired properties. The computation, via the medial axis, can be accomplished in $O(n)$ time [**CSW99**]. The output-sensitive computation of all immobilizations with at most three points satisfying the conditions by Czyzowicz et al. is discussed in Section 4.1.3.

Nguyen [**Ngu88**] described an approach for finding maximal regions on a polygonal and polyhedral parts where contacts can be placed independently while achieving form or force closure. His work dealt with form-closure immobilizations with four contacts in 2D and seven contacts in 3D, and with force-closure immobilizations with two contacts in 2D. The running times of the algorithms is roughly $O(n^c)$, where n is the complexity of part, and c is the number of contacts. The algorithm by Van der Stappen et al. [**vdSWO00**] is capable of reporting the regions admitting independent placement of four contacts providing form closure of a polygon in time $O(n^{2+\varepsilon} + K)$, where K is the number of quadruples of regions.

Several other papers considered the computation of two-contact force-closure immobilizations of planar parts. Chen and Burdick [**CB93**] solved the problem of finding two-contact force-closure immobilizations in the two-dimensional space of all paramterizations of placements of two contacts. The immobilizations correspond to criticalities of an energy function depending on the distance between the contacts. Ponce et al. [**PSF93**] used a cell-decomposition approach to identify pairs of regions—in the spirit of Nguyen's work [**Ngu88**]—on piecewise-smooth objects such that any placement of two points along these regions put the part in force-closure. The approach by Jia [**Jia02**] dissects the boundary of a smooth object into convex and concave pieces and then applies numerical bisection and mraching to find the two-contact immobilizations. The running time of the algorithm has a quadratic dependence on the number of pieces.

Ponce and Faverjon [**PF95**] gave a method for synthesizing triples of regions on a polygon such that any placement of three points along these regions results in force closure. The sufficient conditions for force closure employed by the authors reduces the problem of computing three-contact force-closure to determining the three-dimensional projection of a five-dimensional polytope. The problem of computing frictionless three-contacts immobilizations satisfying Rimon and Burdick's second-order immobility condition [**RB98a, RB98b**] was tackled by Ponce et al. [**PBR95**]. They presented an algorithm for synthesizing all such immobilizations of a part bounded by polynomial splines. The algorithm uses exact cell decomposition to build an explicit description of the set of immobilizations.

Ponce et al. [**PSS$^+$97**] extended the work by Ponce and Faverjon [**PF95**] to quadruples of independent regions on a polyhedron. Their new force closure conditions lead to an algorithm for a specific class of immobilizations. Here the problem is reduced to finding the eight-dimensional projection of an eleven-dimensional polytope.

A substantial body of work, including papers by Ferrari and Canny [**FC92**], Teichmann and Mishra [**TM00**], Markenscoff and Papadimitriou [**MP89**], Mirtich and Canny [**MC94**], and Kirkpatrick et al. [**KMY92**] consider the computation or approximation of the single optimal immobilization according to some quality criterion. Examples of such criteria are the range of the external wrenches that can be balanced by the wrenches induced by the contacts [**FC92**], or the magnitude of the worst-case forces needed to balance any unit force acting on the part's center of gravity [**MP89**]. The problem often leads to determining the immobilization that maximizes the volume of some enclosed shape, e.g., a ball [**KMY92**], of the convex hull or Minkowski sum of the wrenches. Mishra [**Mis94**] provided an overview of results in this direction.

4.1.2. *Output-sensitive computation of form-closure immobilizations.* We consider the problem of computing all form-closure immobilizations of a polygonal part. The availability of *all* grasps of a certain part allows a user—usually a machinist—to select the grasps that best meet specific additional requirements, such as accessibility, which may vary from one operation to another. As the computation of all grasps along a given combination of edges and vertices can be accomplished in constant time [**Ngu88, vdSWO00**], the algorithmic challenge is to efficiently report all combinations of edges and vertices that yield at least one grasp. Since placing a contact at a convex vertex may damage the part, convex vertices are not generally accepted as contact positions, so we restrict the search to combinations of edges and/or concave vertices.

An output-sensitive algorithm to compute, for a simple polygon with n edges, all the edge combinations that have a form-closure grasp with four frictionless contacts was presented by van der Stappen et al. [**vdSWO00**]. The algorithm runs in $O(n^{2+\varepsilon} + K)$ time, where K is the number of edge quadruples reported.

Fewer than four point contacts may suffice for form closure if the object has concave vertices. Informally speaking, such vertices allow to have two contacts at the price of one, as a contact at a concave vertex can be regarded as lying on both incident edges or arcs. Form-closure grasps involving concave vertices were first studied by Gopalakrishnan and Goldberg [**GG02**], who gave an $O(m^2)$ time algorithm to find all K concave vertex pairs that allow a two-contact form-closure grasp, where m is the number of concave vertices of the polygon. We improve this to $O(m^{4/3} \log^{1/3} m + K)$. We also give output-sensitive algorithms for reporting all combinations of one concave vertex and two edges, of two concave vertices and one edge, and of three concave vertices admitting form closure grasps.

We concentrate on pairs of concave vertices first. In the following, we define the segment $\overline{pq}$ to be the *relatively open* segment connecting p and q, that is, the set $\overline{pq} := \{\lambda p + (1-\lambda)q \mid 0 < \lambda < 1\}$. We need a simple geometric lemma.

LEMMA 4.1. *Let w_1, w_2, w_3, w_4 be four points in $\mathbb{R}^3$. The origin O lies in the interior of the convex hull of $w_1, \ldots, w_4$ if and only if there are points $p_{12} \in \overline{w_1 w_2}$ and $p_{34} \in \overline{w_3 w_4}$ such that $O \in \overline{p_{12} p_{34}}$.*

We now build a *screen* Γ in wrench space. In wrench space, the horizontal dimensions x and y represent the direction of the force applied by a wrench, while the vertical dimension z represents the torque that is caused by a wrench. The screen Γ, shown in Figure 8, consists of two vertically infinite slabs, namely $\Gamma_1 := \{(x, 1, z) \mid -1 < x < 1, z \in \mathbb{R}\}$ and $\Gamma_2 := \{(-1, y, z) \mid -1 < y < 1, z \in \mathbb{R}\}$.

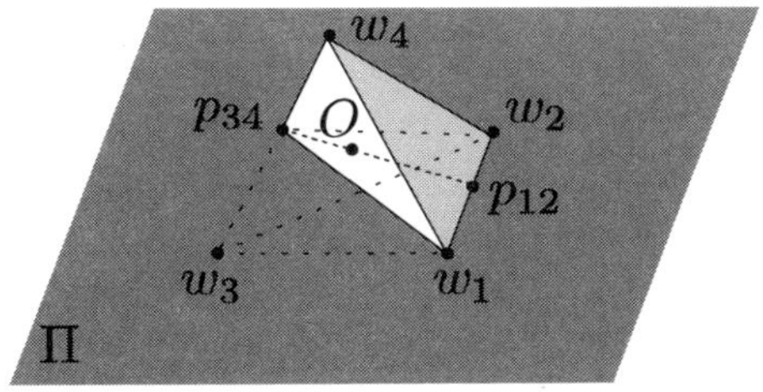

FIGURE 7. The origin O lies in the interior of the convex hull of $w_1, \ldots, w_4$ if and only if there are points $p_{12} \in \overline{w_1 w_2}$ and $p_{34} \in \overline{w_3 w_4}$ such that $O \in \overline{p_1 p_2}$.

We will project wrenches w that do not lie on the z-axis onto Γ as follows. The projection $\pi_i(w)$ of w on Γ_i is the intersection, if it exists, of Γ_i with the line through w and the origin O. If w lies between O and $\pi_i(w)$, we color $\pi_i(w)$ *black*. If O lies between w and $\pi_i(w)$, we color $\pi_i(w)$ *gray*. It is easy to see that for each wrench w, at least one of $\pi_1(w)$ and $\pi_2(w)$ exists. Note that we do not need to consider wrenches that lie on the z-axis, as the point contacts are unable to apply torque to the object without pushing it.

A segment $\overline{w_1 w_2}$ is projected onto Γ by projecting each point $w \in \overline{w_1 w_2}$. The complete projection $\pi(\overline{w_1 w_2})$ consists of at most four segments on Γ: one of each color on each part of the screen.

LEMMA 4.2. *Given an object with four contact wrenches w_1, w_2, w_3, w_4. The object is in form-closure if and only if the interior of a gray part of $\pi(\overline{w_1 w_2})$ intersects the interior of a black part of $\pi(\overline{w_3 w_4})$, or vice versa.*

We now have all the ingredients for an efficient algorithm that reports all pairs of concave vertices that allow a form-closure grasp by placing two frictionless point contacts at these vertices. We assume that the concave vertices have already been identified. For each concave vertex p, we compute the two wrenches $w_1(p)$ and $w_2(p)$ corresponding to it, and project the segment $\overline{w_1(p)w_2(p)}$ onto Γ. Let $s(p) := \pi(\overline{w_1(p)w_2(p)})$ be the projection. By Lemma 4.2, two concave vertices p and q have a form-closure grasp if and only if the interiors of $s(p)$ and $s(q)$ form a gray-black intersection in Γ—see Figure 8.

The family $\{s(p)\}$ consists of at most $4m$ gray and black segments in Γ, where m is the number of concave vertices of P. It remains to compute all gray-black intersections in this set, a problem that can be solved in time $O(m^{4/3} \log^{1/3} m + K)$, using the solution by Agarwal [**Aga90**] with the improvement by Chazelle [**Cha93**].

THEOREM 4.3. *Given a polygon with m concave vertices, all K form-closure grasps formed by two concave vertices can be computed in time $O(m^{4/3} \log^{1/3} m + K)$.*

Form closure may also be achieved by placing three frictionless point contacts, one at a concave vertex p, and one each on two edges e_1 and e_2. We now give an algorithm to report all such triples (p, e_1, e_2). Again, we have four wrenches: $w_1 \in \hat{e_1}$, $w_2 \in \hat{e_2}$, and the two wrenches $w_1(p)$ and $w_2(p)$ corresponding to the concave vertex p. Here, $\hat{e}$ is the set of wrenches corresponding to the possible placement of contacts on the interior of edge e. All such wrenches have a common force vector, so $\hat{e}$ is a vertical segment in wrench space.

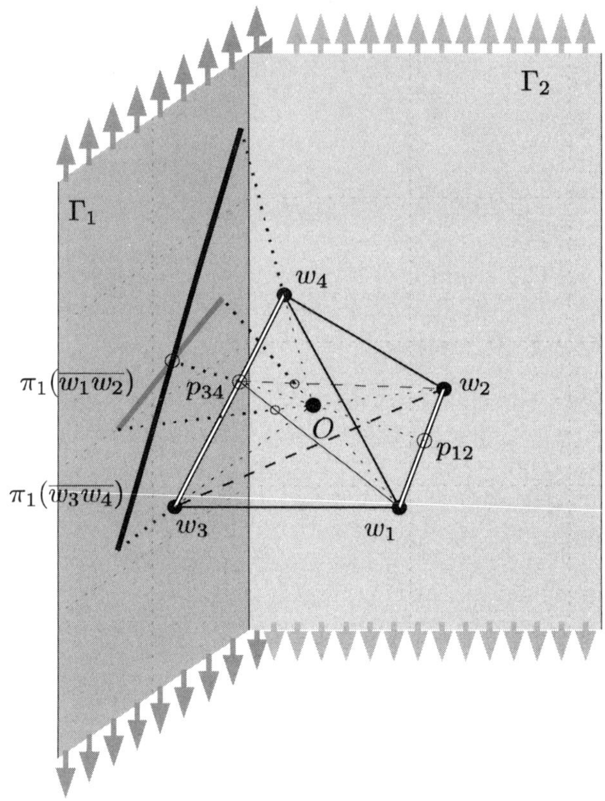

FIGURE 8. $\pi_1(\overline{w_1w_2})$ and $\pi_1(\overline{w_3w_4})$ intersect, and their colors differ. Therefore, the origin O must lie in the interior of the convex hull of $w_1, \ldots, w_4$.

Let $r(e_1, e_2) := \bigcup\{\pi(\overline{w_1w_2}) \mid w_1 \in \hat{e_1}, w_2 \in \hat{e_2}\}$. The region $r(e_1, e_2)$ is the union of at most four relatively open trapezoids with two vertical sides, where each trapezoid is either all gray or all black. For a concave vertex p, let $s(p)$ be as above. By Lemma 4.2, a triple (p, e_1, e_2) allows a form-closure grasp if and only if a black part of $s(p)$ intersects a gray trapezoid of $r(e_1, e_2)$, or vice versa. Observe that the edges of P define $O(n^2)$ trapezoids: at most four trapezoids for each pair of edges.

It remains to solve the following problem: given a set of m line segments and a set of $O(n^2)$ trapezoids, find all intersections between a line segment and a trapezoid. We observe that a segment s intersects a trapezoid r if and only if the midpoint of s lies in r, or s intersects one of the sides of r. We test the two cases separately.

For the first we are going to use a *triangle search structure*, which stores a set P of points in $\mathbb{R}^2$ and supports queries of the following form: given a query triangle Δ, report the points in $\Delta \cap P$. For the second we will use a *segment intersection structure* that stores a set S of line segments in $\mathbb{R}^2$, and supports queries of the form: given a query segment q, report all segments $s \in S$ with $s \cap q \neq \emptyset$. For both data structures, there are several alternatives. However, in this paper, we will stick to data structures based on Matoušek's hierarchial cuttings [**Mat93**], because of the good bounds on the preprocessing time.

We build, in $O(m^2 \log m)$ time, a triangle search structure on the midpoints of the segments: this permits queries with a trapezoid (by decomposing it into triangles), identifying the k points inside the trapezoid in $O(\log^3 m + k)$ time. Furthermore, we build, in $O(m^2 \log^2 m)$ time, a segment intersection structure for segment intersection queries on the segments that represent the concave vertices. Finding all k segments intersecting a given trapezoid boundary takes $O(\log^4 m + k)$ time. To find all the k segments intersecting a given trapezoid, we query both data structures, in total time $O(\log^4 m + k)$. Since there are $O(n^2)$ trapezoids, this takes time $O(n^2 \log^4 m + K)$.

THEOREM 4.4. *Given a polygon with m concave vertices and n edges, all K combinations of one concave vertex and two edges, such that one contact on the vertex and one each on the interior of each edge can put the object in form closure, can be computed in time $O(n^2 \log^4 m + K)$.*

Placing two point contacts at a pair of concave vertices p, q may not achieve form closure. Placing one more contact in the interior of an appropriate edge e, however, can achieve form closure with p and q. Here, we present an algorithm to report all such triples (p, q, e).

Consider a pair of concave vertices p, q that does not achieve form closure. Let w_1, w_2 and w_3, w_4 be the wrenches induced by p and q, respectively, and let $W := \{w_1, w_2, w_3, w_4\}$. From Section 2 we know that the origin O does not lie in the interior of the convex hull of W. An additional contact on an edge contact achieves form closure if and only if O lies in the interior of the convex hull of $W \cup \{w\}$, where w is the wrench induced by the contact.

Let $W' := W \cup \{O\}$. The convex hull of W' is a convex polytope with four or five vertices, one of which is O. Consider a facet f_i incident to O, and let H_i be the open half-space bounded by the supporting plane of f_i not containing W'. If O lies in the interior of the convex hull of $W \cup \{w\}$, for some w, then $w \in H_i$ for all i's. Conversely, if this is true for every facet incident to O, then O does lie in the interior of the convex hull of $W \cup \{w\}$.

It follows that an edge e can achieve form closure together with p and q if and only if the wrench space segment $\hat{e}$ intersects the intersection of three or four half-spaces. The bounding planes of these half-spaces pass through O, so we can again project everything onto a two-dimensional screen. Here, we do not wish to identify wrenches that are symmetric around the origin, so we use a screen $\Gamma' := \{(x, y, z) | \max(|x|, |y|) = 1, z \in \mathbb{R}\}$ enclosing O. We project the n segments $\hat{e}$ onto Γ', build a triangle search structure on their endpoints, and a segment intersection structure on the segments themselves.

For the triangle search structure we use a structure by Matoušek again; however, this time we use the variant that allows us to balance preprocessing and query time (see [**CHvdS04**] for details). This leads to a preprocessing time which is, in any case, $O(n^2 \log^\delta m)$, and a query time that is certainly $O(\log^4 n + k)$.

Before we choose the segment intersection structure, observe that all segments to be stored are vertical. A query segment $q = \overline{q_0 q_1}$, where $q_i = (x(q_i), y(q_i))$, intersects a stored segment $s = \overline{s_0 s_1}$, where $s_i = (x(s), y(s_i))$, if and only if both (i) s_0 lies above $l(q)$ while s_1 lies below $l(q)$ (or the other way around), and (ii) $x(s)$ lies between $x(q_0)$ and $x(q_1)$. Therefore, we can solve our query problem with an order-2 structure. The structure stores tuples (s_0, s_1), and stores the sets P_v associated with internal nodes in order-1 trees sorted by x-coordinate. The complete structure can be constructed in time $O(n^2)$. The query time of an order-2 structure is normally $O(\log^2 n + k)$, but in this case, we cannot just report all contents of the internal nodes found: we have to do a binary search to report only those segments with x-coordinates between $x(q_0)$ and $x(q_1)$. This increases the query time to $O(\log^3 n + k)$.

In total, both data structures are built in $O(n^2 \log^\delta m)$ time. We now consider each pair (p, q) of concave vertices in turn. We compute the wrenches W induced by the two vertices, the convex hull of $W \cup \{O\}$, and the intersection R of the three or four relevant half-spaces. We then compute $R' := R \cap \Gamma'$, a polygonal area of constant complexity. We triangulate R', and find the k segment endpoints inside R' by triangle range queries in time $O(\log^4 n + k)$. Furthermore, we find all k segments intersecting the boundary of R' in time $O(\log^3 n + k)$ by a constant number of segment intersection queries. Since there are $\Theta(m^2)$ pairs of concave vertices, the total running time is $O(n^2 \log^\delta m + m^2 \log^4 n + k)$.

THEOREM 4.5. *Given a polygon with m concave vertices and n edges, all K combinations of two concave vertices and one edge, such that one contact each on each vertex and one contact on the interior of the edge can put the object in form closure, can be computed in time $O(n^2 \log^\delta m + m^2 \log^4 n + K)$.*

It remains to consider triples of concave vertices providing form closure. Three concave vertices provide six wrenches. If the convex hull of four or five of these wrenches contains the origin, then the triple has already been reported implicitly among the triples consisting of a vertex and two edges, or of two vertices and one edge. In fact, the triples not reported by the algorithms for these two cases satisfy a very specific property, making it easy to compute them (see [**CHvdS04**] for details).

THEOREM 4.6. *Given a polygon with m concave vertices and n vertices in total, all K form-closure grasps formed by three concave vertices can be computed in time $O(m^2 \log^3 m + K)$.*

4.1.3. *Output-sensitive computation of second-order immobilizations.* We explain how to find combinations of three edges and combinations of a concave vertex and an edge, that allow a three- or two-point immobility grasp, respectively. We first introduce some notations and definitions used in this section. Let the edges of the simple polygon P be oriented counter-clockwisely around P, that is, P lies locally to the left of each edge. We denote the line orthogonal to an edge e through the start and end point of e by $s_0(e)$ and $s_1(e)$, respectively. Let $\hat{s}(e)$ be the relatively open infinite slab bounded by $s_0(e)$ and $s_1(e)$, that is, the union of all lines that are orthogonal to and intersect the interior of e (see Figure 10). Recall that $H(e)$ is the half-plane bounded by the supporting line of e lying locally to the left of e, that is, locally containing P (see again Figure 2). The following lemma summarizes the conditions by Czyzowicz et al. [**CSU99**] discussed in Section 2.

LEMMA 4.7 (Czyzowicz et al. [**CSU99**]). *There are three point contacts on the interior of three edges e, e', and e'' that immobilize a polygon if and only if:*

(i) $\hat{s}(e) \cap \hat{s}(e') \cap \hat{s}(e'') \neq \emptyset$ *(common normal intersection condition), and*
(ii) $H(e) \cap H(e') \cap H(e'')$ *is a triangle (triangular triple condition).*

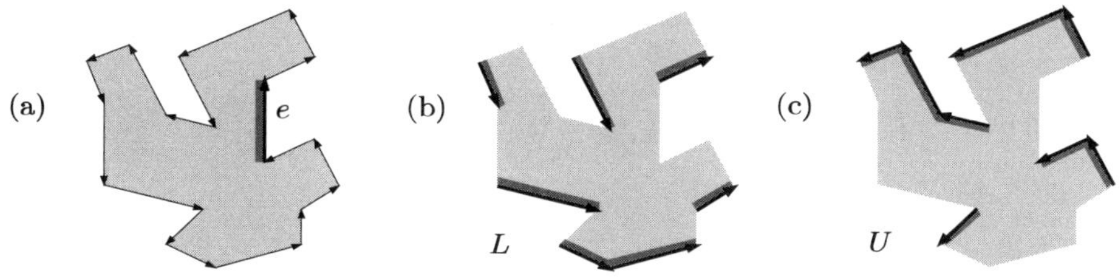

FIGURE 9. (a) A polygon with directed edges, oriented such that e is on the y-axis, pointing upward. (b) The edges in L, and (c) in U.

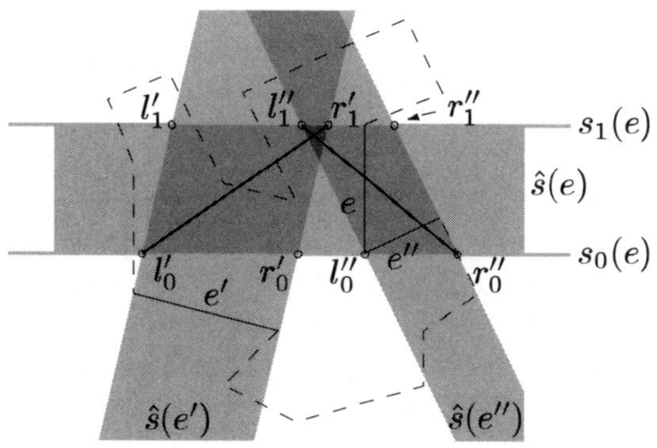

FIGURE 10. Notation for Lemma 4.8.

To find all such edge triples, we take a similar approach as in [**vdSWO00**]. We find all the edge triples that have a common normal intersection; among these, triangular triples will be filtered out. The sketch of the global approach is as follows. For each edge e of P, we build a data structure. It will be queried with each of the remaining $n-1$ edges e', to report all edges e'' such that the triple (e, e', e'') satisfies the conditions of Lemma 4.7.

From now on, we focus on building and searching the data structure for a fixed edge e. We choose a coordinate system such that $l(e)$ is the y-axis, oriented in upward direction. We divide the remaining edges into two groups L(ower) and U(pper); when an edge forms an angle between $-\frac{\pi}{2}$ and $\frac{\pi}{2}$ with the positive x-axis, it is in L, otherwise it is in U (see Fig. 9 (b) and (c)). We omit all vertical edges from L and U, since they could never make a triangular triple with e and a third edge. For $i \in \{0, 1\}$, we define l'_i and r'_i as the x-coordinates of the left and right intersection points of $s_i(e)$ and the slab boundaries of $\hat{s}(e')$. We define l''_0, r''_0, l''_1 and r''_1 for edge e'' likewise—see Figure 10. The following is a necessary and sufficient condition for three edges to have a non-empty common normal intersection region.

LEMMA 4.8. *Two slabs $\hat{s}(e')$ and $\hat{s}(e'')$ have a common intersection with $\hat{s}(e)$ if and only if one of the following is true:*

(i) $l'_0 < r''_0 \wedge l''_1 < r'_1$
(ii) $l''_0 < r'_0 \wedge l'_1 < r''_1$

Suppose $l(e')$ is the line defined by $y = a'x + b'$, and $l(e'')$ is the line defined by $y = a''x + b''$.

LEMMA 4.9. *$H(e) \cap H(e') \cap H(e'')$ is a triangle if and only if one of the following is true:*

(i) $a' < a'' \wedge b' < b'' \wedge e' \in L \wedge e'' \in U$
(ii) $a'' < a' \wedge b'' < b' \wedge e' \in U \wedge e'' \in L$

From Lemmas 4.7, 4.8 and 4.9 it follows that e, e' and e'' allow a three-point immobility grasp if and only if one of the following conditions is satisfied:

(i) $l_1'' < r_1' \wedge r_0'' > l_0' \wedge a'' > a' \wedge b'' > b' \wedge e' \in L \wedge e'' \in U$
(ii) $l_1'' < r_1' \wedge r_0'' > l_0' \wedge a'' < a' \wedge b'' < b' \wedge e' \in U \wedge e'' \in L$
(iii) $l_1' < r_1'' \wedge r_0' > l_0'' \wedge a'' > a' \wedge b'' > b' \wedge e' \in L \wedge e'' \in U$
(iv) $l_1' < r_1'' \wedge r_0' > l_0'' \wedge a'' < a' \wedge b'' < b' \wedge e' \in U \wedge e'' \in L$

Since the roles of e' and e'' are interchangeable, we only need to search for triples satisfying condition (i) or (ii). We can do this with two four-dimensional orthogonal range trees [**dBvKOS97**] as follows. In the first tree, store every edge $e'' \in U$ as a four-dimensional point (l_1'', r_0'', a'', b''). Query this tree with every edge $e' \in L$ for all points in $\langle -\infty, r_1' \rangle \times \langle l_0', \infty \rangle \times \langle a', \infty \rangle \times \langle b', \infty \rangle$. In the second tree, store every edge $e'' \in L$ as a four-dimensional point (l_1'', r_0'', a'', b''). Query this tree with every edge $e' \in U$ for all points in $\langle -\infty, r_1' \rangle \times \langle l_0', \infty \rangle \times \langle -\infty, a' \rangle \times \langle -\infty, b' \rangle$. Every edge e'' reported forms a triple with e and e' such that three point contacts on e, e' and e'' will immobilize the polygon.

A four-dimensional orthogonal range tree can be built in $O(n \log^3 n)$ time using $O(n \log^3 n)$ space, and can be queried in $O(\log^4 n + k)$ time (see Chapter 5.4 in [**dBvKOS97**]). This can be improved to $O(\log^3 n + k)$ query time, with the same building time, using fractional cascading (see Chapter 5.6 in [**dBvKOS97**]). We query each tree with $O(n)$ edges e', for a total building and query time of $O(n \log^3 n + k)$ per tree, and we do this for every edge e, so that the complete search takes $O(n^2 \log^3 n + k)$ time.

THEOREM 4.10. *Given a polygon with n edges, all K edge triples (e, e', e'') such that the polygon can be immobilized by three frictionless point fingers on the interiors of e, e' and e'', can be computed in time $O(n^2 \log^3 n + K)$.*

If we exploit concave vertices, two contacts can immobilize a simple polygon: one at a concave vertex p and the other in the interior of an edge e. When a polygon has n edges and m concave vertices, all such pairs can be reported in time $O(mn)$ by simply checking all vertex-edge pairs. We develop a more efficient specialized algorithm based on the lemma below.

First, we introduce some notations. Let e' and e'' be the edges incident to p. Let n_p' be the inward normal to e', and let n_p'' be the inward normal to e''. Let $\mathrm{Cone}(n_p', n_p'')$ be $\{\lambda' n_p' + \lambda'' n_p'' | \lambda', \lambda'' > 0\}$, that is, the set of all positive linear combinations of n_p' and n_p''. In the same way, let $\mathrm{Cone}^-(n_p', n_p'')$ be the set of all positive linear combinations of $-n_p'$ and $-n_p''$ (see Figure 11(a)). For each edge e, let n_e be the inward normal to e, and let the open simplex $S(e)$ be $\hat{s}(e) \cap H(e)$ (see Figure 11 (b).)

LEMMA 4.11. *Placing two point contacts at a concave vertex p and on an edge e immobilizes a polygon if and only if:*

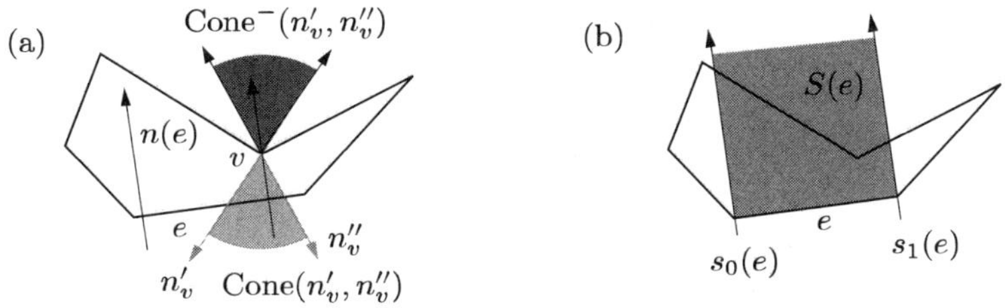

FIGURE 11. (a) Notations of normals of edges and concave vertices. (b) Vertex p is in the simplex $S(e)$ of edge e.

(i) $n_e \in Cone^-(n'_p, n''_p)$, *and*
(ii) $p \in S(e)$.

For any edge e and any concave vertex p, let θ_e, θ'_p and θ''_p be the angles that n_e, $-n'_p$ and $-n''_p$, respectively, make with the positive x-axis. Let θ'_p be the smaller angle of θ'_p and θ''_p, i.e. $\theta'_p < \theta''_p$. Observe that $n_e \in \text{Cone}^-(n'_p, n''_p)$ if and only if one of the following is true:

(i) $\theta''_p - \theta'_p < \pi \wedge \theta_e \in \langle \theta'_p, \theta''_p \rangle$
(ii) $\theta''_p - \theta'_p > \pi \wedge \theta_e \in [-\pi, \theta'_p\rangle \cup \langle \theta''_p, \pi]$.

For a given edge e, we find all concave vertices that have a two-point immobilizing grasp with e. For this, we store the concave vertices in a data structure that stores pairs of the form (I_p, p), where I_p is a one-dimensional interval and p is a point in the plane. Each vertex p with $\theta''_p - \theta'_p < \pi$ is stored once, as a pair $(\langle \theta'_p, \theta''_p \rangle, p)$. Each vertex p with $\theta''_p - \theta'_p > \pi$ is stored twice: once as a pair $(\langle -\infty, \theta'_p \rangle, p)$ and once as $(\langle \theta''_p, \infty \rangle, p)$. We query this data structure with each edge e of P, to report all vertices p stored as a pair (I_p, p) such that $\theta_e \in I_p$ and $p \in S(e)$.

The data structure we use is an order-2 data structure. The top level is a segment tree [**dBvKOS97**] on the intervals I_p. Let X be the set of all begin and end points of intervals I_p to be stored in the tree, in order of increasing value. A segment tree is a balanced binary tree on the intervals between consecutive values from X: each leaf is associated with one such interval. Each internal node v is associated with an interval $I(v)$, which is the union of the intervals of its descendants. With each node v, we associate a data structure $\mathcal{T}(v)$ that stores all pairs (I_p, p) such that I_p contains $I(v)$ but not $I(\text{parent}(v))$. In our case, the data structures $\mathcal{T}(v)$ will be triangle search structures on the points p in the pairs (I_p, p). Again, we use the triangle search structure by Matoušek [**Mat93**]. This data structure storing the m concave vertices allows us to accomplish the n queries within the time bound given in the following theorem. The reader is again referred to [**CHvdS04**] for the analysis of the structure.

THEOREM 4.12. *Given a polygon with n edges and m concave vertices, all K pairs of an edge e and a concave vertex p such that the polygon can be immobilized by two frictionless point fingers on e and at p, can be computed in time $O(n \log^4 m + (mn)^{2/3} \log^{2+\delta} m + K)$.*

4.2. Modular setting. Brost and Goldberg [**BG96**] developed an algorithm that enumerates all form-closure immobilizations of a polygonal part by three locators and one clamp on a square lattice of holes. Recall that if neighboring holes are unit distance apart, the contact provided by a locator is restricted to $\mathbb{Z}\times\mathbb{Z}$ and by a clamp to $(\mathbb{Z}\times\mathbb{R})\cup(\mathbb{R}\times\mathbb{Z})$. The running time of the algorithm is $O(n^4d^5)$, where n is the number of edges of the polygon and d is its diameter (in grid units). Wentink [**Wen98**] gave an algorithm that reports all immobilizations with four clamps in time $O(n^{2+\varepsilon}+Kd^3)$, where K is the number of quadruples of edges for which at least one form-closure immobilization exists.

Overmars et al. [**ORSW95**] and Wentink et al. [**WvdSO97**] extended the common set of locators and clamps with a linear element. They presented an output-sensitive algorithm that enumerates all form-closure immobilizations of a polygon consisting of one linear element, one locator, and one clamp in $O(n(n+p)^{\frac{4}{3}+\varepsilon}+K)$ time, where p is the perimeter of the polygon (in grid units), and K is the size of the output. Another algorithm reports in $O(n(n+p))$ time the immobilizations with two rigidly attached linear elements and a single clamp.

Wallack and Canny [**WC97**] studied a system of two tables, each containing a square lattice of equally-spaced holes equivalent to that described above. The role of the clamp is exchanged for the adjustable distance between the two tables. They described an algorithm that reports all form-closure immobilization with four locators on two tables. The running time is $O(A)$ where A is the number of simultaneous contact placements of the four locators, which is roughly $O(n^4d^6)$ in the worst case according to Wentink [**Wen98**].

Wagner et al. [**WZG95**] proposed a three-dimensional system consisting of four square lattices of holes mounted on a cubic frame in such a way that only two of the lattices are parallel. Seven bendable struts of any integer length can be attached to the holes of the lattice. The authors presented an algorithm for reporting all possible seven-strut form-closure immobilizations.

Sudsang et al. [**SPS97**] studied a device consisting of two parallel plates with locator holes facing each other. The distance between the plates is continuously adjustable and the locators have integer length. They developed an algorithm reporting all four-locator arrangements providing second-order immobility of a polyhedral part with n faces and diameter d. The algorithm runs in time $O(n^4d^8)$ in the worst case.

Wang [**Wan00**] and Wang and Pelinescu [**WP01**] focussed on 3D form-closure immobilization in a point set domain. The possible locations of six locators and one clamp are to be chosen from a discrete set of candidates. The aim of these papers is to avoid the prohibitively large cost of an exhaustive such of all combinations of seven locations. Wang [**Wan00**] used a greedy algorithm searching for a satisfactory rather than optimal solution, while Wang and Pelinescu [**WP01**] chose to improve randomly chosen combinations. Ding et al. [**DLW01**] presented heuristics for selecting fixturing surfaces on a polyhedral part.

Acknowledgements. The author thanks the co-authors of the highlighted papers: Jae-Sook Cheong, Ken Goldberg, Herman Haverkort, Mark Overmars, and Elon Rimon.

References

[Aga90] P. K. Agarwal, *Partitioning arrangements of lines II: Applications*, Discrete & Computational Geometry **5** (1990), 533–573.

[BG96] R. Brost and K. Goldberg, *A complete algorithm for designing planar fixtures using modular components*, IEEE Transactions on Robotics and Automation **12** (1996), 31–46.

[BMK97] D. Baraff, R. Mattikalli, and P. Khosla, *Minimal fixturing of frictionless assemblies: complexity and algorithms*, Algorithmica **19** (1997), 4–39.

[CB93] I.-M. Chen and J.W. Burdick, *Finding antipodal point grasps on irregularly shaped objects*, IEEE Transactions on Robotics and Automation **9** (1993), no. 4, 507–512.

[CG94] J. F. Canny and K. Goldberg, *Risc for industrial robotics: Recent results and open problems*, Proceedings of the IEEE International Conference on Robotics and Automation (ICRA), 1994, pp. 1951–1958.

[Cha93] B. Chazelle, *Cutting hyperplanes for divide-and-conquer*, Discrete & Computational Geometry **9** (1993), 145–158.

[CHvdS04] J.-S. Cheong, H.J. Haverkort, and A.F. van der Stappen, *On computing all immobilizing grasps of a simple polygon with few contacts*, Tech. report, UU-CS-2004-013, Institute of Information and Computing Sciences, Utrecht University (Preliminary version in Proceedings of the 14th Annual International Symposium on Algorithms and Computation (ISAAC), LNCS 2906, Springer Verlag, 2003), 2004.

[CSU99] J. Czyzowicz, I. Stojmenovic, and J. Urrutia, *Immobilizing a shape*, International Journal of Computational Geometry and Applications **9** (1999), 181–206.

[CSW99] F. Chin, J. Snoeyink, and C.A. Wang, *Finding the medial axis of a simple polygon in linear time*, Discrete & Computational Geometry **21** (1999), 405–420.

[CvdSG+04] J.-S. Cheong, A.F. van der Stappen, K. Goldberg, M.H. Overmars, and E. Rimon, *Immobilizing hinged polygons*, Tech. report, UU-CS-2004-004, Institute of Information and Computing Sciences, Utrecht University (Preliminary version in Proceedings of the IEEE International Conference on Robotics and Automation (ICRA), 2002), 2004.

[dBvKOS97] M. de Berg, M. van Kreveld, M. Overmars, and O. Schwarzkopf, *Computational geometry: Algorithms and applications*, Springer-Verlag, 1997.

[DLW01] D. Ding, Y.-H. Liu, and M.Y. Wang, *Automatic selection of fixturing surfaces and fixturing points for polyhedral workpieces*, IEEE Transactions on Robotics and Automation **17** (2001), 833–841.

[FC92] C. Ferrari and J. F. Canny, *Planning optimal grasps.*, Proceedings of the IEEE International Conference on Robotics and Automation (ICRA), 1992, pp. 2290–2295.

[GG02] K. Gopalakrishnan and K. Goldberg, *Gripping parts at concave vertices*, Proceedings of the IEEE International Conference on Robotics and Automation (ICRA), 2002, pp. 1590–1596.

[GG04] ______, *D-space and deform closure: A framework for holding deformable parts*, Proceedings of the IEEE International Conference on Robotics and Automation (ICRA), 2004.

[HLMT90] J. Hong, G. Lafferriere, B. Mishra, and X. Tan, *Fine manipulation with multifinger hands*, Proceedings of the IEEE International Conference on Robotics and Automation (ICRA), 1990, pp. 1568–1573.

[Jia02] Y.-B. Jia, *Curvature-based computation of antipodal grasps*, Proceedings of the IEEE International Conference on Robotics and Automation (ICRA), 2002, pp. 1571–1577.

[KMY92] D. Kirkpatrick, B. Mishra, and C.-K. Yap, *Quantitative steinitz's theorems with applications to multifingered grasping*, Discrete & Computational Geometry **7** (1992), 295–318.

[Lak78] K. Lakshminarayana, *Mechanics of form closure*, Tech. report, ASME 78-DET-32 American Society of Mechanical Engineers, 1978.

[Mas01] M. T. Mason, *Mechanics of robotic manipulation*, MIT Press, Cambridge, MA, 2001.

[Mat93] J. Matoušek, *Range searching with efficient hierarchical cuttings*, Discrete & Computational Geometry **10** (1993), 157–182.

[MC94] B. Mirtich and J. F. Canny, *Easily computable optimal grasps*, Proceedings of the IEEE International Conference on Robotics and Automation (ICRA), 1994, pp. 739–747.

[Mey93] W. Meyer, *Seven fingers allow force-torque closure grasps on any convex polyhedron*, Algorithmica **9** (1993), 278–292.

[Mis91] B. Mishra, *Workholding: analysis and planning*, Proceedings of the IEEE/RSJ International Conference on Intelligent Robots and Systems (IROS), 1991, pp. 53–57.
[Mis94] ———, *Grasp metrics: optimality and complexity*, Algorithms for Robotic Motion and Manipulation (Boston, MA) (K. Goldberg, D. Halperin, J.C. Latombe, and R. Wilson, eds.), A.K. Peters, 1994, pp. 137–166.
[MNP90] X. Markenscoff, L. Ni, and C. H. Papadimitriou, *The geometry of grasping*, International Journal of Robotics Research **9** (1990), 61–74.
[MP89] X. Markenscoff and C. H. Papadimitriou, *Optimal grip of a polygon*, International Journal of Robotics Research **8** (1989), 17–29.
[MSS87] B. Mishra, J. T. Schwartz, and M. Sharir, *On the existence and synthesis of multifinger positive grips*, Algorithmica **2** (1987), 541–558.
[Ngu88] V.-D. Nguyen, *Constructing force-closure grasps*, International Journal of Robotics Research **7** (1988), 3–16.
[Ohw81] M. S. Ohwovoriole, *An extension of screw theory*, Journal of Mechanical Design **103** (1981), 725–735.
[ORSW95] M.H. Overmars, A. Rao, O. Schwarzkopf, and C. Wentink, *Immobilizing polygons against a wall*, Proceedings of the 11th Annual ACM Symposium on Computational Geometry, 1995, pp. 29–38.
[PBR95] J. Ponce, J. Burdick, and E. Rimon, *Computing the immobilizing three-finger grasps of planar objects*, Proc. of the 1995 Workshop on Computational Kinematics, 1995, pp. 281–300.
[PF95] J. Ponce and B. Faverjon, *On computing three-finger force-closure grasps of polygonal objects*, IEEE Transactions on Robotics and Automation **11** (1995), 868–881.
[PSF93] J. Ponce, D. Stam, and B. Faverjon, *On computing force-closure grasps of curved two-dimensional objects*, International Journal of Robotics Research **12** (1993), 263–273.
[PSS+97] J. Ponce, S. Sullivan, A. Sudsang, J-D. Boissonnat, and J-P. Merlet, *On computing four-finger equilibrium and force-closure grasps of polyhedral objects*, International Journal of Robotics Research **16** (1997), 13–35.
[RB95] E. Rimon and J. W. Burdick, *New bounds on the number of frictionless fingers required to immobilize planar objects*, Journal of Robotic Systems **12** (1995), 433–451.
[RB98a] ———, *Mobility of bodies in contact—part I: A second-order mobility index for multiple-finger graps*, IEEE Transactions on Robotics and Automation **14** (1998), 696–708.
[RB98b] ———, *Mobility of bodies in contact—part II: How forces are generated by curvature effects*, IEEE Transactions on Robotics and Automation **14** (1998), 709–717.
[RB99] E. Rimon and A. Blake, *Caging planar bodies by one-parameter two fingered gripping systems*, International Journal of Robotics Research **18** (1999), 299–318.
[Reu76] F. Reuleaux, *The kinematics of machinery*, Macmillan, 1876, Republished by Dover in 1963.
[Rim01] E. Rimon, *A curvature-based bound on the number of frictionless fingers required to immobilize three-dimensional objects*, IEEE Transactions on Robotics and Automation **17** (2001), 679–697.
[Som00] P. Somov, *Ueber gebiete von schraubengeschwindigkeiten eines starren koerpers bei verschiedener zahl von stutzflachten*, Zeitschrift fuer Mathematik und Physik **45** (1900), 245–306.
[SPS97] A. Sudsang, J. Ponce, and N. Srinivasa, *Algorithms for constructing immobilizing fixtures and grasps of three-dimensional objects*, Algorithms for Robotic Motion and Manipulation (Wellesley, MA) (J.-P. Laumond and M. H. Overmars, eds.), A.K. Peters, 1997, pp. 363–380.
[TM00] M. Teichmann and B. Mishra, *Probabilistic algorithms for efficient grasping and fixturing*, Algorithmica **26** (2000), 345–363.
[vdS00] A. F. van der Stappen, *On the existence of form-closure configurations on a grid*, Proceedings of the IEEE/RSJ International Conference on Intelligent Robots and Systems (IROS), 2000, pp. 1237–1242.

[vdSWO00] A. F. van der Stappen, C. Wentink, and M. H. Overmars, *Computing immobilizing grasps of polygonal parts*, International Journal of Robotics Research **19** (2000), 467–479.

[Wan00] M.Y. Wang, *An optimum design for 3d fixture synthesis in a point set domain*, IEEE Transactions on Robotics and Automation **16** (2000), 839–846.

[WC97] A. S. Wallack and J. F. Canny, *Planning for modular and hybrid fixtures*, Algorithmica **19** (1997), 40–60.

[Wen98] C. Wentink, *Fixture planning—geometry and algorithms*, Ph.D. thesis, Department of Computer Science, Utrecht University, 1998.

[WP01] M.Y. Wang and D. Pelinescu, *Optimizing fixture layout in a point set domain*, IEEE Transactions on Robotics and Automation **17** (2001), 312–323.

[WvdSO97] C. Wentink, A. F. van der Stappen, and M. H. Overmars, *Algorithms for fixture design*, Algorithms for Robotic Motion and Manipulation (Wellesley, MA) (J.-P. Laumond and M. H. Overmars, eds.), A.K. Peters, 1997, pp. 321–346.

[WZG95] R. Wagner, Y. Zhuang, and K. Goldberg, *Fixturing parts with seven modular struts*, IEEE International Symposium on Assembly and Task Planning, 1995, pp. 133–139.

[ZG96] Y. Zhuang and K. Goldberg, *On the existence of solutions in modular fixturing*, International Journal of Robotics Research **15** (1996), 646–656.

Institute of Information and Computing Sciences, Utrecht University, PO Box 80089, 3508 TB Utrecht, The Netherlands

E-mail address: `frankst@cs.uu.nl`

DIMACS Series in Discrete Mathematics
and Theoretical Computer Science
Volume **67**, 2005

Geometric Algorithms for Layered Manufacturing

Ravi Janardan and Michiel Smid

ABSTRACT. Layered Manufacturing is a technology that allows physical prototypes of three-dimensional models to be built from their digital representation, as a stack of two-dimensional layers. One of the key problems is the choice of a suitable direction in which the model should be oriented and built, so as to minimize the number of layers, the stair-stepping effect, the volume of the support structures that are generated during the build, or the area of contact between the prototype and the support structures, or a combination of these measures. Other problems of interest include orienting the model so that one or more prescribed facets are not in contact with supports, determining a direction to fill in the individual layers, and decomposing the model into two or more submodels and building them independently so as to reduce the amount of support structures. In this survey, we give an overview of efficient geometric algorithms for these problems. The algorithms use a large variety of techniques from computational geometry, such as convex hulls, Voronoi diagrams, spherical sweep, ray-shooting, Boolean operations on polygons, and arrangements.

1. Introduction

Layered Manufacturing is a fast-growing technology with significant impact on the efficiency of the design process in a broad range of industries; see Jacobs [**Jac92**], Kai and Fai [**KF97**], and Chua *et al.* [**CLL03**]. Layered Manufacturing offers a flexible and cost-effective alternative to traditional methods used in the design phase of physical prototypes. Currently, this technology produces high-quality prototypes in a matter of hours and at low cost. The prototypes can be inspected for flaws and if necessary the design can be modified and the process repeated until the final design has reached the desired quality.

Stereolithography is a widely-used Layered Manufacturing process. The Stereolithography Apparatus consists of a vat of light-sensitive liquid resin, a platform, and a laser; see Figure 1. The input to the process is a surface triangulation of the digital model of the part to be built, in the industry-standard STL-format. This format consists of an unordered list of triangles, each specified by the coordinates of its vertices, and the unit outer-normal of each triangle. (Thus, the digital model is a polyhedron.) The model is sliced into horizontal two-dimensional layers, which are then sent over a network to the Stereolithography Apparatus. The laser traces

1991 *Mathematics Subject Classification.* Primary 68U05; Secondary 68U07.
Key words and phrases. Computational geometry, computer aided design.
The second author was supported by NSERC.

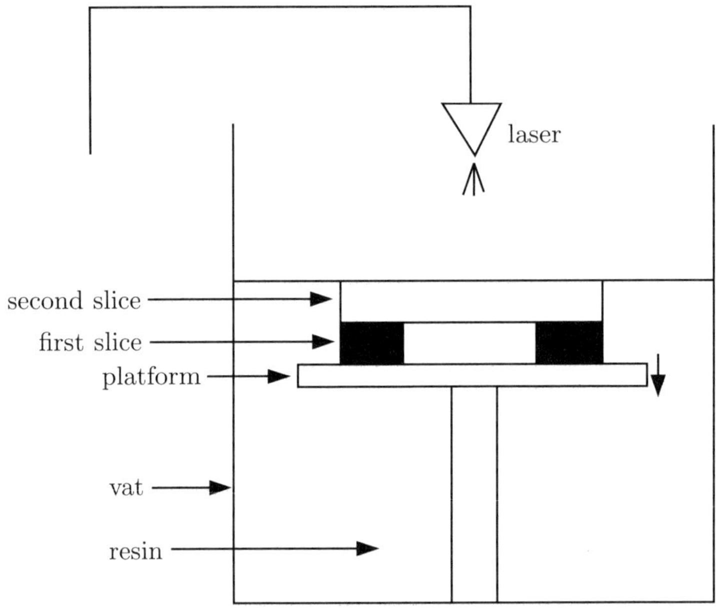

FIGURE 1. *The Stereolithography Apparatus.*

out the contour of each layer (which is a polygon) and then scans the interior in a zig-zag pattern. The exposure to the laser causes the scanned portion of the liquid to harden and form the physical layer. The platform is then lowered by an amount equal to the layer thickness (typically a few thousandths of an inch) and the next layer is then built on top of the previous one. Thus, the three-dimensional prototype is realized eventually as a vertical stack of two-dimensional layers. Ideally each new layer should rest completely on top of the previous one, so that the prototype is self-supporting during the build phase. The complex shape of real-world prototypes, however, often prevents them from being self-supporting. Therefore, during a preprocessing step, the model is analyzed and additional structures called *supports* are created and merged with the description of the model. These support structures are built simultaneously with the prototype and removed later. Figure 1 illustrates the construction of the first two layers of the model shown in Figure 2, along with the supports for the second layer.

The Layered Manufacturing process consists of three phases: preprocessing, building, and postprocessing. Preprocessing includes repairing flaws in the STL file (e.g., gaps between facets, geometric singularities, etc.), deciding upon a suitable initial orientation for the model, computing support requirements, generating and merging a description of the supports into the STL file, and slicing the model and supports. The building phase involves tracing and filling in each layer. Postprocessing includes removal of supports and improving part finish and accuracy.

1.1. Geometric issues. A key step in Layered Manufacturing is choosing an orientation for the model, i.e., the *build direction*. Its choice can impact critically the efficiency and cost of the build process as well as the surface quality of the physical prototype. Below, we mention several competing criteria that need to be

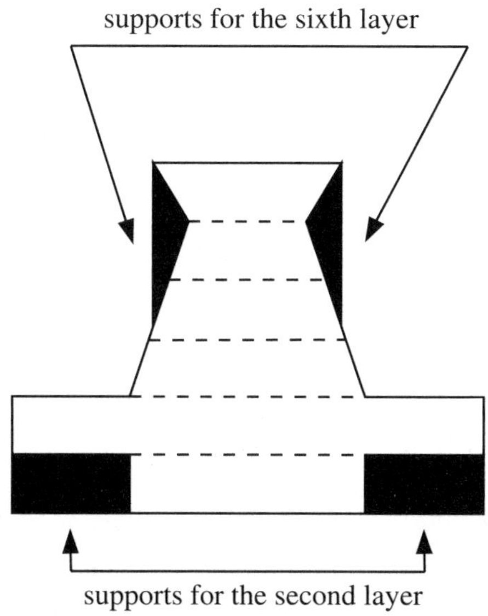

FIGURE 2. *Illustrating support structures, in 2D for convenience.*

addressed when choosing a build direction. Throughout the paper, we denote by $\mathcal{P}$ the (polyhedral) digital model to be built.

1.1.1. *Number of layers.* In the Layered Manufacturing process, the layer thickness is typically measured in thousandths of an inch. Because of this, the number of layers needed to build $\mathcal{P}$ can run into the thousands if $\mathcal{P}$ is oriented along its longest dimension. If we assume the layer thickness to be fixed, then the number of layers for a given build direction, $\mathbf{d}$, is proportional to the smallest distance between two parallel planes that are normal to $\mathbf{d}$ and that enclose $\mathcal{P}$. Hence, orienting $\mathcal{P}$ so as to minimize the number of layers is equivalent to determining a direction $\mathbf{d}$ for which the distance between the enclosing planes is minimum; this minimum distance is called the *width.*

1.1.2. *Stair-stepping effect.* Due to the non-zero layer thickness, the physical prototype will have a stair-stepped finish on each facet f that is not parallel to the build direction $\mathbf{d}$. The degree of stair-stepping on f depends on the angle between the normal of f and $\mathbf{d}$, and it can be mitigated by a suitable choice of $\mathbf{d}$. Bablani and Bagchi [**BB95**] introduce the notion of an *error-triangle* as a way of quantifying the amount of stair-stepping; see Figure 3. They argue that a good build direction is one for which the maximum height of any error-triangle (over all facets of $\mathcal{P}$) is minimized.

1.1.3. *Support structures.* If $\mathbf{d}$ is a direction and f is a facet of $\mathcal{P}$, then we say that f is a *back facet* with respect to $\mathbf{d}$, if the angle between the outer unit-normal $\mathbf{n}_f$ of f and $\mathbf{d}$ is greater than $\pi/2$. If this angle is less than $\pi/2$, then we say that f is a *front facet* with respect to $\mathbf{d}$. Finally, if this angle is equal to $\pi/2$, then we say that f is a *parallel facet* with respect to $\mathbf{d}$.

All back facets of $\mathcal{P}$ will need to be supported. Consider such a back facet f with respect to a build direction $\mathbf{d}$. The *support polyhedron* for f is the closure of the set of all points $p \in \mathbb{R}^3$ such that p is not in the interior of $\mathcal{P}$ and the ray shot

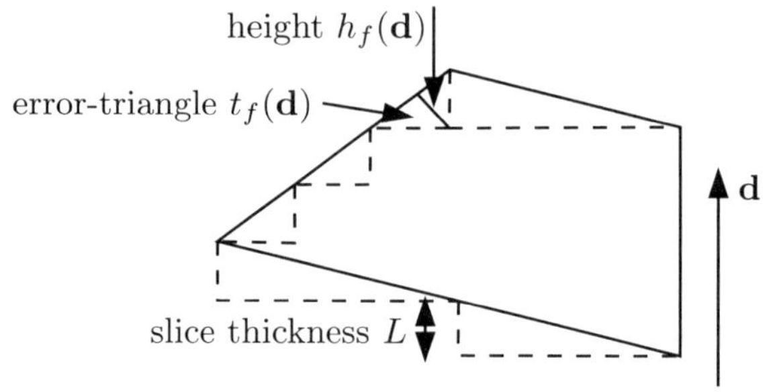

FIGURE 3. *Illustrating stair-stepping. For ease of depiction, the underlying model is assumed to be a prism of uniform cross-section perpendicular to the plane of the paper. The polygon shown is the cross-section of the prism. The model is shown in solid lines, whereas the prototype that is actually built is shown in dashed lines.*

from p in direction $\mathbf{d}$ first enters $\mathcal{P}$ through f. Informally, the support polyhedron of f is bounded from above by f, on the sides by vertical facets that "drop down" from the edges of f, and from below by the platform on which $\mathcal{P}$ rests and/or portions of front facets of $\mathcal{P}$. (Observe that if $\mathcal{P}$ is convex, then it is bounded from below by only the platform.) The *supports* of $\mathcal{P}$ with respect to a build direction is the collection of support polyhedra for all back facets; see Figure 2.

The *contact-area* is the total surface area of $\mathcal{P}$ that is in contact with supports. It includes the area of all the back facets of $\mathcal{P}$ and the areas of those portions of front facets and parallel facets that are in contact with supports.

Observe that for a convex polyhedron, the support structures are relatively simple, in that only back facets are in contact with supports and every point on a back facet is in contact with supports. Furthermore, the support structures extend all the way down to the platform. For a general polyhedron, however, the situation is more complex: In addition to back facets, some front and parallel facets can also be in contact with supports, and the latter two types of facets may be only partially in contact. In addition, supports need not extend all the way down to the platform, but may instead terminate on other parts of the polyhedron. This is illustrated in Figure 2 for the supports for the sixth layer.

A build direction that minimizes the total volume of all supports will translate into a faster build time and less waste of material. Similarly, a build direction that minimizes the contact-area will minimize the portions of the prototype that are affected during the postprocessing stage and help minimize damage to the surface of the prototype during support removal.

1.1.4. *Protecting a facet from being in contact with supports.* The removal of supports from a facet affects the surface quality and accuracy adversely, thereby impacting the functional properties of critical facets, such as, for instance, facets on gear teeth. Therefore, a good build direction is one for which one or more prescribed facets are not in contact with supports.

1.1.5. *Decomposing the model.* Traditionally, a process-planning algorithm for Layered Manufacturing works by viewing the model as a single, monolithic unit. As a result, large models that cannot be accommodated in the workspace cannot be constructed. A possible approach is to decompose the model into a small number of pieces, build each piece separately, and then glue the built pieces together to obtain the physical prototype. Besides the advantage that larger models can be built, the amount of time needed to build a model is reduced in this way. Another crucial advantage is that the support requirements of this decomposition-based approach are often much less than those of the conventional approach. For instance, if a hollow sphere is built the conventional way, supports will be needed in the interior void and below the lower hemisphere. However, if it is decomposed by a horizontal plane through the equator and built as two hemispherical shells, with the upper (resp. lower) shell built in the upward (resp. downward) direction, then supports will be needed only in the region below (resp. above) the upper (resp. lower) shell; this results in an overall reduction in both support volume and contact-area. This motivates the problem of how to decompose a model and build it so that the overall support requirements are minimized.

1.1.6. *Hatching.* The process of printing the layers is called *hatching.* This is essentially a two-dimensional problem, since each layer is a polygon (possibly with holes). The hatching process depends on the specific Layered Manufacturing process being used. In Stereolithography, the tool (i.e., the laser) traces out the boundary of the polygon and then fills in the interior by following paths along equally-spaced parallel lines, switching itself on (resp. off) when it is in the interior (resp. reaches the boundary) of the polygon, and stopping and reversing direction when it reaches the exterior face of the polygon. The path followed by the tool consists of a set of parallel line segments in the interior of the polygon, called *hatching segments*, where each segment has a fixed width determined by the tool-tip. (This width is very small compared to the segment length.) The speed of the hatching process and the life of the tool can be impacted adversely if the number of hatching segments is large. The number of hatching segments is a function of the direction (in the plane) followed by the laser; this direction is called the *hatching direction.* This motivates the problem of computing a hatching direction that minimizes the number of hatching segments.

1.2. Geometric optimization problems. The geometric issues introduced above lead to the problem of computing a build direction (or hatching direction) that optimizes one or more of the criteria. In this paper, we show that, for each of the criteria, the optimization problem can be formulated in purely geometric terms. We then show how techniques from computational geometry can be used to solve these problems efficiently. In particular, we use techniques such as convex hulls, Voronoi diagrams, spherical sweep, ray-shooting, Boolean operations on polygons, and arrangements.

1.3. Notation. Throughout this paper, we denote by $\mathcal{P}$ the polyhedral model of interest, and by n the number of its facets. We assume that the facets of $\mathcal{P}$ are triangles and that its boundary is represented in some standard form, such as, for instance, a doubly-connected edge list (see de Berg *et al.* [**dBvKOS00**]) or a winged-edge structure (see Baumgart [**Bau75**]). If necessary, such a representation can be computed easily from the standard STL representation of $\mathcal{P}$; see McMains and Séquin [**MS99**].

In the problems considered in this paper, we want to compute a build direction $\mathbf{d}$ that optimizes one or more criteria. We will represent directions as points—or unit-vectors—on the *unit-sphere* $\mathbb{S}^2$, i.e., the boundary of the three-dimensional ball centered at the origin and having radius one. The *upper hemisphere* is defined as

$$\mathbb{S}^2_+ := \mathbb{S}^2 \cap \{(x, y, z) \in \mathbb{R}^3 : z \geq 0\}.$$

Similarly, we define the *lower hemisphere* as

$$\mathbb{S}^2_- := \mathbb{S}^2 \cap \{(x, y, z) \in \mathbb{R}^3 : z \leq 0\}.$$

Finally, the *equator* is the intersection of $\mathbb{S}^2$ with the plane $z = 0$.

2. Minimizing the number of layers

Assuming that the layer thickness is fixed, the number of layers for a given build direction $\mathbf{d}$ is proportional to the smallest distance between two parallel planes that are normal to $\mathbf{d}$ and that enclose the model $\mathcal{P}$. We call this smallest distance the *width* of $\mathcal{P}$ *in direction* $\mathbf{d}$, and denote it by $W(\mathbf{d})$. Observe that $W(\mathbf{d}) = W(-\mathbf{d})$. The *width* $W(\mathcal{P})$ of the polyhedron $\mathcal{P}$ is defined as the minimum distance between any two parallel planes that enclose $\mathcal{P}$, i.e.,

$$W(\mathcal{P}) = \min\{W(\mathbf{d}) : \mathbf{d} \in \mathbb{S}^2\}.$$

We consider the problem of computing a build direction $\mathbf{d}$ for which $W(\mathbf{d}) = W(\mathcal{P})$. The problem of computing the width of $\mathcal{P}$ was considered by Houle and Toussaint [**HT88**]. We outline their approach.

First observe that the width of the polyhedron $\mathcal{P}$ is equal to the width of the convex hull $CH(\mathcal{P})$ of $\mathcal{P}$. Therefore, it suffices to consider the convex polyhedron $CH(\mathcal{P})$.

Let v be any vertex and let f be any facet of $CH(\mathcal{P})$. We say that (v, f) is an *antipodal vertex-facet pair* (or *VF*-pair), if the two parallel planes containing v and f, respectively, enclose $CH(\mathcal{P})$. Similarly, two *non-parallel* edges e_0 and e_1 of $CH(\mathcal{P})$ are said to be an *antipodal edge-edge pair* (or *EE*-pair), if the two parallel planes containing e_0 and e_1, respectively, enclose $CH(\mathcal{P})$.

Houle and Toussaint prove that any direction $\mathbf{d}$ that is not associated with some *VF*- or *EE*-pair can be rotated to a direction $\mathbf{d}'$ such that $W(\mathbf{d}') < W(\mathbf{d})$. In other words, any direction minimizing the width of $\mathcal{P}$ is perpendicular to the parallel planes associated with some *VF*- or *EE*-pair. Therefore, the width of $\mathcal{P}$ can be obtained by first computing all *VF*- and *EE*-pairs, and then for each of them computing the distance between the corresponding supporting parallel planes. The smallest distance found is equal to the width $W(\mathcal{P})$ of the polyhedron $\mathcal{P}$.

In order to compute all *VF*- and *EE*-pairs, we consider the *dual graph* G of $CH(\mathcal{P})$, which is the planar graph on the unit-sphere $\mathbb{S}^2$ that is defined as follows. The vertices of G are the outer unit-normals of the facets of $CH(\mathcal{P})$. Two of these vertices define an edge in G if the corresponding facets of $CH(\mathcal{P})$ share an edge. Edges of this dual graph are drawn as great arcs on $\mathbb{S}^2$. Observe that edges (resp. faces) of G are in one-to-one correspondence with edges (resp. vertices) of $CH(\mathcal{P})$. It turns out to be convenient to transform the graph G into a planar graph G' on $\mathbb{S}^2$, by cutting all edges that cross the equator, and "adding" the equator to it.

Let G'_u be the subgraph of G' consisting of all vertices and edges that are in the upper hemisphere $\mathbb{S}^2_+$. Let G^0_ℓ be the subgraph of G' consisting of all vertices and edges that are in the lower hemisphere $\mathbb{S}^2_-$. Finally, let G'_ℓ be the *inverse image* of G^0_ℓ, i.e., the graph obtained by mapping each vertex $\mathbf{v}$ of G^0_ℓ to the vertex $-\mathbf{v}$. Both graphs G'_u and G'_ℓ are in the upper hemisphere $\mathbb{S}^2_+$.

We now show how the graphs G'_u and G'_ℓ can be used to obtain all *VF*- and *EE*-pairs.

2.1. Computing *VF*-pairs. Consider an arbitrary *VF*-pair (v, f). Let f_v be the face of G that corresponds to v, and let $\mathbf{d}_f$ be the vertex of G that corresponds to f. There are two cases to consider. The first case is when $\mathbf{d}_f$ is on or above the equator. Then $\mathbf{d}_f$ is a vertex of G'_u. Let f_v^0 be the face of G_ℓ^0 that is contained in f_v, and let f'_v be the face of G'_ℓ that corresponds to f_v^0. Since the unique planes that support $CH(\mathcal{P})$ at v and f are parallel, vertex $\mathbf{d}_f$ of G'_u is contained in face f'_v of G'_ℓ. The second case is when $\mathbf{d}_f$ is below the equator. In this case, $\mathbf{d}_f$ is a vertex of G_ℓ^0, and $-\mathbf{d}_f$ is a vertex of G'_ℓ. Let f'_v be the face of G'_u that is contained in f_v. Then vertex $-\mathbf{d}_f$ of G'_ℓ is contained in face f'_v of G'_u. We have proved the following result.

LEMMA 2.1. *We can obtain all VF-pairs by performing a point location query with each vertex of G'_u in the graph G'_ℓ, and by performing a point location query with each vertex of G'_ℓ in the graph G'_u. In both these point location problems, all query points are known in advance.*

2.2. Computing *EE*-pairs. Consider two arbitrary edges e_0 and e_1 of $CH(\mathcal{P})$ that form an *EE*-pair. Let g_0 and g_1 be the edges of G that correspond to e_0 and e_1, respectively. We may assume without loss of generality that g_0 is (completely or partially) contained in the upper hemisphere. Hence, g_1 is (again, completely or partially) contained in the lower hemisphere. Let g'_0 be the part of g_0 that is contained in $\mathbb{S}^2_+$. Then g'_0 is an edge of G'_u. Let g_1^0 be the part of g_1 that is contained in $\mathbb{S}^2_-$, and let g'_1 be its inverse image. Then g'_1 is an edge of G'_ℓ. Since the unique planes that support $CH(\mathcal{P})$ at e_0 and e_1 are parallel, the edges g'_0 and g'_1 intersect. Hence, we have proved the following result.

LEMMA 2.2. *We can obtain all EE-pairs by computing all intersections between edges of G'_u and edges of G'_ℓ.*

2.3. The algorithm. Lemmas 2.1 and 2.2 imply that all *VF*- and *EE*-pairs can be obtained from the overlay of the graphs G'_u and G'_ℓ. Since this overlay is defined by $O(n)$ great arcs, it can be computed in $O(n^2)$ time. For details, see Houle and Toussaint [**HT88**]. Below, we give an alternative algorithm and express its running time in terms of (i) the number n of facets of $\mathcal{P}$, (ii) the number h of facets of $CH(\mathcal{P})$, and (iii) the number k of intersections between edges of G'_u and edges of G'_ℓ. Observe that, in the worst case, $h = \Theta(n)$ and $k = \Theta(n^2)$.

The convex hull of the polyhedron $\mathcal{P}$ can be computed in $O(n \log n)$ time, see, e.g., de Berg *et al.* [**dBvKOS00**]. Observe that the graphs G'_u and G'_ℓ have total size $O(h)$. Based on Lemmas 2.1 and 2.2, we obtain all *VF*- and *EE*-pairs by (i) computing for each vertex of one of these graphs, the face of the other graph that contains the vertex, and (ii) computing all intersections between edges of these graphs. Both these problems can be solved in $O(h \log h + k \log h)$ time, using the Bentley–Ottmann sweep algorithm [**BO79**], adapted to great arcs on the unit-sphere. The overall running time of the algorithm is thus $O(n \log n + k \log h)$, which is $O(n^2 \log n)$ in the worst case.

THEOREM 2.3. *Let $\mathcal{P}$ be a polyhedron with n facets. A build direction that minimizes the number of layers can be computed in $O(\min\{n^2, n \log n + k \log h\})$ time, where h is the number of facets of the convex hull of $\mathcal{P}$ and k is the number* *of intersections between edges of the graph G'_u and edges of the graph G'_ℓ. In the worst case, $h = \Theta(n)$ and $k = \Theta(n^2)$.*

An implementation of the algorithm described above, together with extensive experimental results, can be found in Schwerdt *et al.* [**SSMJ99**]. These experiments show that the value of k is usually less than h. That is, the running time on practical instances is much less than the theoretical near-quadratic worst-case upper bound.

Gärtner and Herrmann [**GH01**] present an implementation of an alternative algorithm that finds all *VF*- and *EE*-pairs without using the Bentley–Ottmann sweep technique. They reformulate the width problem as an optimization problem with linear constraints and a non-convex objective function. This new algorithm has a worst-case running time of $\Theta(n^2)$. Experimental results show that it usually outperforms the implementation of [**SSMJ99**].

We finally remark that the asymptotically fastest known algorithm for computing the width of a three-dimensional point set is due to Agarwal and Sharir [**AS96**]; its expected running time is close to $O(n^{1.5})$. The disadvantage of their algorithm is that it computes only one direction that minimizes the width and that it is probably extremely difficult to implement. By contrast, the algorithm described in this section finds *all* such directions. (Houle and Toussaint [**HT88**] have shown that there can be $\Theta(n^2)$ many such directions.) Knowing all directions that minimize the width is useful in situations where one is interested in reconciling multiple design criteria, as in Section 6.

3. Minimizing the stair-stepping effect

In this section, we consider the problem of computing a build direction that minimizes the maximum height of any error-triangle.

Let L denote the layer thickness and, for each facet f of the polyhedron $\mathcal{P}$, let $h_f(\mathbf{d})$ denote the height of the error-triangle $t_f(\mathbf{d})$ for facet f; see Figure 3. Our problem is that of computing a build direction $\mathbf{d}$ that minimizes the maximum error-triangle height $ETH(\mathbf{d}) := \max_f h_f(\mathbf{d})$. We will transform this min-max problem into a max-min problem.

Let $\theta'_f(\mathbf{d})$ be the angle between $\mathbf{d}$ and the outer unit-normal $\mathbf{n}_f$ of f, and let $\theta''_f(\mathbf{d})$ be the angle between $\mathbf{d}$ and $-\mathbf{n}_f$. We define $\theta_f(\mathbf{d}) := \min\{\theta'_f(\mathbf{d}), \theta''_f(\mathbf{d})\}$. Since $h_f(\mathbf{d}) = L\cos\theta_f(\mathbf{d})$, the problem of minimizing $ETH(\mathbf{d})$ is equivalent to computing a build direction $\mathbf{d}$ that maximizes the value of $\min_f \theta_f(\mathbf{d})$. Let us see how the latter problem can be solved. Consider the set

$$S := \{\mathbf{n}_f : f \text{ is a facet of } \mathcal{P}\} \cup \{-\mathbf{n}_f : f \text{ is a facet of } \mathcal{P}\}.$$

Maximizing the value of $\min_f \theta_f(\mathbf{d})$ means that we want to compute a direction $\mathbf{d}$ for which the minimum angle between $\mathbf{d}$ and the elements of S is maximized.

A *cap* on $\mathbb{S}^2$, with *pole* $\mathbf{d}$ and *radius* θ, is the set of all points on $\mathbb{S}^2$ whose distances from $\mathbf{d}$ are less than or equal to θ, as measured along the surface of $\mathbb{S}^2$. Hence, our problem is equivalent to computing a cap whose interior does not contain any element of S and whose radius is maximum, i.e., a *largest empty cap*. The pole of this cap is the desired optimal direction.

Since a largest empty cap must have at least three elements of S on its boundary, it is easy to compute such a cap in $O(n^3)$ time. However, it is possible to derive a significantly faster algorithm, with a running time of $O(n \log n)$, based on the following properties.

(1) Let c be the circle bounding a cap C and let $H(C)$ be the plane such that $c = H(C) \cap \mathbb{S}^2$. If C is empty, then all elements of S lie on the same side of $H(C)$. Conversely, every plane H that intersects $\mathbb{S}^2$ and for which all elements of S lie on the same side of H corresponds to an empty cap.
(2) The larger the cap C is, the closer is $H(C)$ to the origin.
(3) A largest empty cap C must have at least three elements of S on its boundary, i.e., $H(C)$ contains a facet of the convex hull of S.

The discussion above suggests the following algorithm. First compute the set S and its convex hull $CH(S)$. For each facet g of $CH(S)$, determine the plane H_g through g. Among all these planes H_g, compute one that is closest to the origin. Finally, compute the normal $\mathbf{d}$ from the origin to this closest plane. This direction $\mathbf{d}$ has the property that when $\mathcal{P}$ is built in direction $\mathbf{d}$, the maximum error-triangle height $ETH(\mathbf{d})$ is minimized. Observe that the running time of this algorithm is dominated by the $O(n \log n)$ time needed for the convex hull computation; see de Berg *et al.* [**dBvKOS00**].

THEOREM 3.1. *Let $\mathcal{P}$ be a polyhedron with n facets. A build direction that minimizes the maximum error-triangle height can be computed in $O(n \log n)$ time.*

As we saw, the correctness of the algorithm follows from the fact that a direction $\mathbf{d}$ minimizing $ETH(\mathbf{d})$ must be the normal vector of some facet of the convex hull of S. This is equivalent to saying that $\mathbf{d}$ is a vertex of the *spherical Voronoi diagram* of S. (In this Voronoi diagram, distances are measured along $\mathbb{S}^2$.)

The algorithm presented in this section is due to Majhi *et al.* [**MJSG99**]. These authors show how the algorithm can be extended to the case when each facet of $\mathcal{P}$ is assigned a positive weight, to reflect its importance in the model, and the objective is to minimize the maximum weighted error-triangle height. This version of the problem can also be solved in $O(n \log n)$ time. They also show how the sum of the weighted error-triangle heights, taken over all facets, can be minimized in $O(n^2)$ time.

4. Optimization of support structures

Asberg *et al.* [**ABB+97**] give an algorithm that decides, in $O(n)$ time, if there exists a build direction $\mathbf{d}$ for a given polyhedron with n facets, such that no support structures are needed. In this section, we consider the more general problem of minimizing the support structures. For simplicity, we restrict ourselves to the case when we want to minimize the contact-area of the supports. The approach presented here can be generalized to the case when the goal is to minimize the volume of the support structures.

Let $\mathcal{P}$ be a polyhedron with n facets and let $\mathbf{d} \in \mathbb{S}^2$ be a direction. Recall that the *contact-area for build direction* $\mathbf{d}$ is defined as the total area of all points on the boundary of $\mathcal{P}$ that are in contact with supports, when $\mathcal{P}$ is built in direction $\mathbf{d}$. The problem we consider is that of computing a build direction $\mathbf{d}$ for which the contact-area is minimum.

4.1. The convex case. We first show how the case when $\mathcal{P}$ is a convex polyhedron can be solved. The algorithm is due to Majhi *et al.* [**MJSG99**]. Recall the definition of back facets and front facets, see Section 1.1.3. Observe that the contact-area for build direction $\mathbf{d}$ is equal to the total area of all back facets.

The set of all directions $\mathbf{d}$ for which a facet f is a back facet can be represented by an open hemisphere C_f on $\mathbb{S}^2$, whose pole is the point $-\mathbf{n}_f$. We associate with this hemisphere a weight that is equal to the area of f. Our problem is now equivalent to computing a point on $\mathbb{S}^2$ such that the total weight of all hemispheres C_f containing this point is minimized. This problem can be solved by computing and traversing the arrangement defined by the boundaries of all these hemispheres; the time and space requirements will both be $O(n^2)$. Using the topological sweep algorithm of Edelsbrunner and Guibas [**EG89**], the space requirement can be reduced to $O(n)$.

THEOREM 4.1. *Let $\mathcal{P}$ be a convex polyhedron with n facets. A build direction that minimizes the total surface area of $\mathcal{P}$ that is in contact with supports can be computed in $O(n^2)$ time using $O(n)$ space.*

The time bound in Theorem 4.1 is probably optimal, because, as shown in Majhi *et al.* [**MJSG97**], the following closely related problem belongs to the class of *3SUM-hard problems* (see Gajentaan and Overmars [**GO95**]): Given a convex polyhedron, compute a direction with a positive z-component that minimizes the *number* of back facets.

In related work, Agarwal and Desikan [**AD00**] consider the problem of approximating the minimum contact-area. They show that, for a convex polyhedron, the minimum contact-area can be approximated to within a factor of $1 + \epsilon$, in $O((n/\epsilon^3)\log^3 n)$ expected time.

When a three-dimensional model is built using Layered Manufacturing, usually an entire facet is in contact with the platform, because it provides more stability to the part. Majhi *et al.* [**MJSG99**] show how to transform the contact-area problem for this special case to a halfplane range counting problem on weighted points in two-dimensional space. Using a result of Matoušek [**Mat93**], they show that, for a convex polyhedron $\mathcal{P}$, a build direction for which a facet rests on the platform and that minimizes the total surface area in contact with supports, can be computed in close to $O(n^{4/3})$ time. Agarwal and Desikan [**AD00**] show that the approximation version of this problem can be solved in $O((n/\epsilon^2)\log^2 n)$ expected time.

As a final remark, Agarwal and Desikan [**AD00**] show the following: Given a (possibly non-convex) polyhedron $\mathcal{P}$ and a build direction $\mathbf{d}$, the total area of all facets that are in contact with supports can be computed in close to $O(n^{4/3})$ time. Observe, however, that this is not the same as the total contact-area. It is possible that only a portion of a facet is in contact with supports, as discussed in the section below; in this case, only the area of that portion (not the area of the entire facet) should be included in the contact-area.

4.2. The general case. We now consider the contact-area minimization problem for a general (i.e., not necessarily convex) polyhedron $\mathcal{P}$. We remark that this problem is highly non-trivial, even in the two-dimensional case (i.e., when $\mathcal{P}$ is a simple polygon and the two-dimensional equivalent of supports is considered); an (inefficient) algorithm for this case can be found in Majhi *et al.* [**MJS$^+$99**]. In fact, Agarwal and Desikan [**AD00**] argue that it is unlikely that the three-dimensional version can be solved in $o(n^4)$ time. In this section, we outline an algorithm due to Schwerdt [**Sch01**].

The basic approach for solving the problem is by subdividing the unit-sphere $\mathbb{S}^2$ into regions such that for each region I, the combinatorial structure of the supports is the same for all directions $\mathbf{d} \in I$. (This is related to the notion of the *aspect graph* of $\mathcal{P}$, see Bowyer and Dyer [**BD90**] and Plantinga and Dyer [**PD90**].) Given this subdivision of $\mathbb{S}^2$, we derive an expression (as a function of $\mathbf{d}$) for the contact-area for each region I. Finally, for each region I, we minimize the expression over all directions $\mathbf{d} \in I$. The minimum value over all regions gives the minimum contact-area over all $\mathbf{d} \in \mathbb{S}^2$.

Recall that, for any given build direction $\mathbf{d}$, a back facet of $\mathcal{P}$ is completely in contact with supports. On the other hand, a front facet is either not in contact with supports, or is completely or partially in contact. It turns out that the partial contact case is the most difficult to handle.

4.2.1. *Subdividing* $\mathbb{S}^2$. Let f be a front facet that is partially in contact with supports. The boundary of the region on f that is in contact with supports consists of one or more polygons. We call each such polygon a *support polygon*; see Figure 4. Each vertex of a support polygon is either (i) the projection in direction $-\mathbf{d}$ onto f of a vertex of $\mathcal{P}$ (see vertex v in Figure 4), or (ii) the intersection of the projections in direction $-\mathbf{d}$ onto f of two edges of $\mathcal{P}$ (see vertex w in Figure 4). Vertex u in Figure 4 corresponds to a special case of (ii): For this vertex, one of the edges is not a projection, but an edge of f.

As mentioned already, our approach is to subdivide $\mathbb{S}^2$ into regions such that within each region, the expression for the total contact-area does not "change". Directions for which the expression does change will be called *critical directions*; these will form the boundaries of the regions.

Assume we move the build direction $\mathbf{d}$ along $\mathbb{S}^2$. Then $\mathbf{d}$ becomes a critical direction if a front facet becomes a back facet (or vice versa) or the combinatorial structure of some support polygon changes. A facet f changes from front facet to back facet (or vice versa) if $\mathbf{d}$ crosses the great circle on $\mathbb{S}^2$ consisting of all directions that are parallel to f. The combinatorial structure of the support polygons on facet f can change in any one of the following situations:

(1) $\mathbf{d}$ is parallel to f.
(2) A new support polygon appears.
(3) An existing support polygon disappears.
(4) Two support polygons combine into one support polygon.
(5) A support polygon splits into two support polygons.
(6) A new vertex appears on a support polygon.
(7) A vertex on a support polygon disappears.

We now derive necessary conditions for each of these seven cases to occur. If $\mathbf{d}$ is parallel to facet f (case (1) above), then a new support polygon may appear on f or an existing support polygon on f may disappear. Therefore, we consider the plane through f and translate it to the origin. The intersection of this translated plane with $\mathbb{S}^2$ gives all directions that are parallel to f. This covers case (1). Cases (2)–(7) are covered by Cases A and B below.

Case A: The projection in direction $-\mathbf{d}$ onto f of a vertex of $\mathcal{P}$ is on the projection in direction $-\mathbf{d}$ onto f of an edge of $\mathcal{P}$.

In order to obtain all directions in this case, we do the following. For each vertex v and each edge e of $\mathcal{P}$, we consider the wedge consisting of all rays emanating from v that contain some point of e. The directions of all these rays are the critical

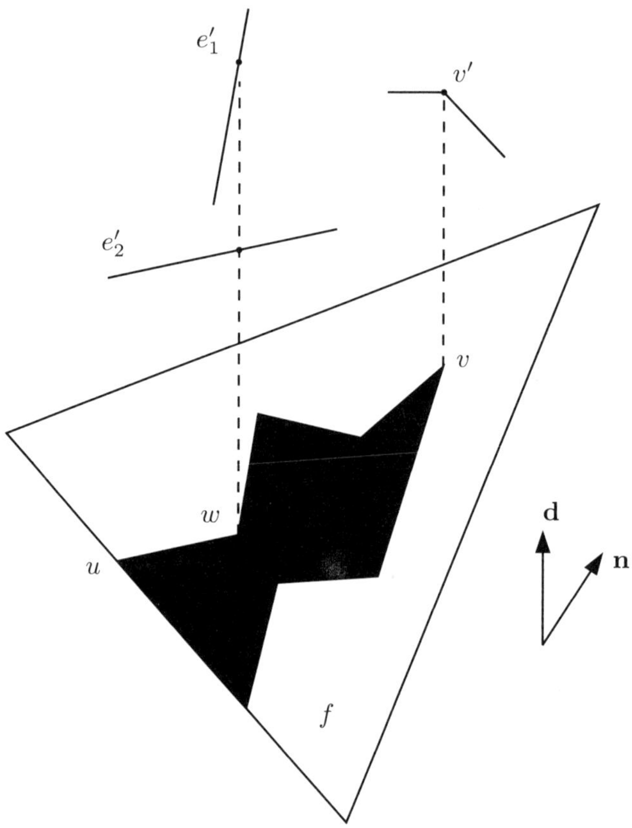

FIGURE 4. *A support polygon on facet f. $\mathbf{n}$ is the outer normal of f. u, v and w illustrate the three different types of vertices of a support polygon. e'_1 and e'_2 are edges and v' is a vertex of the polyhedron $\mathcal{P}$.*

directions for Case A. In this way, we obtain for each vertex-edge pair of $\mathcal{P}$ a great arc on $\mathbb{S}^2$.

Case B: The projections in direction $-\mathbf{d}$ onto f of three edges of $\mathcal{P}$ intersect in one single point.

The critical directions for this case are obtained as follows. We want those directions $\mathbf{d}$ such that a line with direction $\mathbf{d}$ exists that intersects three edges of $\mathcal{P}$. Let (a_1, b_1), (a_2, b_2), and (a_3, b_3) be three edges of $\mathcal{P}$, and let p be a point on one of these edges. Let us assume that p is a point on (a_1, b_1). Then $\mathbf{p} = \mathbf{a_1} + \lambda(\mathbf{b_1} - \mathbf{a_1})$, for some λ with $0 \leq \lambda \leq 1$. Let E be the plane through p and (a_2, b_2), and let E' be the plane through p and (a_3, b_3). Then $\mathbf{d}$ is a critical direction if E and E' intersect in a line having direction $\mathbf{d}$. The planes E and E' have normal vectors $(\mathbf{p} - \mathbf{a_2}) \times (\mathbf{b_2} - \mathbf{a_2})$ and $(\mathbf{p} - \mathbf{a_3}) \times (\mathbf{b_3} - \mathbf{a_3})$, respectively. The direction $\mathbf{d}$ of the intersection of these planes is given by

$$\mathbf{d} = [(\mathbf{p} - \mathbf{a_2}) \times (\mathbf{b_2} - \mathbf{a_2})] \times [(\mathbf{p} - \mathbf{a_3}) \times (\mathbf{b_3} - \mathbf{a_3})].$$

Since each coordinate of the direction $\mathbf{d}$ is a quadratic function in λ, $\mathbf{d}$ is contained in a curve on $\mathbb{S}^2$. (This curve will not be a great arc.)

Over all facets f of $\mathcal{P}$, case (1) gives rise to n great circles on $\mathbb{S}^2$, whereas Case A gives $O(n^2)$ great arcs. For the critical directions in Case B, we obtain $O(n^3)$ curves on $\mathbb{S}^2$, one curve for each triple of edges of $\mathcal{P}$. It can be shown that any two of these curves intersect only $O(1)$ times, and any such curve has only $O(1)$ intersections with any great circle.

The arrangement on $\mathbb{S}^2$ of all these n great circles, $O(n^2)$ great arcs, and $O(n^3)$ curves, subdivides $\mathbb{S}^2$ into $O(n^6)$ regions. The total number of vertices and edges of all these regions is $O(n^6)$. Moreover, the boundary of each such region consists of $O(n^3)$ edges.

4.2.2. *The expression for the contact-area.* We start by deriving expressions for the vertices of the support polygons. Let I be one of the regions in the arrangement of Section 4.2.1, and let f be a facet of $\mathcal{P}$ that is a front facet for all directions in I. Let $\mathbf{n}$ be the outer unit-normal of f, and let S be a support polygon on f. Recall that for all $\mathbf{d} \in I$, the polygon S has the same combinatorial structure. Consider an arbitrary vertex v of S. As we saw before, there are two cases to distinguish.

Case 1: v is the projection in direction $-\mathbf{d}$ onto f of a vertex v' of $\mathcal{P}$.

We can write $v = v' - \lambda\mathbf{d}$, for some $\lambda > 0$. Since the plane through f has the form $\mathbf{n} \cdot \mathbf{x} = c$, for some constant c, we obtain $\mathbf{n} \cdot (v' - \lambda\mathbf{d}) = c$, which can be rewritten as

$$\lambda = \frac{\mathbf{n} \cdot v' - c}{\mathbf{n} \cdot \mathbf{d}}.$$

Hence, we have written v as

$$v = v' - \frac{\mathbf{n} \cdot v' - c}{\mathbf{n} \cdot \mathbf{d}}\,\mathbf{d}. \tag{4.1}$$

Case 2: v is the intersection of the projections in direction $-\mathbf{d}$ onto f of two edges of $\mathcal{P}$.

Let $e_1' = (u_1', v_1')$ and $e_2' = (u_2', v_2')$ be two edges of $\mathcal{P}$ whose projections e_1 and e_2, respectively, onto f in direction $-\mathbf{d}$ intersect in v. Let E_1 be the plane through e_1' and e_1, and let E_2 be the plane through e_2' and e_2. Observe that E_1 contains the points u_1', v_1' and $u_1' - \mathbf{d}$; hence, E_1 has a normal vector $\mathbf{d} \times (v_1' - u_1')$. Therefore, E_1 is given by the equation

$$[\mathbf{d} \times (v_1' - u_1')] \cdot \mathbf{x} = c_1,$$

for some constant c_1. Since u_1' is contained in E_1, we have

$$c_1 = [\mathbf{d} \times (v_1' - u_1')] \cdot u_1' = [\mathbf{d} \times v_1'] \cdot u_1'.$$

Hence, we have obtained the following equation for the plane E_1:

$$[\mathbf{d} \times (v_1' - u_1')] \cdot \mathbf{x} = [\mathbf{d} \times v_1'] \cdot u_1'.$$

In a similar way, we obtain the following equation for the plane E_2:

$$[\mathbf{d} \times (v_2' - u_2')] \cdot \mathbf{x} = [\mathbf{d} \times v_2'] \cdot u_2'.$$

Since v is the intersection of E_1, E_2, and the plane through f, we obtain the system of equations $A_{\mathbf{d}}v = b_{\mathbf{d}}$, where $A_{\mathbf{d}}$ is a 3×3 matrix and $b_{\mathbf{d}}$ is a vector. The coefficients of $A_{\mathbf{d}}$ and $b_{\mathbf{d}}$ are polynomials in $\mathbf{d}$ of degree at most one. Hence, we can write v as

$$v = A_{\mathbf{d}}^{-1} b_{\mathbf{d}}, \tag{4.2}$$

i.e., each coordinate of v is the quotient of two polynomials in $\mathbf{d}$, each having degree at most three. This concludes Case 2.

To summarize, using (4.1) and (4.2), each vertex v of the support polygon S on f can be written in the form

$$v = \begin{pmatrix} a_1(\mathbf{d})/b_1(\mathbf{d}) \\ a_2(\mathbf{d})/b_2(\mathbf{d}) \\ a_3(\mathbf{d})/b_3(\mathbf{d}) \end{pmatrix}, \tag{4.3}$$

where each of a_1, a_2, a_3, b_1, b_2, and b_3 is a polynomial in $\mathbf{d}$ of degree at most three.

We are now ready to express the area of the support polygon S as a function of the build direction $\mathbf{d}$. If the vertices of S are $s_0, s_1, \ldots, s_{k-1}$, then the area of S is equal to

$$\sum_{i=0}^{k-1} \Delta(p, s_i, s_{i+1}), \tag{4.4}$$

where $s_k = s_0$, p is an arbitrary point on f, and $\Delta(p, s_i, s_{i+1})$ is the signed area of the triangle having vertices p, s_i, and s_{i+1}. The sign is positive (resp. negative), if the points p, s_i, and s_{i+1} form a left-turn (resp. right-turn). (See O'Rourke [**O'R98**].) Since

$$\Delta(p, s_i, s_{i+1}) = \pm\frac{1}{2}\,\|(s_i - p) \times (s_{i+1} - p)\|,$$

and since the cross-product on the right-hand side is collinear with $\mathbf{n}$, we can rewrite the expression in (4.4) as

$$\frac{1}{2}\sum_{i=0}^{k-1} \beta_i\,[(s_i - p) \times (s_{i+1} - p)] \cdot \mathbf{n},$$

where each β_i has the appropriate value plus or minus one.

In (4.3), we have shown how to express each vertex s_i of the support polygon. Hence, we can write each term $[(s_i - p) \times (s_{i+1} - p)] \cdot \mathbf{n}$ as a quotient of two polynomials in $\mathbf{d}$, each having degree at most six. In conclusion, we obtain the following expression for the area $A^S(\mathbf{d})$ of the support polygon S:

$$A^S(\mathbf{d}) = \sum_{i=0}^{k-1} \frac{c_i(\mathbf{d})}{d_i(\mathbf{d})},$$

where each of the c_i's and d_i's is a polynomial in $\mathbf{d}$ of degree at most six.

Until now, we considered the area $A^S(\mathbf{d})$ of only one support polygon S on one facet f of $\mathcal{P}$. We have to consider this area for each support polygon on each front facet of $\mathcal{P}$. Let I be one of the regions in the arrangement on $\mathbb{S}^2$, and let K_I be the total number of all vertices of all support polygons on all facets of $\mathcal{P}$ for directions $\mathbf{d}$ in I. Then the total area $A_I(\mathbf{d})$ of all support polygons can be written as

$$A_I(\mathbf{d}) = \sum_{i=1}^{K_I} \frac{f_i(\mathbf{d})}{g_i(\mathbf{d})}, \tag{4.5}$$

where each of the f_i's and g_i's is a polynomial in $\mathbf{d}$ of degree at most six. Observe that $K_I = O(n^2)$.

The algorithm computes the minimum of the expression (4.5) over all directions $\mathbf{d} \in I$, for each of the $O(n^6)$ regions I. The boundary of each such region I is described by $O(n^3)$ curves on $\mathbb{S}^2$. If we denote by $Q(n)$ the time needed to minimize the expression (4.5) within one region I, then we obtain the following result.

THEOREM 4.2. *Let $\mathcal{P}$ be a polyhedron with n facets. A build direction that minimizes the total surface area of $\mathcal{P}$ that is in contact with supports can be computed in $O(n^6 \cdot Q(n))$ time, where $Q(n)$ is the time needed to minimize the expression (4.5).*

It should be clear that the algorithm described in this section is not practical. Therefore, in the next section, we present a heuristic approach for approximating the minimum contact-area.

5. Approximating contact-area of suppports for general polyhedra

Let $\mathbf{d}^*$ be a direction that minimizes the contact-area for $\mathcal{P}$. As seen in the previous section, $\mathbf{d}^*$ appears to be difficult to compute. In this section, we give a simple heuristic to compute a build direction $\hat{\mathbf{d}}$ which approximates $\mathbf{d}^*$. This result is due to Ilinkin *et al.* [**IJS+04**].

Let $CA(\mathbf{d})$ denote the contact-area of $\mathcal{P}$ for a given direction $\mathbf{d}$. We first show how to upper-bound the ratio $CA(\hat{\mathbf{d}})/CA(\mathbf{d}^*)$. Let $BFA(\mathbf{d})$ be the total area of the back facets with respect to $\mathbf{d}$ and let $\mathbf{d}'$ be a direction that minimizes the total area of the back facets of $\mathcal{P}$. Observe that $BFA(\mathbf{d}^*) \leq CA(\mathbf{d}^*)$, since $CA(\mathbf{d}^*)$ also includes possible contact-area on front and parallel facets. Also, by definition of $\mathbf{d}'$, $BFA(\mathbf{d}') \leq BFA(\mathbf{d}^*)$. Therefore,

$$\frac{CA(\hat{\mathbf{d}})}{CA(\mathbf{d}^*)} \leq \frac{CA(\hat{\mathbf{d}})}{BFA(\mathbf{d}^*)} \leq \frac{CA(\hat{\mathbf{d}})}{BFA(\mathbf{d}')}.$$

This result allows us to measure the quality of the approximation provided by our heuristic, for various choices of $\hat{\mathbf{d}}$. In the rest of this section, we show how to compute efficiently $CA(\hat{\mathbf{d}})$ and $BFA(\mathbf{d}')$. We also discuss ways to choose candidate directions $\hat{\mathbf{d}}$ and describe briefly our experimental results. Our emphasis here will be on practical algorithms rather than on the asymptotically most efficient ones.

5.1. Computing contact-area for a fixed direction. The contact-area for a given direction $\hat{\mathbf{d}}$ includes contact-area on front facets, on parallel facets, and on back facets. Handling front and parallel facets is tricky since only parts of these facets may be in contact with supports; handling back facets is straightforward, since they are completely in contact with supports. We describe an algorithm for front facets only; the algorithm for parallel facets is similar.

Our approach is a heuristic based on ray-shooting. Consider a front facet f and let p be a point on f. Point p will be in contact with supports if and only if the ray emanating from p in direction $\hat{\mathbf{d}}$ intersects some other facet of $\mathcal{P}$. The idea is to pick a set of points in the interior of f and identify those points that will be in contact with supports by shooting rays in direction $\hat{\mathbf{d}}$. Let H_f (resp. M_f) denote the set of rays, originating at points on f, that hit a facet of $\mathcal{P}$ (resp. miss all facets of $\mathcal{P}$). Then the area of f that is in contact with supports can be approximated as $|H_f|/(|H_f| + |M_f|)$ times the area of f. As the density of the sample points is increased, the accuracy of the approximation improves.

The sample points are selected through a subdivision process, where each facet is subdivided into two triangular patches and placed at the end of a queue of unprocessed patches. Thus, the subdivision proceeds in a breadth-first fashion and all facets are subdivided to the same depth. Each patch is processed by shooting a ray from its centroid in direction $\hat{\mathbf{d}}$ and recording the result. An iteration of the algorithm corresponds to a complete subdivision of the patches. The algorithm terminates after a predefined number d of iterations or when the change in contact-area from one iteration to the next is not significant, e.g., less than 1%. The running time can be shown to be $O(2^d n^2)$.

To verify the accuracy of the heuristic, we also implemented an exact algorithm; however, it is very slow and we use it only for a one-time verification of the heuristic. (This algorithm involves intersecting the projections of pairs of front and back facets and runs in $O(n^3)$ time.) Our experiments on real-world models showed that the heuristic produced answers that were very close to the exact one, and did so in a fraction of the time. (We used $d = 10$ in our experiments.) Details may be found in [**IJS**+**04**].

5.2. Minimizing the total area of back facets. In Section 4.1, we showed how to minimize the contact-area when $\mathcal{P}$ is convex; in the convex case, this is equivalent to minimizing the total back facet area. It turns out that this algorithm can also be used to minimize the total back facet area for a general polyhedron $\mathcal{P}$. The algorithm runs in $O(n^2)$ time and uses $O(n)$ space (when topological sweep is used). We describe here a simpler algorithm, which runs in $O(n^2 \log n)$ time and uses $O(n)$ space.

Each facet f of $\mathcal{P}$ defines a great circle C_f on $\mathbb{S}^2$, consisting of all points that are at distance $\pi/2$ from the outer unit-normal of f. It can be shown [**IJS**+**04**] that a direction $\mathbf{d}'$ that minimizes the total back facet area coincides with some vertex of the arrangement $\mathcal{A}$ of these great circles. This vertex can be found, without explicitly computing $\mathcal{A}$, as follows.

We pick a great circle C_f, compute its intersection with the other great circles, and sort the intersection points (which are vertices of $\mathcal{A}$) in their circular order around C_f. We pick one of these intersection points and compute the total back facet area for the corresponding direction. We then visit the other intersection points in order and update the total back facet area incrementally. Specifically, suppose that we visit intersection point $\mathbf{d}_2$ from intersection point $\mathbf{d}_1$, and assume that $BFA(\mathbf{d}_1)$ has been computed already. Let $\Delta BFA(\mathbf{d}_1)$ be the total area of the facets that are parallel facets with respect to $\mathbf{d}_1$ but are back facets with respect to $\mathbf{d}_2$. Similarly, let $\Delta BFA(\mathbf{d}_2)$ be the total area of the facets that are parallel facets with respect to $\mathbf{d}_2$ but are back facets with respect to $\mathbf{d}_1$. Then we compute $BFA(\mathbf{d}_2)$ as $BFA(\mathbf{d}_1) + \Delta BFA(\mathbf{d}_1) - \Delta BFA(\mathbf{d}_2)$. The optimum direction $\mathbf{d}'$ is obtained after all great circles C_f have been thus processed.

The running time for each great circle C_f is dominated by the $O(n \log n)$ time to sort the $O(n)$ intersection points. The walk around C_f takes $O(n)$ additional time. Observe that, for each intersection point $\mathbf{d}_1$ on C_f, the facets that are parallel with respect to $\mathbf{d}_1$ can be identified from the great circles that intersect at $\mathbf{d}_1$. Thus, $\Delta BFA(\mathbf{d}_1)$ can be computed in time proportional to the number of great circles that contain $\mathbf{d}_1$, so the time for all intersection points on C_f is $O(n)$. It follows that the running time is $O(n^2 \log n)$. The space used is $O(n)$.

5.3. Choosing candidate directions. We have implemented and tested the following choices for a build direction $\hat{\mathbf{d}}$. In each case, the quality of the heuristic was upper-bounded using the ratio $CA(\hat{\mathbf{d}})/BFA(\mathbf{d}')$, where $\mathbf{d}'$ is computed (just once), as in Section 5.2. A detailed discussion of these choices can be found in [**IJS$^+$04**].

Our choices for $\hat{\mathbf{d}}$ included: (i) a direction which minimizes the total back facet area (i.e., $\mathbf{d}'$); (ii) a direction which maximizes the total area (or, alternatively, the number) of parallel facets; (iii) the principal components of $\mathcal{P}$; and (iv) a direction corresponding to the negation of the outer normal of that facet of the convex hull of $\mathcal{P}$ which contains facets of $\mathcal{P}$ of maximum total area; intuitively, this direction provides high stability to the part as it is built. We tested these choices on several real models obtained from industry and found savings in contact-area ranging from 9% to 83%, when compared to the contact-area for randomly chosen directions.

In related work, Allen and Dutta [**AD95**] also describe heuristics for contact-area minimization. They choose a subset of the normals of the convex hull as candidate build directions. For each orientation, they compute the needed supports and contact-area approximately and pick the one that is smallest. No analysis of the quality of the heuristic is given, however. Support optimization has also been addressed previously by Frank and Fadel [**FF94**] in the framework of an expert system.

6. Multi-criteria optimization

In the previous sections, we have considered different design criteria that are important to optimize. It is not difficult to see that a single build direction will, in general, not optimize multiple design criteria simultaneously. In this section, we consider the problem of computing a build direction that reconciles multiple measures in a meaningful way. The results we present are due to Majhi *et al.* [**MJSS01**]. These authors present several algorithms that optimize different combinations of measures. Their algorithms incorporate the results of the previous sections as building blocks. Moreover, their algorithms are modular in the sense that solutions for different combinations of criteria can be derived quickly by combining the appropriate single-criterion algorithms.

We will consider three types of multi-criteria optimization problems. For each type, we will give one specific example. Further applications can be found in [**MJSS01**].

6.1. Threshold formulation. Let $\mathcal{C}_1$ and $\mathcal{C}_2$ be two generic design criteria. Given two designer-specified values ρ_1 and ρ_2, our goal is to compute a build direction $\mathbf{d}$ such that $\mathcal{C}_1(\mathbf{d}) \leq \rho_1$ and $\mathcal{C}_2(\mathbf{d}) \leq \rho_2$.

Let us assume that $\mathcal{C}_1(\mathbf{d})$ is the maximum error-triangle height $ETH(\mathbf{d})$ for direction $\mathbf{d}$ (see Section 3) and $\mathcal{C}_2(\mathbf{d})$ is the width $W(\mathbf{d})$ of the polyhedron $\mathcal{P}$ in direction $\mathbf{d}$ (see Section 2).

We start by computing a description of all build directions $\mathbf{d}$ for which $ETH(\mathbf{d})$ is less than or equal to ρ_1. Recall from Section 3 that the error-triangle height $h_f(\mathbf{d})$ for facet f satisfies $h_f(\mathbf{d}) = L \cos\theta_f(\mathbf{d})$. Hence we require that $\theta_f(\mathbf{d}) \geq \theta$ for all facets f, where $\theta := \arccos(\rho_1/L)$.

For each facet f of $\mathcal{P}$, the set of all directions $\mathbf{d}$ such that $\theta_f(\mathbf{d}) \geq \theta$ can be represented by the complement of the union of two open caps of radius θ and poles $\mathbf{n}_f$ and $-\mathbf{n}_f$, where $\mathbf{n}_f$ is the outer normal of f. This complement, which we call a *band*, consists of two parallel small circles at a distance of $\pi/2 - \theta$ on either side of the great circle defined by $\mathbf{n}_f$.

Thus, the set of all build directions $\mathbf{d}$ for which $ETH(\mathbf{d}) \leq \rho_1$ is the intersection of these bands, over all facets f of $\mathcal{P}$. This is a set of regions bounded by arcs of small circles. The total size of the regions is $O(n)$ and they can be computed in $O(n \log n)$ time.

After having computed the intersection of all bands, we proceed as follows. For each region I in the intersection, we want to decide if there is a build direction $\mathbf{d}$ in I for which $W(\mathbf{d}) \leq \rho_2$. This can be done by solving the *constrained width problem* within I, i.e., computing a direction $\mathbf{d} \in I$ that minimizes $W(\mathbf{d})$, and then checking if this width is less than or equal to ρ_2. Majhi *et al.* [**MJSS01**] show how this problem can be solved in $O(n^2)$ total time for all regions I.

THEOREM 6.1. *Let $\mathcal{P}$ be a polyhedron with n facets. Given any two real numbers ρ_1 and ρ_2, we can decide in $O(n^2)$ time, if there exists a build direction $\mathbf{d}$ for which the maximum error-triangle height $ETH(\mathbf{d})$ is less than or equal to ρ_1 and the width $W(\mathbf{d})$ is less than or equal to ρ_2.*

6.2. Weighted formulation. Again, let $\mathcal{C}_1$ and $\mathcal{C}_2$ be two generic measures. In the weighted formulation, we are given two designer-specified weights w_1 and w_2, reflecting the relative importance of the two criteria, and we want to compute a build direction $\mathbf{d}$ such that $w_1 \cdot \mathcal{C}_1(\mathbf{d}) + w_2 \cdot \mathcal{C}_2(\mathbf{d})$ is minimized.

We again illustrate this type of problem for the case when $\mathcal{C}_1(\mathbf{d})$ is the maximum error-triangle height $ETH(\mathbf{d})$ for direction $\mathbf{d}$ and $\mathcal{C}_2(\mathbf{d})$ is the width $W(\mathbf{d})$ in direction $\mathbf{d}$.

Let S be the set of directions on $\mathbb{S}^2$ corresponding to the facet normals (and their inverse images) of $\mathcal{P}$, as defined in Section 3, and consider the spherical Voronoi diagram $VD(S)$ of S, where distances are measured along $\mathbb{S}^2$. Also, consider the graphs G'_u and G'_ℓ on the upper hemisphere $\mathbb{S}^2_+$ that were defined in Section 2. We define the graph $\mathcal{G}$ on $\mathbb{S}^2$ to be the overlay of G'_u, G'_ℓ, and their inverse images. Finally, let $\mathcal{A}$ be the arrangement on $\mathbb{S}^2$ obtained by overlaying $VD(S)$ and $\mathcal{G}$.

It can be shown that for each face I in the arrangement $\mathcal{A}$, the expression for the maximum error-triangle height, for directions $\mathbf{d} \in I$, can be written as $ETH(\mathbf{d}) = a_I x + b_I y + c_I z$, where a_I, b_I, and c_I are constants (that depend on the face I) and $\mathbf{d} = (x, y, z)$. Similarly, the expression for the width of $\mathcal{P}$, for directions $\mathbf{d} \in I$, can be written as $W(\mathbf{d}) = a'_I x + b'_I y + c'_I z$, where a'_I, b'_I, and c'_I are again constants that depend on I, and $\mathbf{d} = (x, y, z)$. Hence, within the face I of $\mathcal{A}$, we have to minimize the expression

$$w_1 \left(a_I x + b_I y + c_I z\right) + w_2 \left(a'_I x + b'_I y + c'_I z\right).$$

Since the face I is bounded by great arcs, its boundary is given by a collection of linear constraints on the variables x, y, and z, together with the constraint $x^2 + y^2 + z^2 = 1$. As shown in Majhi *et al.* [**MJSG99**], the optimization problem for I can be solved using Lagrange Multipliers, in time that is proportional to the number of great arcs on the boundary of I. Since the total size of all faces I sums up to $O(n^2)$, we obtain the following result.

THEOREM 6.2. *Let $\mathcal{P}$ be a polyhedron with n facets. Given any two real numbers w_1 and w_2, we can compute, in $O(n^2)$ time, a build direction $\mathbf{d}$ for which $w_1 \cdot ETH(\mathbf{d}) + w_2 \cdot W(\mathbf{d})$ is minimum.*

6.3. Sequential formulation. As before, let $\mathcal{C}_1$ and $\mathcal{C}_2$ be two generic measures. In the sequential formulation, we want to compute, among all build directions $\mathbf{d}$ that minimize $\mathcal{C}_1(\mathbf{d})$, a direction $\mathbf{d}$ that minimizes $\mathcal{C}_2(\mathbf{d})$.

Assume again that $\mathcal{C}_1(\mathbf{d})$ is the maximum error-triangle height $ETH(\mathbf{d})$ for direction $\mathbf{d}$, and $\mathcal{C}_2(\mathbf{d})$ is the width $W(\mathbf{d})$ in direction $\mathbf{d}$.

Consider again the set S of directions on $\mathbb{S}^2$ corresponding to the facet normals (and their inverse images) of $\mathcal{P}$. As we saw in Section 3, the set of all directions $\mathbf{d} \in \mathbb{S}^2$ that minimize $ETH(\mathbf{d})$ is equal to the set of poles of all largest empty caps. Moreover, each such pole is the normal of some facet of the convex hull of S. It follows that the directions $\mathbf{d}$ that minimize $ETH(\mathbf{d})$ (there are $O(n)$ of these), can be computed in $O(n \log n)$ time. For each of these directions $\mathbf{d}$, we have to compute the value of $W(\mathbf{d})$. We can do this by computing and preprocessing the convex hull of $\mathcal{P}$ into the *hierarchical data structure* of Dobkin and Kirkpatrick [**DK85**]; this takes $O(n \log n)$ time. Using this data structure, we can, given any query direction $\mathbf{d}$, compute the extreme vertices in that direction, and hence the value of $W(\mathbf{d})$, in $O(\log n)$ time.

THEOREM 6.3. *Let $\mathcal{P}$ be a polyhedron with n facets, and let D be the set of all build directions $\mathbf{d}$ for which $ETH(\mathbf{d})$ is minimum. In $O(n \log n)$ time, we can compute a build direction $\mathbf{d}$ in D that minimizes the width of $\mathcal{P}$ in direction $\mathbf{d}$.*

7. Protecting facets

Let $\mathcal{P}$ be a polyhedron with n facets, and let f be a fixed facet of $\mathcal{P}$. In this section, we consider the problem of computing a description of all build directions for which facet f is not in contact with supports. Observe that this problem is equivalent to computing all directions in which f can be translated to infinity such that no collision occurs with the rest of the polyhedron. The approach that we sketch in this section is due to Schwerdt *et al.* [**SSJ⁺00**]; see also Nurmi and Sack [**NS89**].

For any point $x \in \mathbb{R}^3$ and for any direction $\mathbf{d} \in \mathbb{S}^2$, we denote by $r_{x\mathbf{d}}$ the ray emanating from x having direction $\mathbf{d}$. For any facet g of $\mathcal{P}$ and for any point $x \in \mathbb{R}^3$, we define

$$R_{xg} := \{\mathbf{d} \in \mathbb{S}^2 : (r_{x\mathbf{d}} \cap g) \setminus \{x\} \neq \emptyset\}.$$

For any facet g of $\mathcal{P}$, we define

$$C_{fg} := \bigcup_{x \in f} R_{xg}.$$

Hence, for each direction $\mathbf{d} \in C_{fg}$, facet f is in contact with supports for build direction $\mathbf{d}$ "because of" facet g. That is, there is a point x on f, such that the ray $r_{x\mathbf{d}}$ emanating from x and having direction $\mathbf{d}$ intersects facet g.

LEMMA 7.1. *Ignoring degenerate cases, the complement of the union of the sets C_{fg}, where g ranges over all facets of $\mathcal{P}$ (except for f), is equal to the set of all directions for which facet f is not in contact with supports.*

Lemma 7.1 implies that our problem can be solved in the following way.

Step 1: For each facet g of $\mathcal{P}$ with $g \neq f$, compute a description of the set C_{fg}.

Step 2: Compute the boundary of the union of the sets C_{fg}, $g \neq f$.

We first consider Step 1. It turns out that there is a simple way to compute a set C_{fg}:

LEMMA 7.2. *Let g be a facet of $\mathcal{P}$. Assume that f and g are disjoint or intersect in a single point. Let*

$$D_{fg} := \{\mathbf{d}_{st} \in \mathbb{S}^2 : s \text{ is a vertex of } f,\ t \text{ is a vertex of } g,\ s \neq t\},$$

where $\mathbf{d}_{st}$ is the point on $\mathbb{S}^2$ having the same direction as the line segment from s to t. Then C_{fg} is the spherical convex hull of the directions in D_{fg}.

Chen and Woo [**CW92**] present an efficient algorithm for computing the spherical convex hull of a finite set of directions. Observe that, since f and g are triangles, the set D_{fg} in Lemma 7.2 consists of at most nine elements.

Next, we consider Step 2. Observe that after Step 1, we have a collection of at most $n-1$ spherically convex polygons C_{fg}, each having at most nine edges. Each edge of such a polygon is a great arc. For each such edge e, let K_e be the great circle that contains e. Using an incremental algorithm, we compute the arrangement on $\mathbb{S}^2$ of the $O(n)$ great circles K_e. (This can be done by adapting the algorithm in, e.g., de Berg *et al.* [**dBvKOS00**], for computing the arrangement of lines in $\mathbb{R}^2$.) By traversing this arrangement, we then remove all vertices and edges that are inside some set C_{fg}.

THEOREM 7.3. *Let $\mathcal{P}$ be a polyhedron with n triangular facets and let f be a facet of $\mathcal{P}$. In $O(n^2)$ time and using $O(n^2)$ space, we can compute a description of all build directions for which f is not in contact with supports.*

Implementing the algorithm described above is a highly non-trivial problem. Part of the difficulty comes from degenerate configurations and numerical problems. Observe that if the facets g_1 and g_2 share an edge, then the spherically convex polygons C_{fg_1} and C_{fg_2} will probably also share an edge. In other words, degenerate configurations are very likely to occur. Details about an implementation, based on LEDA [**MN99**] and its exact rational arithmetic to solve geometric predicates exactly, can be found in Schwerdt *et al.* [**SSJJ03**] and Schwerdt [**Sch01**].

The result of Theorem 7.3 is optimal in the sense that there exists a polyhedron $\mathcal{P}$ that contains a facet f such that the set of all build directions for which f is not in contact with supports has $\Omega(n^2)$ connected components.

The algorithm presented in this section can be extended to the case when k facets $f_1, f_2, \ldots, f_k$ have to be protected from being in contact with supports. Then, both the running time and the space requirement become $O(k^2n^2)$.

8. A decomposition-based approach

Given a polyhedron $\mathcal{P}$ with n facets, our goal is to decompose it into a small number of pieces and then build these pieces individually so that the total support requirements (support contact-area or volume) is minimized. Our notion of a decomposition is to cut $\mathcal{P}$ with a plane H normal to some direction $\mathbf{d}$ and build the pieces lying in each halfspace of H in the direction of the normal of H ($\mathbf{d}$ or $-\mathbf{d}$) lying in that halfspace. The quality of the solution depends on $\mathbf{d}$ and, ideally, we would like to find the direction $\mathbf{d}$ which minimizes the support requirements. As a first step towards this challenging goal, we consider the "fixed-direction" version of this problem, where we assume that the direction $\mathbf{d}$ is a given direction. We show how to compute a plane H, perpendicular to $\mathbf{d}$, which cuts $\mathcal{P}$ into polyhedra

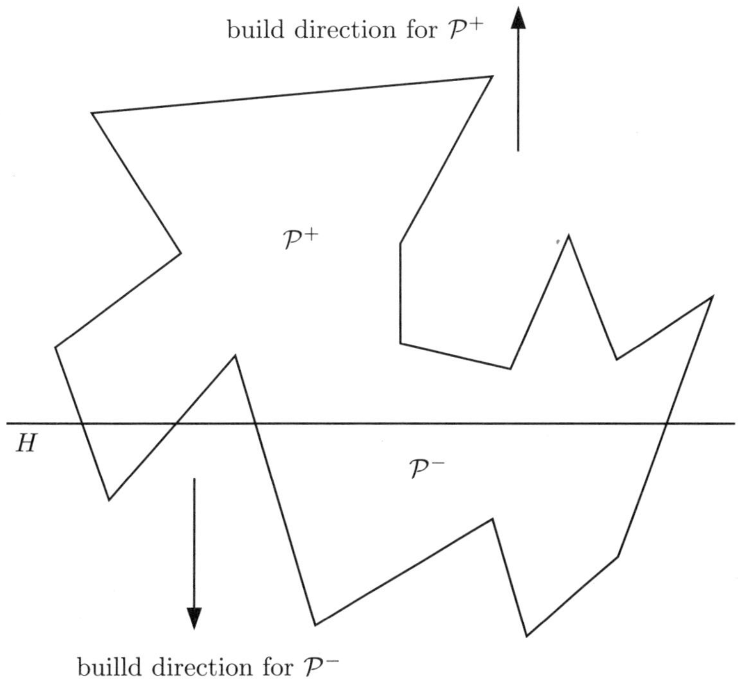

FIGURE 5. *The decomposition-based approach, shown in 2D for convenience. The polyhedron $\mathcal{P}$ is decomposed by a plane H into polyhedra $\mathcal{P}^+$ and $\mathcal{P}^-$, which are then built in the indicated directions.*

$\mathcal{P}^+$ and $\mathcal{P}^-$, lying above and below H, respectively, such that the total support contact-area, or support volume, when $\mathcal{P}^+$ and $\mathcal{P}^-$ are built in directions $\mathbf{d}$ and $-\mathbf{d}$, respectively, is minimized. (See Figure 5.) Since $\mathcal{P}^+$ and $\mathcal{P}^-$ can consist of several connected components, an additional requirement is that the number of connected components in the optimal decomposition be at most a user-specified integer K. We assume for simplicity that $\mathbf{d}$ is the vertical direction.

This problem is considered in Ilinkin *et al.* [**IJM$^+$02**], where an $O(n \log n)$-time algorithm is given for a convex polyhedron and it is shown how this solution can be incorporated into an algorithm for a general polyhedron that runs in $O(n^2 \log n)$ time. Experiments on real-world models show significant reductions in support requirements, e.g., 30% to 86% for contact-area. In what follows, we describe these algorithms briefly, focusing on contact-area minimization, since the approach for support volume minimization is similar.

8.1. Decomposing a convex polyhedron to minimize contact-area. For any position of the plane H, each facet of $\mathcal{P}$ can be classified as *active* or *inactive*, depending on whether or not it is cut by H. An inactive facet is completely contained in either $\mathcal{P}^+$ or $\mathcal{P}^-$ and its contribution to the contact-area is unaffected by small movements in H. An active facet is contained partially in $\mathcal{P}^+$ and partially in $\mathcal{P}^-$ and any movement in H affects its contribution to the contact-area. It can be shown that the contribution of an active facet f to the contact-area is a quadratic function $e_f(h) = a_f h^2 + b_f h + c_f$, where h is the height of H above the xy-plane and a_f, b_f, and c_f are constants depending only on the coordinates of the vertices of f. (However, the expressions for these coefficients depend on which edges of f are intersected by H.)

The algorithm begins by sorting the vertices of $\mathcal{P}$ from bottom to top. It then sweeps a horizontal plane H over $\mathcal{P}$, stopping at each vertex in turn. The total area contributed by the inactive facets (the *inactive-area*) is initialized to the total area of the back facets of $\mathcal{P}$, and the area contributed by the active facets (the *active-area*) is set to zero; the total contact-area is the sum of these two terms. At the current vertex v, the following actions are taken for each facet f incident to v.

If v is the lowest vertex of f, then f changes from inactive to active. An expression of the form $e_f(h)$ is added to the active-area term and, if f is a back facet of $\mathcal{P}$, then the area of f is also subtracted from the inactive-area term. (Parallel facets play no role in the convex case and are ignored.) If v is the highest vertex of f, then f changes from active to inactive. An expression of the form $e_f(h)$ is subtracted from the active-area term and, if f is a front facet of $\mathcal{P}$, then the area of f is added to the inactive-area term (f will henceforth be in $\mathcal{P}^-$ which is built in direction $-\mathbf{d}$, so f will be in contact with supports). If v is the middle vertex of f, then f will continue to be active. The active-area term is updated by subtracting and adding terms of the form $e_f(h)$ (the two terms will be different since H intersects a different edge of f above v than it did below v).

After v is thus processed, a new expression is obtained for the total contact-area. This expression is quadratic in h and does not change for any position of H between v and its successor. The expression is minimized using standard techniques from calculus to obtain the optimum plane in this range. Once all vertices have been processed, the overall optimum plane is known. The total time to process a vertex is proportional to its degree, so the total time for all vertices is $O(n)$. Including the initial sorting, the overall time is $O(n \log n)$.

8.2. Decomposing general polyhedra. The algorithm of Section 8.1 for convex polyhedra cannot be applied directly to a general polyhedron $\mathcal{P}$ because of the complex nature of supports here (see Section 1.1.3). Nevertheless, it is possible to solve the problem by applying the algorithm for the convex case to a carefully selected set of triangles derived from the facets of $\mathcal{P}$.

The main idea is to partition each front or back facet of $\mathcal{P}$ into two classes of triangles, called black and gray triangles; one of these classes may be empty. A *black triangle* will always be completely in contact with supports, regardless of the position of the decomposition plane H; therefore, it always contributes a fixed amount, i.e., its area, to the total contact-area. However, a *gray triangle* will contribute anywhere between zero and its area to the total contact-area, depending on the position of H. Parallel facets are partitioned into three classes of triangles, called black, gray, and white; up to two of these classes may be empty. Black and gray triangles are defined as before; a *white triangle* will never be in contact with supports, regardless of the position of H. It follows that only gray facets need to be considered for support minimization purposes. These are handled using the approach in Section 8.1, as discussed later.

We discuss how to identify black and gray triangles for a front facet f. Suppose that $\mathcal{P}$ is built without decomposition in direction $\mathbf{d}$. The footprint of the supports, if any, that are in contact with f is a collection of disjoint polygons. Let b be any triangle in a triangulation of the footprint and let g be any triangle in the triangulation of the region that is the complement of the footprint. We claim that b and g satisfy the definition of a black and a gray triangle, respectively.

Suppose that $\mathcal{P}$ is built with decomposition in direction $\mathbf{d}$. If the decomposition plane H intersects b, then the part b^+ of b above H will be in contact with supports since the portion of $\mathcal{P}^+$ that is above b^+ is the same as the part of $\mathcal{P}$ that was above b^+ if $\mathcal{P}$ were built without decomposition. The part b^- of b that is below H functions as a back facet in $\mathcal{P}^-$ and will also be in contact with supports. Thus, all of b will be in contact with supports. If H is above b, then $b^+ = \emptyset$ and $b^- = b$; if H is below b, then $b^- = \emptyset$ and $b^+ = b$. In either case, the same argument applies. It follows that b is indeed a black triangle.

Next consider g. If H intersects g, then the part g^- of g that is below H will be in contact with supports since it functions as a back facet in $\mathcal{P}^-$. Since no part of g is in contact with supports when $\mathcal{P}$ is built in direction $\mathbf{d}$, it follows that the part g^+ of g, which is in $\mathcal{P}^+$ and also built in direction $\mathbf{d}$, will also not be in contact with supports. If H is below or above g, then g is either not in contact or completely in contact with supports. Thus, the contribution of g to the contact-area will be a function of the position of H, so g is indeed a gray triangle. It is easy to check that this contribution has the same form as in the convex case: if g is inactive, it is zero or the area of g; if g is active, it is a quadratic function of the height of H.

A symmetric discussion applies for back facets. To identify black, gray, and white triangles for a parallel facet f, we build $\mathcal{P}$ without decomposition in directions $\mathbf{d}$ and $-\mathbf{d}$ and identify three types of regions on f: (i) those consisting of points that are in contact with supports for both directions, (ii) those in contact for exactly one direction, and (iii) those not in contact for either direction. These points define three sets of disjoint polygons that partition f. It is easy to argue as before, that the triangles resulting from triangulating the polygons of types (i)–(iii) are, respectively, black, gray, and white triangles on f.

We can compute the black and gray triangles on the front and back facets simultaneously using the well-known cylindrical decomposition method [**Mul94**]. This method in essence partitions the space that is outside $\mathcal{P}$ and "in between" its facets into triangular prisms that are oriented in direction $\mathbf{d}$. From these prisms the footprints of the supports can be inferred easily and the different types of triangles can be computed. The algorithm runs in $O(n^2 \log n)$ time. Parallel facets are handled somewhat similarly. For each parallel facet, the points of types (i)–(iii) are identified by doing a trapezoidal decomposition on a set of line segments obtained by intersecting each non-parallel facet with the plane containing the parallel facet. (Details on these algorithms may be found in [**IJM$^+$02**].)

At this point, we have computed the black, gray, and white triangles for all facets. Recall that only gray triangles are relevant for support minimization. We store the gray triangles from all the facets in a doubly-connected edge list [**dBvKOS00**] and perform a sweep over them, as in Section 8.1, to compute the optimum plane H. Observe that although the algorithm in Section 8.1 is described for convex polyhedra, it does not really rely on convexity and, in fact, is applicable to any set of triangles.

8.3. Controlling the size of the decomposition. Recall that we assumed that $\mathbf{d}$ is the vertical direction. To ensure that there are at most K connected components in the polyhedra $\mathcal{P}^+$ and $\mathcal{P}^-$, a preprocessing step is applied, where the z-axis is partitioned into a sequence $\mathcal{I}$ of disjoint intervals $I_1, I_2, \ldots$, defined by the z-coordinates of the vertices of $\mathcal{P}$, such that for any position of H within each interval I_j, the number of connected components is the same (say k_j). The above algorithm is then run as before, but the minimization is done only within intervals I_j for which $k_j \leq K$.

The sequence $\mathcal{I}$ is identified via a two-step process. First, a sequence $\mathcal{I}'$ of intervals $I'_1, I'_2, \ldots$, is found such that for any position of H within each interval I'_j, the number of connected components of $\mathcal{P}^-$ is the same. This is done by sweeping a horizontal plane upwards over the sorted sequence of vertices of $\mathcal{P}$ and maintaining the connected components of $\mathcal{P}^-$ using a Union–Find–Makeset data structure; see Cormen *et al.* [**CLRS01**]. Specifically, the vertices of the connected components are maintained as disjoint sets and when a new vertex is encountered its incident edges are examined and used to merge components together; a count of these components is maintained with the interval defined by the z-coordinates of the current vertex and its successor. The key observation is that during this process, the connected components of $\mathcal{P}^-$ can only merge, never split, so the above data structure suffices. The second step is symmetric and it identifies, via a downward sweep, a sequence $\mathcal{I}''$ of intervals $I''_1, I''_2, \ldots$, such that for any position of H within each interval I''_j, the number of connected components of $\mathcal{P}^+$ is the same. The desired sequence $\mathcal{I}$ is inferred by scanning and combining the intervals of $\mathcal{I}'$ and $\mathcal{I}''$ suitably. This preprocessing adds only $O(n \log n)$ time to the running time of the overall algorithm. (Details may be found in [**IJM$^+$02**].)

We summarize the discussion in this section in the following theorem.

THEOREM 8.1. *Let $\mathcal{P}$ be a polyhedron with n facets. Given a direction* $\mathbf{d}$ *and an integer K, it is possible to compute a plane H which decomposes $\mathcal{P}$ into at most K connected components such that when the components in the halfspace of H containing* $\mathbf{d}$ *(resp.* $-\mathbf{d}$*) are built in directions* $\mathbf{d}$ *(resp.* $-\mathbf{d}$*), the total contact-area is minimized. The algorithm runs in $O(n^2 \log n)$ time. A similar result also applies for support volume minimization.*

This result assumes a fixed decomposition direction $\mathbf{d}$. A natural, but challenging, extension is finding an optimium decomposition over all directions. Some progress has been made on a two-dimensional version of the problem. A simple polygon is a *terrain* if each point on its boundary and interior can be connected to a single edge, called a *base*, by a line segment which is contained completely inside the polygon. Ilinkin *et al.* [**IJS02**] present algorithms to decompose a simple polygon with a line into two terrains. Their algorithm runs in $O(n \log n)$ time if the two terrains have a common base and in $O(n^2 \log n)$ time otherwise. The motivation for this is that it allows certain classes of polyhedra to be built without supports—specifically, those that are generated by extruding (and possibly scaling) such a decomposable polygon. It is natural to also consider decompositions that minimize the number of terrains. However, Fekéte and Mitchell [**FM01**] have proved that it is NP-complete to decide if a polygon with holes, or a polyhedron of genus zero, can be decomposed into K terrains, where K is part of the input.

Related problems also arise in the context of casting and injection molding. Rosenbloom and Rappaport [**RR94**] give an efficient algorithm to decompose a simple polygon into two terrains with a common base. Bose *et al.* [**BBvK97**] consider the problem of decomposing a polyhedron by a plane and building mold-halves for the two pieces such that the cast polyhedron can be de-molded without obstruction. This is equivalent to deciding if the polyhedron can be decomposed into two polyhedra that are terrains with respect to the common base formed by the dividing plane. Decomposition problems at the interface of Layered Manufacturing and injection molding have been considered recently by McMains [**McM02**].

9. Computing an optimal hatching direction

In this section, we consider the problem of printing, i.e., hatching, the individual layers. Unlike the other problems considered in this paper, this is essentially a two-dimensional problem since each layer is a polygon. We give an overview of an exact algorithm, due to Schwerdt *et al.* [**SSHJ02**], and an efficient heuristic, due to Hon *et al.* [**HJSS03**].

9.1. An exact algorithm. Let P be an n-vertex polygon in the plane, possibly with holes. Let $\mathbf{d}$ be a hatching direction, specified as a unit-vector in the plane. Let δ be the width of the tool-tip (e.g., the laser in Stereolithography). Let $\ell_0(\mathbf{d})$ be the line through the origin in direction $\mathbf{d}$ and let $\mathcal{L}(\mathbf{d})$ be the set of all lines that are parallel to $\ell_0(\mathbf{d})$ and whose distances to $\ell_0(\mathbf{d})$ are integer multiples of δ. Let S_ℓ be the set of line segments of positive length in the intersection of any line $\ell \in \mathcal{L}(\mathbf{d})$ with P and let $H(\mathbf{d}) = \sum_{\ell \in \mathcal{L}(\mathbf{d})} |S_\ell|$. Thus, $H(\mathbf{d})$ is the number of hatching segments for direction $\mathbf{d}$. Our goal is to find a direction $\mathbf{d}$ such that $H(\mathbf{d})$ is minimized.

Observe that $H(\mathbf{d}) = H(-\mathbf{d})$, so it suffices to consider directions $\mathbf{d} = (d_1, d_2)$ for which $d_2 \geq 0$. The algorithm performs a sweep around the unit-circle, starting from the direction $\mathbf{d} = (-1, 0)$ and going clockwise to $\mathbf{d} = (1, 0)$. During this sweep, the value of $H(\mathbf{d})$ changes at certain directions $\mathbf{d}$, called *critical directions*, and we update $H(\mathbf{d})$ at these points. At the end of the sweep, we will have found the optimal hatching direction.

We now state necessary conditions for a direction $\mathbf{d}$ to be critical, i.e., for which the value of $H(\mathbf{d})$ changes. There are two cases:

- $\mathbf{d}$ is such that the subset of lines in $\mathcal{L}(\mathbf{d})$ that intersect P changes. This happens if there is a vertex v of P such that (a) v is extreme in a direction orthogonal to $\mathbf{d}$, and (b) v lies on a line of $\mathcal{L}(\mathbf{d})$, i.e., the distance between v and $\ell_0(\mathbf{d})$ is an integer multiple of δ.
- $\mathbf{d}$ is such that for some line $\ell \in \mathcal{L}(\mathbf{d})$, the set of segments in S_ℓ changes. This happens if there is a vertex v of P such that (a) both neighbors of v are on the same side of the line through v that is parallel to $\ell_0(\mathbf{d})$, and (b) v lies on a line of $\mathcal{L}(\mathbf{d})$.

Define the set $\mathcal{D}$ to consist of all directions $\mathbf{d}$ for which there is a vertex v of P lying on a line of $\mathcal{L}(\mathbf{d})$. Clearly, $\mathcal{D}$ contains all critical directions.

We begin by computing the set $\mathcal{D}$, for directions $\mathbf{d} = (d_1, d_2)$ for which $d_2 \geq 0$, and sort these in clockwise order starting from the direction $(-1, 0)$. Let $\mathbf{d}^0 \prec \mathbf{d}^1 \prec \cdots \prec \mathbf{d}^{m-1}$ be the m distinct directions in $\mathcal{D}$. Observe that $H(\mathbf{d}') = H(\mathbf{d}'')$ for any directions $\mathbf{d}'$ and $\mathbf{d}''$ that are strictly between $\mathbf{d}^i$ and $\mathbf{d}^{i+1}$.

Next we compute $H(\mathbf{d}^s)$ for some direction $\mathbf{d}^s$ such that $\mathbf{d}^{k-1} \prec \mathbf{d}^s \prec \mathbf{d}^k$. We then consider the elements of $\mathcal{D}$ in the order $\mathbf{d}^k, \mathbf{d}^{k+1}, \ldots, \mathbf{d}^{m-1}, \mathbf{d}^0, \ldots, \mathbf{d}^{k-1}$. For each such $\mathbf{d}^i$ that is a critical direction, we first compute $H(\mathbf{d}^i)$ from $H(\mathbf{d})$, for $\mathbf{d}^{i-1} \prec \mathbf{d} \prec \mathbf{d}^i$, and then compute $H(\mathbf{d})$ from $H(\mathbf{d}^i)$, for $\mathbf{d}^i \prec \mathbf{d} \prec \mathbf{d}^{i+1}$. After all directions $\mathbf{d}^i$ have been processed in this way, the minimum $H(\mathbf{d})$ found is reported along with the corresponding $\mathbf{d}$.

Further details on this algorithm can be found in Schwerdt *et al.* [**SSHJ02**]. We summarize the main result in the following theorem.

THEOREM 9.1. *Let P be an n-vertex polygon, possibly with holes. A direction* $\mathbf{d}$ *which minimizes the number of hatching segments $H(\mathbf{d})$ can be computed in time $O(Cn\log(Cn))$. Here $C = 1 + L/\delta$, where L is the maximum distance of any vertex of P from the origin.*

Although the algorithm is conceptually simple, care must be taken to handle many special cases. Indeed there are seventy-two different cases to be considered! Nevertheless, the algorithm has been implemented and tested on layers generated from real-world models. Owing to the use of exact arithmetic based on the LEDA number type `real` [**MN99**], and the dependence on L/δ, the algorithm is fairly slow. However, this algorithm turns out to be crucial in verifying the performance of a fast heuristic which we describe next.

9.2. An efficient heuristic. The idea behind this approach is as follows. Let $\mathbf{d}$ be any hatching direction and let e be any edge of P. The number of times the tool-tip runs into e is roughly equal to the length of the projection of e normal to $\mathbf{d}$ divided by δ, i.e., $|\mathbf{n}_e \cdot \mathbf{d}|/\delta$, where $\mathbf{n}_e$ is the outer normal to e whose length is the same as that of e. (The error in this bound depends on how far away the first and last lines of $\mathcal{L}(\mathbf{d})$ are from the endpoints of e. The smaller the value of δ, the smaller is the error.) Since each such intersection of the tool-tip with e creates an endpoint of a hatching segment, the total number of hatching segments in direction $\mathbf{d}$ is one-half the sum of the projected lengths of the edges divided by δ. Let $\mathcal{S}$ be the set of outer normals of the edges of P, where each normal has the same length as the corresponding edge and begins at the origin. It then follows that an approximate solution to the hatching problem can be obtained by computing a direction $\mathbf{d}$ such that $\sum_{\mathbf{n}_e \in \mathcal{S}} |\mathbf{n}_e \cdot \mathbf{d}|$ is minimized.

The key observation is that $\sum_{\mathbf{n}_e \in \mathcal{S}} |\mathbf{n}_e \cdot \mathbf{d}|$ depends only on the lengths and the orientations of the edges of P and not on how the edges connect to form P. This suggests the possibility of connecting the edges of P in a different order to produce a new polygon Q for which the desired direction can be computed more easily. Indeed, the solution outlined below replaces P with a convex polygon Q and shows that the desired direction $\mathbf{d}$ can be obtained by computing the width of Q.

We first replace all vectors in $\mathcal{S}$ that point in the same direction by a single vector equal to their sum. We then sort these vectors in circular order around the origin and walk through this list to build a chain of vectors as follows; see also Figure 6. Initially, the chain is empty; when we reach the current vector in the walk, we update the chain by adding the current vector so that its tail is at the head of the chain. The chain obtained at the end of the walk will be a closed polygon since P is a closed polygon and each edge of the chain is a vector that is rotated ninety degrees (say, counterclockwise) from an edge of P and has the same length. Moreover, the polygon will be convex since the edges are added in sorted angular order. Denote this convex polygon by Q.

For any direction $\mathbf{d}$, the boundary of Q can be partitioned into two chains by cutting at the two vertices u and v that are extreme in directions $\mathbf{d}$ and $-\mathbf{d}$, respectively; see Figure 6. The projections of the chains in a direction perpendicular to $\mathbf{d}$ will be non-overlapping (except at u and v). Let ℓ_u and ℓ_v be, respectively, the lines through u and v that are perpendicular to $\mathbf{d}$. It follows that $\sum_{\mathbf{n}_e \in \mathcal{S}} |\mathbf{n}_e \cdot \mathbf{d}|$ is

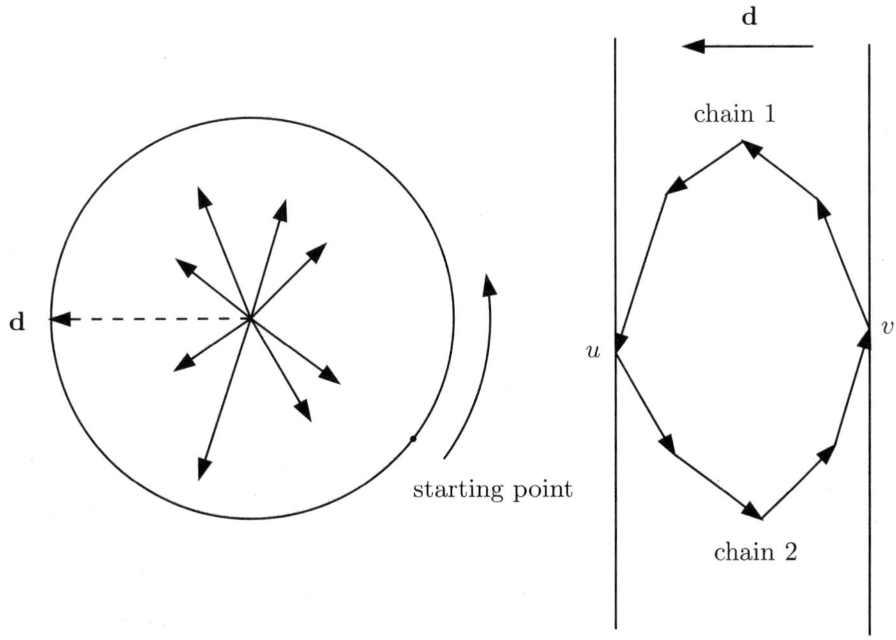

FIGURE 6. *A set of vectors and the resulting convex polygon. The sum of the absolute values of the dot products of the vectors with respect to direction* $\mathbf{d}$ *is twice the width of the convex polygon in the direction perpendicular to* $\mathbf{d}$.

simply twice the distance between ℓ_u and ℓ_v. Thus, the direction $\mathbf{d}$ which minimizes $\sum_{\mathbf{n}_e \in \mathcal{S}} |\mathbf{n}_e \cdot \mathbf{d}|$ is one which realizes the minimum distance between two parallel lines that enclose Q, i.e., the *width* of Q. We can, therefore, solve the problem by computing the width of Q, which can be done in $O(n \log n)$ time and using $O(n)$ space; see Houle and Toussaint [**HT88**]. We note that a similar approach was proposed independently by Sarma [**Sar99**] in the context of path-planning for milling.

THEOREM 9.2. *Let* P *be a simple* n*-vertex polygon, possibly with holes. Let* $\mathbf{n}_e$ *be the outer normal of edge* $e \in P$*, where* $\mathbf{n}_e$ *has the same length as* e *and begins at the origin. Let* $\mathcal{S}$ *be the set of such outer normals for all edges of* P*. Then a direction* $\mathbf{d}$ *such that* $\sum_{\mathbf{n}_e \in \mathcal{S}} |\mathbf{n}_e \cdot \mathbf{d}|$ *is minimized can be found in* $O(n \log n)$ *time and using* $O(n)$ *space.*

We next describe an alternative algorithm which does not require explicit construction of Q. It works for more general sets of vectors than those derived from the normals of P—a fact that will be useful in some other applications that we mention at the end of the section. The algorithm can also be generalized to higher dimensions.

Henceforth, we assume that $\mathcal{S}$ is a set of n arbitrary vectors in the plane, where each vector begins at the origin. We wish to compute a direction $\mathbf{d}$ such that $\sum_{\mathbf{v} \in \mathcal{S}} |\mathbf{v} \cdot \mathbf{d}|$ is minimized. We pick an arbitrary unit-vector $\mathbf{d}$ as a candidate direction and draw a line perpendicular to $\mathbf{d}$ through the origin. This line cuts the plane into two half-planes. The vectors $\mathbf{v} \in \mathcal{S}$ that are in the same closed half-plane as $\mathbf{d}$ generate a non-negative dot-product with $\mathbf{v}$. However, those in the the complement of the above half-plane generate a negative dot product with $\mathbf{d}$. We can correct for this by reflecting these vectors through the origin. Let this new set of vectors be $\tilde{\mathcal{S}}$.

All the vectors $\tilde{\mathbf{v}}$ in $\tilde{\mathcal{S}}$ lie in the same closed half-plane as $\mathbf{d}$. Thus, $\sum_{\mathbf{v} \in \mathcal{S}} |\mathbf{v} \cdot \mathbf{d}|$ reduces to $\sum_{\tilde{\mathbf{v}} \in \tilde{\mathcal{S}}} \tilde{\mathbf{v}} \cdot \mathbf{d}$, which can be rewritten as $\left(\sum_{\tilde{\mathbf{v}} \in \tilde{\mathcal{S}}} \tilde{\mathbf{v}}\right) \cdot \mathbf{d}$. If no vector of $\tilde{\mathcal{S}}$ is on the cutting line, we can rotate $\mathbf{d}$ away from $\sum_{\tilde{\mathbf{v}} \in \tilde{\mathcal{S}}} \tilde{\mathbf{v}}$, thereby decreasing $(\sum_{\tilde{\mathbf{v}} \in \tilde{\mathcal{S}}} \tilde{\mathbf{v}}) \cdot \mathbf{d}$. We can do this until one of the vectors $\tilde{\mathbf{v}}$ is on the cutting line perpendicular to $\mathbf{d}$. Any further movement of $\mathbf{d}$ will cause $\tilde{\mathbf{v}}$ to go to the other side of the cutting line. Thereafter, the contribution of (the reflection of) $\tilde{\mathbf{v}}$ will cause $(\sum_{\tilde{\mathbf{v}} \in \tilde{\mathcal{S}}} \tilde{\mathbf{v}}) \cdot \mathbf{d}$ to increase. Thus, the position of the cutting line that coincides with one of the input vectors must be a local minimum for $(\sum_{\tilde{\mathbf{v}} \in \tilde{\mathcal{S}}} \tilde{\mathbf{v}}) \cdot \mathbf{d}$.

We update $\sum_{\tilde{\mathbf{v}} \in \tilde{\mathcal{S}}} \tilde{\mathbf{v}}$ efficiently by visiting the vectors in a circular order, from $\mathbf{d}$ to $-\mathbf{d}$. Specifically, each vector $\tilde{\mathbf{v}}$ has associated with it two regions, separated by the line perpendicular to $\tilde{\mathbf{v}}$. In the walk, whenever we pass this line, the associated vector's contribution to the sum changes sign. If $\tilde{\mathbf{v}}_i$ is the associated vector, we subtract $2\tilde{\mathbf{v}}_i$ from $\sum_{\tilde{\mathbf{v}} \in \tilde{\mathcal{S}}} \tilde{\mathbf{v}}$: one copy to take it off from the sum, and another copy to insert it back in with a negative sign. At each event, we use the newly updated vector sum to re-calculate $(\sum_{\tilde{\mathbf{v}} \in \tilde{\mathcal{S}}} \tilde{\mathbf{v}}) \cdot \mathbf{d}$. Since the update can be done in $O(1)$ time, we can find the minimum in $O(n)$ time given the circular list of vectors. The total running time is dominated by the time to prepare the list itself, in this case the $O(n \log n)$ time for sorting. This yields the following result, which generalizes Theorem 9.2.

THEOREM 9.3. *Let $\mathcal{S}$ be a set of n arbitrary vectors in the plane, where each vector begins at the origin. A direction $\mathbf{d}$ which minimizes $\sum_{\mathbf{v} \in \mathcal{S}} |\mathbf{v} \cdot \mathbf{d}|$ can be computed in $O(n \log n)$ time and using $O(n)$ space.*

This algorithm has been implemented and tested on layers obtained from real-world models (using a pessimistically large value of 0.1 inches for δ) and the results have been compared to those obtained by the exact algorithm from Section 9.1. It was found that in all cases, the number of hatching segments found by the heuristic was within a small additive constant of the number found by the exact algorithm; moreover, the heuristic was about three to five orders of magnitude faster. A conservative analysis of the heuristic shows that under some reasonable assumptions, the number of hatching segments produced by the heuristic is at most $1+3/k$ times the minimum number, for some integer $k \geq 1$. (Details of this analysis and the experiments may be found in [**HJSS03**].)

We note that it is straightforward to extend the above method to compute an approximation to an optimal hatching direction for all layers of a three-dimensional model: We simply run the algorithm on the set of normals of the edges of the polygons from all layers.

We close by mentioning some other applications of the projection-minimization technique [**HJSS03**]. Each of these problems can be solved using the second algorithm described above in $O(n \log n)$ time and using $O(n)$ space.

Weighted hatching: The idea here is to assign weights to different edges of the polygons, where the weights reflect the importance of the edge to the form or function of the part. The goal then is to find a direction $\mathbf{d}$ which minimizes the sum of the weights of the edges of P that are hit by the lines in $\mathcal{L}(\mathbf{d})$.

Hatching in two directions: The idea here is to improve the strength of the part by hatching it along two prescribed directions that have some fixed angular relationship to one another. The goal is to find a pair of such directions which minimizes the total number of hatching segments.

Minimizing stair-step error in 2D: The non-zero width δ of the tool-tip implies that the hatched polygon is not an exact replica of the original polygon P; instead it has a stair-stepped appearance. Analogous to the three-dimensional case discussed in Section 3, one can define an error-triangle on each edge of the polygon and then minimize the sum of the stair-step errors on all edges of P by finding a suitable hatching direction $\mathbf{d}$.

Finally, we note that Theorem 9.3 can be generalized to $d > 2$ dimensions, using time $O(n^{d-1} \log n)$ (resp. $O(n^{d-1})$) and space $O(n)$ (resp. $O(n^{d-1})$).

10. Final remarks

We have presented an overview of recent work on the application of techniques from computational geometry to several design optimization problems in Layered Manufacturing. Our goal has been to cover the topics in sufficient depth, so that the interested reader can use this as a starting point for further research.

There are several interesting avenues for further research. For instance, can the minimum support volume for general polyhedra be approximated? A challenge in extending the approach presented for contact-area is the absence of a suitable criterion for evaluating the quality of any proposed heuristic for support volume minimization. Another challenging problem is to compute a decomposition of a general polyhedron which is globally optimal, i.e., one which minimizes the contact-area or support volume, when taken over all possible directions $\mathbf{d} \in \mathbb{S}^2$. Similarly, it would be interesting to compute a hatching direction that is optimal over all possible orientations of the model. Also useful would be other decomposition modalities; for instance, decomposing the model with three planes that are spaced 120° apart. Yet another problem is stair-step error minimization in the presence of adaptive slicing, where the layer thickness is varied dynamically in the vicinity of fine features in the model. The methods presented in this paper assume a fixed layer thickness and it would be interesting to generalize them to the case where the layer thickness lies in some prescribed range. Finally, since the STL format is inherently redundant and error-prone, it would be interesting to consider alternatives (e.g., analytic surfaces such as quadrics, see Farouki and König [**FK96**]) and the effect of these formats on the algorithms presented here.

We close with a partial list of other interesting aspects of Layered Manufacturing that we have not discussed here. This includes automatic repair of STL files (see Bøhn [**Bøh95**] and Barequet [**Bar97**]), methods for support structure elimination through selective thickening of walls of the model (see Allen and Dutta [**AD98**]), new modeling techniques based on voxels (see Chandru *et al.* [**CMP95**]), and design issues associated with building a complete software front-end for Layered Manufacturing (see Barequet and Kaplan [**BK98**]). We refer the reader to Jacobs [**Jac92**], Kai and Fai [**KF97**], Chua *et al.* [**CLL03**], and Stucki *et al.* [**SBE95**] for more information on these and related topics.

References

[ABB+97] B. Asberg, G. Blanco, P. Bose, J. Garcia-Lopez, M. Overmars, G. Toussaint, G. Wilfong, and B. Zhu, *Feasibility of design in stereolithography*, Algorithmica **19** (1997), 61–83.

[AD95] S. Allen and D. Dutta, *Determination and evaluation of support structures in layered manufacturing*, Journal of Design and Manufacturing **5** (1995), 153–162.

[AD98] ———, *Wall thickness control in layered manufacturing for surfaces with closed slices*, Computational Geometry: Theory and Applications **10** (1998), 223–238.

[AD00] P. K. Agarwal and P. K. Desikan, *Approximation algorithms for layered manufacturing*, Proceedings of the 11th ACM-SIAM Symposium on Discrete Algorithms, 2000, pp. 528–537.

[AS96] P. K. Agarwal and M. Sharir, *Efficient randomized algorithms for some geometric optimization problems*, Discrete & Computational Geometry **16** (1996), 317–337.

[Bar97] G. Barequet, *Using geometric hashing to repair CAD models*, IEEE Computational Science & Engineering **4** (1997), no. 4, 22–28.

[Bau75] B. G. Baumgart, *A polyhedron representation for computer vision*, Proceedings of the AFIPS National Computer Conference, vol. 44, AFIPS Press, Alrington, Va., 1975, pp. 589–596.

[BB95] M. Bablani and A. Bagchi, *Quantification of errors in rapid prototyping processes and determination of preferred orientation of parts*, Transactions of the 23rd North American Manufacturing Research Conference, 1995.

[BBvK97] P. Bose, D. Bremner, and M. van Kreveld, *Determining the castability of simple polyhedra*, Algorithmica **19** (1997), 84–113.

[BD90] K. W. Bowyer and C. R. Dyer, *Aspect graphs: An introduction and survey of recent results*, International Journal of Imaging Systems and Technology **2** (1990), 315–328.

[BK98] G. Barequet and Y. Kaplan, *A data front-end for layered manufacturing*, Computer–Aided Design **30** (1998), 231–243.

[BO79] J. L. Bentley and T. A. Ottmann, *Algorithms for reporting and counting geometric intersections*, IEEE Transactions on Computers **C-28** (1979), 643–647.

[Bøh95] J. H. Bøhn, *Removing zero-volume parts from CAD models for layered manufacturing*, IEEE Computer Graphics and Applications **15** (1995), no. 6, 27–34.

[CLL03] C. K. Chua, K. F. Leong, and C. S. Lim, *Rapid prototyping: Principles and applications*, World Scientific, Singapore, 2003.

[CLRS01] T. H. Cormen, C. E. Leiserson, R. L. Rivest, and C. Stein, *Introduction to algorithms*, 2nd ed., MIT Press, Cambridge, MA, 2001.

[CMP95] V. Chandru, S. Manohar, and C. Prakash, *Voxel-based modeling for layered manugacturing*, IEEE Computer Graphics and Applications **15** (1995), no. 6, 42–47.

[CW92] L.-L. Chen and T. C. Woo, *Computational geometry on the sphere with application to automated machining*, Journal of Mechanical Design **114** (1992), 288–295.

[dBvKOS00] M. de Berg, M. van Kreveld, M. Overmars, and O. Schwarzkopf, *Computational geometry: Algorithms and applications*, 2nd ed., Springer-Verlag, Berlin, 2000.

[DK85] D. P. Dobkin and D. G. Kirkpatrick, *A linear algorithm for determining the separation of convex polyhedra*, Journal of Algorithms **6** (1985), 381–392.

[EG89] H. Edelsbrunner and L. J. Guibas, *Topologically sweeping an arrangement*, Journal of Computer and System Sciences **38** (1989), 165–194.

[FF94] D. Frank and G. Fadel, *Preferred direction of build for rapid prototyping processes*, Proceedings of the 5th International Conference on Rapid Prototyping, 1994, pp. 191–200.

[FK96] R. Farouki and T. König, *Computational methods for rapid prototyping of analytic solid models*, Rapid Prototyping Journal **2** (1996), no. 3, 41–48.

[FM01] S. P. Fekéte and J. S. B. Mitchell, *Terrain decomposition and layered manufacturing*, International Journal of Computational Geometry & Applications **11** (2001), 647–668.

[GH01] B. Gärtner and T. Herrmann, *Computing the width of a point set in 3-space*, Proceedings of the 13th Canadian Conference on Computational Geometry, 2001, pp. 101–103.

[GO95] A. Gajentaan and M. H. Overmars, *On a class of $O(n^2)$ problems in computational geometry*, Computational Geometry: Theory and Applications **5** (1995), 165–185.

[HJSS03] M. C. Hon, R. Janardan, J. Schwerdt, and M. Smid, *Minimizing the total projection of a set of vectors, with applications to layered manufacturing*, Computer-Aided Design **35** (2003), 57–68.

[HT88] M. E. Houle and G. T. Toussaint, *Computing the width of a set*, IEEE Transactions on Pattern Analysis and Machine Intelligence **PAMI-10** (1988), 761–765.

[IJM+02] I. Ilinkin, R. Janardan, J. Majhi, J. Schwerdt, M. Smid, and R. Sriram, *A decomposition-based approach to layered manufacturing*, Computational Geometry: Theory and Applications **23** (2002), 117–151.

[IJS02] I. Ilinkin, R. Janardan, and M. Smid, *Terrain polygon decomposition, with application to layered manufacturing*, Proceedings of the 8th International Computing and Combinatorics Conference, Lecture Notes in Computer Science, vol. 2387, Springer-Verlag, 2002, pp. 381–390.

[IJS+04] I. Ilinkin, R. Janardan, M. Smid, E. Johnson, P. Castillo, and J. Schwerdt, *Approximating contact-area of supports in layered manufacturing*, Technical Report TR–04–001, Department of Computer Science and Engineering, University of Minnesota, Minneapolis, MN, 2004, Abbreviated version appears in Proceedings of the 16th Canadian Conference on Computational Geometry, 2004.

[Jac92] P. F. Jacobs, *Rapid prototyping & manufacturing: Fundamentals of stereolithography*, McGraw-Hill, New York, 1992.

[KF97] C. C. Kai and L. K. Fai, *Rapid prototyping: Principles and applications in manufacturing*, John Wiley & Sons, 1997.

[Mat93] J. Matoušek, *Range searching with efficient hierarchical cuttings*, Discrete & Computational Geometry **10** (1993), 157–182.

[McM02] S. McMains, *Double sided layered manufacturing*, Proceedings of the Japan-USA Symposium on Flexible Automation, 2002, pp. 269–272.

[MJS+99] J. Majhi, R. Janardan, J. Schwerdt, M. Smid, and P. Gupta, *Minimizing support structures and trapped area in two-dimensional layered manufacturing*, Computational Geometry: Theory and Applications **12** (1999), 241–267.

[MJSG97] J. Majhi, R. Janardan, M. Smid, and P. Gupta, *On some geometric optimization problems in layered manufacturing*, Tech. Report TR–97–002, Department of Computer Science, University of Minnesota, 1997.

[MJSG99] ______, *On some geometric optimization problems in layered manufacturing*, Computational Geometry: Theory and Applications **12** (1999), 219–239.

[MJSS01] J. Majhi, R. Janardan, M. Smid, and J. Schwerdt, *Multi-criteria geometric optimization problems in layered manufacturing*, International Journal of Mathematical Algorithms **2** (2001), 201–225.

[MN99] K. Mehlhorn and S. Näher, *LEDA: A platform for combinatorial and geometric computing*, Cambridge University Press, Cambridge, U.K., 1999.

[MS99] S. McMains and C. Séquin, *A coherent sweep plane slicer for layered manufacturing*, Proceedings of the 5th ACM Symposium on Solid Modeling and Applications, 1999, pp. 285–295.

[Mul94] K. Mulmuley, *Computational geometry: An introduction through randomized algorithms*, Prentice Hall, Englewood Cliffs, NJ, 1994.

[NS89] O. Nurmi and J.-R. Sack, *Separating a polyhedron by one translation from a set of obstacles*, Proceedings of the 14th International Workshop on Graph-Theoretic Concepts in Computer Science, Lecture Notes in Computer Science, vol. 344, Springer-Verlag, 1989, pp. 202–212.

[O'R98] J. O'Rourke, *Computational geometry in C*, 2nd ed., Cambridge University Press, Cambridge, UK, 1998.

[PD90] H. Plantinga and C. R. Dyer, *Visibility, occlusion, and the aspect graph*, International Journal of Computer Vision **5** (1990), 137–160.

[RR94] D. Rappaport and A. Rosenbloom, *Moldable and castable polygons*, Computational Geometry: Theory and Applications (1994), 219–233.

[Sar99] S. Sarma, *The crossing function and its application to zig-zag tool paths*, Computer-Aided Design **31** (1999), 881–890.

[SBE95] P. Stucki, J. Bresenham, and R. Earnshaw, *Computer graphics in rapid prototyping technology*, IEEE Computer Graphics and Applications **15** (1995), no. 6, Edited special issue.

[Sch01] J. Schwerdt, *Entwurf von Optimierungsalgorithmen für geometrische Probleme im Bereich Rapid Prototyping und Manufacturing*, Ph.D. thesis, Department of Computer Science, University of Magdeburg, Magdeburg, Germany, 2001.

[SSHJ02] J. Schwerdt, M. Smid, M. C. Hon, and R. Janardan, *Computing an optimal hatching direction in layered manufacturing*, International Journal of Computer Mathematics **79** (2002), 1067–1081.

[SSJ$^+$00] J. Schwerdt, M. Smid, R. Janardan, E. Johnson, and J. Majhi, *Protecting critical facets in layered manufacturing*, Computational Geometry: Theory and Applications **16** (2000), 187–210.

[SSJJ03] J. Schwerdt, M. Smid, R. Janardan, and E. Johnson, *Protecting critical facets in layered manufacturing: implementation and experimental results*, Computer-Aided Design **35** (2003), 647–657.

[SSMJ99] J. Schwerdt, M. Smid, J. Majhi, and R. Janardan, *Computing the width of a three-dimensional point set: an experimental study*, ACM Journal of Experimental Algorithmics **4** (1999), Article 8.

Department of Computer Science & Engineering, University of Minnesota, Minneapolis, MN 55455, U.S.A.

E-mail address: janardan@cs.umn.edu

School of Computer Science, Carleton University, Ottawa, Ontario, Canada K1S 5B6

E-mail address: michiel@scs.carleton.ca

DIMACS Series in Discrete Mathematics
and Theoretical Computer Science
Volume **67**, 2005

A Process Planning Framework for Multi-Direction Layered Deposition

Prabhjot Singh and Debasish Dutta

ABSTRACT. Layered Manufacturing (LM) techniques build a part by adding thin layers of material. In this process overhangs need to be supported by sacrificial supports, resulting in an increase in the build time, wastage of material and requiring costly post-processing. Emerging LM processes with the capability to deposit material along multiple directions have the potential for support-less part fabrication. In this paper we analyse the process planning issues in multi-directional LM. An overview of our research to solve some of the process planning tasks is presented. Key features of our solution methodology include a part decomposition algorithm and the use of spherical maps to select direction vectors for material deposition. Directions for future research and potential applications of multi-directional LM are suggested.

1. Introduction

Layered Manufacturing (LM) processes have developed the ability to deposit a variety of materials with an increased emphasis on the fabrication of functional prototypes. There is a vast body of literature covering various aspects of Layered Manufacturing processes. The interested reader is directed to [**15**] for a survey of LM processes and the Proceedings of the annual Solid Freeform Fabrication Proceedings hosted by the University of Texas, Austin.

In this paper we will concern ourselves with a specific set of LM processes which can deposit material along multiple directions. The development of these processes has been motivated by two salient characteristics of LM processes, namely the need for support structures to deposit overhanging features (Ref. Figure 1a) and the so called staircase effect. The staircase effect concerns the approximate construction of surfaces which are not aligned along the build (deposition) direction and is qualified by the cusp height (Ref. Figure 1b). The cumulative effect of these is a longer build time, material wastage, deterioration of surface quality and time consuming post processing.

Multi-Direction Layered Deposition (MDLD) systems have the ability to deposit material along multiple directions. A part is decomposed into smaller subvolumes which are built along multiple build directions. This enables support-less layered deposition and a better control over the cusp height.

Key words and phrases. Layered Manufacturing, Multiple Degrees of Freedom, Slicing, Offset Surfaces, Mechanism Design.

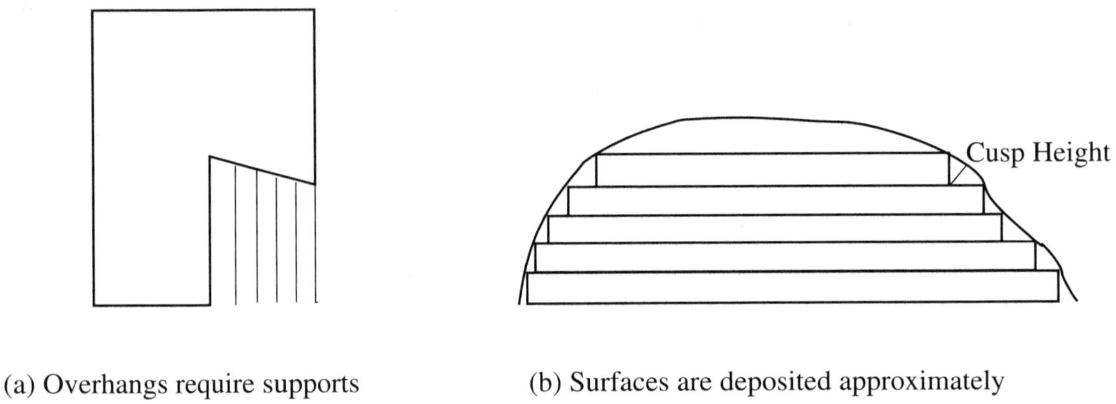

FIGURE 1. Process Characteristics of Layered Manufacturing

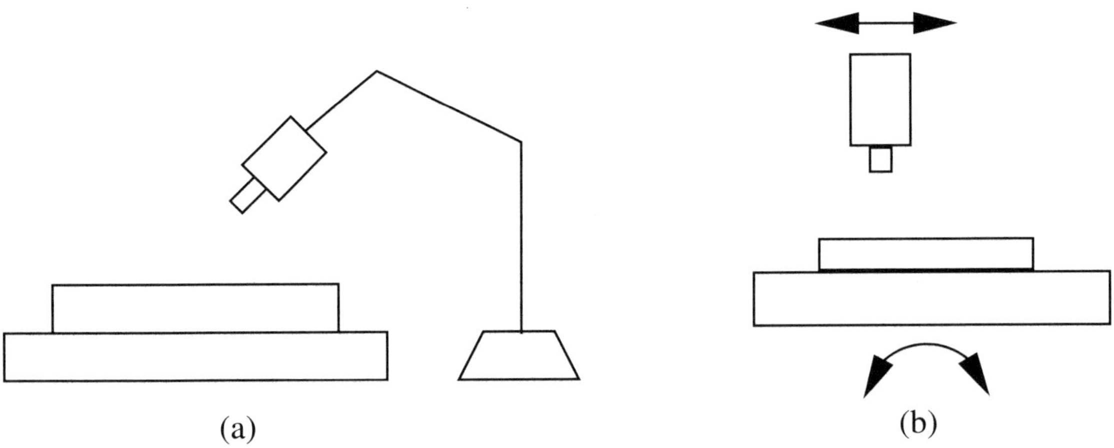

FIGURE 2. (a) Deposition nozzle mounted on a robot arm (b) Multi-Direction Deposition configuration with a rotating deposition table and a translating deposition nozzle - Schematic

Process planning refers to the generation of tool paths and selection of process parameters to build an object using a manufacturing process. The reader is referred to [**15, 10**] for a survey of process planning tasks in LM. The use of multiple deposition directions necessitates additional process planning for MDLD. This paper focuses on the process planning challenges in MDLD and an overview of our research efforts to address them. To better understand the tasks involved, we begin by briefly summarizing the capabilities of the available MDLD techniques and their process characteristics. The physics of the deposition processes is not discussed in detail; instead the focus is on deriving the process characteristics which have a bearing on the associated process planning. This survey also motivates the categorization of the process planning tasks.

2. Multi-Directional Layered Deposition Processes

Multi-direction Layered Deposition processes have developed independently and employ a variety of deposition modes for part fabrication. The degree of freedom in choosing a deposition direction varies from one process to another.

Processes specifically designed to have MDLD capabilities include Direct Metal Deposition (DMD) [**18**], Five Axis Rapid Metal Forming [**13**], Laser Chemical Vapor Deposition (LCVD) [**23**] and 3D Welding [**4**]. Both DMD and the Rapid Metal Forming process utilize a laser cladding process with CO_2 and Nd:YAG lasers respectively. A high power laser generates a small melt pool on the substrate while metal powder is injected through a concentric nozzle. MDLD is realized by the use of a deposition table with translational and rotational degrees of freedom. The Rapid Metal Forming process uses a 5-axis CNC milling machine in addition to the deposition process. This improves the surface quality while the part is being fabricated.

LCVD [**23**] is an adaptation of the traditional Chemical Vapor Deposition (CVD) process. In this process the substrate is placed in a chamber with a supply of CVD reagent gases. The substrate is heated close to its melting point; thereafter a laser is used to locally heat a spot on the surface. This initiates a thermal decomposition of the reactant gases resulting in the deposition of material at the same spot. MDLD is realized by moving the substrate relative to the laser beam. Since the mode of deposition is via a local chemical reaction, a variety of substrate shapes are possible.

The 3D welding process [**4**] was one of the earliest attempts at creating an MDLD system using a welding torch mounted on a highly articulated robot arm. The weld electrode provides the deposition material.

Some of the MDLD processes use 2.5D LM machines retrofitted with specialized kinematics permitting limited MDLD capabilities. Examples are the Double Sided Layered Manufacturing [**19**] and the Multi-Orientation Deposition (MOD) [**30**] processes. Double-Sided Layered Deposition uses a traditional Fused Deposition Modeling (FDM) machine with an additional fixturing mechanism which permits the deposition of the part along two directions. The part to be built is divided into two halves with a parting plane. The bottom half is deposited first. It is then flipped over and the remainder is deposited on its backside. The MOD process uses two deposition nozzles aligned perpendicular to each other. The deposition table in MOD has 2 translational (along x and y axes) and one rotational degree of freedom (along the z axis). This permits the support-less deposition of annular regions using offsets of a base layer.

We have considered a simplified model for MDLD machines in which the either the part or the deposition table can be re-oriented along any direction. Our solution methodology can be adapted to serve a process with limited capabilities. These limitations are usually manifested through reduced mobility to re-orient the part and potential collisions. Where appropriate, we have included guidelines for potential modifications.

3. Process Planning Challenge and Tasks

The objective of MDLD is support-less part fabrication and improved part quality. The overhang angle [**2**] between two contiguous layers is the most important process constraint in MDLD process planning. It limits deposition without

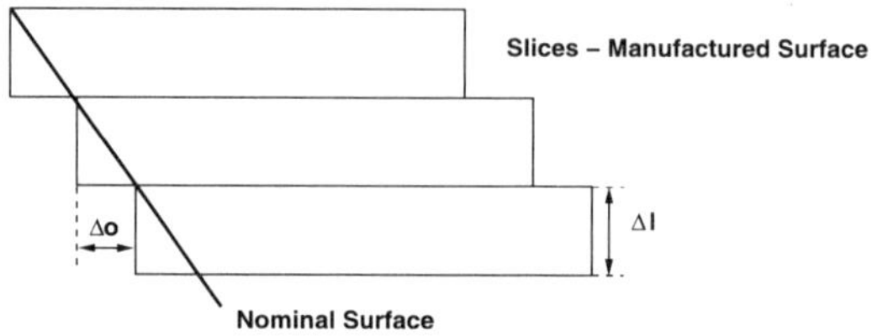

FIGURE 3. Relationship between the overhang angle, the slice thickness, the surface normal and the build direction.

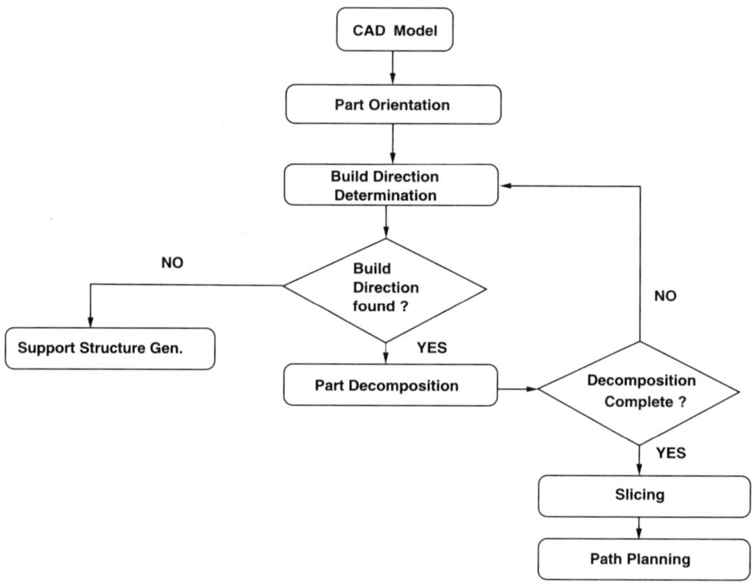

FIGURE 4. Sequence of Process Planning Operations for MDLD

sacrificial supports. We are primarily concerned with the overhang angle between two contiguous layers.

In LM literature [**2**], the overhang is defined as the maximum of the minimum distance between a point of the $i-1$ layer (L_{i-1}) and the corresponding point on the ith layer (L_i). The overhang angle is given by $\frac{\Delta o}{\Delta l}$, where Δo is the overhang and Δl is the slice thickness. The angle, θ, between the surface normal and the build direction and the overhang, Δo (with Δl begin the slice thickness) are related as:

$$\Delta o \approx \Delta l \cdot \cot\theta \tag{3.1}$$

Figure 3 shows the relationship between the layer thickness, the surface normal, the build direction and the overhang angle.

Figure 4 provides an overview of the process planning tasks in MDLD. The first task in MDLD is setting the orientation of the part relative to the deposition table. This is followed by two computational tasks which comprise the core of MDLD process planning. The first of these is determining the extents of the part volume which can be deposited along a certain build direction. The second task is the estimation of a new build direction in the event a part cannot be deposited completely along one direction. The change in build direction is realized by either re-orienting the part using a multi-axis deposition table or by moving the nozzle mounted on a multi-axis robot arm. In either case the direction of deposition is perpendicular to the deposited layers. Together, these tasks are used to progressively decompose the part into smaller volumes which are deposited along the associated build directions. These tasks are carried out recursively. In the ideal case part decomposition enables support-less fabrication. However, real world MDLD systems are constrained by limited freedom of motion and the likelihood of collisions during the deposition process. Depending on the part geometry, sacrificial supports may be required. The volume of sacrificial supports can be minimized by appropriately choosing an initial orientation of the part relative to the deposition table.

After part decomposition, build direction determination and support structure generation are completed the sequence of part volume deposition is established. This is followed by slicing and the generation of deposition path patterns (path planning).

In the following sections, we present an overview of our research in the development of algorithms for the following process planning tasks:

- Part Orientation.
- Part Decomposition.
- Build Direction Determination.
- Support Structure Generation.
- Sub-volume Sequencing.
- Slicing and Deposition Path Generation.

Where literature is not available, we list the likely computational problems in the development of solutions to the respective process planning task. We have also included pointers for further research to improve the presented algorithms.

4. Part Orientation

Part orientation in LM literature refers to orienting the part relative to the deposition table. The computational challenge in this process planning task is the search for a build direction along which either the build time is minimized or the part's surface quality is maximized. A summary of the literature concerning LM part orientation is provided in [**10**].

In the context of MDLD, this task is complex due to multiple build directions assigned to smaller part volumes. Setting up metrics such as improved part quality or reduction in build time must take into account the kind of part decomposition algorithms employed. Furthermore, the choice of initial orientation can also influence collisions between the deposited layers and the MDLD mechanism.

Literature on MDLD part orientation is scant. One of the efforts is presented in [**9**] with the object of minimizing the number of build directions. This, according to the authors will improve surface quality at the interface of the part volumes. The computational problem is shown to be NP-hard for 3D parts of genus 0. Factors such as collisions are not considered.

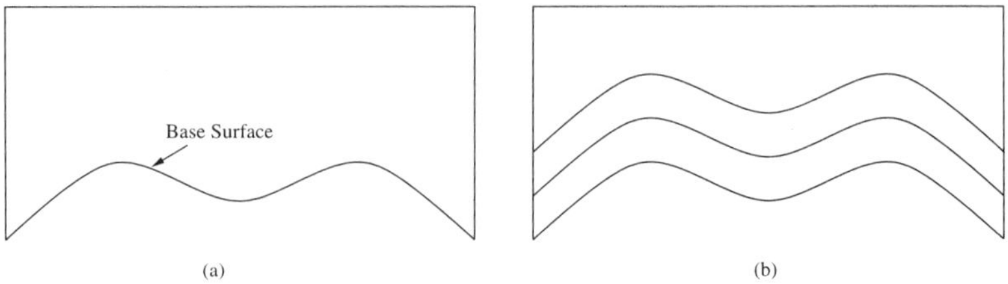

FIGURE 5. (a) Unbuildable Part Volume (b) Offset slices used to deposit the part volume

5. Part Decomposition

The input to this task are a part volume and an associated build vector. We shall refer to a part volume that can be deposited without supports along a chosen build direction as a *buildable volume*. The remainder is classified as *unbuildable*. In this task the input part volume is decomposed into buildable and unbuildable sub-volumes.

Before describing the challenge and solution underlying part volume decomposition, we discuss two relevant process characteristics of MDLD namely the deposition of non-planar slices and the effect of collisions on support-less part deposition.

MDLD processes have the ability to deposit non-planar slices. These are offsets of a base surface. Uniform slice thickness is maintained as they are parallel to each other. Figure 5 shows the geometry of these slices. Figure 5a show the base surface in red. The part is deposited using offsets of this base surface (ref. Figure 5b). The build direction is along the surface normal. Such slices are also referred to as transitional walls [**31**]and conformal layers [**23**] in literature.

The second process characteristic of MDLD is the effect of collisions during the deposition process. Collisions occur due to the finite volume of the MDLD apparatus and limitations in machine kinematics. To avoid collisions, especially for smaller part sub-volumes, support structures are deposited. Most research efforts employ concepts such as the visibility maps of surfaces [**11**] to detect collisions.

The algorithm for part volume decomposition involves two steps. In the first step, unbuildable surface regions are identified. In the second stage the part sub-volumes corresponding to the unbuildable surface regions are computed. Offset and planar slices are considered separately.

5.1. Planar Slices. The first objective in part volume decomposition is to identify unbuildable surface features. An oriented surface (S) is deemed to be unbuildable along a build direction, $\overline{B}$ if $\forall p \in S, \angle(N(p), \overline{B}) > \theta_{max}$. $N(p)$ is the directed surface normal at p and θ_{max}, the maximum permissible overhang angle. In our analysis the overhang angle is restricted to 90 degrees. This is a simplification of the problem statement. The identification and the demarcation of such surface features is done using silhouette edges.

Silhouette edges [**5, 28**] are defined as the loci of points on a surface on which the surface normal is at a constant inclination of 90 degrees with a given vector $\overline{V}$. For a C^1 continuous non-planar parametric surface silhouette edges either form closed loops or start and terminate at the boundary of the surface patch. Consequently, silhouette edges bound the open subset of the surface where

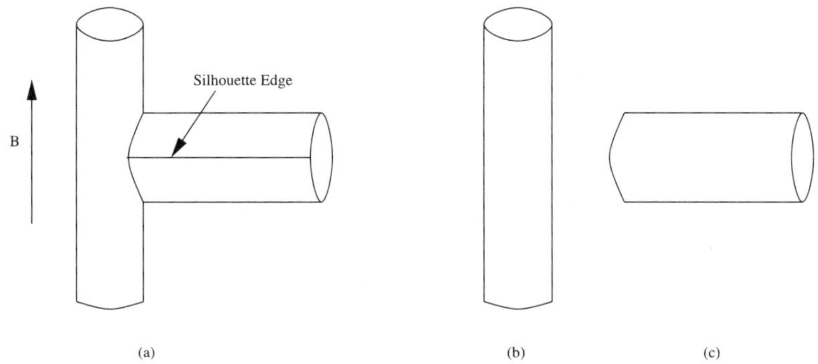

FIGURE 6. (a) Model with silhouette curves (b) Part which can be built (c) Part which cannot be built along B

$\angle(N(p), \overline{V}) < 90$. If $\overline{V} = \overline{B}$, the silhouette edges delineate the unbuildable and buildable surface features.

The second task in part volume decomposition is the identification of unbuildable sub-volumes. These sub-volumes are derived from unbuildable surface regions. The faces of the part are split into buildable and unbuildable surface features using these edges. For a given build direction, $\overline{B}$ and an unbuildable surface feature $S_{unbuild}$, the corresponding unbuildable sub-volume is given by $V_{unbuild}(\overline{B}, S_{unbuild})$ in Equation 5.1. In Equation 5.1, t is a sufficiently large scalar, $\|t\overline{B}\| = \infty$

$$V_{unbuild}(\overline{B}, S_{unbuild}) = V_{sweep} \bigcap V \tag{5.1}$$

$$V_{sweep} = S_{unbuild} \oplus t\overline{B} \tag{5.2}$$

In equation 5.2, $\oplus$ denotes the Minkowski sum operation. The identification of the unbuildable volume is aided by the *maximal boundary property* of silhouette edges. This property implies that the boundary of the projection of a face along a viewing vector $\overline{V}$ is either the projection of its silhouette edges or the bounding edges of the face itself. As such the sweep of these unbuildable surface features will include all regions on the part's boundary which are oriented at an angle greater than 90 degrees with the build direction. The unbuildable sub-volume is obtained by intersecting the swept solid (V_{sweep}) with the input part volume (V). Figures 6b,c show the silhouette-edge based decomposition for the part shown in Figure 6a.

5.2. Offset Slices. Offset slices are required in the following cases:

(1) Change in the build direction in MDS. Figure 7b shows the decomposition (for MDS) of the part shown in Figure 7a, the base of volume V_2 is non-planar. Depositing planar slices aligned along B_2 leads to a variable slice thickness. This can be avoided by depositing a transitional region using offset slices which creates a (nominally) planar base for the deposition of slices along the new direction of build. Since the base and its offsets are parallel, uniform slice thickness is maintained in the transition zone. An example is shown in Figure 8, where a transition zone is created to aid a smooth change between the build directions B_1 and B_2.

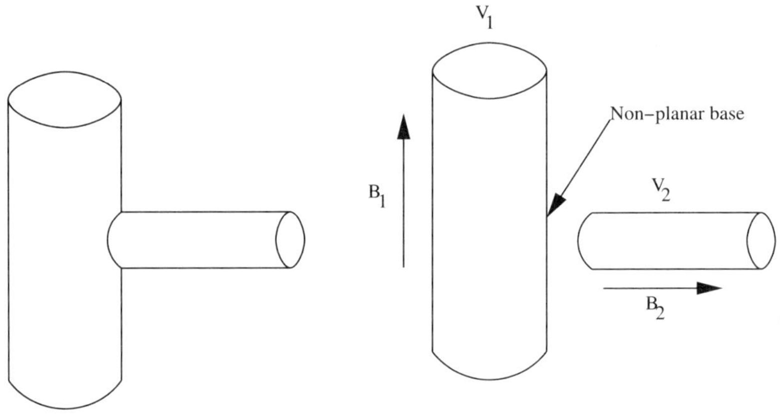

FIGURE 7. (a) Part (b) MDS decomposition, V_2 has a non-planar base which cannot be deposited using planar slices.

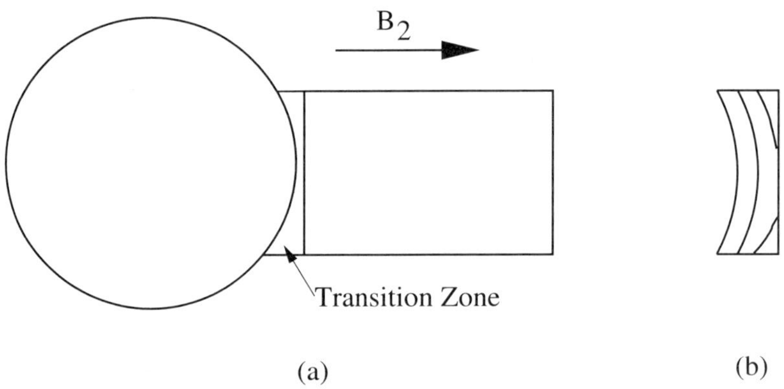

FIGURE 8. (a) Decomposition - with the Transition zone shown in dark (b) Enlarged view of the transition zone deposited using offset slices

(2) In certain cases, a support-free build direction cannot be found when using planar slices. Annular regions present one such case.
(3) Use in a combined NC-LM set-up. This refers to *local deposition* in which a custom feature is deposited on a machined part using LM. LM plays a complementary role, by depositing features which may otherwise be unmachinable. Alternately the use of LM aids design changes or part repair.

In each of the three scenarios the deposition of offset slices is done in conjunction with planar slices.

5.2.1. *Build Direction.* In using offset slices, the notion of a build direction is more subtle. The offset slice is obtained by uniformly translating each point on the base surface along the surface normal (at that point). Therefore, the build direction at a point on the offset slice is the corresponding surface normal. Consequently, the overhang angle is the angle between the part surface normal and the corresponding surface normal to the offset of the base surface at their point of intersection. Determining the corresponding normal to the base surface is a non-trivial task and is one of the key challenges in computing the unbuildable part volume.

5.2.2. *Part Decomposition with Offset Slices.* In our analysis, we restrict the maximum overhang angle (θ_{max}) to 90° degrees. Note that this is the angle between the normal to the surface being deposited and the offset of the base at a particular point of intersection. Assuming the surface $S(u, v)$ is continuously offset to create slices for a volume V_{unproc}, the family of offset slices is given by the intersection of $O(u, v, d)$ and ∂V_{unproc}. Where $O(u, v, d)$ [**12**] is defined as:

$$O(u, v, d) = S(u, v) + d{\cdot}N(u, v) \tag{5.3}$$

. In Equation 5.3, d represents the offset distance and $N(u, v)$ is the unit normal to the base $S(u, v)$ computed at the point (u, v).

Equation 5.4 represents the intersection of the nominal base surface with the boundary of V_{unproc}.

$$S_{int}(s, t) : S(u_1, v_1) + d{\cdot}N(u_1, v_1) - \partial V_{unproc}(u_2, v_2) = 0 \tag{5.4}$$

The boundary of V_{unproc} is represented by B-Spline or NURBS surfaces. From Equation 5.4, a functional dependence of u_1, v_1 and d on the parameters u_2 and v_2 can be found. The normal vectors to ∂V_{unproc} and the corresponding offset of the base surface at their point of intersection can be calculated. We are interested in getting the loci of points on V_{unproc} at which the angle between the corresponding normals is equal to the maximum permitted overhang angle (θ_{max}). This condition can be expressed as:

$$\angle(\mathbf{N}_d^{offset}(u_1, v_1), \mathbf{N}^{\partial \mathbf{V}_{unproc}}(u_2, v_2)) - \theta_{max} = 0 \tag{5.5}$$

5.2.3. *Challenges in Offset Computation.* Offsets of rational curves and surfaces may not be rational [**12**]. As a consequence it is not possible to exactly represent offset curves (surfaces) using a rational polynomial basis such as the Bernstein Basis. Further, singularities or cusps are observed in curve offset computation [**8**]. At cusp points the normal vector and the curvature are undefined. Cusps signify the start of self-intersections in the offsets. The reader is referred to [**17**] for a discussion on the development of singularities in surface offsets.

Most methods to evaluate offset curves and surfaces [**24**] try to determine trimmed offsets at a prescribed distance from the base surface (curve). These methods are primarily aimed at generating contour-parallel CNC toolpaths by computing offsets at discrete intervals. Since Equations 5.4 and 5.5 require a correlation between parameters u_1, v_1, u_2, v_2 and d, these methods are not applicable.

5.2.4. *Solution Methodology.* We have used the level sets method [**22**] to solve the equation set underlying unbuildable part volume computation. The level set method is based on advances in Computational Fluid Mechanics (CFD). The generation of cusps or shocks in fluid flow is avoided by using a physically motivated entropy condition. The offsets of a 3D surface are traced as zero or level sets of an evolving 4D surface. We present the relevant results of the level set method developed by [**22**]. The computation of surface offsets follows the algorithm developed in [**14**]. This methodology presented in the following sections assumes that the base surface is closed, although it may be modified to handle open surfaces.

5.2.5. *Level Set Preliminaries for Offset Surface Computation.* The level set method of [**22**] introduces a smooth function $\phi(x, y, z, d)$, such that the surface $S(u, v)$ is given by $\phi(x, y, z, 0) = 0$. Offsets of $S(u, v)$ are the level sets of ϕ, i.e. the offset at a distance d is $\phi(x, y, z, d) = 0$. An Eulerian formulation for the motion is set up by differentiating $\phi(x, y, z, d) = 0$ as:

$$\phi_d + \|\nabla\phi\| = 0 \tag{5.6}$$

In 1D, the differentiation of Equation 5.6 w.r.t. x along with the substitution of $u = \phi_x$ and $G(u) = (u^2)^{\frac{1}{2}}$ gives the Hamilton-Jacobi (HJ) equation which can be written as:

$$u_d + [G(u)]_x = 0 \tag{5.7}$$

The numerical solution to the weak form of the HJ equation is given by:

$$\frac{u_i^{n+1} - u_i^n}{\Delta d} = \frac{g(u_i^n, u_{i+1}^n) - g(u_{i-1}^n, u_i^n)}{\Delta x} \tag{5.8}$$

In Equation 5.8 $u_i^n = u(i\Delta x, n\Delta d)$ and $g(u, u) = G(u)$ is the consistency condition. To avoid the formation of shocks or cusps [**22**] proposes the following definition for g in 1D:

$$g_{EO}(u_i^n, u_i^{n+1}) = (max(u_i^n, 0)^2 + min(u_i^{n+1}, 0)^2) \tag{5.9}$$

From Equation 5.7 and $u = \phi_x$ it follows that the 1D weak entropy solution for ϕ is given by:

$$\phi_i^{n+1} = \phi_i^n - \Delta d \cdot g_{EO}(D_-\phi_i^n, D_+\phi_i^n) \tag{5.10}$$

where

$$\begin{aligned} D_-\phi_i^n &= \frac{\phi_i^n - \phi_{i-1}^n}{\Delta x} \\ D_+\phi_i^n &= \frac{\phi_{i+1}^n - \phi_i^n}{\Delta x} \end{aligned} \tag{5.11}$$

In 3D, g_{EO} is:

$$\begin{aligned} {g_{EO}}^2 = ((max(D_-^x\phi_{ijk}^n, 0))^2 + (min(D_+^x\phi_{ijk}^n, 0))^2 \\ +(max(D_-^y\phi_{ijk}^n, 0))^2 + (min(D_+^y\phi_{ijk}^n, 0))^2 \\ +(max(D_-^z\phi_{ijk}^n, 0))^2 + (min(D_+^z\phi_{ijk}^n, 0))^2) \end{aligned} \tag{5.12}$$

The evolution of ϕ in 3D is given by:

$$\phi_{ijk}^{n+1} = \phi_{ijk}^n - \Delta d \cdot g_{EO}(D_-^x\phi_{ijk}^n, D_+^x\phi_{ijk}; D_-^y\phi_{ijk}^n, D_+^y\phi_{ijk}; D_-^z\phi_{ijk}^n, D_+^z\phi_{ijk}) \tag{5.13}$$

The initial condition in Equation 5.13 is based on the approach taken in [**14**] which uses a rectilinear grid to offset (2D) curves. The function ϕ_{ij}^0 is C^0 continuous and defines a tight neighborhood around the curve. Analogously, in 3D, if the edge length of the rectilinear grid be C then the function ϕ_{ijk}^0 is defined as:

(5.14)

$$\phi(x, y, z, 0) = \begin{cases} \min[+d((x, y, z), S(u, v)), C] & (x, y, z) \in exterior(S(u, v)) \\ \max[-d((x, y, z), S(u, v)), -C] & (x, y, z) \in interior(S(u, v)) \\ 0 & (x, y, z) \in S(u, v) \end{cases}$$

The contours at an offset distance are extracted using a marching squares algorithm. In the case of surfaces, the zero set of ϕ can be extracted using the marching cubes algorithm.

The normal vector to an offset surface is $\frac{\nabla\phi}{\|\nabla\phi\|}$. To avoid discontinuities in the normal vector at cusps, [**25**] proposes the average of directional normals, refer Equation 5.15. It follows that the normal vector, n_{ijk} at a grid point is obtained by normalizing n^*_{ijk}.

$$n^*_{ijk} = \sum_l \sum_m \sum_n \frac{(D^{l\cdot x}_{ijk}, D^{m\cdot y}_{ijk}, D^{n\cdot z}_{ijk})}{\sqrt{(D^{l\cdot x}_{ijk})^2 + (D^{m\cdot y}_{ijk})^2 + (D^{n\cdot z}_{ijk})^2}} \tag{5.15}$$

$$l = (1, -1), m = (1, -1), n = (1, -1)$$

5.2.6. *Identification of Unbuildable Surface Features.* The evolution of the function ϕ is computed at the nodes of a uniform 3D grid. Therefore, the application of the level set method to identify unbuildable surface features requires the discretization of the base and part surfaces using the same grid. The function $\phi(x, y, z, 0)$ is defined by Equation 5.14 with $S(u, v)$ being the base surface. Grid elements which intersect with the part boundary are identified. A first order approximation, i.e. planar approximation, of the part boundary is used in these grid elements.

Our algorithm takes in the part and base boundaries (P and B respectively) as input. A uniform grid, **G**, with edge length, C is used to compute the evolution of the function ϕ. The output from the algorithm is in the form of grid cells, C_u, in which the overhang angle is violated. The base surface is approximated by the distance function ϕ^0_{jkl} as defined in Equation 5.14. Normals of ϕ^0_{jkl}, n^0_{jkl} are computed at the nodes of **G** using Equation 5.15. The average of n^0_{jkl}, $normal_{av}$ is computed in each grid cell. If $\|normal_{av}\| > 0$, the grid cell is added to the list G_B and $\|normal_{av}\|$ to the list $Norm_B$. The grid **G** is also used to construct a first order approximation of the part boundary. Grid cells which intersect with **G** are listed under G_P. A planar approximation (1st order) of P, represented by the normal vector $Norm_P[i]$ is computed in each $G_P[i]$. The algorithm uses G_P, G_B, $Norm_B$ and $Norm_P$ to identify C_b. First the overlapping cells in G_B and G_P are identified. The angle, θ, between $Norm_B$ and $Norm_P$ is computed in the overlapping cells. If θ exceeds the overhang angle (90°), the cells are added to the list C_u, otherwise they are added to C_b. All overlapping cells are removed from the list G_P as they represent intersection with offsets of the base surface. The algorithm continues to offset the base surface till $G_P = \emptyset$. The lists G_B and $Norm_B$ are reformulated after each offsetting operation. Grid cells, listed in C_u, are divided into face connected components.

5.2.7. *Comments.* The identification of unbuildable surface regions on 3D surfaces requires the discretization of the base and the part boundaries. The advantage in using the level set method is that the computation of N_d^{offset} (the surface normal of the base offset at a distance d) does not require an explicit representation of the corresponding offset surface. Traditional offsetting algorithms approximate surface offsets as freeform B-Spline surfaces. Further, the formation of cusps is handled using a physically motivated entropy condition. Our implementation of the level sets methodology uses a uniform grid. Computationally efficient methods [**25**] which use non-uniform grid elements are known in literature

5.2.8. *Part Volume Decomposition.* We motivate the construction of the unbuildable part volumes using Figure 9a. For clarity, we have shown a 2 dimensional analogy. The overhang is exceeded at the point P, i.e. it is unbuildable. Let the normal to the offset surface at P be N_P^O. Since P lies on an offset of the base

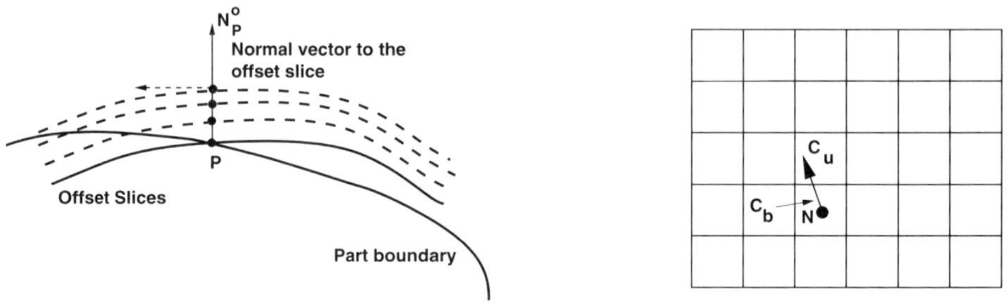

(a) Boundary between buildable and unbuildable regions.

(b) Construction of the unbuildable part volume

FIGURE 9. Computation of the unbuildable part volume

curve, all further offsets of P are along the direction given by N_P^O. Let R_P be the ray originating at P and parallel to N_P^O.The angle between the normals to the offsets and the normal to the ray R_P is 90 degrees. Therefore, R_P is the boundary between the buildable and unbuildable regions of the part. The average normal vector, $normal_{av}$ is computed in each unbuildable grid cell $C_{unbuild}$ using Equation 5.15. A ray originating at the center of $C_{unbuild}$ is extended along $normal_{av}$ till it intersects another cell (refer Figure 9b). This cell, C_{new}, is also labeled unbuildable. Further unbuildable cells are identified using the average cell normal of C_{new}. The overall process is repeated till all the unbuildable cells have been identified. The unbuildable volume, thus obtained, is bounded as the grid is finite. The boundary grid elements are extracted and the marching cubes algorithm is used to reconstruct the surface of the unbuildable region. The intersection of this solid with the part surface gives the unbuildable part volume.

5.3. Future Research. The algorithms presented above restrict the overhang angle to 90 degrees. This allowed us to use silhouette edges. Future research will likely focus on the development of algorithms for angles greater than 90 degrees. Another important thrust area could be the development of algorithms for polygonal (STL) models. The use of such polygonal models is widespread in the Rapid Prototyping industry.

6. Build Direction Determination

The input to this task is an unbuildable part volume. The purpose of this task is the selection of a build direction for the input part.

Prior to the assignment of a build direction an unbuildable volume will be referred to as *unprocessed* or V_{unproc} and subsequent to the assignment, *processed* or V_{proc}. We take a two stage approach to determine the build direction for an unprocessed volume:

(1) Establish the set of all feasible build directions for a given V_{unproc}.
(2) Evaluate the members of this set against a metric to arrive at a suitable build direction.

In order to expedite the above, we use well known mappings on the sphere namely the Gauss [**27**]and Visibility maps [**29**]. The Gauss map is a transformation from an orientable surface in $\mathcal{E}^3 \to \mathcal{S}^2$, which associates each point on the surface with its oriented normal vector. $\mathcal{S}^2$ represents the unit sphere. A related concept

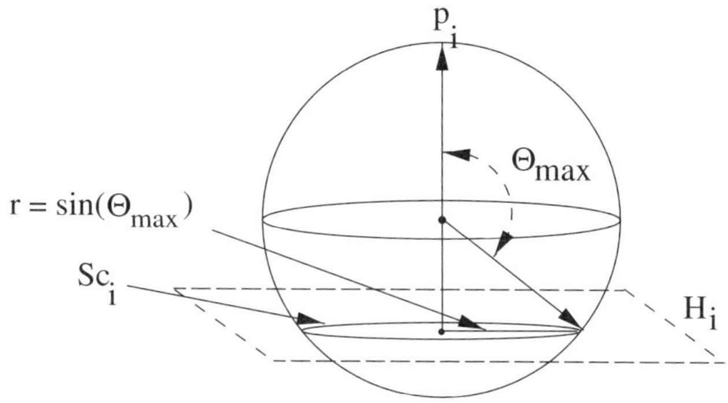

FIGURE 10. The Build Map

is the spherical visibility map, it describes how much of a surface is accessible to a line segment based at a point p on the surface and directed along the surface normal at p. It is also called the *accessibility map* in literature.

On similar lines, the set of all feasible build directions for a surface can be described by a spherical map. We call this the *build map* of the surface. The set of all build directions along which a point $p \in S$ can be built is described as:

$$\{X \in \mathcal{S}^2 | (X \bullet n_p \geq cos(\theta_{max})\} \tag{6.1}$$

This definition (ref. Equation 6.1) of the build map assumes the slices of the CAD model to be planar. The build map is computed for each face of the unprocessed sub-volume except the base surface. The cumulative build map of all these faces gives the set of feasible build directions for the unprocessed sub-volume. The base surface is excluded since it lies on a part volume which has already been deposited along a selected build direction.

The exact build map of the surface is obtained by the intersection of the build maps of each point on the surface. Clearly, this is computationally infeasible. Specialized techniques developed [**6**] to compute convex spherical polygons (such as the visibility maps) cannot be applied to the build map which is non-convex. The notion of a convex set on a unit sphere needs to clarified at this stage. Janardan et al. [**11**] define a spherical disc $D(P, r)$ with pole P and radius r, where $0 \leq r \leq \frac{\Pi}{2}$ as the set of points $X \in \mathcal{S}^2 : D(P,r) = \{X \in \mathcal{S}^2 | d(P, X) \leq r\}$. $d(P, X)$ being the angular distance between the points P and X. A spherical polygon is convex if the radius $r \leq 90^o$. The build map is a spherical disc of (angular) radius $r = \theta_{max}$ with point p_i as the pole. Since θ_{max} is greater than 90° the build map is non-convex.

An approximate scheme to evaluate the build map starts by computing the Gauss map of the surface (S). The discrete evaluation of the Gauss map - $G(S)$ is a point set on $\mathcal{S}^2$. It represents the unit normals at various points on the surface. Since the Build map is a function of the unit surface normal, the corresponding build map, $B(S)$ is obtained by intersecting the build map associated with each point in $G(S)$. The build map of each $p \in G(S)$ is a quadratic surface. $B(S)$ is the intersection of $\forall B(p), p \in G(S)$ and is therefore computationally expensive.

Since the radius of $B(p_i)$ ($p_i \in G(S)$) is greater than 90°, it is bounded by a small circle, Sc_i. The radius of Sc_i is $sin(\theta_{max})$. Let the half plane containing Sc_i be H_i. Figure 10 shows the relationship between Sc_i, H_i and θ_{max}.

The volume of the unit open 3-ball (B^3) bounded by H_i and $B(p_i)$ is $clo(B^3 \cap H_i$, where $clo(B^3)$ is the closure of B^3 and H_i is the half space $h_i > 0$. Since both H_i and $clo(B^3)$ are closed convex sets, the resulting set is also convex. The build map for p_i can be expressed as, $\partial(clo(B^3 \cap H_i) \cap \mathcal{S}^2$. Consequently, the build map of the surface, $B(S)$ can be written as:

$$B = \partial((\bigcap_{i=0}^{n} H_i) \cap clo(B^3)) \cap \mathcal{S}^2 \tag{6.2}$$

The sequence of intersections is not important in calculating intersections, the resultant expression for $B(S)$ is reduced to n intersections of half-spaces and only one intersection with $\mathcal{S}^2$. The intersection of half-spaces which are bounded by linear surfaces is computationally less expensive than the intersection of quadratic surfaces. Hence Equation 6.2 represents a more efficient expression to calculate the build map of the surface.

6.1. Heuristics. Heuristics are required to select a build direction for a part sub-volume (V_{unproc}), when its build map is an empty set. These heuristics identify a set of build directions for a subset of V_{unproc}. Since only a part of V_{unproc} can be built along any selected direction, the remainder needs to be processed further. This section discusses two heuristics which can be employed when the build map is an empty set. Before doing so, the notion of a *base surface* and an *adjacent processed sub-volume* are introduced. We describe the *base surface* to be the common surface between two volumes V_1 and V_2. V_1 and V_2 in the present context are decomposed sub-volume of the original CAD model. At any stage of the decomposition, a part volume (P) is split into a processed (V_{proc}) and an unprocessed sub-volume (V_{unproc}). Since V_{proc} and V_{unproc} share a common base surface, V_{proc} is referred to as an adjacent processed sub-volume (AP) of V_{unproc}.

The heuristics and the scenarios for their use are as follows:

(1) The build map of the unprocessed sub-volume is empty but that is non-empty in the neighborhood of its base surface (shared with at least one of its adjacent processed sub-volumes). A build direction is chosen from the build map of this neighborhood. This build direction is used to further disjunction the part into buildable and unbuildable sub-volumes. Figure 11a shows a part which has an empty build map. The neighborhood (ref. Figure 11b) of the unprocessed sub-volume V_2 in the immediate vicinity of its base surface with V_1 has a non-empty build map (ref. Figure 11c) which is exploited to select a new build direction.

(2) The build map in the neighborhood of the base surface is also empty. In this case, we consider the possibility of using offset slices. If the use of offset slices is not applicable, supports are used.

6.2. Build Direction Selection. The considerations in the selection of a build direction from the build map are the surface quality and collision free deposition. In LM literature [**1**], surface quality has been quantified in terms of the *cusp height*. The cusp height is defined as:

$$h = \begin{cases} z_T|cos\theta| & \text{for } -1 < cos\theta < 1, \\ 0 & \text{if } cos\theta = 1 \text{ or } cos\theta = -1. \end{cases} \tag{6.3}$$

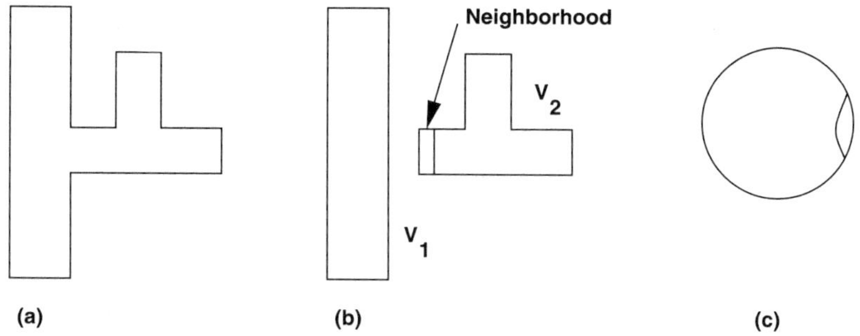

FIGURE 11. (a) Model (b) Decomposition (c)Build Map of the neighborhood

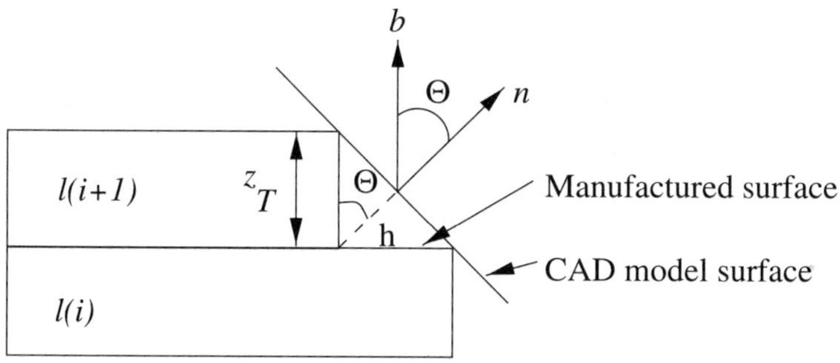

FIGURE 12. Cusp Height

where, $|cos\theta| = |\overline{n \bullet b}|$. $\overline{n}$ is the surface normal, $\overline{b}$ is the build direction and z_t is the slice thickness. The cusp height is a measure of the stepped approximation of a surface as is evident from Figure 12.

For a faceted approximation of a CAD model, the overall surface quality be estimated in terms of its *average weighted cusp height*, H_{av}. If the number of facets be N and the area of the ith facet be $A_{facet}(i)$, the average weighted cusp height is expressed as:

$$H_{av} = \frac{\sum_{i=0}^{N} h(i) \times A_{facet}(i)}{\sum_{i=0}^{N} A_{facet}(i)} \tag{6.4}$$

where $h(i)$ is the cusp height of the ith facet as defined by Equation 6.3. A lower value of H_{av} implies a better surface quality and vice-versa. The selected build direction for a part sub-volume in MDS minimizes H_{av}. The constraint on the minimization is provided by the collision avoidance criterion. This constraint also entails the solution to the problem of finding nozzle orientation along the deposition path. Collision detection has been studied extensively, specially in the context of 4/5 axis CNC machining [**16, 7**]. The task of collision free deposition can be analyzed on similar lines - the details are not presented here.

6.3. Nozzle Orientation. The selection of the deposition nozzle orientation is an open problem and is quite likely to be process specific. In the simplest case the nozzle can be oriented along the chosen build direction when depositing planar slices. In the case of offset slices, the nozzle orientation is governed by the normal to the slice surface. In a realistic scenario, the finite dimension of the nozzle can lead to collisions. The obstacles could be layers of material which have already been deposited or the surrounding apparatus. Any solution to the nozzle orientation problem must propose a set of collision-free nozzle orientations at each point along the deposition path. In addition, many laser based deposition processes have limitations on the maximum deviation (of the nozzle) from the deposition direction.

6.4. Support Structure Generation. Collisions and the resulting lack of a feasible build direction may lead to the deposition of supports. However, collisions do not always necessitate support generation.

Given an adjacent processed sub-volume P_1 and a sub-volume P_2 which has to be deposited along a build direction $\overline{B_{P_2}}$, possible collisions can be categorized as:

(1) Collisions between P_1 and P_2 below and with the common base surface between them. Distance being measured along $\overline{B_{P_1}}$.
(2) All other collisions.

The existence of a base surface is a prerequisite for the deposition of any part sub-volume. As a result, collisions with the base surface lead to the deposition of supports. Collisions falling in the second category can be avoided by simultaneous deposition of P_1 and P_2.

At each stage of the part volume decomposition, a processed and an unprocessed sub-volume are created. The decomposition resembles a binary tree [**3**]. At each stage the *parent* volume yields at most 2 *child* volumes. Since the original part is the progenitor of the decomposed volumes, it is the *root* of the decomposition tree. Processed volumes are not divided further, they are the *leaves*. A processed (unprocessed) volume at the ith stage of the decomposition is referred to as $V_{proc}(i)$ $(V_{unproc}(i)).V_{proc}(i)$ shares a *child-parent* relationship with at least one $V_{unproc}(i-1)$. The total number of stages in the decomposition is the *height* of the decomposition tree. This tree like structure of the decomposition process forms the basis of our support structure generation algorithm.

The algorithm for the deposition of supports is summarized as follows: The input to the algorithm is the decomposition tree, build directions $(\overline{B})$ for each processed sub-volume and *depth* - denoting a particular level in the decomposition tree. The algorithm is recursive. The seed *depth* is the depth $(depth_{max})$ of the tree. The algorithm checks for collisions at each level. If collisions are reported at a level, the algorithm creates supports, $sup(depth - 1)$ along $\overline{B}(depth - 1)$. $V_{proc}(depth - 1)$ is suitably modified. Since the deposition is along $\overline{B}(depth - 1)$, supports, $sup(i), i = depth$ to $depth_{max}$ are not required and are eliminated from the modified $V_{proc}(depth - 1)$. The decomposition tree is suitably modified and the recursive process continues while $depth > 0$. At the end of the recursion, supports are generated to ensure collision-free deposition. Figure 13 shows an example of the support generation process. Collisions occur during the deposition of $V_{unproc-2}(= V_{proc-3})$ at depth 3. As a result, the union of V_{proc-2}, V_{proc-3} and sup_2 is deposited along B_2 without collisions. Supports are required at only one stage of the decomposition tree.

Algorithm 1 Algorithm for Support_Structure_Generation

Require: Decomposition Tree, $\overline{B}$, *depth*
Ensure: Supports

```
if check_for_collisions(V_proc(depth), V_proc(depth − 1)) then
  if collisions_in_simultaneous_deposition(V_proc(depth), V_proc(depth − 1)) then
    for i = depth to depth_max do
      V_temp = V_temp ∪ (V_proc(i) − sup(i))
    end for
    sup(depth -1) = create_supports(V_temp, B̄(depth − 1))
    V_proc(depth − 1) = V_proc(depth − 1) ∪ sup(depth − 1) ∪ V_temp
    depth_max = depth
    Support_Structure_Generation(Decomposition Tree, B̄, depth)
  else
    V_proc(depth − 1) = V_proc(depth) ∪ V_proc(depth − 1)
    depth = depth − 1
    Support_Structure_Generation(Decomposition Tree, B̄, depth)
  end if
else
  depth = depth -1
  Support_Structure_Generation(Decomposition Tree, B̄, depth)
end if
```

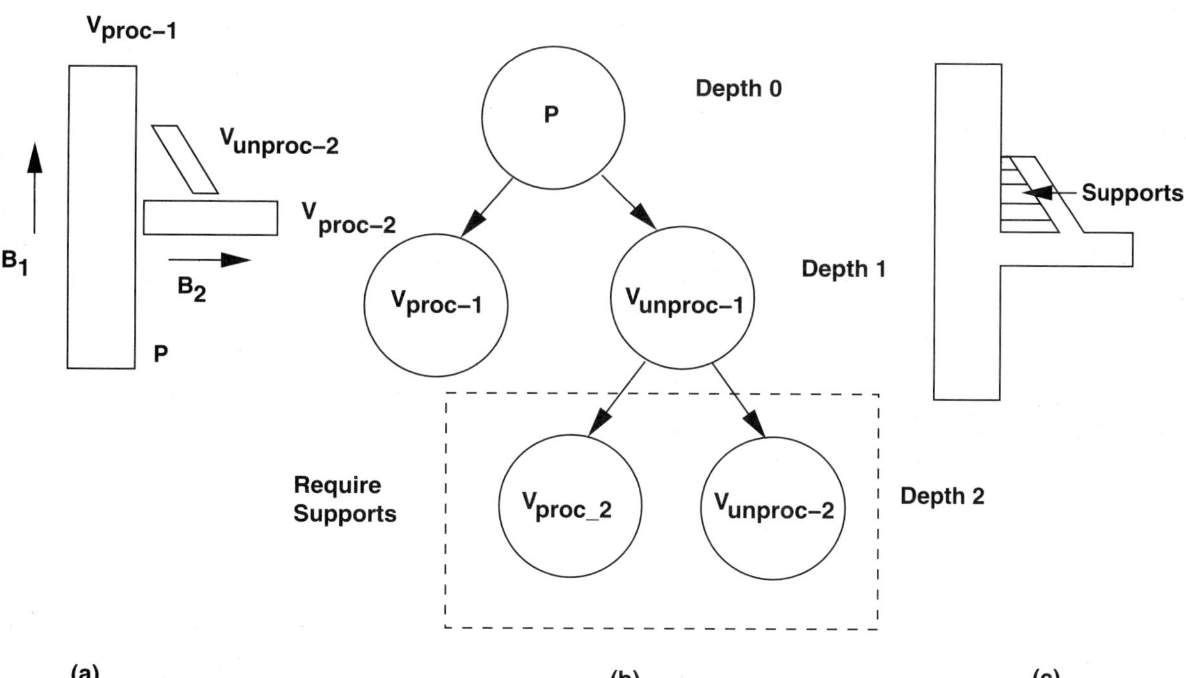

FIGURE 13. (a) The Model - P (b) Decomposition Tree (c) Supports

6.5. Sequencing. Collision avoidance and the existence of a base surface are the defining constraints on volume deposition sequencing. The decomposition tree is used to arrive at a sequence of the decomposed sub-volumes. It is assumed that the support structure (where required) is already available. Sub-volumes and their corresponding supports are assumed to be one processed sub-volume. The base surfaces of sub-volumes at depth i lie on one or more processed sub-volumes at depth i-1. A child sub-volume cannot be deposited till at least the base surface on the parent has been deposited. Therefore processed sub-volumes are sequenced in the ascending order of their depths in the decomposition tree. Sub-volumes at the same depth may be assigned any sequence as long as collisions are avoided.

6.6. Future Research. Future research efforts may focus on improving the computational efficiency of the build map construction. The main challenge is likely to be posed by the non-convex nature of the build map. Efficient search methods for the selection of a build direction from the build map is another important area of research. In our research we have used the average weighted cusp height for build direction selection. This in turn gives a build direction best suited for the entire part volume. Research may be directed at developing metrics which take into consideration sub-regions of the part surface instead of the whole.

7. Slicing and Deposition Path Planning

Multiple degrees of freedom in MDLD permit the deposition of both planar and non-planar slices. Planar slices are deposited along their respective build directions. The path planning for material deposition within a slice proceeds along the same lines as 2.5D LM (refer [**15**] for details). The planar slice is oriented along the build direction assigned to its corresponding part sub-volume. The deposition paths for an offset slice are generated by intersecting it with planes aligned along the build direction of the corresponding base surface. Recall that such slices are offsets of the base surface. The distance between two contiguous planes is equal to the slice thickness. The number of the planes depends on the span of the base surface along its build direction.

8. Examples

The first example shows the use of planar slices in MDLD. Figure 14 shows a CAD model which is sliced into buildable sub-volumes using the algorithm explained in previous sections. The orientation of the part was assumed to be known. The seed build direction was chosen to be the +Z axis. The build directions were selected using the build map concept discussed in section 6.2. The resulting slices are shown in Figure 15.

The second example shows the use of offset slices to decompose an unprocessed sub-volume into its buildable and unbuildable constituents. Figure 16 shows the input CAD model. The initial build direction was selected to be the +Z axis. The cylindrical volume in the part comprises the buildable sub-volume whereas the flange comprises the unbuildable sub-volume. The flange is then built using the offsets of the cylindrical base surface. The hole in the flange prevents the deposition of the entire sub-volume using offset slices. Figure 17 shows the computation of the unbuildable volume. The angle between the offsets to the base surface and the

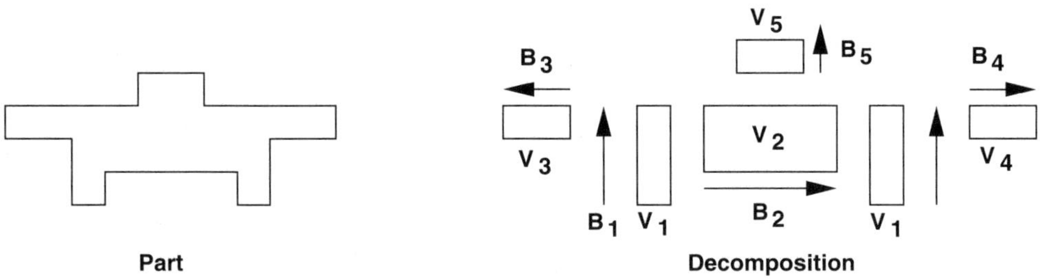

FIGURE 14. Multi-Direction Deposition - Decomposition of a Part into sub-volumes

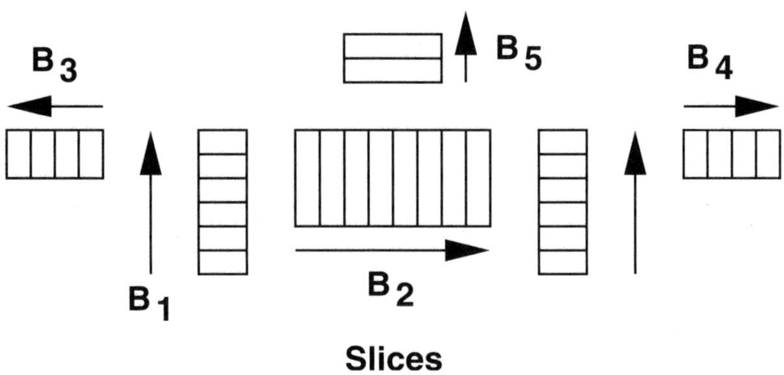

FIGURE 15. Multi-Direction Slicing

(hole) part surface exceeds 90 degrees. As a result the region ABCD cannot be built by using offsets of the cylindrical base surface (refer Figure 17).

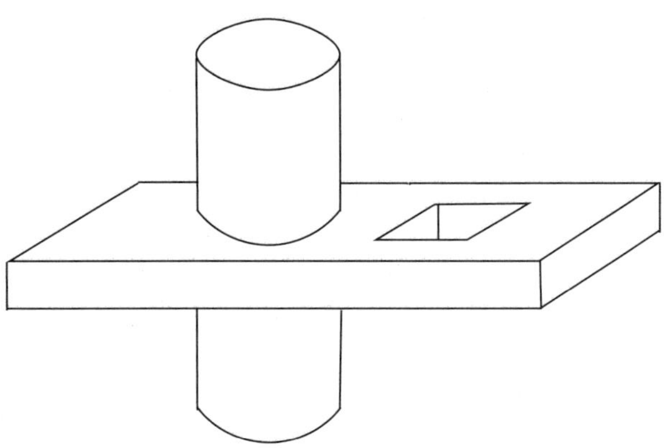

FIGURE 16. (a) The input CAD model

9. Applications

In this section we present two potential areas of collaborative research using the MDLD research framework. These are part repair and the development of layered

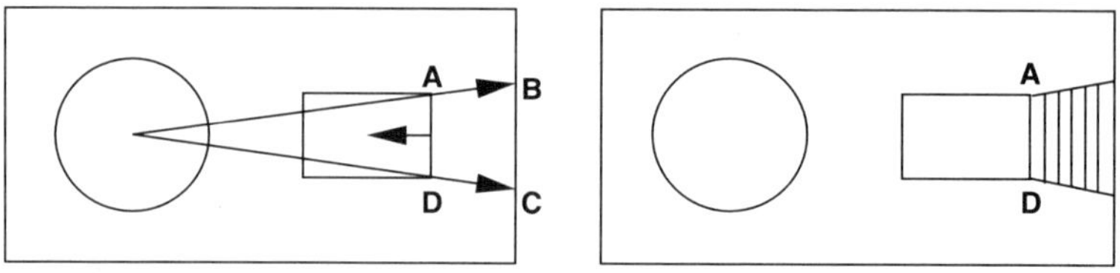

FIGURE 17. The unbuildable part volume - ABCD

manufacturing machines customized for a part family. Each application is discussed in turn.

9.1. Part Repair. Wear and breakage is common in engineering components. In many instances, repair is more economical than replacement. In such instances it is possible to use MDLD to reconstruct the damaged regions in the part. Repair of tubine airfoils represent a potential application. Airfoil sections can chip or wear with continuous use. In such cases layered deposition can be used to repair the blades. Since the base surface on which material is deposited is non-planar, the slices are likely to be offsets of the base. Further development in this application domain could incorporate CNC machining along with local deposition. Developments in the capabilities of layered manufacturing to deposit multi-material and heterogeneous materials could also find application.

9.2. Customized Kinematics. MDLD necessitates a mechanism with more than 3 degrees of freedom. A full 6 degree of freedom mechanism is expensive and is very often redundant for customers engaged in the design and fabrication of specific geometries. Recognizing this, recent research efforts such as [**19, 30**] have proposed simplified kinematics for multi-direction deposition systems. While they have a distinct advantage over 2.5D LM systems, the fixed configuration of the nozzles/deposition table restricts support-less deposition to a few part geometries.

We believe it is possible to take a systematic approach to the design of customized kinematics for MDLD. The design of machine kinematics could quite possibly be driven by a part family which the customer intends to build. The part family may be parametric, i.e. the geometry of the constituent parts is controlled by geometric or dimensional constraints. Alternately a part family may be a grouping of disparate parts.

We have made an initial effort [**26**] to develop customized kinematics driven by a parametric part family. The compliant parallel kinematic machine (CPKM) design methodology [**20, 21**] was used to construct the customized kinematics. We briefly explain the concepts underlying CPKMs before elaborating on our research.

The CPKM methodology synthesizes mechanisms using dual number algebra and screw theory. We expect the synthesized mechanism to have a high accuracy because it uses compliant joints which are clearance and friction free mechanical components. A CPKM is comprised of a top plate (end effector or work holding), a bottom plate, a constraining leg and active legs. The top plate holds the workpiece and the bottom plate is attached to the MDLD machine. The constraining leg is a serial kinematic chain that connects the two plates and defines the total degrees of freedom of the deposition table. The active legs are attached after the constraining leg is configured so that the machine can be activated.

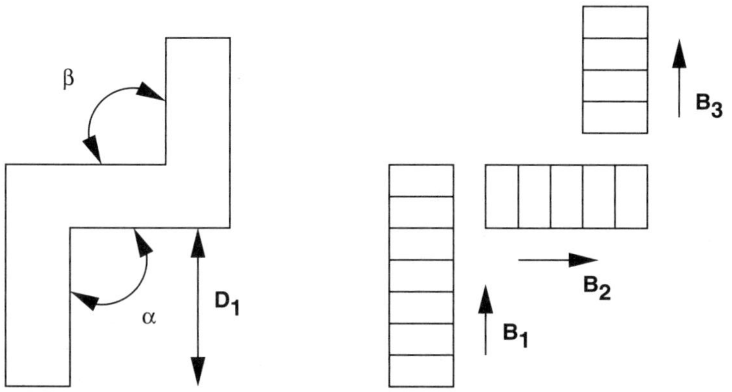

FIGURE 18. An example parametric part family

Figures 18, 19 and 20 illustrate the development of customized kinematics using CPKMs for a geometric part family. We have considered a parametric part family in our research. Members of the part family were generated from a master model by varying the angular and dimensional constraints as shown in Figure 18. There were two principal tasks in our approach. The first task is the process planning underlying multi-direction layered deposition of a given member of the part family. The input part geometries were decomposed into buildable part volumes and associated build vectors. Uniform, planar slices aligned with the corresponding build vectors were created. This is shown in Figure 18. The dependence of the build direction selection on the shape parameters defining the part family is established. In the second task, slice geometry and build directions were used to derive the machine kinematics using the CPKM methodology [**21**]. The deposition table of the MDLD machine was given only rotational degrees of freedom. The construction of the serial kinematic chain composing the constraining leg is driven by the variation in the build direction vectors. The construction of the constraining leg was driven by a library of kinematic configurations (refer Figure 19). To build the part family shown in Figure 18, the constraining leg was assigned 1 rotational degree of freedom. Three translational degrees of freedom were allocated to the deposition nozzle. The resulting mechanism is shown in Figure 20.

10. Summary

In this paper we have reviewed the process planning tasks involved in multi-direction layered deposition. MDLD consolidates the advantages offered by layered manufacturing by limiting the need for supports. Many MDLD processes are under development. We also presented a summary of our research aimed at solving some of the geometry related process planning tasks in MDLD. Key tasks in our research included the decomposition of an input part geometry into smaller volumes, each of which could be built along a chosen build direction. Another important aspect of MDLD is the possibility of depositing offset slices. The challenges in the use of these slices were discussed and a solution framework was put forth. Examples in support of our research were presented.

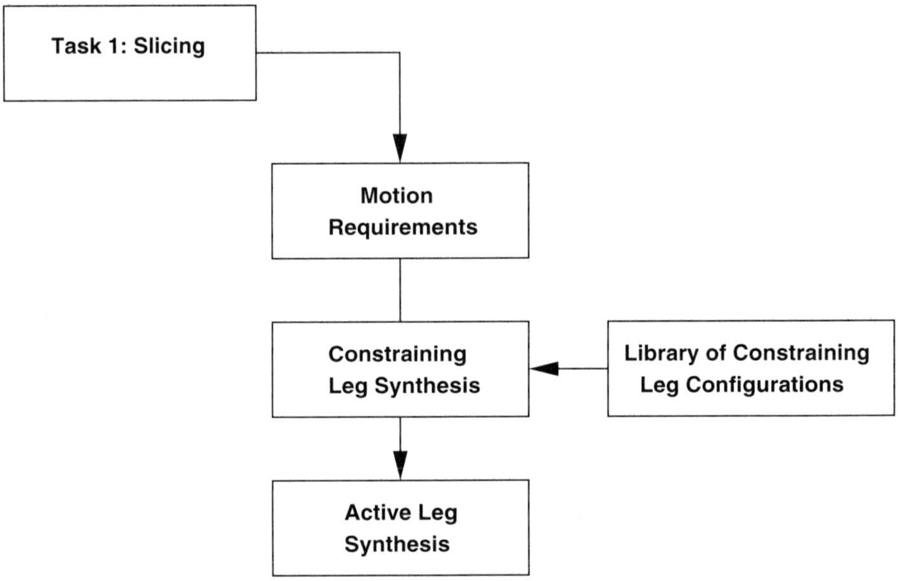

FIGURE 19. The tasks underlying the construction of customized kinematics for MDLD

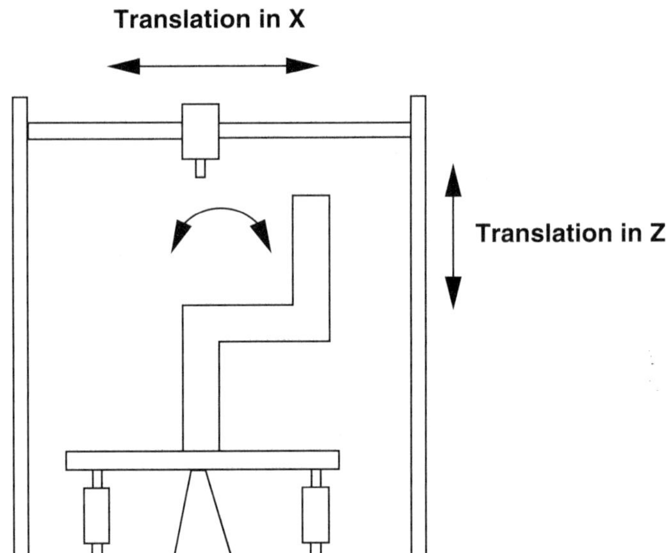

FIGURE 20. Customized kinematics to the multi-directional layered deposition of the part family shown in Figure 18

References

1. P. Alexander, S. Allen, and D. Dutta, *Part Orientation and Build Cost Determination in Layered Manufacturing*, Computer-Aided Design **30** (1998), 343–358.
2. S. Allen and D. Dutta, *Wall thickness control in layered manufacturing for surfaces with closed slices*, Computational Geometry: Theory and Applications **10** (1998), 223–238.

3. T. H. Cormen, C. E. Leiserson, and R. L. Rivest, *Introduction to Algorithms*, McGraw-Hill Book Company, 1995.
4. P. Dickens, M. Pridham, R. Cobb, I. Gibson, and G. Dixon, *Rapid Prototyping using 3D Welding*, Solid Freeform Fabrication Symposium, 1992.
5. G. Elber, *Computation of Isoclines of Freeform Parametric Surfaces*, Preprint (1999).
6. G. Elber and E. Cohen, *Arbitrarily Precise Computation of Gauss Maps and Visibility Sets for Freeform Surfaces*, The 3rd ACM/IEEE Symposium on Solid Modeling and Applications, 1995.
7. ———, *A Unified Approach to Accessibility in 5-Axis Freeform Milling Environments*, Computer-Aided Design **31** (1999), 795–804.
8. R. Farouki and C. Neff, *Analytic Properties of Plane Offset Curves*, Computer Aided Geometric Design **7** (1990), 83–89.
9. S. Fekete and J. Mitchell, *Terrain Decomposition and Layered Manufacturing*, International Journal of Computational Geometry and Applications **11** (2001).
10. I. Gibson, *Software Solutions for Rapid Prototyping*, Professional Engineering Publishing Limited, 2002.
11. P. Gupta, R. Janardan, R. Majhi, and T. Woo, *Efficient Algorithms for Workpiece Orientation in 4- and 5- Axis NC Machining*, Computer-Aided Design **28** (1996), 577–587.
12. J. Hoschek and D. Lasser, *Computer Aided Geometric Design*, A. K. Peters, 1993.
13. J.H. Kao, J.M. Pinnilla, and F. Prinz, *Evolution of Process Planning for Solid Freeform Fabrication Processes*, AUTOFACT-98, Detroit, Michigan, 1998.
14. R. Kimmel and A. Bruckstein, *Shape offsets via level sets*, Computer-Aided Design **25** (1993), 154–162.
15. P. Kulkarni, A. Marsan, and D. Dutta, *A Review of Process Planning Techniques in Layered Manufacturing*, Rapid Prototyping Journal **6** (1999).
16. Y. S. Lee and T. C. Chang, *2-Phase Approach to Global Tool Interference Avoidance in 5-Axis Machining*, Computer-Aided Design **27** (1995), 715–729.
17. T. Maekawa, *An overview of Offset Curves and Surfaces*, Computer-Aided Design **31** (1999), 165–173.
18. J. Mazumder, J. Koch, K. Nagarathnam, and J. Choi, *Direct Metal Deposition of H13 Tool Steel for 3-D Components*, ICALEO, 1997.
19. S. McMains, *Double Sided Layered Manufacturing*, Proceedings of the Japan-USA Symposium on Flexible Automation, 2002.
20. Y. Moon and S. Kota, *Automated Synthesis of Mechanisms using Dual-Vector Algebra*, Mechanism and Machine Theory **37** (2002), 143–166.
21. ———, *Design of Compliant Parallel Kinematic Machines*, Proceedings of AMSE DETC, 2002.
22. S. Osher and J. Sethian, *Fronts Propogating with Curvature Dependent Speed: Algorithms Based on Hamilton-Jacobi Formulations*, Journal of Computational Physics **79** (1988), 12–49.
23. J. Park and D. Rosen, *Issues in Process Planning for Laser Chemical Vapor Deposition*, Proceedings of the Solid Freeform Fabrication Symposium Proceedings, 2002.
24. B. Pham, *Offset Curves and Surfaces: a brief survey*, Computer-Aided Design **24** (1992), 223–229.
25. J. Sethian, *Level Set Methods and Fast Marching Methods - Evolving Interfaces in Computational Geometry, Fluid Mechanics, Computer Vision, and Material Science*, Cambridge University Press, 1999.
26. P. Singh, Y. Moon, D. Dutta, and S. Kota, *Design of a customized multi-directional layered deposition system based on part geometry*, Proceedings of the 14th Annual Solid Freeform Fabrication Symposium, 2003.
27. D. J. Struik, *Lectures on Classical Differential Geometry*, Addison-Wesley Publishing Co., 1961.
28. Y. Tokuyama and S. Bae, *An Approximate Method for Generating Draft on a Freeform surface*, Visual Computer **15** (1999), 1–8.
29. T. Woo, *Visibility Maps and Spherical Algorithms*, Computer-Aided Design **26** (1994), 6–16.
30. Y. Yang, J. Fuh, H. Loh, and Y. Wong, *Multi-orientation Deposition for Support-less Layered Manufacturing Process*, Pre-print.

31. J. Zhang, J. Ruan, and F Liou, *Process Planning for a Five-Axis Hybrid Rapid Manufacturing Process*, Proceedings of the Eleventh Annual Solid Freeform Fabrication Symposium, 2000.

Department of Mechanical Engineering, University of Michigan, Ann Arbor, Michigan, 48105
E-mail address: psinghz@umich.edu

Department of Mechanical Engineering, University of Michigan, Ann Arbor, Michigan, 48105
E-mail address: dutta@umich.edu

DIMACS Series in Discrete Mathematics
and Theoretical Computer Science
Volume **67**, 2005

Machinability: Geometric Reasoning for Cutting

Taejung Kim and Sanjay E. Sarma

ABSTRACT. Machining is a process by which we remove unnecessary material from a workpiece to get some desired shape. The shape of the workpiece at any instant is determined by the cumulative of material-removing tool motions until that instant. Planning for machining is difficult because, to prevent the cutting tool from colliding with the workpiece, we need to consider the history of tool motions as we plan the motion. The tool-workpiece collision problem is further complicated because while a portion of the tool is not allowed to hit any part of the workpiece, a special portion of the tool is capable of penetrating the material. This is the cutting portion of the tool, where cutting flutes can contact material. Our approach for tool path generation is to first assign a tool orientation field over space and to later find a collision-free tool motion that conforms to the preassigned orientation field. In the final stages of the tool motion, the unnecessary material needs to be removed completely. We present a condition on the pre-assigned orientation field by which we can guarantee the existence of the collision-free complete tool motion.

1. Introduction

Machining is to transform a block of solid material into a desired form by removing a volume from the original solid using cutting tools. The initial block is called an initial stock or a workpiece and the final resulting form is called the designed part. Diverse mechanical parts are produced in this way. Most consumer products that require dies and molds rely on this process, too.

The motion planning problem for machining can be thought of as a sophisticated version of the robot navigation problem. We think of the cutting tool as the moving robot and the designed part as the obstacle to be avoided by the robot. We list additional features that are unique in machining:

(1) Only a portion of the boundary of the cutting tool is allowed to penetrate the given material. This special portion of the boundary is called the cutting portion. As the cutting process proceeds, the volume swept by the cutting portion is removed from the stock. The non-cutting portion of the tool boundary is not allowed to interfere with the remaining stock material during the entire process.

(2) While robotic path planning seeks a path from a start-point to a finish-point, machining path planning seeks a path which starts with an initial

1991 *Mathematics Subject Classification.* Primary 54C40, 14E20; Secondary 46E25, 20C20.
Key words and phrases. CAD/CAM, machinability, geometry, path planning, collision.

stock and leaves behind the designed part, sweeping **all** the stock material in the exterior of the designed part.

(3) The moving object in machining, which is the cutting tool, can be abstracted as an axially symmetric semi-infinite region.

Loosely put, the designed part is a "hard" obstacle, which must be avoided by the entire cutting tool, and the stock material in the exterior of the designed part is a "soft" obstacle, which must be avoided by the non-cutting portion of the boundary of the cutting tool. The cutting portion must deal with the soft material first and create enough free space so that the rest of the cutting tool can continue to navigate. A rigid body motion of the cutting tool, or simply called a tool path, must be planned carefully so that the cutting tool can avoid both the "hard" and "soft" collisions as described above.

Checking such collisions demands recording the evolution of stock shape, which is a computationally-expensive operation. As far as we know, there is little prior work in this area (apart from solid modeling simulation.) Most previous work focuses on a simpler condition called accessibility, which asks whether the tool can reach points on the boundary of the part in the absence of the soft obstacle. This ignores the evolution of stock shape. An even simpler condition to check is whether the limit case of a long slender tool can reach the boundary of the part. When the tool is infinitesimally slender and infinitely long, in effect, a light ray, reachability is reduced to visibility. The main objective of this paper is to clarify the underlying assumptions on the geometry of cutting tools and parts, which guide motion planners' spatial reasoning. We will approach this problem by asking machinability of a designed part: "using a given cutting tool, is it possible to shape a given designed part by means of machining?" We will develop **history-independent** sufficient conditions for machinability with which we simplify the original problem for a class of limited but practical geometries of cutting tools and parts.

Outline. In the following section, we briefly introduce CNC machine tools and review previous work in this field. In Section 3, we set a framework for our analysis and limit our attention to a major subclass of tools, called monotonic tools. In Section 4, we define machinability in an increasingly strict manner and clarify the relationships among them. In Section 5, we develop a sufficient condition for complete-machining, which is the main theorem of this paper. In Section 6, we show the applicability of the theory developed in this paper to 5-axis roughing path planning in the context of visibility-based path planning. In Section 7, we conclude.

2. Background

In this paper, we restrict our attention to an important mode of machining, called milling. Computer-numerically-controlled (CNC) milling-machine tools are essentially robots of 3, 4 or 5 degrees of freedom. An n-axis machine tool is a machine that has n actuators. A 3- and a 5-axis machine tool are shown in Figure 1.

In milling, a cutting tool is attached to a spindle which consists of a motor, a casing and a tool holder. Spindles are much larger entities than the cutting tools. A spindle in turn is attached to a machine structure, which is an even larger structure

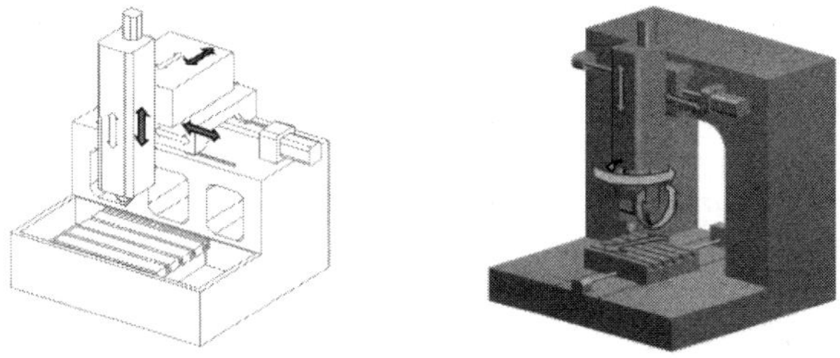

FIGURE 1. A 3-axis and a 5-axis machine tool.

with several degrees of freedom. Because the cutting tool spins in a relatively high speed compared to the "bulk" motion, we regard the axi-symmetric envelope generated by the spin as the cutting tool boundary. Because of this symmetry, 5 degrees of freedom out of the 6 in its rigid body motion are enough to describe the necessary configurations of the cutting tool. When the machine tool has more than 5 actuators, we say that the machine tool is redundant. On the other hand, many available machine tools have less than 5 actuators, which can generate only a restricted class of tool motions. Typical 3-axis machines can generate translations as shown in Figure 1. Since our objective is to clarify the "ultimate" machinability, we focus on 5-axis machining, which fully explores all possible tool motions.[1]

A milling process of a part is usually decomposed into two stages, respectively called roughing and finishing. The goal of roughing is to produce a near-net shape part. In a finishing stage, most remaining material is further removed typically with smaller cutting tools to sculpture the fine detail. This paper focuses mainly on the roughing process because the history-related complexity is greatly reduced in the finishing stage.

There has been a significant amount of research in the automation of CNC tool path generation, some of which is parallel to robotics literature. A comprehensive list of machining path planning literature can be found in [1]-[4]. Visibility related concepts from robotics have found applications in machining, though in a different way than in robotic path planning. This is because accessibility to a point by a cutting tool with 5 degrees of freedom is similar to stabbing the point with a ray of light. Seminal work in the area of visibility and visibility maps was done by Chen, Tang and Woo [5]-[7]. They introduced the concept of visibility cones for facets on a workpiece, which can be mapped on to the unit sphere using "G-maps." The same authors also show how G-maps can be extended with a central projection to manipulate access information and minimize setups. The idea of G-maps was used by Wuerger and Gadh [8] to evaluate separability of dies. Lee [9] uses a convex hull based approach to approximate local visibility. Suh and Kang [10] used a table to store the information on accessibility. Elber and Cohen [11] presented an approach to surface accessibility, which maps convex surfaces to a plane. Obstacles to the surface are also mapped into the plane, and tool path generation is carried out in a Cartesian world. Accessibility in various forms has been explored [12]-[18]. Morishige *et. al.* [19] present an approach for tool path generation using a three dimensional configuration space. The work presented here also draws inspiration,

[1]In theory, the workpiece can be manually adjusted in a 3-axis machine to "simulate" 5-axis machining although it is clearly not practical.

if not direct results, from the large body of literature in robotic path planning [20]-[24].

In recent years, the problem of determining tool accessibility has been approached from the perspective of computer graphics. In a series of publications, we developed the concepts of graphics-based visibility determination, which makes our visibility-based path generation method feasible in practice [17], [25]-[27]. Spitz *et. al.* [28] present a similar method for performing accessibility analysis using computer graphics hardware. The emphasis of their work is on computing accessible directions for tactile probes used in 3D digitization with coordinate measuring machines.

Previous literature, for example [25], assumes that if for every point on the boundary of a designed part, there exists a tool posture which contacts the point and which is not in collision with the designed part anywhere else, then collision-free tool postures exist over the exterior of the part. Second, the literature assumes that the existence of such a posture over the exterior implies that there is a cutting path to bring it there. In general, these assumptions are incorrect. We will show later the conditions under which such simplifying assumptions are valid.

3. Parts and Tools

In this section, we set a framework for our analysis and limit our attention to a major subclass of tools.

3.1. Parts. A *workspace* or *scene* is defined as the Euclidean 3-space. A *designed part* is a connected 3-manifold with its boundary bounded by a closed surface, being embedded in the scene. A larger volume of material, which we call a *stock*, is converted into the designed part through a machining process. It is general enough to require the stock to be a compact 3-manifold in the scene. The part must be enclosed by the stock. A setup is defined as a pair $(\mathcal{S}, \mathcal{P})$ where $\mathcal{S}$ is a stock and $\mathcal{P}$ is a part in it. A setup is shown in Figure 2. The volume in the workspace that does not belong to the stock is referred to as the *initial free space*. The material which is in the stock but not in the part is the material that must be removed, and we refer to it as the *delta-volume* of the setup. We define the delta-volume of a setup unambiguously as the closure of $\mathcal{S}\backslash\mathcal{P}$, where $\mathcal{S}$ is the stock and $\mathcal{P}$ is the part of the setup $(\mathcal{S}, \mathcal{P})$. The closure of the complement $\mathcal{P}^c$ of a part $\mathcal{P}$ in a scene is referred to as the *transparent region*.

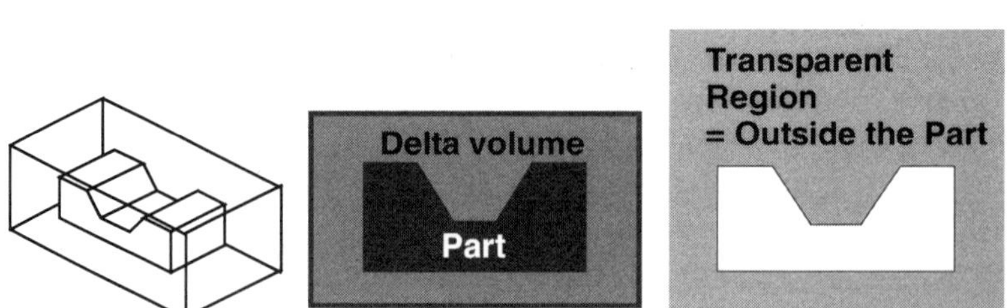

FIGURE 2. A part to be machined.

3.2. Cutting Tools. The readers are referred to Figure 3-(a) for the following definitions on cutting tools. By a cutting tool or tool, we will, in fact, mean a tool-spindle assembly. We will treat cutting tools as if they extended to infinity and they were axi-symmetric. The *bottom* of a tool is the intersection of the axis of symmetry and the boundary of the tool, from which the level of a point in the tool is measured along the axis of symmetry. No portion of the tool is allowed to be below the bottom level. In the implementation stage, we will disregard the part of the tool beyond a plane at a certain level, which we refer to as the *stem surface*. The level of the stem surface is chosen such that the geometry beyond this plane is not involved in the cut. A vector parallel with the axis of symmetry pointing towards the stem surface from the bottom is referred to as the *orientation* of the tool. A tool has a connected closed region of non-vanishing area called the *cutting portion* of the tool on its boundary. The remaining region on the boundary is referred to as the *non-cutting portion*. We require that the cutting portion should be a surface of revolution and that at least one point in the cutting portion be at the bottom level of the tool. It is only the cutting portion of a cutting tool that is allowed to make contact with the uncut material, and sweep of this portion removes material. In summary, a cutting tool, simply called a tool, is an axi-symmetric semi-infinite rigid body, whose boundary is composed of two axi-symmetric regions. The two regions are referred to as the cutting portion and the non-cutting portion of the tool, respectively.

In general, tools may have a "complicated" cutting portion as shown in Figure 3-(a) where the cutting portion faces downwards, sideways and upwards. This permits the tool to cut overhung surfaces for specialized features such as slots. The use of such tools is, however, rare, and we consider a major subclass of tools which we will refer to as *monotonic tools*. A tool is said to be *monotonic* if at any point on the boundary of the tool, there exists a ray based at the point such that (1) it is parallel with the line of symmetry of the tool and (2) it is contained entirely in the tool. A monotonic tool is shown in Figure 3-(b). Most cutting tools used in machining, such as end mills, faces mills and drills can be modeled as monotonic tools. Henceforth, we will limit our discussion to monotonic tools. Later in Section 4, we will make it clear why monotonicity is an important condition which makes path planning simpler.

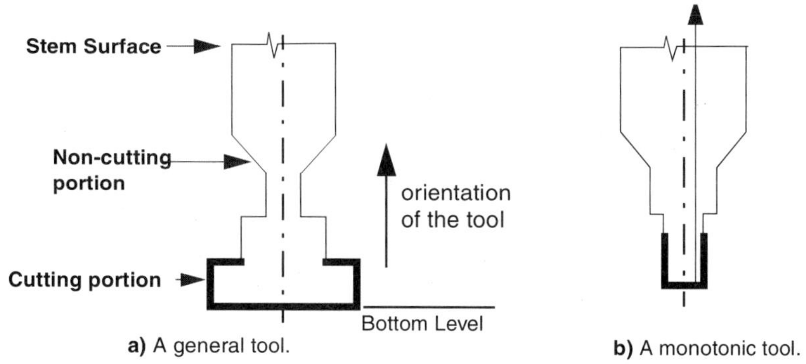

FIGURE 3. Tools.

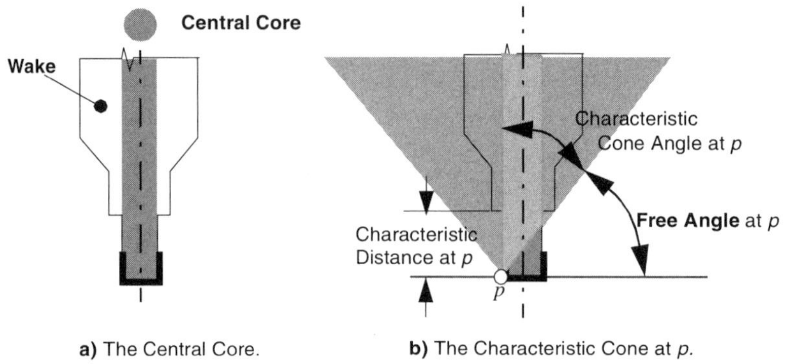

FIGURE 4. Terms for monotonic tools.

For later use, we define additional terms related to monotonic tools, especially the free angle of a tool. Figure 4 provides intuitive definitions. The *central core* of a tool is the volume in the tool swept out by the cutting portion of the tool when it is translated upwards to infinity, in a direction parallel to the axis of the tool. The points of a tool that do not belong to its central core constitute what we call the *wake* of the tool. A cone whose vertex is located at a point p on a cutting portion of a tool is referred to as the *wrapping cone* of the tool at the point p, if (1) the axis of the cone is parallel to the line of symmetry of the tool and (2) the cone, as a volume, covers the wake of the tool. A wrapping cone of a tool at a point is said to be a *characteristic cone* at the point if no wrapping cone at the point has a smaller cone angle than the wrapping cone of the tool. The *characteristic cone angle* of a tool at a point is the angle between a generatrix and the axis of the characteristic cone of the tool at the point. Its complementary angle with respect to the right angle will be referred to as the *free angle* at the point in the tool as shown in Figure 4-(b). The greatest lower bound of the distance between a point p on the cutting portion of a tool and the wake of the tool is referred to as the *characteristic distance* at the point p in the tool. Most tools have a non-vanishing characteristic distance and a non-negative free angle at every point on the cutting portion.

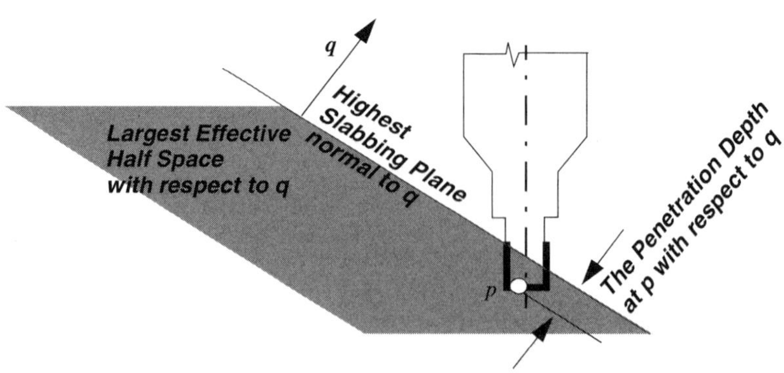

FIGURE 5. A tool penetrative with respect a vector **q**.

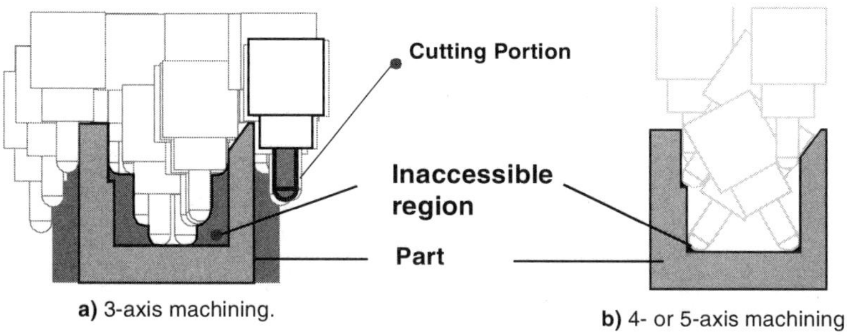

FIGURE 6. Accessibility.

We define a more special subclass of tools and related terms for later use, especially the penetration depth of a tool. See Figure 5. A monotonic tool is said to be *penetrative with respect to a vector* $\mathbf{q}$, if there is a plane $\mathcal{L}$ normal to the vector $\mathbf{q}$ such that one of the two half spaces, whose boundary is $\mathcal{L}$, is disjoint with the non-cutting portion of the tool and contains at least one point in the cutting portion. If so, the plane $\mathcal{L}$ is referred to as a *slabbing plane* normal to $\mathbf{q}$ and the half space is referred to as an *effective half space* with respect to $\mathbf{q}$. The effective half space is required to be pointed by $-\mathbf{q}$. The closure of the union of all the effective half spaces with respect to $\mathbf{q}$ is referred to as the *largest effective half space* with respect to $\mathbf{q}$ and its boundary plane is referred to as the *highest slabbing plane* normal to $\mathbf{q}$. The intersection of the largest effective half space with respect to $\mathbf{q}$ with the cutting portion of the tool is referred to as the *effective cutting portion* with respect to $\mathbf{q}$. The *penetration depth* at a point p in the effective cutting portion with respect to a vector $\mathbf{q}$ is defined as the distance between the point p and the highest slabbing plane normal to the vector $\mathbf{q}$. These are shown in Figure 5.

4. Machinability

In this section we define various machinability conditions, and clarify the relationships among them.

Typically, machine tools are designed to provide up to 5 degrees of freedom. The difference between 3-axis and 4- or 5-axis machining is shown in Figure 6, which shows that 4- or 5-axis machines can remove more delta volume in a single setup than 3-axis machines can, generating a larger "accessible" region. In the literature, including several papers that we have written, the term "accessibility" is used as a general term for the ability to perform collision-free cutting. However, the concept of avoiding collisions is subtle as pointed out in the introduction. We present increasingly strict definitions below and avoid using the term accessibility because of the ambiguity associated with this term in the literature.

First, we define machinability in the point-wise manner.

Visibility: A point p in the transparent region of a part is said to be *visible* if there exists a ray emitted from the point p to a point at infinity that does not intersect the interior of the part. In the computational geometry community, this definition is strictly referred to as *weak-external-visibility* [30], [31]. However, we will not use the more elaborate term because the only visibility involved in this paper is weak-external-visibility.

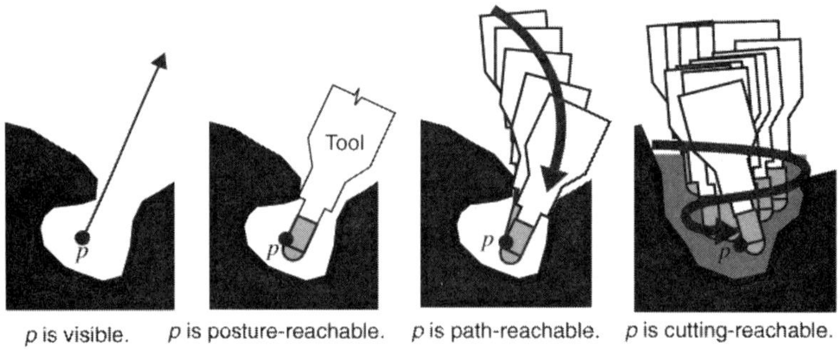

FIGURE 7. Machinability.

Posture-reachability: At each instant during machining, a *posture* of a tool is determined by a rigid body transformation from a certain configuration of the tool. A point p in the transparent region of a part is said to be *posture-reachable* by a tool if there is a posture of the tool satisfying the following criteria: (1) a point in the cutting portion of the tool in the posture coincides with the point p and (2) the tool in the posture and the interior of the part are mutually disjoint. If so, we will also say that we can *place* the tool at the point in the posture.

Path-reachability: A point p in the transparent region of a part is said to be *path-reachable* by a tool if there is a rigid body motion[2] of the tool from infinity such that (1) at the final posture of the motion, the tool is placed at the point p and (2) at each instant during the motion, the tool and the interior of the part are disjoint with each other.

Cutting-reachability: For a prescribed rigid body motion of a tool, the *free space at any instant* is the union of the initial free space with the volume swept by the cutting portion of the tool up to the instant. If at each instant during a motion, the tool lies in the free space at the instant and the tool is disjoint with the interior of the part, the motion is said to be a *legal cutting path*. A point p in the transparent region of a part is said to be *cutting-reachable* by a tool if there is a legal cutting path that carries the tool, which starts at infinity, to a posture in which the point p lies on the cutting portion of the tool. Note that this is an even stricter condition than path-reachability.

The definitions are shown in Figure 7. Visibility is the weakest but most canonical test of machinability because it postulates the use of a needle-like slender tool. Furthermore, it is easiest to check. Posture-reachability takes account of the thickness of tools, and it is essentially a "static" test. Path-reachability emphasizes that there should be a collision free motion for machinability in addition to the static test at the final posture. Cutting-reachability is the strictest condition, taking account of the subtleties related to the fact that the shape of uncut material evolves during machining. Most prior research has dealt with visibility and some has dealt with posture-reachability, *c.f.* [28] and [29].

[2] A rigid body motion is a time-parametrized family of rigid body transformations or postures.

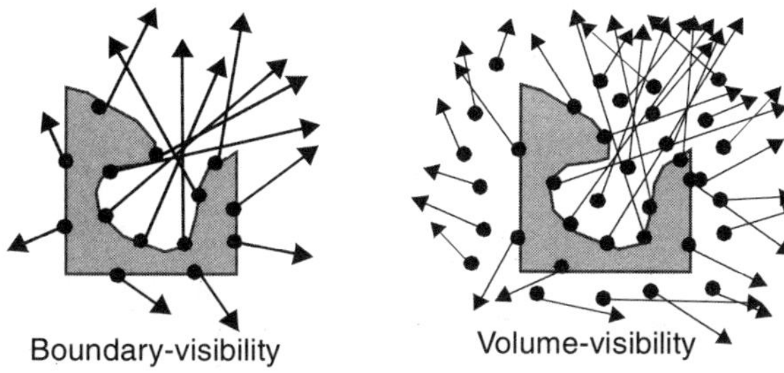

FIGURE 8. Boundary- and volume-visibility.

Based on the above point-wise definitions, we define machinability conditions over particular regions.

Volume-visibility: If all the points in the transparent region of a part are visible, then we say that the part is *volume-visible.*

Boundary-visibility: If all the points on the boundary of a part are visible, we say that the part is *boundary-visible* as shown in Figure 8.

Volume-reachability: If all the points in the transparent region of a part are posture-reachable by a tool, we say that the part is *volume-reachable* by the tool.

Boundary-reachability: If all the points on the boundary of a part are posture-reachable by a tool, we say that the part is *boundary-reachable* by the tool.

Complete-machinability: If all the points in the transparent region of a part are cutting-reachable by a tool, the part is said to be *completely machinable* by the tool.

The following propositions between the machinability conditions are immediate from their definitions:

THEOREM 4.1. *For every point p in the delta volume of any setup,*

$$p \textit{ is cutting-reachable} \Rightarrow p \textit{ is path-reachable} \Rightarrow p \textit{ is posture-reachable} \Rightarrow p \textit{ is visible.}$$

THEOREM 4.2. *For any part $\mathcal{P}$,*

(1) *$\mathcal{P}$ is completely-machinable$\Rightarrow\mathcal{P}$ is volume-reachable$\Rightarrow\mathcal{P}$ is boundary-reachable.*
(2) *$\mathcal{P}$ is volume-reachable$\Rightarrow\mathcal{P}$ is volume-visible.*
(3) *$\mathcal{P}$ is volume-visible$\Rightarrow\mathcal{P}$ is boundary-visible.*
(4) *$\mathcal{P}$ is boundary-reachable$\Rightarrow\mathcal{P}$ is boundary-visible.*

Figure 9-(a) shows that the implication, *posture-reachability$\Rightarrow$visibilty*, holds for monotonic tools, to which we have limited our discussion. The following less apparent relation is also due to our use of monotonic tools.

THEOREM 4.3. *For every point p in the delta volume of any setup,*

$$p \textit{ is path-reachable} \iff p \textit{ is posture-reachable.}$$

PROOF. The implication *path-reachability$\Rightarrow$posture-reachability*, is immediate from the definitions as mentioned in Theorem 4.1. We only need to prove its converse. Assume that the point p is posture-reachable by a tool. Then, there is a posture of the tool, placed at the point p, without any intersection with the interior

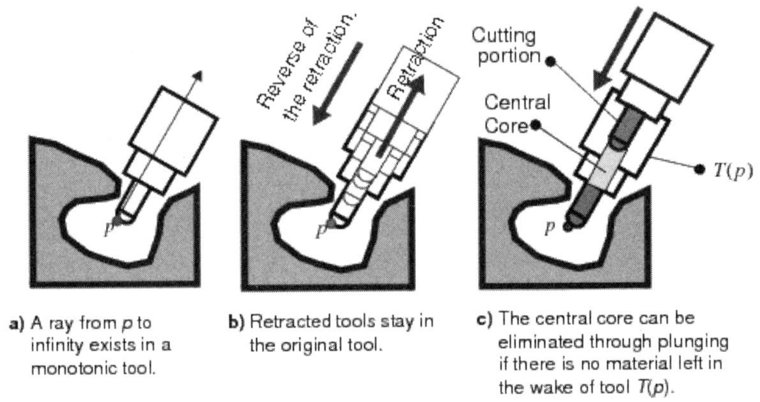

FIGURE 9. Consequences of monotonicity of tools.

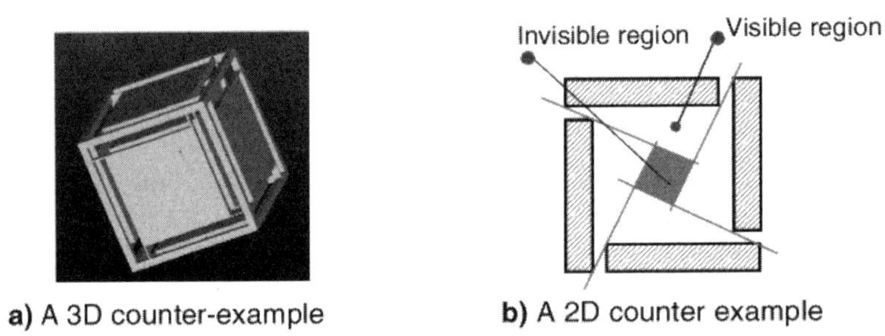

FIGURE 10. Boundary-visibility does not guarantee volume-visibility: *All the points on the boundary are visible but a region is not.*

of the part. Now, consider retraction of the tool along its axis. Because the tool is monotonic, the retracted tools stay in the original tool placed at p, implying that the retracted tools are also disjoint with the interior of the part. The reverse of the retracting translation satisfies all the requirements for the path in the definition of path-reachability. This is shown in Figure 9-(b). □

Since we only consider monotonic tools, the theorem above permits us to limit our consideration to posture-reachability.

For later use, we present the following theorems, which sounds obvious.

THEOREM 4.4. *Every point in the outside of the convex hull of any stock is cutting-reachable.*

PROOF. Let $\mathcal{C}$ be the convex hull of the stock. Let p be an arbitrary point that does not belong to $\mathcal{C}$. Because of the convexity of $\mathcal{C}$, we can construct a plane passing the point p in such a way that one of the two closed half spaces generated by the plane is disjoint with the interior of the stock, which we call empty. Since every tool is semi-infinite and has a point on the cutting portion in the bottom level, the tool can be placed at p in such a way that the tool lies completely in the

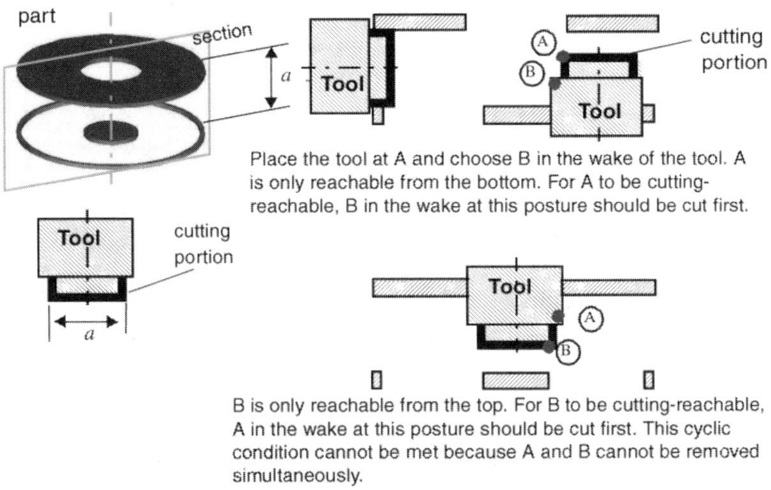

FIGURE 11. Volume-reachability does not guarantee complete-machinability: *A connected part should have been drawn. However, this figure is more instructive.*

empty half space and its axis is normal to the plane. This posture can be reached along a legal cutting path from infinity, which is simply a translation along the vector normal to the plane. □

THEOREM 4.5. *Let $\mathcal{T}$ be a monotonic tool and p be a point in the delta-volume of a setup which is not cutting-reachable but is posture-reachable by $\mathcal{T}$. By $T(p)$, we denote a copy of the tool $\mathcal{T}$ placed at p in a certain posture, which is embedded in the scene. Then, in the region occupied by the wake of $T(p)$, there is a point which is not cutting-reachable.*

PROOF. Assume that all the points in the wake of $T(p)$ are cutting-reachable. Then, we can "eliminate" all the points in the wake of $T(p)$ "legally." As a result of the elimination, the stock material remaining in the region occupied by $T(p)$ is only in its central core. The remaining material can be further removed by a translation from infinity along the axis of $T(p)$. This is shown in Figure 9-(c). By this motion, the tool $\mathcal{T}$ is finally placed in the posture of $T(p)$. This implies the cutting-reachability of p and contradicts our assumption that p is not cutting-reachable. Therefore, there should be at least one point which is not cutting-reachable in the wake of $T(p)$. □

A corollary of Theorem 4.5 is that if we use a tool without a wake, then volume-reachability implies complete-machinability.[3] However, the converses of Theorem 4.1 and Theorem 4.2 are not generally true except for the equivalence of posture- and path-reachability confirmed in Theorem 4.3. We present two important counter-examples against those converses. In Figure 10, we show a part

[3]Most work in machining deals with tools without a wake, and consequently the question on cutting-reachability has seldom been asked. Exceptions include [32]. We mention that the collision-free condition suggested in [32] is a condition that is subtly different from our cutting-reachability. This is because their "cutting portion" is defined as a volume while ours is defined as a surface on the boundary of a tool.

which is not volume-visible but is boundary-visible.[4] We can also show that the part is not volume-reachable but is boundary-reachable by a tool because we can regard a ray as a special tool. That is, boundary-reachability does not guarantee volume-reachability. Figure 11 shows a volume-reachable part with a given monotonic tool and shows why this part is not completely-machinable by the tool. The example shows that even the volume-reachability of the part does not guarantee its complete-machinability.

On the other hand, it is also observed that the counter-examples look rather "artificial" and will seldom occur in practice. In the following section, we establish certain conditions under which the "history-related" complete-machinability is reduced to a simpler condition, posture-reachability, for restricted but practical geometries of tools and parts.

5. Sufficient Conditions for Complete Machinability

We discuss the possibility of reducing complete-machinability into a more manageable form. Our approach for tool path generation is to first assign a tool orientation field over the delta volume and to later find a collision-free tool path that conforms to the preassigned orientation field. We present a condition on the preassigned orientation field. When this condition is met, we can guarantee for the second step of our approach that there exists a collision-free tool path that completely removes the delta volume. In other words, we provide a means to answer the query on complete-machinability by examining only posture-reachability which is a much simpler "static" test.

First, we define a few terms related to this condition. The readers are referred to Figure 12 for these definitions.

A cover $\{A_k, k = 1, 2, ..., N\}$ of a region A is said to be *non-excessive* if $\cup_k A_k = A$ and A_k is a simply connected closed set. Consider a setup $(\mathcal{S}, \mathcal{P})$, a tool $\mathcal{T}$ and a non-excessive cover $\{\Delta_k, k = 1, 2, ..., N\}$ of the delta volume Δ of the setup. The union of the initial free space of the setup and the region $\Delta_k \in \{\Delta_k\}$ is referred to as the *kth transparent region* of the setup by the cover. A map $T_k : \Delta_k \to SE(3)$ is called a *posture assignment* in $\Delta_k \in \{\Delta_k\}$ for the tool $\mathcal{T}$ if for every $p \in \Delta_k$, the wake of $T_k(p)(\mathcal{T})$ is in the kth transparent region and p is in the cutting portion of $T_k(p)(\mathcal{T})$. Note that $T_k(p)$ is a rigid body transformation while $T_k(p)(\mathcal{T})$ is the region occupied by the transformed tool. However, we will sometimes abbreviate $T_k(p)(\mathcal{T})$ into $T_k(p)$. A pre-roughing plan in $\Delta_k \in \{\Delta_k\}$ for a tool is a pair $(T_k, \mathbf{q}_k^o)$ where T_k is a posture assignment in Δ_k for the tool and $\mathbf{q}_k^o$ is a vector.

Consider a pre-roughing plan $(T_k, \mathbf{q}_k^o)$ in $\Delta_k \in \{\Delta_k\}$ for a tool $\mathcal{T}$. We call $\mathbf{q}_k^o$ the *kth slabbing direction.* The orientation of $T_k(p)$ $(p \in \Delta_k)$ is symbolically referred to as $\mathbf{q}_k(p)$. The angle between the orientation $\mathbf{q}_k(p)$ $(p \in \Delta_k)$ and the slabbing direction $\mathbf{q}_k^o$ will be referred to as the *deviation angle* at p in Δ_k for the pre-roughing plan. The free angle minus the deviation angle at $p \in \Delta_k$ in $T_k(p)$ is called the *angular margin* at the point p for the pre-roughing plan.

A pre-roughing plan $(T_k, \mathbf{q}_k^o)$ in $\Delta_k \in \{\Delta_k\}$ is said to be *proper* if both $\{A_k(p) | p \in \Delta_k\}$ and $\{D_k(p) | p \in \Delta_k\}$ have positive lower bounds where $A_k(p)$ is the angular margin and $D_k(p)$ is the characteristic distance at p in $T_k(p)$ as defined in Section

[4] A similar observation on the difference between volume-visibility and boundary-visibility is made in [34], where the author modifies the traditional art gallery problem to the one stationing the guards who will be hired to watch the entire walls of the gallery, but not the entire floor.

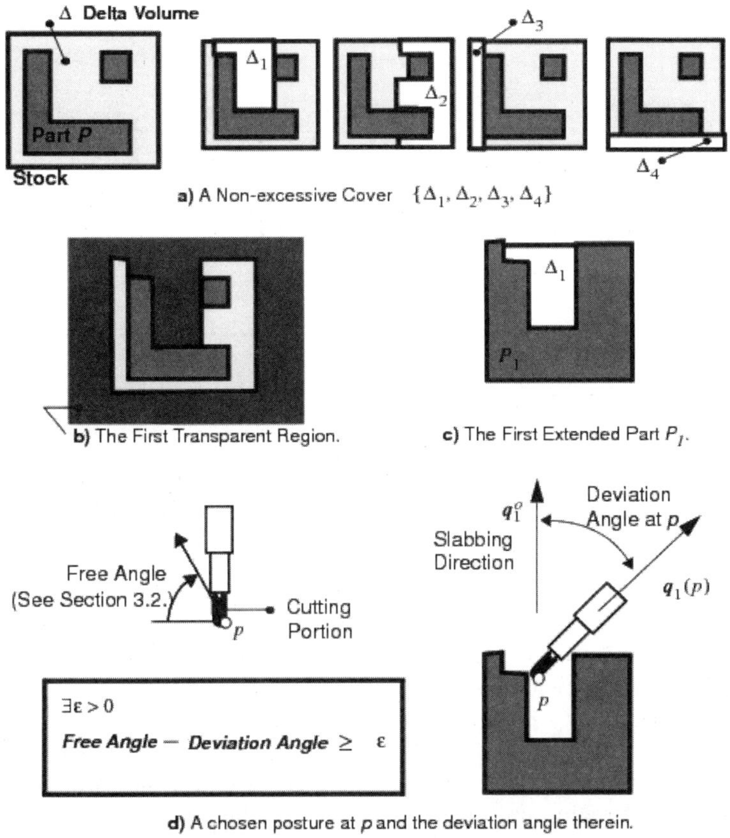

FIGURE 12. The deviation angle theorem.

3.2. A pre-roughing plan $(T_k, \mathbf{q}_k^o)$ in $\Delta_k \in \{\Delta_k\}$ is said to be *penetrative* if (1) for every point $p \in \Delta_k$, $T_k(p)$ is penetrative with respect to the slabbing direction $\mathbf{q}_k^o$ and (2) $\{B_k(p) | p \in \Delta_k\}$ has a positive lower bound where $B_k(p)$ is the penetration depth at p in $T_k(p)$ with respect to the slabbing direction $\mathbf{q}_k^o$. (See Section 3.2 for the related definitions.) A pre-roughing plan $(T_k, \mathbf{q}_k^o)$ is said to be *continuous* if T_k is continuous.

Roughly speaking, a pre-roughing plan is proper if the free angle is greater than the deviation angle. Figure 12-(a) shows a setup and a cover of its delta volume. Figure 12-(b) shows the 1st transparent region of the setup by the cover. The closure of the complement of the kth transparent region is referred to as the kth extended part, and symbolically as P_k. This is shown in Figure 12-(c). Note that if each extended part is completely-machinable, then the original part is completely machinable. The deviation angle for a slabbing direction $\mathbf{q}_1^o$ is shown in Figure 12-(d).

THEOREM 5.1 (Deviation Angle Theorem). *Consider a part $\mathcal{P}$ in a convex stock $\mathcal{S}$. Let $\{\Delta_k, k = 1, 2, ..., N\}$ be a non-excessive cover of the delta-volume of*

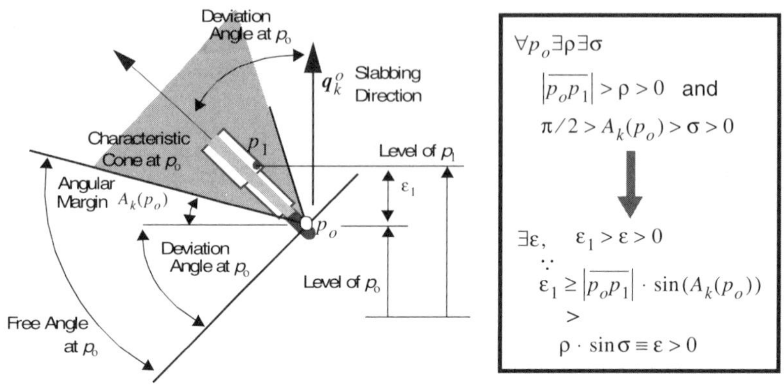

FIGURE 13. The level of p_1 is higher than that of p_0.

the setup $(\mathcal{S}, \mathcal{P})$. For a given monotonic tool, if there is a proper pre-roughing plan in each $\Delta_k \in \{\Delta_k\}$, the part in the setup is completely-machinable by the tool. [5]

PROOF. We mention that Theorem 4.5 in the previous section is a key step of our proof. Let $(T_k, \mathbf{q}_k^o)$ be the proper roughing plan in $\Delta_k \in \{\Delta_k\}$. Here, we measure the level of a point in the kth element Δ_k of the cover along the slabbing direction $\mathbf{q}_k^o$. Assume that the part $\mathcal{P}$ is not completely machinable. Then, in a $\Delta_k \in \{\Delta_k\}$, there must be a point p_0 which is not cutting-reachable. Then in the wake of $T_k(p_0)$, there must be a point, say p_1, which is not cutting-reachable according to Theorem 4.5. p_1 should be also in the stock because if not, it is cutting-reachable according to Theorem 4.4. In addition, p_1 should be in Δ_k because the wake of $T_k(p_0)$ is in the kth transparent region due to the definition of the posture assignment $(T_k, \mathbf{q}_k^o)$. Using elementary geometry, we can show that the level of p_1 is greater than that of p_0 by a positive amount, say ε_1—we substitute the proof of this statement with the drawing in Figure 13. Since p_1 is not cutting-reachable, there must be a point p_2 in the wake of $T_k(p_1)$, which is not cutting-reachable. The level of p_2 must be greater than that of p_1 by a positive amount, say ε_2, because of the same reason. In this manner, we can construct a sequence $\{p_1, p_2, ...\}$ of non-cutting-reachable points in Δ_k and another sequence $\{\varepsilon_1, \varepsilon_2, ...\}$ of positive numbers. Both sequences should continue indefinitely. Because $(T_k, \mathbf{q}_k^o)$ is proper, we can show that the sequence $\{\varepsilon_1, \varepsilon_2, ...\}$ must have a positive lower bound—we substitute the proof of this statement with the box in Figure 13. The level of the nth point p_n $(n > 0)$ in $\{p_1, p_2, ...\}$ is greater than that of p_0 simply by the amount of the summation, $\varepsilon_1 + \varepsilon_2 + ... + \varepsilon_n \equiv S_n$. Since $\{\varepsilon_1, \varepsilon_2, ...\}$ has a positive lower bound, the summation S_n diverges as n goes infinity. This implies that for a certain $m > 0$, p_m is so far from p_0 that p_m is outside the stock. p_m is not cutting-reachable by definition but Theorem 4.4 implies that p_m must be cutting-reachable. This is a contradiction and proves our theorem. □

The above theorem provides a means to check whether there exists a legal cutting path for a given posture field, T_k, but it provides neither a specific tool

[5]A related algorithmic issue is to find an appropriate non-excessive cover of the delta-volume, which can be approached from a technique similar to the cone-visibility decomposition described in [33].

path nor a practical path. In rough machining, it is a usual practice to remove the delta-volume layer by layer. When this is possible, we say that the part is millable in the conventional manner. The question is then what is the condition that guarantees this practice. Below, we will show that the justification is made if we use a "penetrative tool."

First, we define what we mean by conventional millability with unambiguous terms. For a given path $p : [0,T] \to E^3$ in the scene E^3, a continuous rigid body motion of a tool during the time interval $[0,T]$ is said to *prescribe* the path $p : [0,T] \to E^3$ if the cutting portion of the moving tool at each instant $t \in [0,T]$ contains the point $p(t)$ in the path. Let $\mathcal{U}$ be a simply-connected closed region in the delta-volume of a setup $(\mathcal{S},\mathcal{P})$. Consider a one-parameter family of mutually parallel planes. There must be a common normal vector of the planes in the family. The level of a plane in the family can be measured along the common normal vector. Let $\mathcal{L}(c)$ be the plane in the family at a level $c \in R$. The range of the levels of the planes in the family that intersect the region $\mathcal{U}$ must be a certain closed interval, say $[a,b]$. By $\mathcal{S}(c)$ ($c \in [a,b]$), we denote the closed region formed by removing the subregion of $\mathcal{U}$ above the plane $\mathcal{L}(c)$ from the stock $\mathcal{S}$. The region $\mathcal{U}$ is said to be *millable in the conventional manner* along the family $\{\mathcal{L}(x)|x \in R\}$ of planes if there exists a certain positive real number $\varepsilon > 0$ such that for every $c \in [a,b]$ and for every continuous path $p : [0,T] \to E^3$ lying in the plane $\mathcal{L}(c)$, there exists a legal continuous rigid body motion for the setup $(\mathcal{S}(c+\varepsilon),\mathcal{P})$, a portion of which is a rigid body motion that prescribes the path $p : [0,T] \to E^3$ starting from the outside of the stock. If the delta volume of a set up is covered by regions and each region is millable in the conventional manner, we say that the part in the setup is *millable in the conventional manner.*

In the following theorem, we show a condition that guarantees the conventional millability.

THEOREM 5.2 (Conventional Millability). *Let $\mathcal{P}$ be a part in a convex stock $\mathcal{S}$. Let $\{\Delta_k, k = 1,2,...,N\}$ be a non-excessive cover of the delta-volume of the setup $(\mathcal{S},\mathcal{P})$. For a given monotonic tool, if there is a continuous and penetrative pre-roughing plan in each Δ_k, the part $\mathcal{P}$ in the setup is millable in the conventional manner by the tool.*

PROOF. Let $(T_k, \mathbf{q}_k^o)$ be the continuous penetrative pre-roughing plan in Δ_k. Let ε be a lower positive bound of $\{B_k(p)|p \in \Delta_k\}$ where $B_k(p)$ is the penetration depth at p in $T_k(p)$ with respect to the slabbing direction $\mathbf{q}_k^o$. We take this lower bound as the symbol ε appearing in the definition of conventional millability. Let $\{\mathcal{L}_k(x)|x \in R\}$ be the family of planes normal to the slabbing direction $\mathbf{q}_k^o$, where $\mathcal{L}_k(x) \cap \Delta_k \neq \oslash$ only for $x \in [a_k, b_k]$. Let c be an arbitrary level in $[a_k, b_k]$. For every path $p : [0,T] \to \mathcal{L}_k(c) \cap \Delta_k$, there is a continuous rigid body motion prescribing the path, namely $t \to T_k(p(t))$. For the legality of an arbitrary path in the plane at the level c for the setup $(\mathcal{S}(c+\varepsilon),\mathcal{P})$, it is enough to indicate that (1) the non-cutting portion of $T_k(p(t))$ is above the highest slabbing plane of $T_k(p(t))$ with respect to the slabbing direction $\mathbf{q}_k^o$, (2) there is no delta-volume of the setup $(\mathcal{S}(c+\varepsilon),\mathcal{P})$ in Δ_k that meets the non-cutting portion of $T_k(p(t))$ above the highest slabbing plane during the motion (See Figure 5) and (3) there is a cutting-reachable point in the plane for the setup $(\mathcal{S}(c+\varepsilon),\mathcal{P})$. The existence of a cutting-reachable point is proved by the retraction argument similar to the one given in Theorem 5. □

In this section, we have seen how it is possible to guarantee complete machinability of a part by simply constructing collision-free postures of a tool over the delta-volume. Using the result, we have also shown the condition under which the commonly-practiced layer-by-layer roughing strategy is justified.

6. Visibility-based Planning

As shown in Theorem 4.1 and Theorem 4.2, visibility is only a necessary condition for complete-machinability. However, visibility is also a good approximation of machinability because tools are typically narrow and long. Our approach to path planning for 5-axis roughing is to assign a field of visible rays over the delta volume and to use it as an initial guess for feasible tool postures. We mention that the millability conditions can be applied to make a good initial guess in this context. We summarize our approach:

(1) Sample enough number of points from the delta volume of a given part.
(2) For each sample point, find all the rays emitted from the sample point that do not intersect the interior of the part. We refer to such rays as feasible rays.
(3) For each sample point,
   (a) Choose a ray among the feasible rays.
   (b) Align a given tool with the chosen feasible ray and locate the tool tip at the sample point.
   (c) If the tool interferes with the part in the chosen posture, adjust the tool to a nearby posture, until the sample point is confirmed to be posture-reachable by the adjustment.
(4) Check the **conventional millability** or the **deviation angle condition** developed in the previous section to guarantee the complete-machinability with a tool path compatible with the selected postures.
(5) Interpolate the chosen postures in a smooth fashion.

The suggested approach is "brute-force"; however, by leveraging well-known hardware and algorithmic techniques, we achieve this computation in a scalable and efficient way. In Step 3-(c), we use a very rapid collision detection algorithm developed for this purpose [37], [38]. In Step 2, we use a hardware solution in computing all the feasible rays. In computer graphics, dedicated hardware solutions such as the Z-buffer have greatly improved the resolution and the cost of visibility computation. First, we discretize the boundary of the part by tessellating it. To each individual triangle, we assign a unique color of the $16,777,216$ ($\sim 16 \times 10^6$) colors available on a 24 bit color graphics card. Second, we discretize a sphere that represents a set of viewing directions. We now view the solid model of the part from each of the discretized viewing directions. We check whether a triangle is visible from a particular direction by looking for the color of that triangle in the rendered image. This computation could have been performed in software, but would be very slow. Essentially, the graphics buffer performs the hidden surface removal in our approach. The result is a set of feasible rays at each sample point on the boundary of the part. For the sample points on the delta-volume not on the boundary, we use feasible rays emitted from the boundary. We create a coarse voxelization covering the delta-volume. For each triangle on the boundary of the part, we check the directions it is visible from, and draw a ray in each such direction. We then intersect that ray with the voxels encountered. We mark each voxel encountered

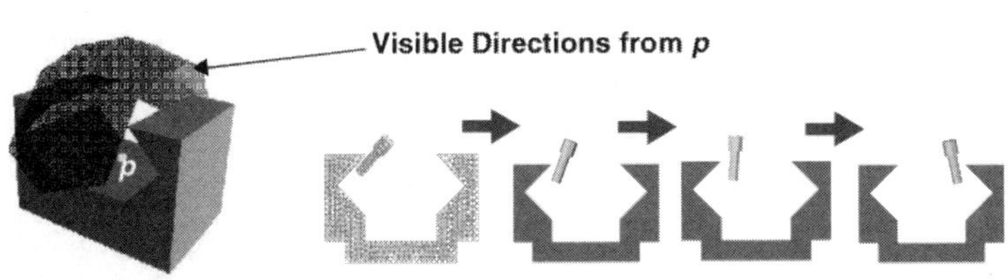

a) A Roughing tool path in a Layer

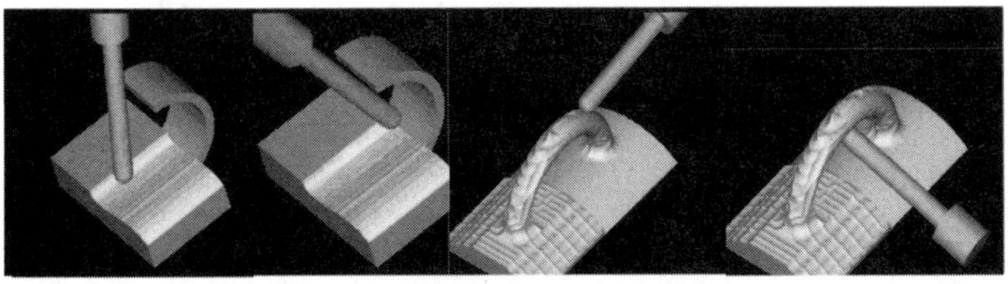

b) Finishing tool paths

FIGURE 14. Roughing and finishing tool paths.

as visible from that direction. The details of the above approach can be found in [25]-[27] except for Step 4. These techniques were recently granted a US patent [39]. A roughing path generated with this method is shown in Figure 14-(a).

The result of Step 2 in the above procedure is a rather complete picture of the feasible configuration space for motions of a "narrow" tool. This information can be used for other purposes such as 3- and 4-axis setup determination [17] *c.f.* [6], [35], [36] and 5-axis finishing [27]. Examples of finishing tool paths are shown in Figure 14-(b).

7. Conclusions

The collision checking problem related to machining is history-dependent and therefore difficult. We developed a sufficient condition for complete-machinability that could be checked without considering a specific instance of a tool path. We showed also the application of the theory to 5-axis roughing in the context of visibility-based path planning.

References

[1] Kim, T. and Sarma, S. E., *Optimal sweeping paths on a 2-manifold: a new class of optimization problems defined by path structures*, IEEE Transactions on Robotics and Automation, Vol 19 (4), Aug. 2003.

[2] Dragomatz, D. and Mann, S., *A classified bibliography of literature on NC milling path generation*, Computer-Aided Design, Vol. 29, No. 3, pp. 239-247, 1997.

[3] Choi, B.K. and Jerard, R.B. *Sculptured Surface Machining, Kluwer Academic Publishers*, ISBN 0-412- 78020-8. 1998.

[4] M. Held, *On the Computational Geometry of Pocket Machining*, LNCS, Vol. 500. Springer–Verlag, 1991.

[5] Chen. L-L and Woo, T. C. *Computational geometry on the sphere with application to automated machining*, Transactions of ASME, Vol. 114: 288-295.

[6] Woo, T. *Visibility maps and spherical algorithms*, Computer Aided Design 26(1). 1994.

[7] K. Tang, T. Woo, and J. Gan. *Maximum intersection of spherical polygons and workpiece orientation for 4- and 5-axis machining*, Journal of Mechanical Design, Vol. 114, 1992, pp. 477-485.

[8] Wuerger, D. and Gadh, R. *Virtual prototyping of die design algorithmic, computational and practical considerations*, Proceedings of the Computer Aided Concurrent Design Symposium, ASME Design Engineering Technical Conferences, Boston. 1995.

[9] Lee, Y.S., and Chang, T.C., *Two-Phase Approach to Global Tool Interference Avoidance in 5-axis Machining*, Computer Aided Design, Vol. 27, No. 10, pp. 715-729. 1995.

[10] Suh, S.-H. and Kang J.-K., *Process planning for multi-axis NC machining of free surfaces*, International Journal of Production Research, Vol. 33, No. 10, pp. 2723-2738, 1995.

[11] Elber, G. and Cohen, E. *Arbitrarily precise computation of gauss maps and visibility sets for freeform surfaces*, ACM Transactions on Solid Modeling, 271-279. 1995.

[12] Spyridi, A. J. and Requicha, A. A. G., *Accessibility analysis for automatic inspection of mechanical parts by coordinate measuring machines*, Proceedings of the IEEE International Conference on Robotics and Automation, Pages 1284-1289, Cincinnati, Ohio, May 1990.

[13] Spyridi, A. J. and Requicha, A. A. G., *Accessibility analysis for polyhedral objects*, Engineering systems with Intelligence: Concepts, Tools and Applications, pp. 317-324, Dordrecht Holland, Kluwer Academic Publishers, Inc., 1991.

[14] Henderson, M. R., Chell, A. R. and Hubele, N. F., *Feature based manufacturing evaluation using statistical process control*, Proceedings of the 1996 NSF Design And Manufacturing Grantees Conference. 1996.

[15] Limaiem, A. and ElMaraghy, H. A., *A general method for accessibility analysis*, Proceedings of the 1997 IEEE International conference on Robotics and Automation, pp. 2346-2351, Albuquerque, New Mexico. April, 1997.

[16] Tangelder, J., Vergeest, J. and Overmars, M., *Computation of voxel maps containing tool access directions for machining free-form shapes*, Proceedings of the ASME DETC/DFM, 1996.

[17] Laxmiprasad, P. and Sarma, S. E., *A Feature Free Approach to 5-Axis Path Generation*, Proceedings of the ASME Design Engineering Technology Conference, Sacramento, CA, September 1997.

[18] S. Dhaliwal, S.K. Gupta, and J. Huang, *Computing exact global accessibility cones for polyhedral object*, ASME Design for Manufacturing Conference, Baltimore, Maryland, September 2000.

[19] Morishige, K., Takeuchi, Y. and Kase, K., *Tool Path Generation Using C-Space for 5-Axis Control Machining*, Journal of Manufacturing Science and Engineering, February 1991, Vol. 121, 144-150.

[20] Lozano-Perez, T. *Automatic planning of manipulator transfer movements*, IEEE Transactions on Systems, Man and Cybernetics, SMC-10(11): 681-698. 1981.

[21] Lozano-Perez, T., *Spatial planning: A configuratin space approach*, IEEE Transactions of Computers, C-32(2): 108-120. 1983.

[22] Udupa, S., *Collision detection and avoidance in computer controlled manipulators*, Ph. D. Thesis, Department of Electrical Engineering and Computer Science, California Institute of Technology, 1977.

[23] Latombe, J. C., *Robot motion planning*, Kluwer Academic Pres. 1991.

[24] Gupta K. and Pobil, A. P., *Practical motion planning in robotics: current approaches and future directions*, John Wiely & Sons, NY, USA, 1998.

[25] Balasubramaniam, M., Laxmiprasad, P., Sarma, S.E.and Shaikh, Z. *Generating 5-axis NC roughing paths directly from a tessellated representation*, Computer-aided Design, 32 (4) pp. 261-277. 2000.

[26] Balasubramaniam, M., *Automatic 5-axis NC tool path generation*, Ph. D. thesis, Department of Mechanical Engineering, Massachusetts Institute of Technology, 2001.

[27] Balasubramaniam, M., Sarma, S. E., and Marciniak, K., *Collision-free finishing toolpaths from visibility data*, Computer-aided design, Vol. 35, pp. 359-374, 2003.

[28] Spitz, S. N. and Requicha, A. A. G., *Accessibility Analysis using computer graphics hardware*, IEEE Transactions on Visualization and Computer Graphics, Vol. 6, No. 3, July/Sep 2000.

[29] Glaeser, G., Wallner, J. and Pottmann, H., *Collision-free 3-axis milling and selection of cutting tools*, Computer-Aided Design 31 (3), pp. 225-232, 1999.

[30] Bhattacharya, B. K., Das, G. and Mukhopadhyay, A. and Narrasimhan, G., Section 1, Introduction, *Optimally computing a shortest weakly visible line segment inside a simple polygon*, Computational Geometry, Vol. 23, pp.1-29, 2002.

[31] Hernando, C., Houle, M. E. and Hurtado, F., *On local transformation of polygons with visibility properties*, Theoretical Computer Science, Vol. 289, pp. 919-937, 2002.
[32] Tangelder, J., Vergeest, J. and Overmars, M., *Interference-free NC machining using spatial planning and Minkowski operations*, Computer-Aided Design, Vol. 30, No. 4, pp. 277-286, 1996.
[33] Elber, G. and Zussaman, E., *Cone visibility decomposition of freeform surfaces*, Computer-aided design, Vol. 30, No. 4, pp. 315-320, 1998.
[34] Laurentini, A., *Guarding the walls of an art gallery*, The Visual Computer, 15:265-278, 1999.
[35] Tang, K., Chen, L. and Chou, *Optimal workpiece setups for 4-axis numerical control machining based on machinability*, Computers in Industry, Vol. 37, pp. 27-41, 1998.
[36] Elber, G., *Accessibility in 5-axis milling environment*, Computer-aided design, Vol. 26, No. 11, pp. 796-802, 1994.
[37] Ho, S., Sarma, S. E. and Adachi, Y., *Real-time interference analysis between a tool and an environment*, Computer-Aided Design, Vol. 33, pp. 935-947, 2001.
[38] Balasubramaniam, M., Ho S., Sarma, S. E. and Adachi, Y., *Generation of collision-free 5-axis tool paths using a haptic device*, Computer-Aided Design, Vol. 34, pp. 935-947, 2002.
[39] Sarma, S. E., Laxmiprasad, P. and Balasubramaniam, M. *Tool path generator for computer aided manufacturing*, US Patent Number 6311100.

School of Engineering, Dankook University, Hanam-dong, San 8, Seoul, Korea
E-mail address: taejungkim@dankook.ac.kr

Department of Mechanical Engineering, MIT Room 35-014, 77 Mass. Ave., Cambridge, MA 02139
E-mail address: sesarma@mit.edu

DIMACS Series in Discrete Mathematics
and Theoretical Computer Science
Volume **67**, 2005

Zig-Zag Tool Path Generation for Sculptured Surface Finishing

Debananda Misra, V Sundararajan, and Paul K. Wright

ABSTRACT. Sculptured surfaces are relatively difficult to machine due to their complex geometry. Sculptured surface machining primarily has two stages, the first being roughing and the other being finishing. The current work focuses on generating tool paths for the finishing operation. Given any arbitrary sculptured surface, G&M machine codes have been generated to machine that surface using ball-end milling cutters on a 3-axis CNC machine. Tool path planning has been done on the offset surface. Iso-planar zig-zag toolpaths, due to their robustness and simplicity, have been generated. The generated tool paths comply to a certain tolerance and surface finish specified on the surface. Tool holder collision detection has been done to help select the appropriate tool. The input is a solid model of the surface in .SAT format while the output is a text file containing the machine code. Examples are presented to validate the algorithm.

1. Introduction

Sculptured surfaces are extensively used in designing complex parts such as automobile bodies, turbine blades and packages for electronic hand-held devices. Broadly, sculptured surface machining is done in two stages, 1. *roughing* and 2. *finishing*. The majority of the material is removed in the roughing stage using larger tool(s) while the left over material is removed in the finishing stage. Typically, flat end milling cutters are used for roughing operations [**CCTW03, HC98, LKH94**] while ball end milling cutters are used for finishing operations.

Due to the inherent nature of the motion of the tool, tool paths are always a series of straight lines/arcs whereas the actual sculptured surface is a surface of varying slope and curvature. Thus, the sculptured surface is approximated by a series of straight lines and arcs for machining, and the machining goal is to get as close as possible to the design surface. The current commercially available CAD/CAM systems for generating cutter location(CL) data for machining of such sculptured surfaces require some important human decisions, such as determination of the precise interval between successive toolpaths. After the cutter location(CL) data are generated, post processing is done to get machine executable codes for

Key words and phrases. Sculptured Surface, Inverse Tool Offset, Feed-forward Marching, Tool Holder Collision.

The first author was supported in part by NSF Grant #DMI-0070128.

actual production. A tool path interval that is too large can result in a rough surface while one that is too small can increase the machining time, making the process inefficient. Due to the complex geometry of the surface, tool body and tool holder interference with the surface pose many constraints on tool path generation. Smaller tool lengths and the vertical motion of the tool might make some regions inaccessible in the case of 3-axis CNC machining.

The current work is aimed at providing a systematic approach to tool path generation of sculptured surfaces using 3-axis CNC machines. Tool paths have been generated for finishing operation only. Tool paths for the roughing stage are assumed to have been previously generated while a set of cutter location points are planned on the offset surface. Importance is given in generating cutter location points that will meet the tolerance requirement while maintaining a certain surface finish.

The layout of the paper is as follows: Sec. 2 reviews previous approaches to the same problem. Sec. 3 lists some of the definitions and notations used while Sec. 4 gives an overview of some of the standard tool path generation methods. Sec. 5 describes the offset generation method employed, followed by a section describing the tool path generation scheme in the current approach. Sec. 7 presents the tool holder collision detection algorithm. Sec. 8 and 9 show the resulting implementation of the algorithm with some example of machined parts.

2. Previous Work

In one of the earliest methods of machining sculptured surfaces, Sata et al. [**SKOH81**] chose small incremental iso-parametric curves as the tool paths. However, the non-uniform relationship between the parametric coordinates and the corresponding physical coordinate made tool path planning difficult. Huang et al. [**HO92**] and Li and Jerard [**LJ92**] thus proposed an algorithm that generated tool paths based on the non-constant parameter tool contact curves. At about the same time, Suh and Lee [**SL90**] and Hwang [**Hwa92**] presented methods to determine the tool paths by calculating, at each path, the smallest tool path interval and using it as a constant offset in the next tool path. However, there still remained an unpredictable scallop height in the part surface, which caused either surface roughness or inefficient machining. A tool path strategy that kept the scallop height constant was thus presented by Suresh and Yang [**SY94**]. Their work led to a reduction of cutter location data; however reduction of cutter location data may not guarantee a proportional shortening in machining time. Feng et al. [**FL02**] also presented an approach for constant scallop height tool path generation. Sarma and Dutta [**SD97**] proposed a method to generate tool paths within any specified distribution of scallop heights on the manufactured surface. Tool paths were generated as offsets of an initial curve on the design surface. Choi et al. [**CKJ97**] generated tool paths using a configuration space (C-Space) approach. The geometric data describing the design surface and stock surface were transformed to the C-Space elements. Sarma [**Sar99**] developed an algorithm to pick the direction of minimum number of switch-backs for a zig-zag path on a polygon. This lead to reduction in machining time of planar surfaces. An offset parametric CNC tool path planning was developed by Kim and Kim [**KK95**] . Their idea was to first obtain an offset surface of a desired part surface and then obtain toolpaths on the offset parametric plane. Bi-cubic surface patch and surface sub-division were used

TABLE 1. Basic Definitions

Term	Definition	Notation Used
CC-Point	Point where tool touches the design surface	C_i
CL-Point	Tool center location	L_i
Feed-forward Distance	Distance the tool travels in a single continuous tool motion	δ_{cc} : along CC-points δ_{cl} : along CL-points
Side-step Distance	Distance between two consecutive parallel tool paths	ϵ_{cc} : along CC-points ϵ_{cl} : along CL-points
Feed-forward error	Deviation of the tool path from the actual design surface	τ
Scallop Height	Height of material left between two consecutive passes of the tool	h

to generate the offset surface accurately. A new tool path scheduling method to improve the accuracy and efficiency of 3-axis surface machining was presented by Chih and Lin [**CCL01**]. In their approach, the compromise between the efficiencies in the feed-forward segmentation and the path-interval segmentation was based on the selection of scallop height limit. The proposed method included compensation for the tool location to control the surface error and the segmentation efficiency. To the present authors' knowledge, none of the above approaches proposed solutions for tool holder collision detection. This detection is a contribution of this paper.

3. Basic Definitions and Notations Used

Table 1 lists some of the basic definitions and notations that are being followed through the rest of the paper.

4. Tool Path Generation Methods

The main objective of tool path generation (TPG) is to compute a sequence of cutter location points (CL-points) from the design surface. Various investigators [**CJ98, KC00, SD97**] have given a detailed description and classification of various tool path generation methods. Tool-path generation methods are classified as either the *CC-based method* or the *CL-based method* depending on the type of tool path generation surface.

CC-Based Method. In the Cutter Contact(CC)-based method, tool paths are generated by sampling a sequence of CC-points from the part surface and then each CC-point is converted to a cutter location(CL)-point. Tool path generation is done on the design surface. In general this method can be classified into 3 main categories as (1) *parametric method*; (2) *drive surface method*; (3) *guide plane method* depending on the path planning techniques. These are often called as the *"conventional"* tool path generation methods.

- *Parametric Method* computes the tool paths (Figure 1(a)) in the parametric domain and then maps it back to the design/part surface. Iso-parametric machining, one of the earliest techniques [**SKOH81**], involves machining along iso-parameter lines. Uniformly spaced parametric lines in the parametric domain are non-uniformly spaced on the part surface due to non-uniform transformation between Parametric and Euclidean spaces. This results in varying scallop height and thus non-uniform surface finish.
- *Drive Surface Method* computes toolpaths (Figure 1(b)) by intersecting the design surface with a series of planes called drive surfaces. The curves generated by intersection are then used to generate accurate tool paths. If the intersecting planes are parallel then it is known as iso-planar machining. However, the orientation of the intersecting plane can be changed to achieve optimal conditions of machining. This method is very robust and can handle complex surfaces reliably.
- *Guide Plane Method* plans the tool paths (either in the form of straight line or contours) first on a 2D plane and then maps it back (Figure 1(c)) to the design surface. For 3-axis milling, the plane perpendicular to the tool axis is mainly used as the guide plane. The major strength of this method is that the shape of the region to be cut on the part surface can be considered during planning the tool path layout on the guide plane.

CL-Based Method. In the Cutter Location(CL)-based method the CL-surface is used as a path generation surface. In this approach, the CL surface has to be first generated from the design surface. Surface finish criteria are met on the CL surface. In most of the previous work in this area [**KK95, CKJ97**], the offset surface was approximated as the CL-surface. *CL-based methods have been employed in the current work.* The offset surface is first generated from the design surface and tool paths are planned on the offset surface. Zig-zag tool paths, parallel to each other, are planned according to the machining parameters.

5. Offset Surface Generation

The tool path generation strategy adopted in this paper is based on the offset of the design surface. Thus, the final surface finish is greatly affected by the accuracy of the offset surface. Offset surface generation has been done in two steps which are described below.

Discretization of the surface. Discretization of the surface involves sampling the design surface with a certain resolution and storing the z co-ordinate values associated with the sampled (x, y) co-ordinates. Since the surface is one of varying slope and curvature, it is desired to get denser points at steep regions than in flat regions. In order to do this, surface-plane intersection has been done to get the initial curves on the surface. The intersecting planes have been chosen to be vertical and parallel to each other. The intersecting curves have been sampled at *constant arc length* to the get the (x, y, z) coordinates of the discretized points. The smaller the arc length, the finer are the points obtained on the surface.

Figure 2(a) shows a design surface while Figure 2(b) shows the top view of its discretized model. The dark areas in figure 2(b) correspond to steep regions and the light areas correspond to flat regions on the surface.

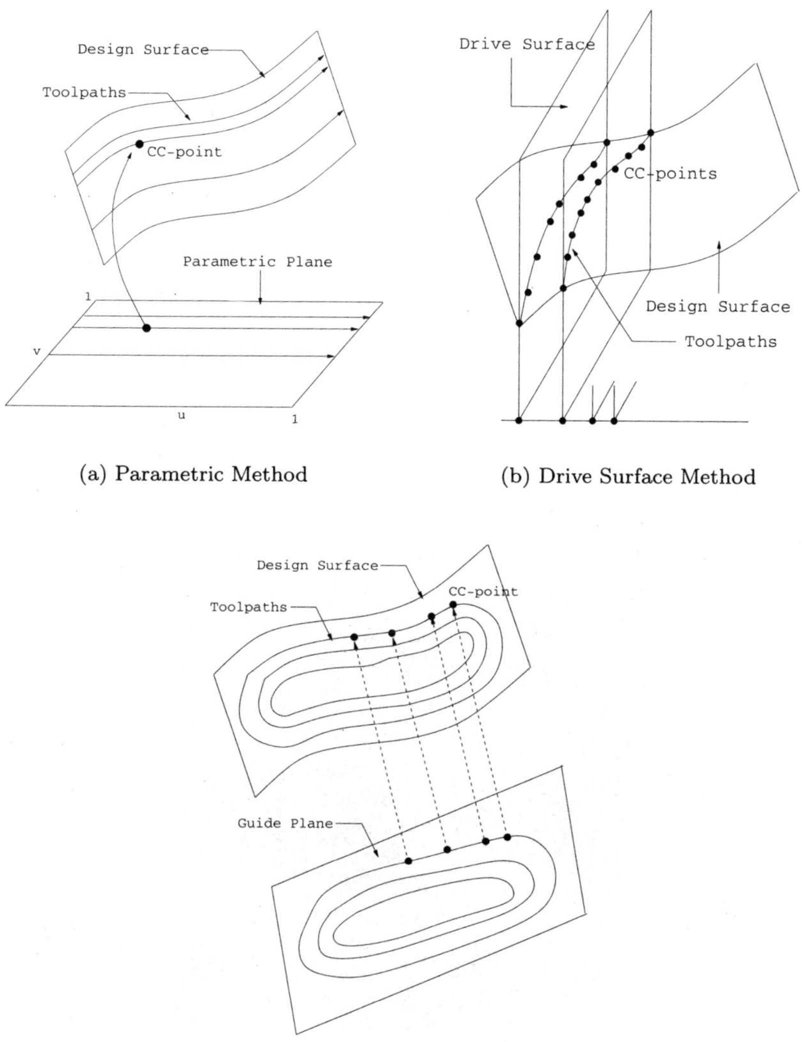

(a) Parametric Method

(b) Drive Surface Method

(c) Guide Plane Method

FIGURE 1. CC-Based Tool Path Generation Methods [based on refs. [**SD97, CJ98, KC00**]]

Inverse Tool Offset. The Inverse Tool Offset (ITO) algorithm [**CJ98, TSAO89, KKS88**] has been used to generate the offset surface. The algorithm states that *the offset surface can be obtained by taking the envelope of the inverse of the tool in the z-direction, with the center of the inverted tool moving on the design surface* (Figure 3). The discretization methodology adopted in this paper helps in getting accurate results for the ITO algorithm. For every discretized point on the design surface, the ITO algorithm gives a tool center location so that the tool is tangent

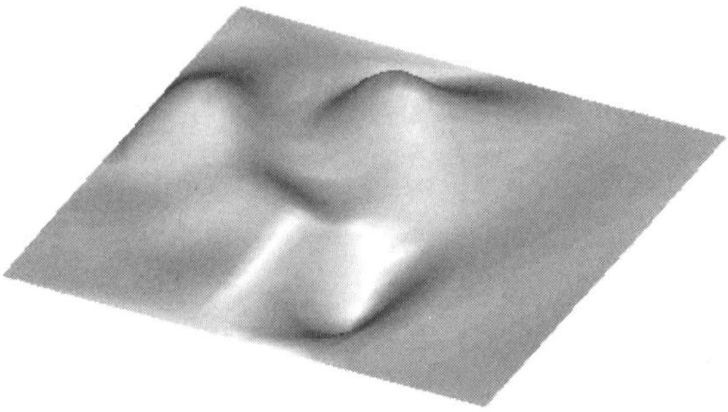

(a) Design Surface

(b) Discretized Surface (top view)

FIGURE 2. Surface Discretization using constant arc length method

to the surface. The algorithm works by considering all potential tool body collision points for each discretized point on the surface. The main advantage of the algorithm lies in generating gouge free offset points [**CKJ97**].

6. Tool Path Generation Strategy

In this paper zig-zag toolpaths have been generated for the finishing operation. In the case of the zig-zag tool paths it is desired that the tool moves in a straight line in the feed-forward direction. Planning tool motion on the design surface guarantees straight line motion of the tool contact point, but the tool center may not move

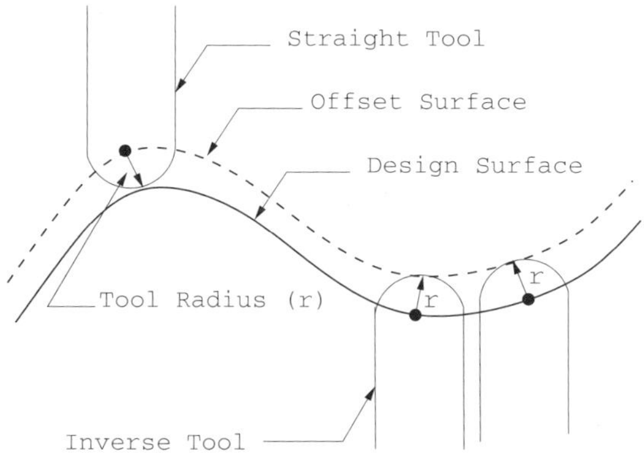

FIGURE 3. Inverse Tool Offset Algorithm

in a straight line. On the other hand, tool path planning on the offset surface ensures straight line motion of the tool center. Hence, in the current approach, CL points have been directly planned on the offset surface generated through the ITO. Depending on the direction of motion of the tool, tool motion can be divided into two categories.

(1) *Feed-forward Marching*
(2) *Side-step Marching*

Offset surface generation, described in the previous section, gives a list of offset points (O_k) for a list of points on the design surface (D_k). The offset points however do not meet the criterion of feed-forward or side-step marching. Hence, the objective is to find intermediate CC-points (C_k) on the surface and the corresponding offset points, which will be the required CL-points (L_k), that will meet the machining criterion of specified design-tolerance.

Feed-forward Marching. Most of the cutting operation is done in the feed-forward direction. The tool moves in straight line between two successive CC-points. The deviation of the tool path from the actual surface defines the error (Figure 4(a)) in the feed-forward direction. The maximum error produced should be less then the design tolerance. The feed-forward distance that meets that tolerance requirement is given by equation 6.1 (chapter V of Choi and Jerrard [**CJ98**]).

$$\delta_{cl} = 2[2\tau(r + \rho_f) - \tau^2]^{\frac{1}{2}} \tag{6.1}$$

δ_{cl}	:=	Distance between two consecutive CL-points in the feed-forward direction
r	:=	Tool Radius
ρ_f	:=	Curvature at the CC-point in feed-forward direction

Figure 4(b) shows the variation of feed-forward distance with change in tolerance for two different tools. The curvature has been kept the same for both cases. As can be seen from the graph, a lower limit on error will demand a smaller feed-forward distance and thus more number of CL-points while a higher tolerance will give less number of CL-points. Also, for a given feed-forward distance a tool of higher radius will give less error than a tool of smaller radius. The following lines summarize the steps followed for feed-forward marching.

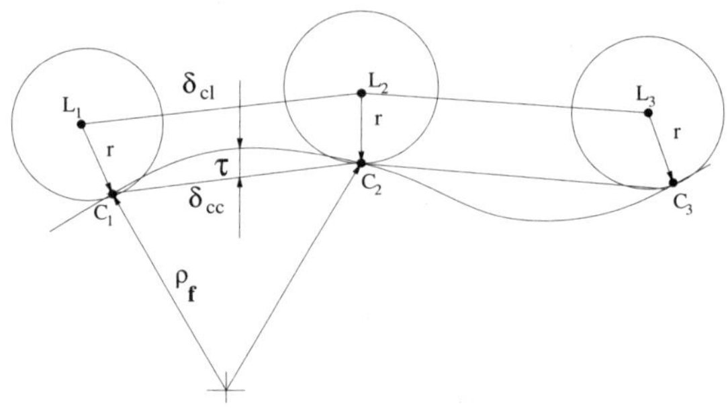

(a) Error in feed-forward direction

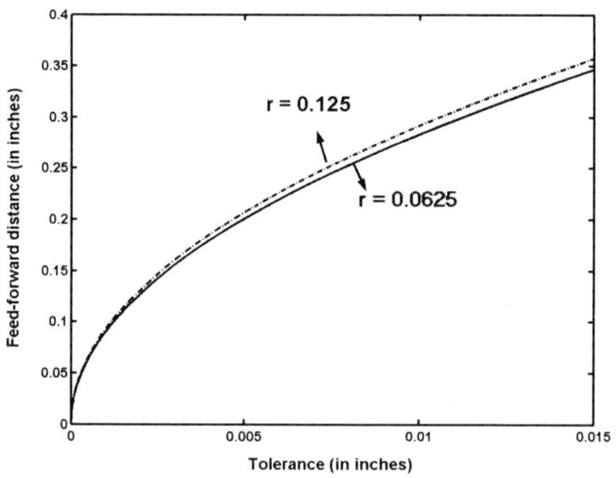

(b) Plot of Feed-forward Distance Vs Error

FIGURE 4. Feed-forward Marching

ALGORITHM - 1

- Start current pass of the tool
  - $i, k = 0$. Here, CL-point(L_0) = Offset point(O_0).
  - For a CL-point (L_i) calculate the feed-forward distance(δ_{cl_i}).
  - Get the distance(d_k) of the next offset point(O_k) from L_i along the current pass in the feed-forward direction.
  - Increment k till $d_k < \delta_{cl_i} < d_{k+1}$.
  - Linearly interpolate between O_k and O_{k+1} to get an intermediate offset point(O_{intm}). Offset corresponding point (C_{intm}) on the design surface vertically below it to get the next CL-point (L_{i+1}).
  - increment i.
- end current pass

Side-step Marching. Side-step marching is the motion of the tool in a direction orthogonal to the feed-forward direction. Referring to Figure 5(a), $\epsilon_{cl} \times \cos(\theta)$ is the path interval where ϵ_{cl} is given by equation 6.2 (Chapter V of Choi and Jerrard [**CJ98**]). The maximum path interval at a CL-point is limited by the cusp height specified on the surface. After the CL-points for a single pass have been obtained by feed-forward marching, the minimum path interval of all those CL-points has been taken as the path interval for that specific pass. This ensures that the cusp height stays within the specified limit at all points on the surface.

$$\epsilon_{cl} = \frac{\sqrt{4(r+\rho_s)^2(\rho_s+h)^2 - [\rho_s^2 + 2\rho_s r + (\rho_s+h)^2]^2}}{|\rho_s + h|} \tag{6.2}$$

ϵ_{cl} := Distance between two consecutive parallel tool paths along the CL-points
r := Tool Radius
ρ_s := Curvature at the CC-point in side-step direction

Figure 5(b) shows the variation of side-step distance with cusp height for two different tool radius. Again, the curvature has been kept the same for both cases. It can be seen from the graph that for a given side-step distance and curvature, a tool of larger radius will generate cusps of smaller height and thus will achieve better surface finish. The following steps summarize the method to obtain the path interval for a particular pass.

ALGORITHM - 2

- Start current pass of the tool
  - For every CL-point(L_i) calculate the side step distance (ϵ_{cl_i}).
  - Obtain the neighboring offset points in the feed-forward direction on the next pass on the offset surface in the side-step direction.
  - Interpolate between the neighboring points to obtain the parallel offset point to L_i in the side-step direction. Compare this with L_i to get the slope(θ_i).
  - Calculate $\epsilon_{cl_i} \times \cos(\theta_i)$ to get the path interval. increment i.
- end current pass.
- Find the minimum path interval of all the CL-points.

7. Tool Selection and Tool Holder Collision Detection

Tool selection is done based on the accessibility of the tool to all regions of the surface. The minimum curvature(ρ_{min}) of all the discretized points on the surface is obtained. The tool with radius r, less than ρ_{min} has been selected for machining operation.

Tools of small diameter are used for the finishing operation. Typically, tools of smaller diameter are shorter in length. Hence, in some cases even if the radius of the tool is bigger than the largest curvature of the surface at the point, the tool cannot enter into the region because of shorter tool body length. This is due to the tool holder which collides with the surface. Thus, tool holder collision poses a constraint to finish machining. In the current approach the tool holder, along with the tool, has been abstracted as a series of stacked cylinders (Figure 6). Collision detection has been done both for the tool body and tool holder. The process to detect the potential collision at a certain CL-point (x_0, y_0, z_0) for a certain tool of

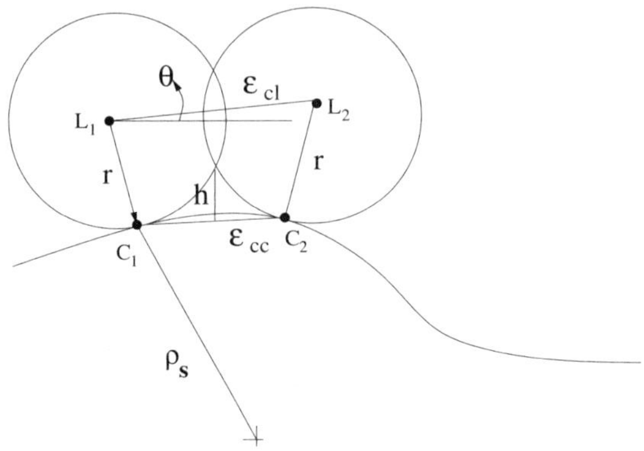

(a) Cusp height in side-step direction

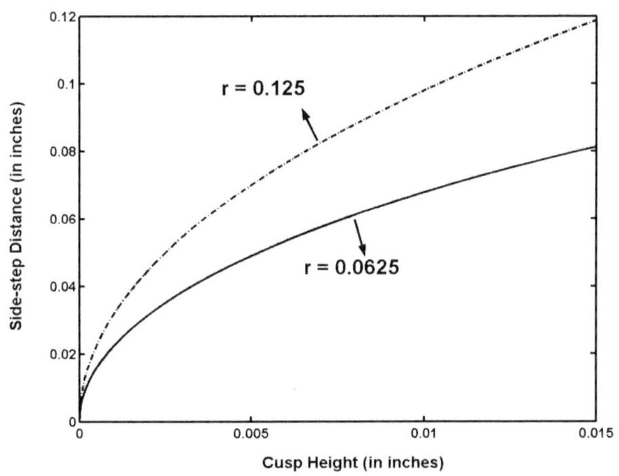

(b) Plot of Side-step Distance Vs Cusp Height

FIGURE 5. Side-step Marching

radius r involves finding all the points (x_i, y_i, z_i) on the discretized surface that lies within the tool holder assembly. Since the tool holder collision detection is done on the discretized surface, the technique is more intensive at steep regions than flat regions. The discretization resolution can be chosen according to the minmum diameter of the tools available for finish machining thus accounting for tool selection along with tool path planning. One of the limitations of the method is that both tool and tool holder collision are done based on the same grid which is computationly expensive. The tool holder collision detection can be done on a grid with lower resolution which can be later refined for higher resolution grids for tool collision detection which requires a finer grid.

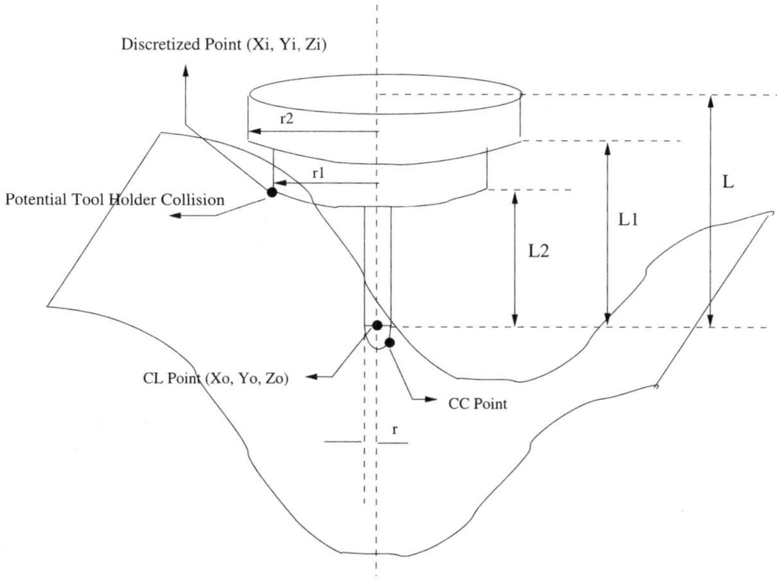

FIGURE 6. Tool Holder Collision Detection

Equation 7.1, 7.2 and 7.3 describes the set of equations to check tool holder collisions.

$$(x_i - x_0)^2 + (y_i - y_0)^2 + (z_i - z_0)^2 < r^2 \quad \forall \quad z_i < z_0 \tag{7.1}$$

$$(x_i - x_0)^2 + (y_i - y_0)^2 < r^2 \quad \forall \quad z_0 \le z_i < z_0 + L \tag{7.2}$$

$$(x_i - x_0)^2 + (y_i - y_0)^2 < r_j^2 \quad \forall \quad z_0 + \sum_{k=0}^{j-1} L_k < z_0 + \sum_{k=1}^{j} L_k \tag{7.3}$$

$$(j = 1 - 4, L_0 = L)$$

8. Implementation

The complete implementation procedure for generating zig-zag toolpaths is given below.

IMPLEMENTATION STEPS

- Input .SAT file
  - Surface-plane intersection to obtain parallel curves on the surface.
  - Constant arch length marching on the curves to obtain discretized model of the surface.
  - Select appropriate tool based on accessibility determination.
  - Inverse Tool Offset to obtain the discretized offset surface.
  - Begin starting feed.
    - ∗ Do feed forward marching on the discretized offset points (*Algorithm 1*).
    - ∗ Find the side step distance (*Algorithm 2*).
    - ∗ Do surface-plane intersection to get the discretized points on the surface.

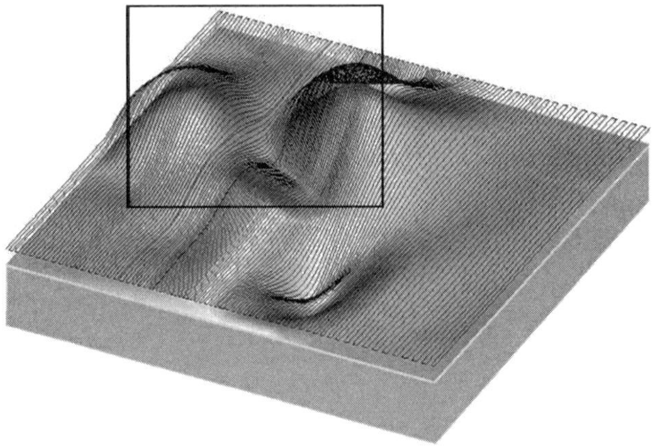

(a) Example part

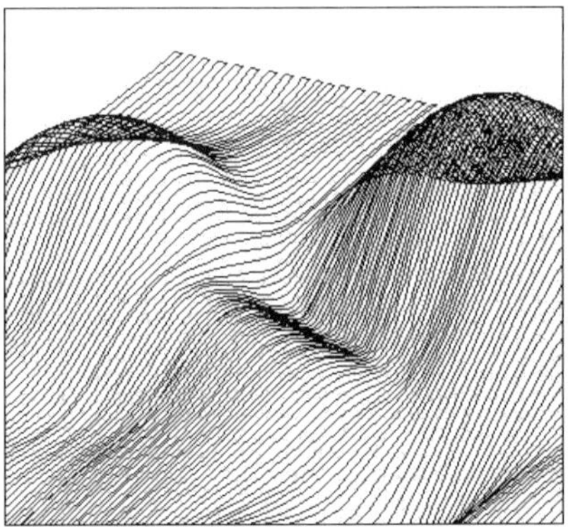

(b) Toolpaths at a specific region

FIGURE 7. Example part and generated tool path

    * Offset to get discretized offset points.
  - end of surface.
  - Tool holder collision detection at the CL-points.
- Post process data to output .NC file.

Figure 7(a) shows a design surface and the zig-zag tool paths generated to machine the surface. Figure 7(b) shows a magnified square region (Figure 7(a)) of the same tool paths. A maximum scallop height of 0.0015in and feed-forward error of 0.0006in has been used for this case. Ball-end milling cutter of 0.125in diameter was used for machining. Denser tool paths can be observed in steeper regions (Figure 7(b)) to keep the scallop height with in specified limits. Figure

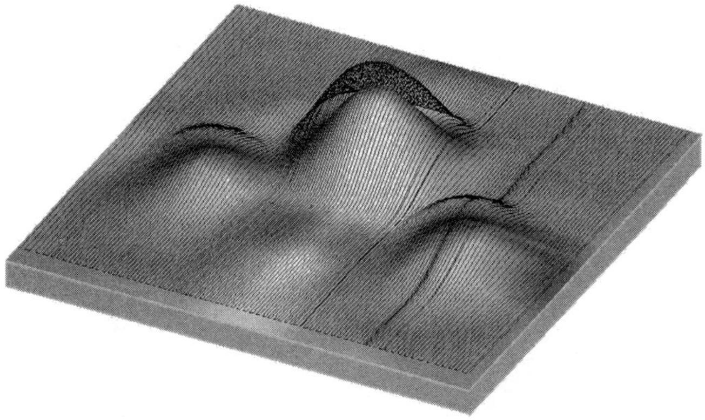

(a) Example part

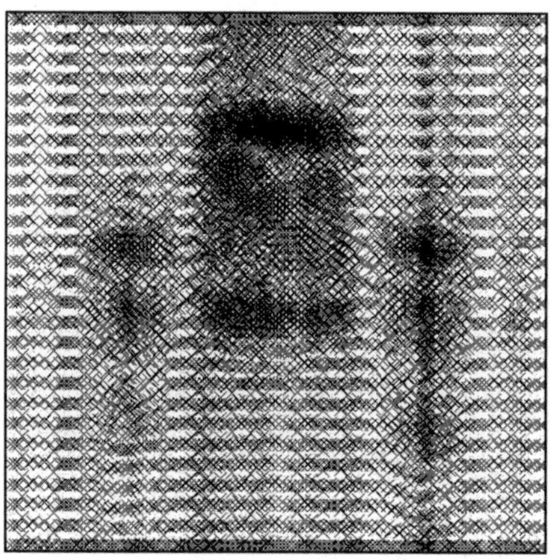

(b) CL-points for example part

FIGURE 8. Example part and generated CL-points

8(a) shows another example part and the tool paths obtained. A maximum scallop height of 0.0012in and an allowable tolerance of 0.0008in has been used for this case. Ball-end milling cutter of 0.125in diameter was used for machining. Figure 8(b) shows top view of the CL-points obtained for the same part. Closely spaced CL-points have been obtained in steep regions. Figure 9 shows a machined surface using a 0.25 diameter ball-end milling cutter. A maximum scallop height of 25μ inch and an allowable tolerance of 0.0004in was used. ACIS© Solid Modelling Kernel has been used for all solid modelling operations.

FIGURE 9. Machined surface using 0.25in diameter cutter

9. Conclusions

In this paper, a CL-based approach was taken to generate tool paths for finishing operation of sculptured surfaces. An inverse tool offset algorithm was used to generate the discretized offset surface. *Tolerance requirements* and *maximum allowable scallop height* were used as the limiting conditions for CL-data generation in the feed-forward and the side-step direction respectively. The conclusions of this method can be summarized as follows

- *Zig-zag tool path* planning on the offset surface ensured that the CL-points were placed on a straight line thus guaranteeing straight line motion of the tool.
- Parallel curves on the surface were obtained by surface-plane intersection. Discretization of the surface was done by *constant arc length* marching on those curves. This gave dense points in steep regions as compared with those in flat regions. The accuracy of the inverse tool offset algorithm was dependent on the resolution of the discretized surface.
- It was shown that the problem of tool holder collision can be successfully tackled by abstracting the tool body and holder as a series of stacked cylinders
- Linear interpolation and inverse tool offset provided an efficient method for feed-forward marching and side-step direction. However, for accurate results, the resolution for discretized offset surface needs to be properly chosen.

10. Acknowledgments

This work done in collaboration with Prof. Jami Shah and his students at Arizona State University. We would also like to thank Profs. Carlo Sequin, Profs. Sara McMains, Dr. Ganping Sun, Palani Kumaresan and Dr. Roshan D.Souza. for their support through out the work.

References

[CCL01] L. Chih-Ching and R-S. Lin, *An improved method for scheduling the tool paths for three-axis surface machining*, International Journal of Machine Tools and Manufacture **41** (2001), pp 133–147.

[CCTW03] K. W. Chan, W. K. Chiu, S. T. Tan, and T. N. Wong, *A high-efficiency rough milling strategy for mold core machining*, Journal of Engineering Manufacture **217** (2003), no. B3, pp335–348.

[CJ98] B.K Choi and R.B Jerrard, *Sculptured surface machining: Theory and applications*, First ed., Kluwer Academic Publishers, The Netherlands, 1998.

[CKJ97] B. K. Choi, D. H. Kim, and R. B. Jerrard, *C-space approach to tool-path generation for die and mold machining*, Computer Aided Design **29** (1997), no. 9, pp657–669.

[FL02] Hsi-Yung Feng and Huiwen Li, *Constant scallop height tool path generation for three-axis sculptured surface machining*, Computer Aided Design **34** (2002), pp647–654.

[HC98] J. S. Hwang and T Chang, *Three-axis machining of compound surfaces using flat and filleted endmills*, Computer Aided Design **30** (1998), no. 8, pp641–647.

[HO92] Y. Huang and H. James Oliver, *Non-constant parameter nc tool path generation on sculptured surfaces*, ASME Journal of Computer in Engineering **1** (1992), pp411–418.

[Hwa92] J. Hwang, *Interference-free tool-path in nc machining of parametric compund surfaces*, Computer Aided Design **24** (1992), no. 12, pp667–677.

[KC00] B. H. Kim and B.K Choi, *Guide surface based tool path generation in 3-axis milling:an extension of guide plane method*, Computer Aided Design **32** (2000), pp191–199.

[KK95] K. I. Kim and K. Kim, *A new machine strategy for sculptured surfaces using offset surface*, International Journal of Production Research **33** (1995), no. 6, pp1583–1697.

[KKS88] T. Kondo, T. Kishinami, and K. Saito, *Machining system based on inverse offset method*, Journal of Japan Society of Precision Engineering **54** (1988), no. 5, pp971–976.

[LJ92] S. X. Li and R. B. Jerard, *Non-isoparametric three-axis nc tool-path generation for finish machining of sculptured surfaces*, Geometric Modeling for Product Realization, Selected and Expanded Papers from the IFIP TC5/WG5.2 Working Conference on Geometric Modeling, Rensselaerville, NY, USA, 27 September - 1 October, IFIP Transactions, vol. B8, 1992, pp. 251–265.

[LKH94] K. Lee, J. K. Kim, and S. E. Hong, *Generation of toolpath with selection of proper tools for rough cutting process*, Computer Aided Design **26** (1994), no. 11, pp822–831.

[Sar99] S.E. Sarma, *The crossing function and its application to zig-zag tool paths*, Computer Aided Design **31** (1999), no. 14, pp881–890.

[SD97] R. Sarma and D. Dutta, *The geometry and generation of nc tool paths*, Journal of Mechanical Design **119** (1997), pp253–258.

[SKOH81] T. Sata, F. Kimura, N. Okada, and M. Hosaka, *A new method of nc interpolator for machining the sculptured surface*, Annals of the CIRP **30** (1981), no. 1, pp369–372.

[SL90] Y. S. Suh and K. Lee, *Nc milling tool-path generation for aribtrary pockets defined by sculptured surfaces*, Computer Aided Design **22** (1990), no. 5, pp273–283.

[SY94] K. Suresh and D. C. H. Yang, *Constant scallop height machining of free-form surfaces*, Journal of Engineering for Industry **116** (1994), pp253–259.

[TSAO89] Y. Takeuchi, M. Sakamoto, Y. Abe, and M Orita, *Developement of a personal cad/cam system for mold manufacture based on solid modeling techniques*, Annals of the CIRP **38** (1989), no. 1.

Integrated Manufacturing Lab, Department of Mechanical Engineering, University of California, Berkeley , California 94720
Current address: GE Global Research Center, Bangalore, India 560066
E-mail address: Debananda.Misra@ge.com

Integrated Manufacturing Lab, Department of Mechanical Engineering, University of California, Berkeley, California 94720
E-mail address: vsundar@kingkong.me.berkeley.edu

Integrated Manufacturing Lab, Department of Mechanical Engineering, University of California, Berkeley, California 94720
E-mail address: pwright@kingkong.me.berkeley.edu
URL: http://me.berkeley.edu/~pwright

DIMACS Series in Discrete Mathematics
and Theoretical Computer Science
Volume **67**, 2005

Implicitization exploiting sparseness

Ioannis Z. Emiris and Ilias S. Kotsireas

ABSTRACT. We review toric elimination theory in order to exploit sparseness in the implicitization problem. More specifically, we show how information on the support of the toric resultant, as well as degree bounds in terms of the mixed volume, can be used to predict the support of the implicit equation. The same approach also determines certain coefficients in this equation. We illustrate three implicitization methods, namely those based on resultant perturbations, the eigenvalue method and moving curves, which are accelerated by the knowledge of the implicit support.

1. Introduction

The problem of switching from a rational parametric representation to an implicit, or algebraic, representation of a curve, surface, or hypersurface lies at the heart of several algorithms in computer-aided design and geometric modelling. Given are rational parametric expressions

$$x_i = \frac{p_i(t)}{q(t)} \in K(t) = K(t_1, \ldots, t_n), \quad i = 0, \ldots, n, \tag{1.1}$$

over some field K of characteristic zero. The implicitization problem consists in computing the smallest algebraic surface in terms of $x = (x_0, \ldots, x_n)$ containing the closure of the image of the parametric map $t \mapsto x$. The most common case is for curve and surface implicitization, namely when $n = 1$ and $n = 2$ respectively.

Several algorithms exist for this problem, cf. e.g. [**AS01, Buc88, Bus01a, BCD03, CGKW00, DE03, Dok01, GV97, MC92, Sed90, Wan04**]. Recently there has been a revival of the subject, with new tools from algebraic geometry contributing to significant advances in this area. A comprehensive survey of all algorithms is beyond our scope. See [**BE97, Hof89, Kot04, SZ02**] for surveys of implicitization algorithms.

In [**EK03**] we used various tools from toric (or sparse) elimination theory, in order to predict the support of the implicit equation by using information on the support of the toric (or sparse) resultant, by considering the extreme monomials

1991 *Mathematics Subject Classification.* Primary: 65D17; Secondary 14Q10.

Key words and phrases. Implicitization, base point, toric elimination, polynomial support.

This work was supported in part by a grant from NSERC. The first author is also partially supported by PYTHAGORAS, a project under the EPEAEK program of the Greek Ministry of Education and Religions.

as described in [**GKZ94, Stu94**]. In addition, certain coefficients of the implicit equation can be determined. The computed support depends on the sparseness of the parametric expressions and is much tighter than the one predicted by degree arguments. Our Maple implementation illustrates many cases in which we obtain the exact support. We refer to this method as IPSOS. In this paper we discuss a nontrivial example by Fröberg and Dickenstein, which exhibits significant sparseness, which is optimally exploited by IPSOS. Furthermore, IPSOS is suitable for *generic implicitization* because the formulation of the input polynomials in terms of Newton polytopes (and, subsequently, mixed volumes) depends only on which terms appear with nonzero coefficients.

The motivation for exploiting sparseness in implicitization comes mainly from algorithms based on evaluation and interpolation. There are several algorithms that would benefit from IPSOS. In the sequel, we focus and illustrate three different approaches.

The first approach concerns algorithms based on perturbed resultant matrices, which yield the implicit equation even in the presence of base points, e.g. [**DE03, MC92**]. The problem reduces to sparse interpolation, which is substantially accelerated when we can accurately predict the output support. More specifically, the algorithm of Ben-Or and Tiwari requires a number of evaluations which is linear in the bound on the support cardinality [**Zip93**].

The second motivation for our work ([**CGKW00**]) treats parametric families of curves and (hyper)surfaces. The method can handle base points, and works both symbolically and numerically. The method looks for an implicit equation of a particular degree at a time. This implies that any information on the degree of the implicit equation accelerates execution. More importantly, the method constructs a symmetric singular square matrix and computes a basis of its nullspace. The dimension of this matrix equals the number of monomials in the implicit equation, which is in principle $\binom{m+n}{m}$, where the number of parametric equations is n and the algorithm seeks an implicit equation of degree m. When it is possible to constrain the monomials that appear in the implicit equation, then we diminish the size of the matrices.

Thirdly, we sketch the application of the IPSOS algorithm to the method of moving lines and curves, which expresses implicit equations in compact determinantal forms. This method (see [**BCD03, CSC98, SGD97**]) uses a family of lines (or, more generally, curves) that follow the given parametric curve or surface, in order to set up an overdetermined linear system. A determinant of the matrix of this linear system is the implicit equation. It is possible to compute the determinant by evaluating it at a sufficient number of points, as mentioned above. Knowledge of the monomials that appear in the implicit equation speeds up this computation.

IPSOS could also be combined with other resultant-based implicitization methods, where the final computation of the implicit polynomial can be accelerated when knowing its support. Examples include [**AS01, Bus01a, BJ03, DE01b, Khe02, Wan04, ZG00**].

This paper is organized as follows. First we give a survey of toric elimination theory tools. Second we apply these tools to predict the support of the implicit equation and certain of its coefficients. Then we survey briefly three implicitization methods. In section 4 we present implicitization by perturbed resultants. In section 5 we present the eigenvalue method for implicitization. In section 6 we present an implicitization method using moving curves. All three methods are illustrated with fully worked out examples. These three methods can be improved using the knowledge of the support of the implicit equation.

2. Toric elimination theory

In this section we study systems of $n+1$ polynomials in n variables, by means of algebraic elimination theory and resultants. Different resultants exist depending on the space of the roots we wish to characterize, namely projective, affine, toric, residual [**BEM01, CLO98, EM99, GKZ94, Stu02**]. Projective resultants (also known as classical) were historically the first to be studied and characterize the existence of projective roots. We shall focus on toric resultants below; for background information see [**CLO98, EM99, GKZ94, Stu02**]. Residual resultants were more recently introduced in order to study roots in the difference of two varieties.

Toric resultants express the existence of toric roots. Formally,

$$f_0, \ldots, f_n \in K[x_1^{\pm 1}, \ldots, x_n^{\pm 1}], \tag{2.1}$$

where each f_i corresponds to generic point $c_i = (c_{i1}, \ldots, c_{im_i})$ in the space of polynomials with support $A_i \subset \mathbb{Z}^n$. This space is identified with projective space $\mathbb{P}_K^{m_i-1}$. Then system (2.1) can be thought of as point $c = (c_0, \ldots, c_n)$. Let Z denote the Zariski closure, in the product of projective spaces, of the set of all c such that the system has a solution in $(\overline{K}^*)^n$. Note that Z is an irreducible variety.

A technical assumption is that, without loss of generality, the affine lattice generated by $\sum_{i=0}^n A_i$ is n-dimensional. This lattice is identified with $\mathbb{Z}^n$ possibly after a change of variables, which can be implemented by computing the appropriate Smith's Normal form [**Stu94**].

The *toric (or sparse) resultant* $R = R(A_0, \ldots, A_n)$ of system (2.1) is a polynomial in $\mathbb{Z}[c]$. If $\operatorname{codim}(Z) = 1$ then R is the defining irreducible polynomial of the hypersurface Z. If $\operatorname{codim}(Z) > 1$ then $R = 1$. R is homogeneous in the coefficients of f_i with $\deg_{f_i} R = \mathrm{MV}_{-i}$, where the latter stands for the *mixed volume* of $A_0, \ldots, A_{i-1}, A_{i+1}, \ldots, A_n$. Recall that, given a system of n polynomials in n variables, the mixed volume of the corresponding Newton polytopes bounds the number of the system's common isolated roots in $(\overline{K}^*)^n$.

The vanishing of $R(A_0, \ldots, A_n)$ is a necessary and sufficient condition for the existence of roots in the toric variety X, corresponding to the Minkowski (or vector) sum of the $n+1$ Newton polytopes [**Cox95, GKZ94**].

The simplest case, where the classical projective and toric resultants coincide, is that of a linear system of $n+1$ equations in n variables.

EXAMPLE 2.1. The bilinear system $f_i = c_{i0} + c_{i1}s + c_{i2}t + c_{i3}st$, $i = 0, 1, 2$, has toric resultant equal to

$$R = \det \begin{bmatrix} c_{i0} & c_{i1} & c_{i2} & c_{i3} & 0_3 & 0_3 \\ 0_3 & c_{i0} & 0_3 & c_{i2} & c_{i1} & c_{i3} \end{bmatrix},$$

where the column vectors satisfy $c_{ij}, 0_3 \in \mathbb{Q}^{3\times 1}, i = 0, 1, 2; j = 0, 1, 2, 3$ and assuming the matrix $[c_{ij}]_{i,j \geq 0}$ is regular. One can show that the projective (classical) resultant vanishes identically in this case.

Let us define the initial form $\mathrm{In}_\omega(F)$ of a multivariate polynomial F in k variables, with respect to some functional $\omega : \mathbb{Z}^k \to \mathbb{R}$. This is the sum of all terms in

F which maximize the inner product of ω with the corresponding exponent vector. Let us define

$$k := |A_0| + \cdots + |A_n| \subset \mathbb{N},$$

then ω defines a *lifting* function on the input system, by lifting every support point $a \in A_i$ to $(a, \omega(a)) \in \mathbb{Z}^n \times \mathbb{R}$. This generalizes the linear lifting used in [**CE00**]. The lower hull facets of the lifted Minkowski sum correspond to maximal cells of an induced coherent mixed subdivision of Q. If ω is sufficiently generic, then this subdivision is tight; in the sequel, we assume the mixed subdivision is both coherent and tight and denote it by Δ_ω; these concepts are detailed in [**CLO98, GKZ94, Stu02**]. The cells of Δ_ω have a unique, optimal representation as the Minkowski sum of faces from the Newton polytopes. A cell is mixed if all of its summands have dimension at most 1; it is i-mixed if and only if it is the sum of n edges and a single vertex from A_i.

If $a_i \in A_i$ is the vertex summand of an i-mixed cell, then the corresponding coefficient in f_i is denoted by c_{ia_i}. We recall the assumption that the A_i span $\mathbb{Z}^n$.

THEOREM 2.2. *The initial form of the toric resultant R with respect to a generic ω equals the monomial*

$$\mathit{In}_\omega(R) = \prod_{i=0}^{n} \prod_{\sigma} c_{ia_i}^{\mathrm{Vol}(\sigma)}, \tag{2.2}$$

where Vol$(\cdot)$ *denotes ordinary Euclidean volume and the second product is over all i-mixed cells σ in the mixed subdivision Δ_ω.*

For a detailed proof of this theorem, see [**Stu94**]. This proof can be obtained from the toric resultant matrix construction, by means of the subdivision-based algorithm of [**CE00**]. For a generic vector ω, the initial form $\mathrm{In}_\omega(R)$ corresponds to a vertex of the Newton polytope of the resultant R. It is precisely the vertex with inner normal ω. So, by varying the lifting ω, we can compute all vertices of this Newton polytope. The integer points in this polytope yield a superset of the resultant's support.

A bijective correspondence exists between the extreme monomials and the configurations of the mixed cells of the A_i. So, it suffices to compute all distinct mixed-cell configurations, as discussed in [**MC00, MV99**]. Another (simpler) means of reducing the number of relevant mixed subdivisions is by bounding the number of cells in the relevant subdivisions. This bound is usually straightforward to compute in small dimensions (e.g. when $n = 2, 3$) and reduces drastically the set of mixed subdivisions. For instance, when studying the implicitization of a biquadratic surface, we can determine that the number of cells in the relevant subdivisions is 8. The total number of mixed subdivisions is 19728, whereas those with 8 cells are 62.

In certain special cases, we can be more specific about the Newton polytope of the toric resultant. First, its dimension equals $k - 2n - 1$ [**GKZ94, Stu94**]. Certain corollaries follow: For essential support families (defined in [**Stu94**]), a 1-dimensional Newton polytope of R is possible if and only if all polynomials are binomials. The only resultant polytope of dimension 2 is the triangle; in this case the support cardinalities must be 2 and 3. For dimension 3, the possible polytopes are the tetrahedron, the square-based pyramid, and polytope $N_{2,2}$ given in [**Stu94**]; the support cardinalities are respectively 2, 2 and 3.

A corollary of Theorem 2.2 (and of its proof) is proven in [**GKZ91, CE00, Stu94**]:

COROLLARY 2.3. *The coefficients of all extreme monomials are in* $\{-1, 1\}$.

Sturmfels [**Stu94**] also specifies, for all extreme monomials, a way to compute their precise coefficients. But this requires computing several coherent mixed subdivisions, and its benefit in terms of computational complexity is not clear.

The so-called *Cayley trick* introduces a new point set $C := \{(z, a_{0j}) : a_{0j} \in A_0\} \cup \{(e_i, a_{ij}) : i = 1, \ldots, n, \ a_{ij} \in A_i\} \subset \mathbb{Z}^{2n}$, where $z = (0, \ldots, 0) \in \mathbb{N}^n$ is the zero vector and $e_i = (0, \ldots, 0, 1, 0, \ldots, 0) \in \mathbb{N}^n$ has a unit at the i-th position and $n-1$ zeroes. Regular are those triangulations that can be obtained by projection of a lifted triangulation.

THEOREM 2.4. *The problem of computing all mixed subdivisions of* $A_0, \ldots, A_n \subset \mathbb{Z}^n$ *is equivalent to computing all regular triangulations of the set* C *defined above, which contains* $|A_0| + \cdots + |A_n|$ *points.*

EXAMPLE 2.5. Consider the univariate system:

$$f_0 = c_{00} + c_{01}x, \ f_1 = c_{10} + c_{12}x^2.$$

The Cayley trick goes as follows: The points in the set C appear in the columns of matrix

$$\begin{bmatrix} 0 & 0 & 1 & 1 \\ 0 & 1 & 0 & 2 \end{bmatrix}.$$

There are two possible triangulations of these points, namely

$$\left(\begin{bmatrix} 0 \\ 0 \end{bmatrix}, \begin{bmatrix} 0 \\ 1 \end{bmatrix}, \begin{bmatrix} 1 \\ 2 \end{bmatrix}\right), \left(\begin{bmatrix} 0 \\ 0 \end{bmatrix}, \begin{bmatrix} 1 \\ 0 \end{bmatrix}, \begin{bmatrix} 1 \\ 2 \end{bmatrix}\right),$$

and

$$\left(\begin{bmatrix} 0 \\ 0 \end{bmatrix}, \begin{bmatrix} 0 \\ 1 \end{bmatrix}, \begin{bmatrix} 1 \\ 0 \end{bmatrix}\right), \left(\begin{bmatrix} 0 \\ 1 \end{bmatrix}, \begin{bmatrix} 1 \\ 0 \end{bmatrix}, \begin{bmatrix} 1 \\ 2 \end{bmatrix}\right).$$

Efficient algorithms (and implementations) exist for computing all regular triangulations of a point set [**Ram01**].

3. Predicting the implicit support

The approach of [**EK03**] is to exploit toric elimination theory in order to predict the extreme monomials i.e., the vertices of the Newton polytope of the toric resultant R. This enables us to predict the support of the implicit equation of a parametric (hyper)surface. Our approach is called IPSOS, an acronym for Implicit Polynomial Support Optimized for Sparseness [**EK03**]. We produce a superset of the monomials in the support of the implicit equation of the input.

The output support scales with the sparseness of the parametric polynomials and is much tighter than the one predicted by degree arguments. In many cases, the exact support of the implicit equation is obtained, as seen in Table 1, by applying our Maple program. Moreover, it is possible to specify certain coefficients in this equation. The motivation comes mainly from implicitization algorithms which apply interpolation.

Consider the polynomials

$$f_i(t) = p_i(t) - x_i q(t)$$

from (1.1), where we ignore the specific values of the coefficients. This is an interesting feature of the algorithm, namely that it considers the monomials in the parametric equations but not their actual coefficients. This shows that the algorithm is suitable for use as a preprocessing off-line step in CAGD computations, where one needs to compute thousands of examples with the same support structure in real time. This handles the implicitization of (multiparametric) families of (hyper)surfaces, indexed by one or more parameters.

Of course, the generic resultant coefficients are eventually specialized to functions of the x_i. Then, bounds on the implicit degree in the x_i are applied, in order to reduce the final support set. The required computation involves obtaining all partial mixed volumes MV_{-i} for $i = 0, \ldots, n$, and hence the implicit degree separately in the x_i variables [**EK03**]. Let us illustrate the idea for plane curves and for surfaces. When we intersect a plane curve with a generic straight line, we obtain generically a number of points equal to the total degree of the implicit equation. Then, we substitute x_0 and x_1 by (1.1), clear out denominators and compute the degree of the resulting equation in t. This will be the total degree of the sought implicit equation.

We examine IPSOS on some examples.

EXAMPLE 3.1. We consider the Folium of Descartes, shown in Figure 1, with parametric expression:

$$x = \frac{3t^2}{t^3+1}, \quad y = \frac{3t}{t^3+1}. \tag{3.1}$$

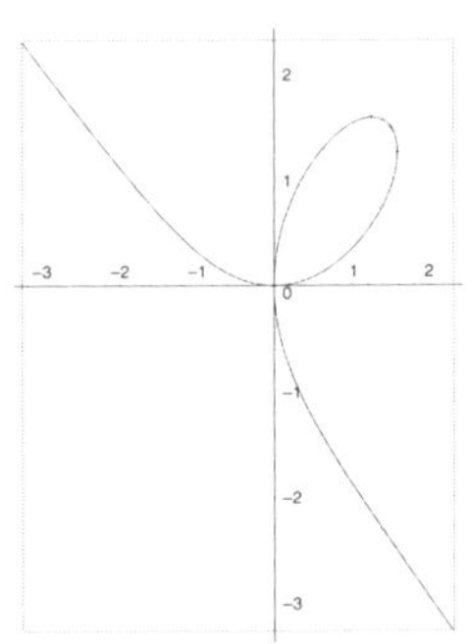

FIGURE 1. The Folium of Descartes

The output monomials are $\{y^3, x^3, x^3\,y^3, x\,y, y^2\,x^2\}$. After applying the degree bound $d = 3$ we obtain the support $\{y^3, x^3, x\,y\}$, which is optimal, since the implicit equation is

$$x^3 + y^3 - 3\,x\,y = 0.$$

EXAMPLE 3.2. An example in 3 dimensions comes from [**Buc88**]; the surface is drawn in Figure 2. The parametric expressions are:

$$x = s\ t,\ y = s\ t^2,\ z = s^2.$$

In order to apply toric elimination theory, we consider polynomials

$$f_0 = c_{00} - c_{01}st,\ f_1 = c_{10} - c_{11}st^2,\ f_2 = c_{20} - c_{21}s^2.$$

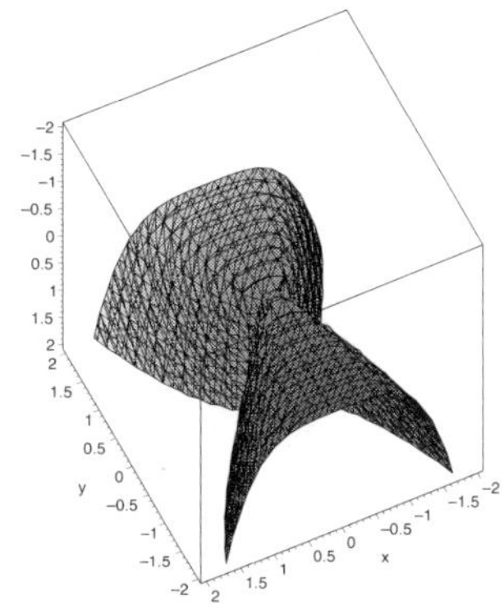

FIGURE 2. The surface in Example 3.2.

There are two possible mixed subdivisions, each containing exactly three maximal cells, all of which are mixed; see Figure 3. To apply degree arguments in order to

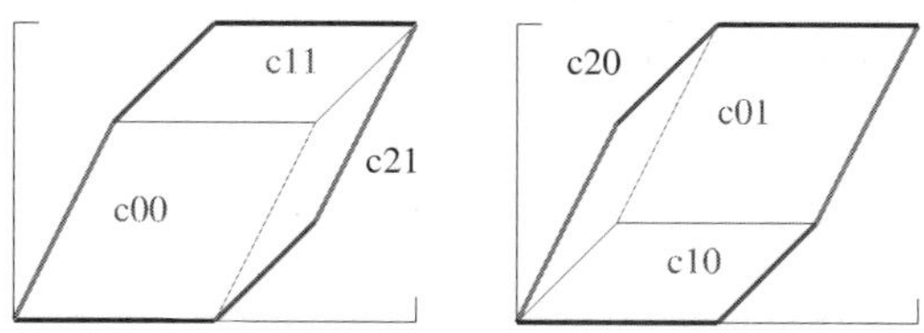

FIGURE 3. Mixed cells in the subdivisions, with vertex summands shown.

establish a bound on the degree, we take two generic planes $a_1x + b_1y + c_1z = d$ and $a_2x + b_2y + c_2z = d$ and substitute x, y, z by their parametric expressions. The resulting system of equations has support $\{(0,0),(2,0),(1,1),(1,2)\}$ and the Newton polygon is a triangle with volume equal to 2. Hence $MV = 4$, which bounds the total implicit degree. Using this degree bound, the computed support is optimal and the implicit equation is

$$x^4 - y^2z = 0.$$

EXAMPLE 3.3. Let us consider a system attributed to Fröberg, also known as the Fröberg-Dickenstein example, see [**Dic04**]:

$$x = t^{48} - t^{56} - t^{60} - t^{62} - t^{63}, \ y = t^{32}.$$

The Minkowski sum is the segment $Q_0 + Q_1 = [0, 95]$. One type of triangulations, obtained from a non-linear lifting, divides it to the following 3 cells (which are segments):

$$(Q_0' + 0), (a + Q_1), (Q_0'' + 32), \text{ where } Q_0' = [0, a], Q_0'' = [a, 63], \tag{3.2}$$

and $a \in \{48, 56, 60, 62, 63\} \subset A_0$. This triangulation is associated to extreme monomial $\pm y^a c_{0,a}^{32} c_{1,32}^{63-a}$, each factor corresponding to the respective cell, where $c_{0,a}$ is the coefficient of t^a in f_0 and $c_{1,32}$ is the coefficient of t^{32} in f_1. Since the last two factors are specialized to ± 1, every triangulation (3.2) yields an extreme monomial $\pm y^a$.

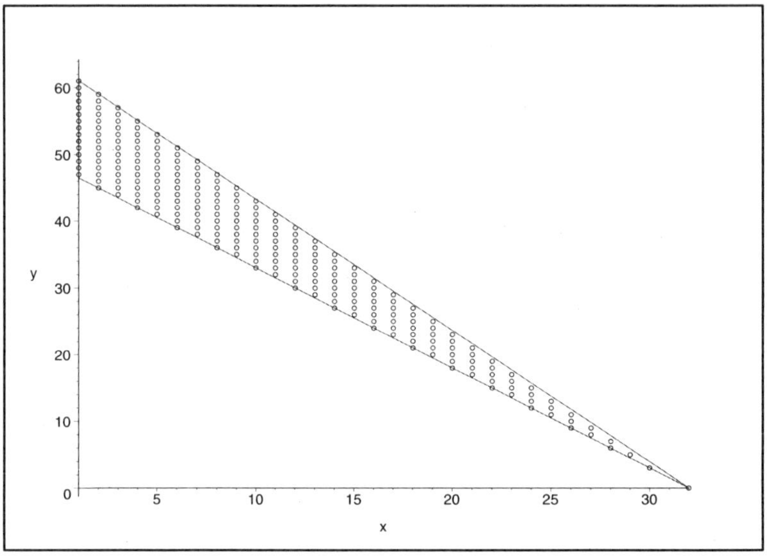

FIGURE 4. Toric resultant support for the Fröberg-Dickenstein example.

The triangulation $(0+Q_1)$,(Q_0+32) is a special case of triangulation (3.2), for $a = 0 \in A_0$. This special triangulation, which is induced from a linear lifting, yields support point $\pm x^{32}$, using an argument similar to the one above.

We can characterize the extreme monomials when all coefficients are indeterminate, but only certain of these remain extreme when we consider the specialized input coefficients. Thus, we obtain the triangle with vertices $(32,0),(0,48)$ and $(0,63)$. Equivalently, it is delimited by the y-axis and the lines

$$y = -(3/2)x + 48 \text{ and } y = -(63/32)x + 63,$$

as shown in Figure 4. Counting the points with integer coordinates inside (and on the sides) of the triangle, we see that there are 257 such points, which is seen to be optimal by actually computing the resultant.

Table 1 shows the results obtained by IPSOS. The general number of monomials represents the number of monomials obtained by defining the support using only degree arguments. In the case of the Fröberg-Dickenstein example, bounding the degree in x and y gives 1057 monomials.

EXAMPLE 3.4. The well-known bicubic surface still represents a challenge for our current implementation:

$x = 3\ t\ (t-1)^2 + (s-1)^3 + 3\ s,$

$y = 3\ s\ (s-1)^2 + t^3 + 3\ t,$

$z = -3s(s^2-5s+5)t^3 - 3(s^3+6s^2-9s+1)t^2 + t(6s^3+9s^2-18s+3) - 3s(s-1).$

We computed 737129 regular triangulations (using TOPCOM) [**Ram01**]. For illustration purposes, we show one of them (the last one):

```
{2,3,4,7,13},{3,4,5,7,13},{3,5,6,7,13},{3,6,9,13,14},{6,9,12,13,14},
{3,6,9,14,15},{6,9,12,14,15},{6,12,13,14,16},{6,12,14,15,16},
{6,12,15,16,17},{3,6,9,15,18},{6,9,12,15,18},{6,12,15,17,18},
```

problem	**input degree**	**implicit degree**	**general #monomials**	**#monomials from** IPSOS
Unit Circle	2	2	6	3
Example 3.1	3	3	10	3
Example 3.3	63	63	1057	257
Example 3.2	1,2	4	35	2
Example 4.3	3	5	56	4
Example 2.1	1,1	2	10	9

TABLE 1. Predicting the implicit support. In these examples, the prediction is always completely accurate.

```
{3,9,15,18,19},{3,6,9,18,19},{6,9,12,18,19},{6,12,16,17,20},
{6,12,17,18,20},{3,6,9,19,23},{6,9,12,19,23},{6,12,19,22,23},
{6,12,22,23,24},{6,12,23,24,25},{3,6,9,23,26},{6,9,12,23,26},
{6,12,23,25,26},{0,2,4,7,13},{3,6,7,9,13},{6,12,18,19,22},
{6,12,18,20,24},{6,7,9,12,13},{6,12,18,22,24}.
```

The size of the file is 383 MBytes. This underlines the fact that we should not compute all regular triangulations but only the mixed-cell configurations.

4. Perturbed resultant

This section discusses implicitization by resultant perturbations.

Implicitization is equivalent to eliminating all parameters t from the polynomial system

$$f_i(t) = p_i(t) - x_i q(t),\ i = 0, \ldots, n,$$

regarded as polynomials in t; here we use notation as in (1.1). The resultant is well-defined for this system, and shall be a polynomial in x, equal to the implicit expression, provided that it does not vanish and the parametrization is generically one-to-one. Otherwise, the resultant is a power of the implicit equation.

EXAMPLE 4.1. Let us take the implicit equation of the parametrized surface with

$$p_0 = t_1 + 3t_2 + t_2^2, p_1 = t_1 + t_2 + 2t_1t_2, p_2 = t_1 + 4t_2, q = t_1 + 2t_2 + t_1^2.$$

The toric resultant construction of [**DE01b**] requires that all Newton polygons be scaled copies of each other; hence, we would take the union of the supports, then specialize the missing coefficients to zero. The matrix has dimension 7 and determinant of degree 9 in the symbolic coefficients, whereas the matrices of [**CE00, ZG00**] have dimensions (and determinant degree) equal to 12 and 15 respectively.

The toric resultant has total degree 7, hence all three expressions here contain superfluous factors with respect to the toric resultant, albeit of different degrees. However, when the symbolic coefficients are specialized to those of the given polynomials, the determinants of the matrices from [**CE00, DE01b**] are precisely the irreducible implicit equation, of degree 3. A similar construction can be applied with Bezoutian matrices, which yields a 4×4 matrix, whose determinant is directly of degree 3.

More subtle is the case where the resultant is identically zero, which happens precisely when there exist values of t, known as *base points*, for which the f_i vanish for all x. Base points forming a component of codimension 1 can be easily removed by canceling common factors in the numerator and denominator of the rational expressions for the x_j's. But higher codimension presents a harder problem.

Besides cases where the (toric) resultant vanishes, another problem with non-generic coefficients is that the resultant matrix may be identically singular. Avoiding degeneracies is an important problem, whose relevance extends beyond the question of implicitization with base points.

Symbolic infinitesimal perturbation is a standard method for handling degeneracies in the case of resultants, see [**Hof89, Sed90, MC92, Man94**]. In the classical context, Canny [**Can90**] perturbed each f_i by adding a new factor $\epsilon x_i^{d_i}$, where $i = 1, \ldots, n$, and f_0 by adding ϵ, where ϵ is a positive infinitesimal indeterminate. Rojas proposed a perturbation scheme for toric resultants in [**Roj99**] which yields a perturbed resultant of low degree in ϵ but is, nonetheless, rather expensive to compute.

In [**DE01a**], a toric projection operator is defined; this generalized the resultant, in a sense, since it projects the common roots to a single dimension. This scheme generalizes [**Can90**], and requires virtually no extra computation besides the matrix construction. It is computed by perturbing the subdivision-based matrix of [**CE00**] so that, after specialization, this operator is not identically zero but vanishes on roots in the proper components of the variety, including all isolated roots.

Suppose we have a family $p := (p_0(x) \ldots, p_n(x))$ of Laurent polynomials such that $\mathrm{supp}(p_i) \subseteq A_i$, and $R(p_0, \ldots, p_n) \neq 0$. The Toric Generalized Characteristic Polynomial (p-GCP) is

$$C_p(\epsilon) := R\left(f_0 - \epsilon p_0, \ldots, f_n - \epsilon p_n\right).$$

The coefficient of $C_p(\epsilon)$ of lowest degree in ϵ is a suitable projection operator. In fact, the polynomials p_i may have random coefficients and support including precisely those monomials of f_i which appear on the diagonal of the toric resultant matrix. The perturbation has been implemented in Maple library `Multires`[1].

EXAMPLE 4.2. We consider the following system in order to illustrate the toric resultant perturbation in a general setting:

$$\begin{aligned} f_0 &= 1 + x_1 x_2 + x_1^2 x_2 + x_1, \\ f_1 &= c_{11} y + c_{12} x^2 y^2 + c_{13} x^2 y + c_{14} x, \\ f_2 &= 1 + x_2 + x_1 x_2 + x_1. \end{aligned} \tag{4.1}$$

First, we study the corresponding generic system, which is further discussed in [**CE00**]. The mixed volumes are $\mathrm{MV}(Q_0, Q_1) = 4$, $\mathrm{MV}(Q_1, Q_2) = 4$, $\mathrm{MV}(Q_2, Q_0) = 3$, so the toric resultant's total degree is 11. Compare this with the Bézout numbers of these subsystems: $8, 6, 12$; hence the projective resultant's total degree is 26. The Newton matrix M constructed by the subdivision-based algorithm in [**CE00**] has dimension 15; the greedy version of [**CE00**] produces a matrix with dimension 14. The incremental algorithm of [**EC95**] yields a matrix of dimension 13.

For the specialization in (4.1) the toric resultant vanishes for all c_{1j} since the variety $V(f_0, f_1)$ has positive dimension: it is formed by the union of the isolated

[1] `http://www-sop.inria.fr/galaad/logiciels/multires.html`

point $(1, -1)$ and the line $\{-1\} \times \mathbb{C}$. For a specific lifting and matrix construction, the trailing coefficient in the perturbed determinant is that of ϵ^2 and equals

$$-(c_{12}c_{13})(c_{14} - c_{11} + c_{12} - c_{13})(c_{14} + c_{11} - c_{12} + c_{13}).$$

So we can recover in the last two factors the value of f_1 at the isolated zero $(1, -1)$ and the point $(-1, -1)$ in the positive-dimensional component.

The toric perturbation method applies directly to implicitization, since the projection operator will contain, as an irreducible factor, the implicit equation [**DE03**]. The extraneous factor has to be removed by factorization. Distinguishing the implicit equation from the latter is possible by using the parametric expressions to generate points on the implicit surface.

EXAMPLE 4.3. Let us consider the de-homogenized version of a system defined in [**Bus01a**]:

$$p_0 = t_1^2,\ p_1 = t_1^3,\ p_2 = t_2^2,\ q = t_1^3 + t_2^3.$$

It has one base point, namely $(0,0)$, of multiplicity 4. The toric resultant here does not vanish, so it yields the implicit equation

$$x_2^3x_1^2 - x_0^3x_1^2 + 2x_0^3x_1 - x_0^3.$$

But under the change of variable $t_2 \to t_2 - 1$ the new system has zero toric resultant. The determinant of the perturbed 27×27 resultant matrix has a trailing coefficient which is precisely the implicit equation. The degree of the trailing term is 4, which equals in this case, the number of base points in the toric variety counted with multiplicity.

The subdivision-based approach of [**CE00**] can be coupled with the existence of a minor in the Newton matrix that divides the determinant so as to yield the exact toric resultant [**D'A02**]. This generalizes Macaulay's famous quotient formula that yields the exact projective resultant [**Mac02**].

EXAMPLE 4.4 (Continued from Example 2.1). The bilinear system $f_i = c_{i0} + c_{i1}x_1 + c_{i2}x_2 + c_{i3}x_1x_2,\ i = 0, 1, 2$, despite its apparent simplicity, does not admit an optimal toric resultant matrix, when we apply the subdivision-based algorithm [**CE00**]. In contrast, the greedy variant may yield an optimal matrix and the incremental algorithm of [**EC95**] produces the optimal 6×6 matrix in Example 2.1. It is possible to construct the following 9×9 numerator matrix, using the subdivision-based algorithm:

$$M = \begin{bmatrix} c_{00} & c_{01} & c_{02} & c_{03} & 0 & 0 & 0 & 0 & 0 \\ c_{10} & c_{11} & c_{12} & c_{13} & 0 & 0 & 0 & 0 & 0 \\ c_{20} & c_{21} & c_{22} & c_{23} & 0 & 0 & 0 & 0 & 0 \\ 0 & 0 & 0 & c_{00} & c_{01} & c_{02} & 0 & 0 & c_{03} \\ 0 & c_{10} & 0 & c_{12} & c_{13} & 0 & c_{11} & 0 & 0 \\ 0 & 0 & c_{20} & c_{21} & 0 & c_{23} & 0 & c_{22} & 0 \\ 0 & c_{20} & 0 & c_{22} & c_{23} & 0 & c_{21} & 0 & 0 \\ 0 & 0 & c_{10} & c_{11} & 0 & c_{13} & 0 & c_{12} & 0 \\ 0 & 0 & 0 & c_{10} & c_{11} & c_{12} & 0 & 0 & c_{13} \end{bmatrix} \quad \begin{matrix} f_0 \\ f_1 \\ f_2 \\ x_1x_2f_0 \\ x_1f_1 \\ x_2f_2 \\ x_1f_2 \\ x_2f_1 \\ x_1x_2f_1 \end{matrix}$$

It is now possible to define a denominator matrix M', of dimension 3, which is a submatrix of M. It is defined by the rows indexed by polynomials $f_1, f_2, x_1x_2f_1$ and the respective columns; these correspond precisely to the integer points in non-mixed cells. The ratio of the determinants yields precisely the toric resultant.

5. Eigenvalue Method

The eigenvalue method for implicitization is applicable to curves, surfaces and hypersurfaces given by very general classes of parametric equations, as well as multiparametric families (generic implicitization) The implicitization problem is reduced to the computation of the basis of the nullspace of certain structured matrices, called Implicitization matrices; see [**CGKW00, Kot04**]. In [**Che98**] a similar method is used for approximate implicitization of rational curves.

Here we illustrate the method for the case of curves. Suppose a curve is given by parametric equations

$$x = f(t),\ y = g(t)$$

and denote by m the (total) degree of the sought implicit equation. Reliable estimates for the value of m, can be given by IPSOS (rather than general algebraic geometry arguments). The method requires the generation of the $\binom{2m+2}{2}$ monomials in x, y up to total degree $2m$, then the computation of the corresponding $\binom{2m+2}{2} = (2m+1)(m+1)$ integrals:

$$\underbrace{\int_a^b 1\ dt,}_{\deg=0}\ \underbrace{\int_a^b x\ dt, \int_a^b y\ dt,}_{\deg=1}\ldots,\ \underbrace{\int_a^b x^{2m}\ dt, \ldots, \int_a^b y^{2m}\ dt,}_{\deg=2m}$$

where x and y are replaced by $f(t)$ and $g(t)$ respectively and the interval $[a, b]$ is chosen with the aim to avoid the singularities (if any) of f and g. Now we define the vector of monomials in x, y up to total degree m,

$$\ell_m = [1, x, y, x^2, xy, y^2, \ldots, x^m, x^{m-1}y, \ldots, xy^{m-1}, y^m]$$

and construct the matrix

$$\ell_m^t \ell_m = \begin{bmatrix} 1 & x & y & \cdots & x^m & \cdots & y^m \\ x & x^2 & xy & \cdots & x^{m+1} & \cdots & xy^m \\ y & xy & y^2 & \cdots & x^m y & \cdots & y^{m+1} \\ \vdots & \vdots & \vdots & \vdots & \vdots & \vdots & \vdots \\ x^m & x^{m+1} & x^m y & \cdots & x^{2m} & \cdots & x^m y^m \\ \vdots & \vdots & \vdots & \vdots & \vdots & \vdots & \vdots \\ y^m & xy^m & y^{m+1} & \cdots & x^m y^m & \cdots & y^{2m} \end{bmatrix} \tag{5.1}$$

We construct a new matrix by replacing each entry $x^i y^j$ of the matrix (5.1) by the result of the corresponding integral. Thus we get an Implicitization matrix with scalar elements and we compute a basis for its nullspace. If there is an implicit equation of degree m, then it will be given by the inner product of a nullvector by ℓ_m.

Matrix (5.1) is symmetric and of dimension $\binom{m+2}{2}$; it contains only $\binom{2m+2}{2}$ different elements, i.e., much less than the total number of its elements. Hankel-like structural properties of implicitization matrices are established in [**KL03**].

EXAMPLE 5.1 (Cont'd from Example 3.1). Degree argument for the parametric equations of the Descartes' folium, allow us to conclude that the sought implicit equation is of degree 3. Using the straightforward eigenvalue implicitization method would require to form a 10×10 implicitization matrix $M = v^t \cdot v$ where

$$v = [1, x, y, x^2, xy, y^2, x^3, x^2y, xy^2, y^3].$$

However, using the information on the support of the implicit equation from IPSOS it suffices to form a 3×3 implicitization matrix $M = v^t \cdot v$ where $v = [xy, x^3, y^3]$. We perform the integrations symbolically with respect to t over the interval $[0, 2]$ to avoid the root $t = -1$ of the denominator of equations (3.1). We obtain matrix

$$\begin{bmatrix} \frac{2}{3}\pi\sqrt{3} + \frac{2}{3}\ln(3) + \frac{8}{81} & \frac{7}{6}\pi\sqrt{3} + \frac{7}{6}\ln(3) - \frac{2}{9} & \frac{5}{6}\pi\sqrt{3} + \frac{5}{6}\ln(3) + \frac{14}{27} \\ \frac{7}{6}\pi\sqrt{3} + \frac{7}{6}\ln(3) - \frac{2}{9} & \frac{7}{3}\pi\sqrt{3} + \frac{7}{3}\ln(3) - \frac{1564}{1215} & \frac{7}{6}\pi\sqrt{3} + \frac{7}{6}\ln(3) + \frac{754}{1215} \\ \frac{5}{6}\pi\sqrt{3} + \frac{5}{6}\ln(3) + \frac{14}{27} & \frac{7}{6}\pi\sqrt{3} + \frac{7}{6}\ln(3) + \frac{754}{1215} & \frac{4}{3}\pi\sqrt{3} + \frac{4}{3}\ln(3) + \frac{1136}{1215} \end{bmatrix},$$

with rank equal to 2. Therefore its nullspace is of dimension 1. We indeed find with Maple a nullspace basis composed of the column vector $[-3, 1, 1]$. Upon multiplying by $v = [xy, x^3, y^3]$ we obtain the implicit equation of the Descartes folium as:

$$x^3 + y^3 - 3\,x\,y = 0.$$

It is important to notice that the nullspace of the original 10×10 implicitization matrix cannot be computed symbolically in Maple 8 after one day on a Pentium III CPU and one had to resort to numerical computations to actually obtain the implicit equation. Therefore IPSOS results in a quite sizable optimization.

EXAMPLE 5.2 (Cont'd from Example 3.2). The degree argument allows us to conclude that the sought implicit equation is of degree 4. Using the eigenvalue implicitization method would require to form a 35×35 implicitization matrix $M = v^t \cdot v$ where

$$v = [1, x, y, x^2, xy, y^2 \ldots x^4 \ldots y^4].$$

However, using IPSOS it suffices to form a 2×2 implicitization matrix $M = v^t \cdot v$ where $v = [x^4, y^2 z]$. We perform the double integrations symbolically, over the interval $[0, 1]$ for s and t successively and we obtain matrix

$$\begin{bmatrix} \frac{512}{81} & \frac{512}{81} \\ \frac{512}{81} & \frac{512}{81} \end{bmatrix}$$

with rank 1 and nullspace basis $[-1, 1]$. Upon multiplying by $v = [x^4, y^2 z]$ we obtain the implicit equation as:

$$x^4 - y^2\,z = 0.$$

6. Moving curves

The method of moving curves for implicitization (see, e.g., [**BCD03, CSC98, SGD97**]) uses a family of lines or curves that follow the curve or surface, in order to set up an overdetermined linear system. A determinant of the matrix of this linear system is the implicit equation. It is possible to compute the determinant by evaluating it at a sufficient number of points. Knowledge of the monomials that appear in the implicit equation speeds up this computation.

We illustrate the simplest case, where moving lines are used to implicitize a parametric curve. Let a rational curve be given by the parametric equations

$$x_0 = \frac{p_0(t)}{q(t)},\quad x_1 = \frac{p_1(t)}{q(t)}.$$

It is assumed that $\gcd(p_0(t), q(t)) = 1$ and $\gcd(p_1(t), q(t)) = 1$. We begin by seeking all moving lines

$$(6.1) \qquad (A_{n-1}x_0 + B_{n-1}x_1 + C_{n-1})t^{n-1} + \cdots + (A_0x_0 + B_0x_1 + C_0) = 0$$

that follow the curve. Substituting into (6.1) the rational functions for x_0 and x_1 and, by clearing out denominators, we obtain the equation

(6.2)

$$(A_{n-1}p_0(t) + B_{n-1}p_1(t) + C_{n-1}q(t))t^{n-1} + \cdots + (A_0p_0(t) + B_0p_1(t) + C_0q(t)) = 0.$$

If (6.1) is a moving line that follows the rational curve, then the polynomial (6.2) must be identically zero. If $\deg p_0(t) = \deg p_1(t) = \deg q(t) = n$, this is a polynomial of degree $2n-1$ in t. Setting all its coefficients to 0, gives rise to a homogeneous linear system of $2n$ equations in the $3n$ unknowns $A_{n-1}, B_{n-1}, C_{n-1}, \ldots, A_0, B_0, C_0$. Such a system has (at least) n linearly independent solutions, for $j = 1, \ldots, n$:

$$(A_{j,n-1}x_0 + B_{j,n-1}x_1 + C_{j,n-1})t^{n-1} + \cdots + (A_{j,0}x_0 + B_{j,0}x_1 + C_{j,0}) = 0.$$

It has been proven that the implicit equation of the curve is given by the determinant

$$\begin{vmatrix} A_{1,n-1}x_0 + B_{1,n-1}x_1 + C_{1,n-1} & \ldots & A_{1,0}x_0 + B_{1,0}x_1 + C_{1,0} \\ \vdots & \vdots & \vdots \\ A_{n,n-1}x_0 + B_{n,n-1}x_1 + C_{n,n-1} & \ldots & A_{n,0}x_0 + B_{n,0}x_1 + C_{n,0} \end{vmatrix}.$$

EXAMPLE 6.1. To illustrate the Descartes Folium example with the method of moving curves we consider the generic moving line of degree $n-1=2$. The equation corresponding to (6.2) is of degree $2n-1=5$ and equals:

$$t^5 C_2 + (C_1 + 3A_2)t^4 + (C_0 + 3A_1 + 3B_2)t^3 + (3A_0 + 3B_1 + C_2)t^2 + (3B_0 + C_1)t + C_0$$

Equating all the coefficients to zero we get a homogeneous linear system of $2n = 6$ equations in the $3n = 9$ unknowns $A_2, B_2, C_2, \ldots, A_0, B_0, C_0$. We find that this linear system has at least 3 linearly independent solutions. We set random values for any 3 unknowns, in order to find 3 specific linearly independent solutions that we place into a matrix:

$$\begin{array}{ccccccccc} A_0 & B_0 & C_0 & A_1 & B_1 & C_1 & A_2 & B_2 & C_2 \end{array}$$
$$\begin{bmatrix} 2 & -3 & 0 & 4 & -2 & 9 & -3 & -4 & 0 \\ 5 & -4 & 0 & -1 & -5 & 12 & -4 & 1 & 0 \\ -3 & -4 & 0 & -4 & 3 & 12 & -4 & 4 & 0 \end{bmatrix}$$

We verify in Maple that the rank of this matrix is 3 and therefore we have found 3 linearly independent solutions. We view these solutions as polynomials in t, and construct the corresponding determinant of linear forms

$$\begin{vmatrix} -3x - 2y & 4x + 3y + 6 & -2x - 4y \\ -4x - 5y & -x + 4y + 15 & -5x + y \\ -4x + 3y & -4x + 4y - 9 & 3x + 4y \end{vmatrix}$$

which is equal (up to a constant factor) to the implicit equation:

$$x^3 + y^3 - 3xy = 0.$$

This last step would rely on IPSOS, in order to accelerate the evaluation-interpolation procedure of determinant computation. Here it is interesting that IPSOS has provided the optimal support; in other words, it tells us that several monomials, that could in principle appear, have zero coefficient. This becomes crucial in larger examples.

7. Conclusion

In this paper we show that the knowledge of the support of the implicit equation may result in significant optimizations in different implicitization algorithms. Predicting the support of the implicit equation can be done efficiently with toric elimination theory, and we have referred to this method as IPSOS; we showed its applicability to 3 implicitization algorithms. IPSOS should be able to accelerate other implicitization algorithms as well.

We have mentioned certain open questions, related to the more efficient implementation of IPSOS, as well as the *a priori* computation of the extreme coefficients of the implicit equation. The application of toric elimination theory into the problem of implicitization is a promising avenue that will surely yield new and interesting results. Recent efforts in this direction include [**DK04, Kra02**].

References

[AS01] F. Aries and R. Senoussi, *An Implicitization Algorithm for Rational Surfaces with no Base Points*, J. Symbolic Computation, 2001, **31**(4) pp. 357–365.

[BE97] C.L. Bajaj and S. Evans, *Splines and geometric modeling*, Handbook of Discrete and Computational Geometry (J.E. Goodman and J. O'Rourke, eds.), CRC Press, Boca Raton, Florida, 1997, pp. 833–850.

[Buc88] B. Buchberger, *Applications of Gröbner bases in non-linear computational geometry*, Mathematical Aspects of Scientific Software (J.R. Rice, ed.), IMA Volumes in Math. & Appl., vol. 14, Springer, 1988, pp. 59–87.

[Bus01a] L. Busé, *Residual resultant over the projective plane and the implicitization problem*, Proc. ACM Intern. Symp. on Symbolic & Algebraic Comput., 2001, pp. 48–55.

[BCD03] L. Busé, D. A. Cox and C. D'Andrea, Carlos, *Implicitization of surfaces in $\mathbb{P}^3$ in the presence of base points.* Journal of Algebra and its Applications, **2** (2003) pp. 189–214.

[BEM01] L. Busé, M. Elkadi, and B. Mourrain, *Residual resultant of complete intersection*, J. Pure & Applied Algebra **164** (2001), 35–57.

[BJ03] L. Busé and J.-P. Jouanolou, *On the closed image of a rational map and the implicitization problem*, J. Algebra **265** (2003), 312–357.

[Can90] J. Canny, *Generalised characteristic polynomials*, J. Symbolic Computation **9** (1990), 241–250.

[CE00] J.F. Canny and I.Z. Emiris, *A subdivision-based algorithm for the sparse resultant*, J. ACM **47** (2000), no. 3, 417–451.

[Che98] F. Chen, *Approximate Implicitization of Rational Curves.* Chinese J. Computers, **21** (1998), pp. 855–859. (In Chinese, Personal Communication)

[CGKW00] R.M. Corless, M.W. Giesbrecht, I.S. Kotsireas, and S.M. Watt, *Numerical implicitization of parametric hypersurfaces with linear algebra.* Artificial intelligence and symbolic computation, (Madrid, 2000), Springer, Berlin, (2001), pp. 174–183.

[Cox95] D. Cox, *The homogeneous coordinate ring of a toric variety*, J. Algebraic Geom. **4**(1) (1995), pp. 17–50.

[CLO98] D. Cox, J. Little, and D. O'Shea, *Using algebraic geometry*, Graduate Texts in Mathematics, no. 185, Springer-Verlag, New York, 1998.

[CSC98] D.A. Cox, T.W. Sederberg, and F. Chen *The moving line ideal basis of planar rational curves.* Computer Aided Geometric Design, **15** (1998), pp. 803–827.

[D'A02] C. D'Andrea, *Macaulay-style formulas for the sparse resultant*, Trans. of the AMS **354** (2002), pp. 2595–2629.

[DE01a] C. D'Andrea and I.Z. Emiris, *Computing sparse projection operators*, Symbolic Computation: Solving Equations in Algebra, Geometry, and Engineering. Contemporary Mathematics, **286**, AMS (Providence, Rhode Island), (2001), pp. 121–139.

[DE01b] C. D'Andrea and I.Z. Emiris, *Hybrid resultant matrices of bivariate polynomials*, Proc. Annual ACM Intern. Symp. on Symbolic and Algebraic Computation (London, Ontario), ACM Press, (2001), pp. 24–31.

[DE03] C. D'Andrea and I.Z. Emiris, *Sparse Resultant Perturbations.*, Algebra, Geometry, and Software Systems, M. Joswig and N. Takayama, eds., Mathematics and Visualization, Springer-Verlag, Berlin, (2003), pp. 93–107.

[Dic04] A. Dickenstein, Residues and resultants, In *Proc. of CIMPA Graduate School on Polynomial Systems*, July 2003, Buenos Aires, Argentina. A. Dickenstein and I.Z. Emiris, Eds.

[DK04] C. D'Andrea and A. Khetan, *Implicitization of rational surfaces using toric varieties.* Preprint, `http://arXiv:math.AG/0401403`, 2004.

[DS02] A. Dickenstein and B. Sturmfels, *Elimination theory in codimension two*, J. Symbolic Computation **34**(2) (2002), pp. 119–135.

[Dok01] T. Dokken, *Approximate implicitization*, Mathematical methods for curves and surfaces (Oslo 2000), Innov. Appl. Math., Vanderbilt Univ. Press, Nashville, (2001), pp. 81–102.

[EC95] I.Z. Emiris and J.F. Canny, *Efficient incremental algorithms for the sparse resultant and the mixed volume*, J. Symbolic Computation **20** (1995), no. 2, 117–149.

[EK03] I. Z. Emiris and I. S. Kotsireas, *Implicit polynomial support optimized for sparseness*, Proceedings of ICCSA'2003 (Montreal, Canada) (V. Kumar et al, ed.), LNCS 2669, (2003), pp. 397–406.

[EM99] I.Z. Emiris and B. Mourrain, *Computer algebra methods for studying and computing molecular conformations*, Algorithmica, Special Issue on Algorithms for Computational Biology **25** (1999), 372–402.

[GKZ91] I.M. Gelfand, M.M. Kapranov, and A.V. Zelevinsky, *Discriminants of polynomials in several variables and triangulations of Newton polytopes*, Leningrad Math. J. **2** (1991), no. 3, 449–505, (Translated from *Algebra i Analiz* 2, 1990, pp. 1–62).

[GKZ94] I.M. Gelfand, M.M. Kapranov, and A.V. Zelevinsky, *Discriminants, resultants and multidimensional determinants*, Birkhäuser, Boston, 1994.

[GV97] L. Gonzalez-Vega, *Implicitization of parametric curves and surfaces by using multidimensional Newton formulae*, J. Symbolic Computation **23** (1997), no. 2-3, 137–151, Parametric algebraic curves and applications (Albuquerque, NM, 1995).

[Hof89] C.M. Hoffmann, *Geometric and solid modeling*, Morgan Kaufmann, 1989.

[Khe02] A. Khetan. Determinantal formula for the Chow form of a toric surface, Proc. ACM Intern. Symp. on Symbolic & Algebraic Comput., (2002). Final version to appear in *J. Symb. Computation.*

[Kot04] I.S. Kotsireas. *Panorama of methods for exact implicitization of algebraic curves and surfaces.* Geometric Computation, World Scientific Press, Lecture Notes Series on Computing - Vol. 11, 2004, F. Chen, D. Wang (Eds.) pp. 126–155.

[KL03] I.S. Kotsireas and E.S.C. Lau, *Implicitization of polynomial curves*, Proceedings of ASCM'2003, Beijing, P.R. China (Z. Li and W. Sit, eds.), (2003), pp. 217–226.

[Kra02] R. Krasauskas, *Toric surface patches: Advances in geometrical algorithms and representations*, Adv. Comput. Math. **17** (2002), no. 1-2, pp. 89–113.

[Mac02] F.S. Macaulay, *Some formulae in elimination*, Proc. London Math. Soc. **1** (1902), no. 33, pp. 3–27.

[Man94] D. Manocha, *Solving systems of polynomial equations*, IEEE Comp. Graphics and Appl., Special Issue on Solid Modeling (1994), pp. 46–55.

[MC92] D. Manocha and J.F. Canny, *Algorithm for implicitizing rational parametric surfaces*, Comput. Aided Geom. Design **9** (1992), no. 1, pp. 25–50.

[MC00] T. Michiels and R. Cools, *Decomposing the secondary Cayley polytope*, Disc. & Comput. Geometry **23** (2000).

[MV99] T. Michiels and J. Verschelde, *Enumerating regular mixed-cell configurations*, Disc. & Comput. Geometry **21** (1999), no, 4, pp. 569–579.

[Ram01] J. Rambau, *TOPCOM*, 2001, www.zib.de/rambau/TOPCOM.

[Roj99] J.M. Rojas, *Solving degenerate sparse polynomial systems faster*, J. Symbolic Computation, Special Issue on Elimination **28** (1999), pp. 155–186.

[Sed90] T.W. Sederberg, *Techniques for cubic algebraic surfaces*, IEEE Comp. Graphics & Appl. (1990), pp. 14–25.

[SGD97] T. Sederberg, R. Goldman, and H. Du. *Implicitizing rational curves by the method of moving algebraic curves.* J. Symbolic Computation, (1997) **23**(2-3), pp. 153–175.

[SZ02] T.W. Sederberg, and J. Zheng, *Algebraic methods for computer aided geometric design*, Handbook of computer aided geometric design, (G. Farin, J. Hoschek, M.-S. Kim, Eds) North-Holland, Amsterdam, (2002), pp. 363–387.

[Stu94] B. Sturmfels, *On the Newton polytope of the resultant*, J. of Algebr. Combinatorics **3** (1994), 207–236.

[Stu02] B. Sturmfels, *Solving systems of polynomial equations*, CBMS Regional Conference Series in Math., no. 97, AMS, Providence, RI, 2002.

[Wan04] D. Wang, *A Simple Method for Implicitization of Rational Curves and Surfaces*, J. Symbolic Computation, (2004) **38**, pp. 899–914.

[ZG00] M. Zhang and R.N. Goldman, *Rectangular corner cutting and Sylvester A-resultants*, Proc. ACM Intern. Symp. on Symbolic & Algebraic Comput., (2000), pp. 301–308.

[Zip93] R. Zippel, *Effective polynomial computation*, Kluwer Academic Publishers, Boston, 1993.

DEPARTMENT OF INFORMATICS & TELECOMMUNICATIONS, NATIONAL UNIVERSITY OF ATHENS, PANEPISTIMIOPOLIS, 15771 GREECE

E-mail address: `emiris@di.uoa.gr`

DEPARTMENT OF PHYSICS AND COMPUTER SCIENCE, WILFRID LAURIER UNIVERSITY, 75 UNIVERSITY AVENUE WEST WATERLOO, ONTARIO N2L 3C5 CANADA

E-mail address: `ikotsire@wlu.ca`

DIMACS Series in Discrete Mathematics
and Theoretical Computer Science
Volume **67**, 2005

The Exact Rational Univariate Representation for Detecting Degeneracies

John Keyser, Koji Ouchi, and J. Maurice Rojas

ABSTRACT. Detecting degenerate object configurations is an important part of a robust boundary evaluation system. The most efficient current methods for exact boundary evaluation rely on a general position assumption, and fail in the presence of degeneracies. To address this problem, we describe here a method, based on the rational univariate reduction (RUR), for computing roots of systems of multivariate polynomials with rational coefficients. Our method enables exact algebraic computations for coordinates of the roots and works even if the underlying set of roots is positive dimensional. We describe several applications, with special emphasis on geometric modeling. In particular, we describe a new implementation that has been used successfully on certain degenerate boundary evaluation problems.

1. Introduction

1.1. Motivation. Like many geometric computations, solid modeling operations often suffer from problems related to robustness. These robustness problems, due to unexpected conditions in the input or computation, can cause serious program failures. The two major sources of such robustness problems are those due to numerical error, and those due to degenerate configurations of data.

Exact computation provides a method for eliminating the problems due to numerical error. Note that exact computation (where enough precision is used to guarantee that every decision made is correct) does not necessarily mean exact arithmetic (where every computation is done to full precision). In particular, it is possible to apply exact computation for algebraic points and curves. Unfortunately, exact geometric computations, particularly when curved objects are involved, can be prohibitively slow. To combat this, a number of speedup techniques are used to improve efficiency while maintaining exactness. Some of these techniques continue to assume general position of objects, meaning that while robustness problems due to numerical error are eliminated, those due to degenerate configurations of data remain or are made worse.

A key aspect of dealing with degenerate data is detecting when degenerate situations arise. If we know the data is not degenerate, we can make use of general position assumptions to achieve faster exact computations. On the other hand, if

This work was supported in part by NSF Grants DMS-0138446 and CCR-0220047.

a degenerate situation is detected, we would like to be able to handle it smoothly, performing all computations smoothly enough to be easily handled by subsequent special-case code.

The work presented here is aimed at providing a computational tool that will aid in the detection of degenerate cases. We have expanded on the concept of the Rational Univariate Reduction (RUR) (also called the Rational Univariate Representation) and present an exact implementation of this. The RUR allows us to solve for roots of systems of polynomials, even for degenerate configurations. Solving a system of polynomial equations is a key operation in geometric algorithms, and particularly boundary evaluation. In geometric algorithms, polynomials are commonly used to describe the shape of objects and their relations. Points are described as intersections of curves and/or surfaces, or the common roots of a system of polynomial equations. Implementation of robust geometric algorithms thus requires exact computation over coordinates of the common roots of a system of polynomials.

1.2. Summary. We propose a new development of *exact* arithmetic over the coordinates of the points in the zero set of a system of polynomial equations with rational coefficients.

Fix a *square* system of polynomials, i.e., a system of n polynomials in n variables. Assume temporarily the zero set of the system is of *dimension zero*, i.e., the system has only finitely many common roots. Then, there exist *univariate* polynomials h and $h_1, \ldots, h_n$ with rational coefficients s.t. the finite zero set of the system is represented as [**BPR03**]

$$\{(h_1(\theta), \ldots, h_n(\theta)) \mid \theta \in \mathbb{C} \text{ with } h(\theta) = 0\}.$$

This representation for the zero set of the system is called the *Rational Univariate Representation (RUR)*.

We implement an algorithm for computing the *exact* RUR for the zero set of a square system of polynomial equations, meaning that we compute all the rational coefficients of univariate polynomials in the RUR to full precision. Then, we develop *exact* arithmetic over numbers of the form $e(\theta_1, \ldots, \theta_m)$ where e is a polynomial with rational coefficients and θ_i is an algebraic number. Putting these together, we can compute the *exact* RUR for the zero set of a (not necessarily square) system and have *exact* arithmetic over the coordinates of the points in the zero set of a system of polynomial equations with rational coefficients. Furthermore, we can make our method succeed even if the zero set of the system has some positive dimensional component and in fact can detect such cases.

We compute the exact RUR by using the *toric resultant.*

It is shown that some modifications of the method allow us to compute a super set of the zero set of a not necessarily square system in $\mathbb{C}^n$. A resultant-based method fails when the resultant for a system becomes zero. The latter happens, for instance, when the zero set of the system has some positive dimensional components. To handle these cases, we use the resultant perturbation technique[**DE03**]. Different points on positive dimensional components of the zero set of the system are picked up when different perturbations are applied. Thus, if we can perform *exact* comparison over the points expressed in the *exact* RUR then we could develop a generic algorithm to detect whether or not the zero set of the system has some positive dimensional components.

There exist implementations of both numerical and exact algorithms for computing the RUR for the zero set of a system of polynomial equations. The exact algorithm requires some normal form computation, typically, Gröbner basis computation. The toric resultant based method is presumably more efficient than the Gröbner basis method. Both have complexity exponential in n, but, Gröbner basis methods may have quadratic exponents in the worst case. Furthermore, unlike the toric resultant, the Gröbner basis is discontinuous with respect to perturbations in the coefficients of the input polynomials.

We develop exact arithmetic over the points in the zero set of a system of polynomial equations expressed in the RUR via a *root-bound approach*. The root-bound approach has been successfully used for the exact sign determination of *real* algebraic numbers of the form $e(\theta_1, \ldots, \theta_n)$ where θ_i is a rational number or a *real* root of a univariate polynomial with rational coefficients and e is an algebraic expression involving $\pm, *, /$ and $\sqrt[k]{\ }$. The sign of such a number can be obtained from the location (on the real line) of some approximation of the number to a certain precision that is controlled by the root-bound for the number. We apply this method to the exact sign determination of the real and imaginary parts of the coordinates of the points expressed in the exact RUR. In fact, our implementation can use LEDA [**MN99**] or CORE [**KLY99**], the libraries that support the root-bound approach for the exact sign determination of real algebraic numbers. The non-trivial part of our development is computing approximations for the real and imaginary parts of all the roots of a univariate polynomial with rational coefficients to any given precision. No algebraic mehtod exists for doing so; instead we use a numerical method.

In order to show the ability of our method, we have developed an exact geometric solid modeler that detects degenerate configurations. Subsequent processing [**OK04**] will be needed to actually handle these degenerate cases.

Thus, the key results of this paper are:

- We describe the computation of the RUR based on the toric resultants and see the resulting RUR is exact.
- We describe the exact RUR can be used to perform exact computation over algebraic numbers.
- We describe now the exact RUR can be used to detect degeneracies appearing solid modeling.
- We show an implementation with the run-time of experiments.

The rest of this paper is organized as follows:

Section 2 gives a brief summary of related work.

Section 3 gives a detailed description of the RUR and how it is implemented.

Section 4 describes ways in which the RUR approach can be applied to various geometric problems.

Section 5 gives examples and experimental results from our implementation of the RUR method.

Section 6 concludes with a discussion of useful avenues for future work.

2. Related Work

2.1. Robust Geometric Computation. The need for robustness in geometric algorithms has been known for a long time [**HHK89, DY95, Yap97**]. Exact computation as a way to eliminate numerical errors appearing in solid modeling has also been discussed for years [**Hof89**]. Much of the earlier work deals with polyhedral solids though some addresses curved solids [**SI89, Yu91, MOBP94, For97**]. The Effective Computational Geometry for Curves and Surfaces (ECG) project is currently developing a large set of exact geometric algorithms [**BFMS99, BFMS00, DFMT02, MVY02**].

2.2. RUR. Representing roots of a system of polynomial equations in the RUR was described by Kronecker [**Kro31**] but started to be used in computer algebra only recently [**GV97, Roj99, Rou99**].

If a system is zero-dimensional, i.e., if a system has only finitely many roots, then the RUR for the zero set of the system can be computed via the multiplication table method [**Rou99, GVRR99, BPR03**]. The method has been extended so that it finds all the isolated roots as well as at least one point from each *real* positive dimensional component [**ARD02**]. A standard implementation of the multiplication table method requires some normal form computation, usually Gröbner basis computation.

A Gröbner-free algorithm to compute the RUR for a zero dimensional system has been proposed [**GLS01**]. The most recent work even handles some systems with multiple roots [**Lec02**]. However, because of their iterative nature, it is hard to apply these algorithms to exact computation.

2.3. Resultants Based Polynomial Solving. The *toric resultant* (or the *sparse resultant*) for a system of $n+1$ polynomial equations in n variables is a polynomial in the coefficients of the polynomials in the system whose vanishing implies the existence of common roots in $(\mathbb{C}^*)^n$ [**Stu94, CLO98, CE00, Stu02, Emi03, Roj03**]. The subdivision based algorithm for computing the toric resultant for a system has been described [**Stu94, Emi96, CE00, Stu02, Emi03**], as have improvements of the algorithms [**EC95, D'A02, Emi02, Khe03**]. The "toric" theory for the solutions of a system of polynomial equations gives a tighter bound on the number of isolated roots of a system known as the BKK bound [**Ber75, Kus76, Kho78, Roj94, Roj04**], and thus lower degree resultants compared to traditional bounds such as Bezout's.

A standard resultant based method for solving a system of polynomial equations, the u-resultant based method, fails if the zero set of the system has some positive dimensional component. In that case, the Generalized Characteristic Polynomial (GCP) based method is used. The homogeneous Generalized Characteristic Polynomial (GCP) [**Can90**] works for homogeneous systems that have some positive dimensional components unless there are some multiple roots at the point at infinity. The toric perturbation based method works even if the system has some multiple roots at the point at infinity [**Roj97, Roj99**]. A potentially more efficient version of the GCP has also been proposed [**DE01, DE03**].

Solving a positive dimensional system means constructing an irreducible decomposition of the zero set of a system of polynomial equations [**SW96**]. For bivariate systems, irreducible decomposition is practical via computation of the gcd of polynomials. For general cases, numerical irreducible decomposition has been developed [**SV00, SVW01, SVW03**]. This approach obtains geometric information by numerically computing homotopies to generic points lying on positive dimensional components, and is thus hard to implement exactly.

2.4. Root Bound. The root-bound approach to exact sign determination for *real* algebraic numbers has been implemented in the libraries LEDA [**BMS96, MN99**] and CORE [**KLY99**]. Several improvements on root-bounds have also been reported [**BFMS97, BFMS00, BFM$^+$01, LY01, PY03**].

The sign of the real algebraic number (given as roots of a univariate polynomial with rational coefficients) can be determined algorithmically (e.g. Sturm's method). However, the sign of the real and imaginary parts of a complex algebraic numbers cannot be determined algorithmically. Instead, a numerical method must be used. Research on numerical methods for computing all the roots of a univariate polynomial has a long history [**Abe73, Bin96, BCSS97, BF00**], but none fits our purpose.

2.5. MAPC and ESOLID. MAPC [**KCMK00**] is a library that manipulates exact computation for two-dimensional real algebraic points and curves. ESOLID [**Key00, KCF$^+$04**] is a robust geometric modeler built on top of MAPC. ESOLID is currently the only system we know of that supports exact boundary evaluation for solids with curved surfaces. For efficiency, MAPC assumes curves are non-singular and ignores some singular intersections such as tangential intersections. Thus, ESOLID works only when provided solids that are in general position. Addressing this deficiency has been a goal of our work.

3. Exact Rational Univariate Representation

Consider a system of m polynomials $f_1, \ldots, f_m$ in n variables with rational coefficients. Let Z be the set of all the common roots of the system. Fix a finite subset Z' of Z. Every coordinate of the points in Z' is represented as a univariate polynomial with rational coefficients evaluated at some (complex) root of the other univariate polynomial with rational coefficients [**BPR03**]. That is, there exist $h, h_1, \ldots, h_n \in \mathbb{Q}[T]$ s.t.

$$Z' = \{(h_1(\theta), \ldots, h_n(\theta)) \in \mathbb{C}^n \mid \theta \in \mathbb{C} \text{ with } h(\theta) = 0\}.$$

This representation for Z' is called the *Rational Univariate Representation (RUR)* for Z'.

We describe a method to compute the exact RUR for some finite subset Z' of Z which contains at least one point from every irreducible component of Z. In particular, if Z is zero-dimensional then $Z' = Z$.

3.1. The Exact RUR for Square Systems. In this section, we describe a strictly exact implementation of the toric resultant based algorithm for computing the exact RUR for the roots of a square system having non-zero coordinates. The polynomial h in the RUR is derived from the toric perturbation [**Roj99**] which is a generalization of the "toric" u-resultant.

3.1.1. *Toric Resultants.* Let f be a polynomial in n variables $X_1, \ldots, X_n$ with rational coefficients. Define the *support* of f to be the finite set A of exponents of all the monomials appearing in f corresponding to non-zero coefficients. Thus, A is some finite set of integer points in n-dimensional space $\mathbb{R}^n$. Then

$$f = \sum_{a \in A} c_a X^a, \qquad c_a \in \mathbb{Q}^*$$

where $X^a = X_1^{a_1} \cdots X_n^{a_n}$ for $a = (a_1, \ldots, a_n)$.

Consider a system of $n+1$ polynomials $f_0, f_1, \ldots, f_n$ in n variables $X_1, \ldots, X_n$ with corresponding supports $A_0, A_1, \ldots, A_n$. Then, the system is associated with coefficient vectors $\mathbf{c}_0, \mathbf{c}_1, \ldots, \mathbf{c}_n$ where

$$\mathbf{c}_i = (c_{ia} \in \mathbb{Q}^* \mid a \in A_i) \quad \text{s.t.} \quad f_i = \sum_{a \in A_i} c_{ia} X^a.$$

For such a system, there exists a unique (up to sign) irreducible polynomial

$$\mathrm{Res}_{A_0, A_1, \ldots, A_n}(\mathbf{c}_0, \mathbf{c}_1, \ldots, \mathbf{c}_n) \quad \in \quad \mathbb{Z}[\mathbf{c}_0, \mathbf{c}_1, \ldots, \mathbf{c}_n],$$

called the *toric resultant* or the *sparse resultant*, s.t.

$$\begin{aligned} &\text{the system } (f_0, f_1, \ldots, f_n) \text{ has a common root in } (\mathbb{C}^*)^n \\ \Longrightarrow \quad &\mathrm{Res}_{A_0, A_1, \ldots, A_n}(\mathbf{c}_0, \mathbf{c}_1, \ldots, \mathbf{c}_n) = 0. \end{aligned}$$

We also write $\mathrm{Res}_{A_0, A_1, \ldots, A_n}(f_0, f_1, \ldots, f_n)$ for the toric resultant.

We implement the subdivision based algorithm [**CE00, Emi03**], summarized here. The algorithm constructs a square matrix N, called the *toric resultant matrix* or *Newton matrix*, whose determinant is some multiple of the toric resultant. The elements of N are the coefficients of the input polynomials $f_0, f_1, \ldots, f_n$. The rows and columns of N are labeled by monomials, or equivalently, integer points belonging to the supports $A_0, A_1, \ldots, A_n$ of the input polynomials. The row labeled by $a_r \in A_i$ contains the coefficients of $X^{a_r} f_i$ s.t. the column labeled by a_c contains the coefficient of the monomial term X^{a_c}. Thus, the determinant of N is a homogeneous polynomial in each coefficient vector $\mathbf{c}_i$. It follows that the total degree of the determinant of N, $\deg_{\mathbf{c}_i} \det N$, with respect to the coefficient vector $\mathbf{c}_i$ is well-defined [**CE00, Emi03**].

The integer points used to label the rows and columns of N are chosen according to the *mixed-subdivision* [**Stu94, Stu02**] of the *Minkowski sum* of the convex hulls $Q_0, Q_1, \ldots, Q_n$ of $A_0, A_1, \ldots, A_n$. The number of integer points (and the degree of $\det N$) is predicted in terms of the *mixed-volume* [**CE00, Emi03**] of the Minkowski sum of $Q_0, Q_1, \ldots, Q_n$. More precisely, writing MV_{-i} for the mixed-volume for the Minkowski sum $Q_0 + Q_1 + \cdots + Q_{i-1} + Q_{i+1} + \cdots + Q_n$, it is shown [**CE00**] that

$$\begin{cases} \deg_{\mathbf{c}_0} \det N = \mathrm{MV}_{-0}, \\ \deg_{\mathbf{c}_i} \det N \geq \mathrm{MV}_{-i}, \quad i = 1, \ldots, n. \end{cases} \tag{3.1}$$

The equalities hold if $\det N$ is the sparse resultant [**PS93**].

We develop a strictly exact implementation of the subdivision based algorithm for computing the toric resultant matrix N. Computation of $Q_0, Q_1, \ldots, Q_n$ and the mixed subdivision of their Minkowski sum both reduce to linear programming problems [**CE00**] which are solved by using a standard 2-phase simplex method implemented with multi-precision rational arithmetic.

3.1.2. *Toric Perturbations.* Consider a square system of n polynomials $f_1, \ldots, f_n \in \mathbb{Q}[X_1, \ldots, X_n]$ with supports $A_1, \ldots, A_n$.

Let $A_0 = \{\mathbf{o}, \mathbf{b}_1, \ldots, \mathbf{b}_n\}$ where $\mathbf{o}$ is the origin and $\mathbf{b}_i$ is the i-th standard basis vector in $\mathbb{R}^n$. Also, let $f_0 = u_0 + u_1 X_1 + \cdots + u_n X_n$ where $\mathbf{u} = (u_0, u_1, \ldots, u_n)$ is a vector of indeterminates. Choose n polynomials $f_1^*, \ldots, f_n^* \in \mathbb{Q}[X_1, \ldots, X_n]$ with supports contained in $A_1, \ldots, A_n$ that have only finitely many common roots in $(\mathbb{C}^*)^n$. There is a deterministic way to choose n polynomials $f_1^*, \ldots, f_n^*$ that work [**Roj99**], but this is expensive. In practice, we simply use polynomials with random coefficients. Then, the probability that those polynomials are appropriate is 1, and furthermore, we can detect the cases when these random choices are not generic. The *toric Generalized Characteristic Polynomial* $\mathrm{TGCP}(s, \mathbf{u})$ for the system $(f_1, \ldots, f_n)$ is defined to be the toric resultant for the perturbed system $(f_0, f_1 - s f_1^*, \ldots, f_n - s f_n^*)$:

$$\mathrm{TGCP}(s, \mathbf{u}) = \mathrm{Res}_{A_0, A_1, \ldots, A_n}(f_0, f_1 - s f_1^*, \ldots, f_n - s f_n^*) \quad \in \quad \mathbb{Q}[s][\mathbf{u}].$$

Also, define a *toric perturbation* $\mathrm{Pert}_{A_0, f_1^*, \ldots, f_n^*}(\mathbf{u})$ for the system $(f_1, \ldots, f_n)$ to be the non-zero coefficient of the lowest degree term in $\mathrm{TGCP}(s, \mathbf{u})$ regarded as a polynomial in variable s. We also write $\mathrm{Pert}_{A_0}(\mathbf{u})$ for $\mathrm{Pert}_{A_0, f_1^*, \ldots, f_n^*}(\mathbf{u})$ when $f_1^*, \ldots, f_n^*$ are fixed.

The following theorem (the Main Theorem 2.4 in [**Roj99**]) guarantees that the RUR for a system is obtained from the toric perturbation.

THEOREM 3.1. [**Roj99**]
$\mathrm{Pert}_{A_0}(\mathbf{u})$ is well-defined, i.e. with a suitable choice of $f_1^*, \ldots, f_n^*$, for any system, there exist rational numbers $u_0, u_1, \ldots, u_n$ s.t. $\mathrm{TGCP}(s, \mathbf{u})$ always has a non-zero coefficient. $\mathrm{Pert}_{A_0}(\mathbf{u})$ is a homogeneous polynomial in parameters $u_0, u_1, \ldots, u_n$ with rational coefficients, and has the following properties:

1. *If $(\zeta_1, \ldots, \zeta_n) \in (\mathbb{C}^*)^n$ is an isolated root of the system $(f_1, \ldots, f_n)$ then $u_0 + u_1\zeta_1 + \cdots + u_n\zeta_n$ is a linear factor of $\mathrm{Pert}_{A_0}(\mathbf{u})$.*
2. *$\mathrm{Pert}_{A_0}(\mathbf{u})$ completely splits into linear factors over $\mathbb{C}$. Recalling Z be the zero set of the system (which might be infinite), for every irreducible component W of $Z \cap (\mathbb{C}^*)^n$, there is at least one factor of $\mathrm{Pert}_{A_0}(\mathbf{u})$ corresponding to a root $(\zeta_1, \ldots, \zeta_n) \in W \subseteq Z$.*

Thus, with a suitable choice of $f_1^*, \ldots, f_n^*$, there exists a finite subset Z' of Z s.t. a univariate polynomial $h = h(T)$ in the RUR for $Z' \cap (\mathbb{C}^*)^n$ is obtained from $\mathrm{Pert}_{A_0}(\mathbf{u})$ by setting u_0 to a variable T and specializing $u_1, \ldots, u_n$ to some values. Immediately from (3.1)

COROLLARY 3.2.

$$\deg_{u_0} \mathrm{Pert}_{A_0}(\mathbf{u}) = \mathrm{MV}_{-0}, \tag{3.2}$$

$$\deg_s \mathrm{TGCP}(s, \mathbf{u}) = \sum_{i=1}^{n} \mathrm{MV}_{-i} \leq \dim N - \mathrm{MV}_{-0}. \tag{3.3}$$

3.1.3. *The Exact RUR Computation.* We give our algorithm to compute the exact RUR for the roots having non-zero coordinates of a square system. More details will be given below.

Algorithm RUR_square:
Input: $f_1, \ldots, f_n \in \mathbb{Q}[X_1, \ldots, X_n]$ with supports $A_0, A_1, \ldots, A_n \subseteq \mathbb{Z}^n$.

Output: $h, h_1, \ldots, h_n \in \mathbb{Q}[T]$, forming the RUR for some finite subset of the roots of the system in $(\mathbb{C}^*)^n$.

1: Compute the toric resultant matrix N and MV_{-0} for the system of polynomials with supports $A_0, A_1, \ldots, A_n$.

2:

- **2-1:** Choose generic $u_0, u_1, \ldots, u_n \in \mathbb{Q}$ and $f_1^*, \ldots, f_n^* \in \mathbb{Q}[X_1, \ldots, X_n]$.
- **2-2:** Compute some multiple of $\mathrm{TGCP}(s, \mathbf{u})$.
- **2-3:** If $\mathrm{TGCP}(s, \mathbf{u}) \equiv 0$ then go to 2-1.

3: Compute $h(T) := \mathrm{Pert}_{A_0}(T, u_1, \ldots, u_n)$ where $\mathrm{Pert}_{A_0}(\mathbf{u})$ is the non-zero coefficient of the lowest degree term in (the multiple of) $\mathrm{TGCP}(s, \mathbf{u})$.

4: For $i := 1, \ldots, n$ do:

- **4-1:** Compute $p_i^{\pm}(t) := \mathrm{Pert}_{A_0}(t, u_1, \ldots, u_{i-1}, u_i \pm 1, u_{i+1}, \ldots, u_n)$.
- **4-2:** Compute the square-free parts $q_i^{\pm}(t)$ of $p_i^{\pm}(t)$.
- **4-3:** Compute a linear (as a polynomial in t) gcd $g(t)$ of $q_i^-(t)$ and $q_i^+(2T - t)$.
- **4-4:** Set $h_i(T) := -T - \frac{g_0(T)}{g_1(T)} \bmod h(T)$ where $g_0(T)$ and $g_1(T)$ are the constant term and the linear coefficient of $g(t)$.

Step 1 determines the toric resultant matrix by using the subdivision based algorithm. Some entries are still undetermined. Note step 1 needs to be performed once and only once. The result of step 1 is used in later computation.

Step 2-1 is performed by choosing random values for u_i and coefficients of f_i^*. While the randomly chosen u_i's could turn out not to be generic, this is almost never the case, and is detected in step 2-3 if it is.

At step 2-2, from (3.3), we know a bound on the degree of $\mathrm{TGCP}(s, \mathbf{u})$ (as a polynomial in s), and can therefore compute some multiple of it via interpolation. More precisely, choose $\dim N - \mathrm{MV}_{-0}$ many values for s, specialize the entries of N with those values for s, and interpolate $\det N$ to obtain the non-zero coefficient of the lowest degree term s^l in $\mathrm{TGCP}(s, \mathbf{u})$.

Step 2-3 checks to make sure the generic choice made in step 2-1 was acceptable. It can be shown that step 2 must be performed only once with probability 1.

Step 3 determines the univariate polynomial h in the RUR, introducing the variable T. From (3.2), we know the degree of $\mathrm{Pert}_{A_0}(T, u_1, \ldots, u_n)$ (as a polynomial in T), and can compute it via interpolation. More precisely, we choose $\mathrm{MV}_{-0} - 1$ many values for u_0 along with the one we chose at step 2-2, specialize the elements of N with those values, and interpolate to compute $\mathrm{Pert}_{A_0}(T, u_1, \ldots, u_n)$.

Similar to steps 2-2 and 3, $p_i^{\pm}(t)$ are determined via interpolation at step 4-1.

At step 4-3, we compute the gcd $g(t)$ of $q_i^-(t)$ and $q_i^+(2T - t)$ (as a polynomial in t). We use the fact that $g(t)$ will be linear with probability 1 (dependent on the choices of u_i in step 2-1). In this case, the coefficients of $g(t)$ are the *first subresultants* for $q_i^-(t)$ and $q_i^+(2T - t)$ [**Can90, GV91**] which are defined to be Sylvester resultants for some submatrices of the Sylvester matrix for $q_i^{\pm}$.

The rest of the above involve only basic operations over the ring of polynomials with rational coefficients.

Hence, it is possible to implement the algorithm **RUR_Square** exactly by the use of multi-precision rational arithmetic in order to obtain the *exact* RUR.

3.2. Root Bound Approach to Exact Computation. In this section, we describe our method to support exact computation for the coordinates of the roots of a system, expressed in the exact RUR. Note that in general these numbers are complex. We begin by summarizing the method for finding root bounds for real algebraic numbers proposed in [**BFMS99**] and [**KLY99**]. We then extend this method to find bounds for complex algebraic numbers. Finally, we describe our method to determine the sign of numbers expressed in the exact RUR.

In this section, we assume for simplicity that all the polynomials have integer coefficients. The results are still valid for polynomials with rational coefficients.

3.2.1. *Root Bounds.* Let α be an algebraic number. There exists a positive real number ρ, called a *root bound* ρ for α, which has the following property: $\alpha \neq 0 \Leftrightarrow |\alpha| \geq \rho$. Having a root bound for α, we can determine the sign of α by computing an approximation $\widetilde{\alpha}$ for α s.t.$|\widetilde{\alpha} - \alpha| < \frac{\rho}{2}$, namely, $\alpha = 0$ iff $|\widetilde{\alpha}| < \frac{\rho}{2}$.

We use the root bound introduced by Mignotte [**MŞ99**]: define the Mahler measure (or simply the measure) $M(e)$ of a polynomial $e(T) = e_n \prod_{i=1}^{n} (T - \zeta_i) \in \mathbb{Z}[T]$ with $e_n \neq 0$ as

$$M(e) = |e_n| \prod_{i=1}^{n} \max\{1, |\zeta_i|\}.$$

Moreover, define the *degree* $\deg \alpha$ and *measure* $M(\alpha)$ of an algebraic number α to be the degree and measure of a minimal polynomial for α over $\mathbb{Z}$. Since, over $\mathbb{Z}$, a minimal polynomial for an algebraic number is uniquely determined up to a sign, the degree and measure of α are well-defined. Then, $\frac{1}{M(\alpha)} \leq |\alpha| \leq M(\alpha)$.

Note we don't compute $M(\alpha)$ explicitly, but rather an upper bound for $M(\alpha)$. In `LEDA` and `Core`, tighter bounds are used where case-by-case analysis can find the most effective (known) root bound for each specific case. We currently use a more general (but looser) bound for flexibility.

3.2.2. *Exact Computation for Real Algebraic Numbers.* The root bound approach to exact sign determination for real algebraic numbers has been implemented in the libraries `LEDA` [**MN99**] and `Core` [**KLY99**]. These libraries support exact arithmetic and comparison over real algebraic numbers of the form $e(\xi_1, \dots, \xi_m)$ where e is an expression involving $+, -,\ /$ and $\sqrt[k]{\ }$, and each of $\xi_1, \dots, \xi_m$ is a rational number. Recent extensions [**BFM$^+$01, PY03**] allow each of ξ_i's to be a real root of some univariate polynomial with integer coefficients. To determine whether or not $e = 0$, we compute an approximation $\widetilde{e}$ for e to enough precision s.t. the root bound guarantees the sign.

The root bounds implemented in `LEDA` and `Core` are "constructive", i.e., they are efficiently calculated without actually computing minimal polynomials for numbers, typically using some norms for minimal polynomials. They established recursive rules to bound the degree and measure of the minimal polynomial for a real algebraic number $f(\xi_1, \dots, \xi_m) \circ g(\xi_1, \dots, \xi_m)$ from those for $f(\xi_1, \dots, \xi_m)$ and $g(\xi_1, \dots, \xi_m)$ where f and g are some expressions and $\circ$ is some operator.

`LEDA` and `Core` use *precision-driven computation* [**DY95, BFMS99**] to compute an approximation $\widetilde{e}$ for e to prescribed precision p. Suppose e is a number of the form $e_1 \circ e_2$. Precision-driven computation is applied to e as follows; First, calculate precisions p_1 and p_2 to which e_1 and e_2 will be approximated. Then, compute approximations $\widetilde{e_1}$ and $\widetilde{e_2}$ for e_1 and e_2 to precision p_1 and p_2, respectively. Finally, compute $\widetilde{e_1} \circ \widetilde{e_2}$ to obtain $\widetilde{e}$.

3.2.3. *Exact Computation for Complex Algebraic Numbers.* We want to determine the exact sign of the coordinates of points expressed in terms of polynomials

h_i's evaluated at roots of the polynomial h. In general, roots of h are not real. Thus, the root bounds approach to exact sign determination for real algebraic numbers must adapt to an exact zero-test for complex algebraic numbers.

Let $e(\zeta_1, \ldots, \zeta_m)$ be a complex algebraic number where e is given as an expression involving $+, -, \cdot$ and $/$, and $\zeta_1, \ldots, \zeta_m$ are complex algebraic numbers. In order to test whether or not $e = 0$, we apply recursive rules to the real algebraic numbers

$$e_R(\Re\zeta_1, \ldots, \Re\zeta_m, \Im\zeta_1, \ldots, \Im\zeta_m) \quad \text{and} \quad e_I(\Re\zeta_1, \ldots, \Re\zeta_m, \Im\zeta_1, \ldots, \Im\zeta_m)$$

where e_R and e_I are expressions satisfying

$$\begin{aligned} e(\zeta_1, \ldots, \zeta_m) &= e_R(\Re\zeta_1, \ldots, \Re\zeta_m, \Im\zeta_1, \ldots, \Im\zeta_m) \\ &\quad + \sqrt{-1} e_I(\Re\zeta_1, \ldots, \Re\zeta_m, \Im\zeta_1, \ldots, \Im\zeta_m). \end{aligned}$$

For example, if $e(\zeta) = \zeta^2$ then $e_R(\Re\zeta, \Im\zeta) = (\Re\zeta)^2 - (\Im\zeta)^2$ and $e_I(\Re\zeta, \Im\zeta) = 2\Re\zeta\Im\zeta$.

In order to complete the adaption, the base cases for recursion must be treated. Thus, our remaining task is stated as follows:

Let ζ be an algebraic number specified as a root of a univariate polynomial e with integer coefficients. For real numbers $\Re\zeta$ and $\Im\zeta$, we would like to compute 1) "constructive" bounds for the degrees and measures of $\Re\zeta$ and $\Im\zeta$ and 2) approximations to any prescribed precision.

For 2), Aberth's method [**Abe73**] is used to compute approximations for the (real and imaginary parts of) roots of univariate polynomials. The method is implemented with floating point numbers with multi-precision mantissa in order to obtain an approximation to any given precision.

For 1), we first find univariate polynomials (with integer coefficients) R_e and I_e such that $R_e(\Re\zeta) = I_e(\Im\zeta) = 0$. We then calculate bounds on the degrees and measures of $\Re\zeta$ and $\Im\zeta$ from the degrees and coefficients of R_e and I_e.

PROPOSITION 3.3. *Let ζ be an algebraic number specified as a root of a polynomial $e(T) \in \mathbb{Z}[T]$. Write $\mathrm{SylRes}_U(f, g)$ for the Sylvester resultant of univariate polynomials f and g w.r.t. variable U. Then*

(1) *$\Re\zeta$ is a real algebraic number and a root of*

$$R_e(T) = \sum_{i=0}^{m} 2^i s_i T^i \quad \in \quad \mathbb{Z}[T]$$

where $\sum_{i=0}^{m} s_i T^i = \mathrm{SylRes}_U(e(T-U), e(U))$.

(2) *$\Im\zeta$ is a real algebraic number and a root of*

$$I_e(T) = \sum_{j=0}^{\lfloor \frac{m}{2} \rfloor} 2^{2j} (-1)^j s_{2j} T^{2j} \quad \in \quad \mathbb{Z}[T]$$

where $\sum_{i=0}^{m} s_i T^i = \mathrm{SylRes}_U(e(T+U), e(U))$.

PROOF. Recall that, for f and $g \in \mathbb{Z}[T]$, if α and β are roots of f and g, respectively, then $\alpha \pm \beta$ is a root of $\mathrm{SylRes}_U(f(T \mp U), g(U))$.

If ζ is a root of e then its complex conjugate $\overline{\zeta}$ is also a root of e. Thus, the sum $\zeta + \overline{\zeta} = 2\Re\zeta$ of two roots of e is a root of $\mathrm{SylRes}_U(e(T-U), e(U))$.

Similarly, the difference $\zeta - \overline{\zeta} = 2\sqrt{-1}\Im\zeta$ of two roots of e is a root of $\mathrm{SylRes}_U\left(e\left(T+U\right), e\left(U\right)\right)$.

If 2ξ is a root of $\sum_{i=0}^{m} s_i T^i$ then ξ is a root of $\sum_{i=0}^{m} 2^i s_i T^i$.

If, for $\xi \in \mathbb{R}$, $\sqrt{-1}\xi$ is a root of $\sum_{i=0}^{m} s_i T^i$ then ξ is a root of $\sum_{j=0}^{\lfloor \frac{m}{2} \rfloor} (-1)^j s_{2j} T^{2j}$.

Putting these together, the statements in the proposition hold. □

By Gauß's lemma, if α is a root of a polynomial $e\left(T\right) = \sum_{i=0}^{n} e_i T^i \in \mathbb{Z}\left[T\right]$ with $e_n e_0 \neq 0$ then $\deg \alpha \leq \deg e$ and $M\left(\alpha\right) \leq M\left(e\right)$. By Landau's theorem [**MŞ99**], for $e\left(T\right) \in \mathbb{Z}\left[T\right]$, $M\left(e\right) \leq ||e||_2 = \sqrt{\sum_{i=0}^{n} |e_i|^2}$. Thus, we could use $\deg e$ and $||e||_2$ as "constructive" upper bounds on $\deg \alpha$ and $M\left(\alpha\right)$.

PROPOSITION 3.4. *Following the notation above*

(1) $\deg \Re\zeta \leq \deg R_e \leq \deg^2 e$,
$M\left(\Re\zeta\right) \leq M\left(R_e\right) \leq ||R_e||_2 \leq 2^{2n^2+n} ||e||_2^{2n}$,

(2) $\deg \Im\zeta \leq \deg I_e \leq \deg^2 e$ *and*
$M\left(\Im\zeta\right) \leq M\left(I_e\right) \leq ||I_e||_2 \leq 2^{2n^2+n} ||e||_2^{2n}$.

PROOF. (1) It can be shown [**Yap00**] that, for f and $g \in \mathbb{Z}\left[T\right]$,

$$\deg \mathrm{SylRes}_T\left(f\left(T\right), g\left(T\right)\right) \leq \deg f \deg g \quad \text{and}$$
$$||\mathrm{SylRes}_U\left(f\left(T-U\right), g\left(U\right)\right)||_2 \leq \left(2^{\deg f+1} ||f||_2\right)^{\deg g} ||g||_2^{\deg f}.$$

Thus

$$m = \deg \sum_{j=0}^{m} s_j T^j = \deg \mathrm{SylRes}_U\left(e\left(T-U\right), e\left(U\right)\right) \leq \deg^2 e = n^2$$

and

$$\begin{aligned} \left|\left|\sum_{j=0}^{m} s_j T^j\right|\right|_2 &= ||\mathrm{SylRes}_U\left(e\left(T-U\right), e\left(U\right)\right)||_2 \\ &\leq \left(2^{\deg e+1} ||e||_2\right)^{\deg e} ||e||_2^{\deg e} \\ &= 2^{n^2+n} ||e||_2^{2n}. \end{aligned}$$

Hence

$$\begin{aligned} \left|\left|\sum_{j=0}^{m} 2^j s_j T^j\right|\right|_2 &= \sqrt{\sum_{j=0}^{m} 2^{2j} |s_j|^2} \\ &\leq \sqrt{\sum_{j=0}^{m} 2^{2m} |s_j|^2} \\ &= 2^m \sqrt{\sum_{j=0}^{m} |s_j|^2} \\ &= 2^m \left|\left|\sum_{j=0}^{m} s_j T^j\right|\right|_2 \\ &\leq 2^{n^2} \left(2^{n^2+n} ||e||_2^{2n}\right). \end{aligned}$$

The proof for (2) is similar. □

3.2.4. *Exact Sign Determination for the Exact RUR.* Let e be a rational function in n variables $X_1, \ldots, X_n$ with rational coefficients. Also, let Z' be a finite set of n-dimensional algebraic numbers. Assume we have univariate polynomials $h, h_1, \ldots, h_n$ with integer coefficients s.t. for every point $\left(\zeta_1, \ldots, \zeta_n\right) \in Z'$, $\left(\zeta_1, \ldots, \zeta_n\right) = \left(h_1\left(\theta\right), \ldots, h_n\left(\theta\right)\right)$ for some root θ of h. We would like to determine whether or not $e\left(\zeta_1, \ldots, \zeta_n\right) = 0$ exactly.

Algorithm Exact_Sign:
Input: $e \in \mathbb{Q}\left(X_1, \ldots, X_n\right)$ and $h, h_1, \ldots, h_n \in \mathbb{Z}\left[T\right]$.

Output: The exact signs of the real and imaginary parts of $e(h_1(\theta), \ldots, h_n(\theta))$ for every root θ of h.

1: Set up bivariate rational functions r_R and r_I with integer coefficients s.t.
$$e(h_1(\theta), \ldots, h_n(\theta)) = r_R(\Re\theta, \Im\theta) + \sqrt{-1}r_I(\Re\theta, \Im\theta).$$

2: Recursively compute the root bounds for r_R and r_I. The base cases are given in Proposition 3.4.

3: Recursively compute approximations for $r_R(\Re\theta, \Im\theta)$ and $r_I(\Re\theta, \Im\theta)$ to a certain precision such that the root bounds allow us to determine their signs. The base case, i.e. computing approximations for $\Re\theta$ and $\Im\theta$ to a certain precision, is done by Aberth's method.

Algorithm **Exact_Sign** is an irregular application of the root bound approach. We must compute approximations of the real and imaginary parts of a number up to their root bounds, which usually costs a lot. However, if the numbers are repeatedly used in an application, this high precision representation can be reused, thus providing a significant advantage.

3.3. Exact Computation for the Exact RUR. In this section, we present our exact algorithm to compute the exact RUR for the affine roots of a system.

3.3.1. *Affine Roots.* Recall that Theorem 3.1 describes only those roots having non-zero coordinates. It can be shown [**Roj99, Roj00**] that, if $A_1, \ldots, A_n$ are replaced by $\{\mathbf{o}\} \cup A_1, \ldots, \{\mathbf{o}\} \cup A_n$ then the RUR for some finite *superset* Z'' of Z' is found. Those extra points in $Z'' \setminus Z$ can be removed simply by testing, for each $x \in Z''$, whether or not x is a root of the original system, i.e., $f_1(x) = \cdots = f_n(x) = 0$. The last test reduces to the exact sign determination problem: for every root ζ of h in the exact RUR for Z'', test whether or not $f_i(h_1(\zeta), \ldots, h_n(\zeta)) = 0$ for $i = 1, \ldots, n$.

3.3.2. *Non-square Systems.* Consider a system of m polynomials $f_1, \ldots, f_m$ in n variables with rational coefficients. We would like to compute the exact RUR for some finite subset Z' of the set Z of the roots of the system. If $m \neq n$ then we construct some square system s.t. the set of the roots of the square system is a super set of the zero set of the original system.

If $m < n$ then construct a square system by adding $m - n$ copies of f_m to the input system.

Otherwise, $m > n$. Let
$$g_i = c_{i1}f_1 + \cdots + c_{im}f_m, \qquad i = 1, \ldots, n,$$
where $c_{11}, \ldots, c_{nm}$ are generic rational numbers. It can be shown that there exist $c_{11}, \ldots, c_{nm} \in \mathbb{Q}$ s.t. every irreducible component of the set $\overline{Z}$ of all the common roots of $g_1, \ldots, g_n$ is either an irreducible component of the set Z of all the common roots of $f_1 \ldots, f_m$ or a point [**Roj00**]. We have already seen that we can compute the exact RUR for points in some finite subset $\overline{Z}'$ of $\overline{Z}$ s.t. $\overline{Z}'$ contains at least one point from every irreducible component of $\overline{Z}$. Then, we can construct a finite subset Z' of Z containing at least one point on every irreducible component of Z by simply testing, for each $x \in \overline{Z}'$, whether or not x is a root of the original system, i.e., $f_1(x) = \cdots = f_m(x) = 0$. The last test reduces to the exact sign determination problem. In implementations, $c_{11}, \ldots, c_{nm}$ can be chosen randomly. A choice is generic with probability 1.

3.3.3. *The Exact RUR.* Our exact algorithm to compute the RUR is thus as follows:

Algorithm: RUR:
Input: $f_1, \ldots, f_m \in \mathbb{Q}[X_1, \ldots, X_n]$.
Output: $h, h_1, \ldots, h_n \in \mathbb{Q}[T]$, forming the RUR for some finite subset of the roots of the system.
1: Form a square system $(g_1, \ldots, g_n)$.
1-1: If $m < n$ then form a square system

$$g_1 = f_1, \ldots, g_m = f_m, \; g_{m+1} = \cdots = g_n = f_m.$$

1-2: If $m = n$ then $g_i = f_i, \; i = 1, \ldots, n$.
1-3: If $m > n$ then form a square system

$$g_i = c_{i1} f_1 + \cdots + c_{im} f_m, \qquad i = 1, \ldots, n$$

where $c_{11}, \ldots, c_{nm}$ are generic rational numbers.
2: For $i := 1, \ldots, n$ do:
2-1: Set A_i to be the support of g_i. If g_i does not have a constant term then $A_i = \{\mathbf{o}\} \cup A_i$.
3: Use Algorithm **RUR_square** to compute $h, h_1, \ldots, h_n \in \mathbb{Q}[T]$ forming the RUR for some finite subset $\overline{Z}'$ of the set $\overline{Z}$ of the affine roots of $g_1, \ldots, g_n$.
4: If steps 1-3 and/or 2-1 are executed then use Algorithm **Exact_Sign** to test whether or not

$$f_i(h_1(\theta), \ldots, h_n(\theta)) = 0, \qquad i = 1, \ldots, n$$

for every root θ of h.

Exactness immediately follows from the fact that both **RUR_square** and **Exact_Sign** are implemented exactly.

4. Applications

In this section, we describe some applications of the RUR to various geometric problems. A number of geometric problems involve solving systems of polynomials, and thus algebraic number computations. We first present some algorithms to support exact computation over algebraic numbers, and then focus on how those can be used in specific situations that arise during boundary evaluation in CAD applications.

4.1. Exact Computation for Algebraic Numbers.

4.1.1. *Positive Dimensional Components.* We present a generic algorithm to determine whether or not the set of roots of a given system of polynomials with rational coefficients has positive dimensional components.

Recall that Algorithm **RUR** finds the exact RUR for some finite subset Z' of Z which contains at least one point from every irreducible component of Z. If Z is infinite (i.e., has positive-dimensional components) then Z' depends on the polynomials $f_1^*, \ldots, f_n^*$ used to perturb the input system.

Suppose two distinct executions of Algorithm **RUR** find two finite subsets Z_1' and Z_2' of Z and their exact RUR's:

$$Z_k' = \left\{ \left(h_1^{(k)}(\theta_k), \ldots, h_n^{(k)}(\theta_k) \right) \mid h^{(k)}(\theta_k) = 0 \right\}, k = 1, 2.$$

If ζ is an isolated point in Z then $\zeta \in Z_1' \cap Z_2'$, and thus $\exists \theta_1$ and $\theta_2 \in \mathbb{C}$ s.t.

$$\zeta = \left(h_1^{(1)}(\theta_1), \ldots, h_n^{(1)}(\theta_1)\right) = \left(h_1^{(2)}(\theta_2), \ldots, h_n^{(2)}(\theta_2)\right).$$

Hence, $Z_1' \setminus Z_2' \neq \emptyset$ implies that Z has some positive dimensional components. We can compare Z_1' and Z_2' pointwise using Algorithm **Exact_Sign**.

Algorithm Positive_Dimensional_Components:
Input: $f_1, \ldots, f_m \in \mathbb{Q}[X_1, \ldots, X_n]$ and a small positive integer `Max_Counts`.
Output: *True* if the set Z of the roots of the system $(f_1, \ldots, f_m)$ has some positive dimensional components.
1: `Count` $:= 0$, `Has_Pos_Dim_Compo` $:=$ *Probably_False*.
2: While `Count` $<$ `Max_Counts` and `Has_Pos_Dim_Compo` $=$ *Probably_False* do:
  2-1: Use Algorithm **RUR** to compute the exact RUR for some finite subsets Z_1' and Z_2' of Z. Increment `Count`.
  2-2: Use Algorithm **Exact_Sign** to compare Z_1' and Z_2' pointwise. If they are not the same then `Has_Pos_Dim_Compo` $:=$ *True*.

If Z has a positive dimensional component and polynomials f_i^* in Algorithm **RUR** are chosen generically, then almost always $Z_1' \setminus Z_2' \neq \emptyset$. Thus, `Max_Counts` is usually set to be 2.

4.1.2. *Root Counting.* We present an algorithm to count the number of the common roots of a given system of polynomials with rational coefficients.

Algorithm **Positive_Dimensional_Components** detects whether or not the set of the roots has positive-dimensional components. If the system has only finitely many roots then the number of roots of the system is the same as the number of the roots of the univariate polynomial h in the RUR.

Algorithm: Root Counting:
Input: $f_1, \ldots, f_m \in \mathbb{Q}[X_1, \ldots, X_n]$.
Output: The number of the roots of the system $(f_1, \ldots, f_m)$.
1: Use Algorithm **Positive_Dimensional_Components** to test whether or not the set Z of the roots of the system $(f_1, \ldots, f_m)$ has some positive dimensional components. If so then return ∞.
2: Otherwise, we may assume Z is finite. At step 1 of Algorithm **Positive_Dimensional_Components**, we computed the exact RUR for Z. Return the number of roots of the polynomial h in the RUR for Z, namely, $\deg h$.

4.1.3. *Real Roots.* We present an algorithm to compute the real roots of a given system of polynomials with rational coefficients.

Assume that the system has only finitely many roots. We first compute the exact RUR for the roots of the system, and then use Algorithm **Exact_Sign** to determine the sign of the imaginary parts of the roots.

Algorithm Real_Roots:
Input: $f_1, \ldots, f_m \in \mathbb{Q}[X_1, \ldots, X_n]$.
Output: The set R of the real roots of the system $(f_1, \ldots, f_m)$. Each root is represented in the RUR as well as an interval of any given precision that contains 1 and only 1 root.

1: Use Algorithm **RUR** to compute the exact RUR for the set Z of the roots of the system:

$$Z = \{(h_1(\theta), \ldots, h_n(\theta)) \mid h(\theta) = 0\}.$$

2: Set $R := \emptyset$.
3: For $i := 1, \ldots, n$ do:
3-1: Set up expressions h_{Ri} and h_{Ii} satisfying

$$h_i(\theta) = h_{Ri}(\Re\theta, \Im\theta) + \sqrt{-1} h_{Ii}(\Re\theta, \Im\theta).$$

3-2: Use Algorithm **Exact_Sign** to determine the sign of $h_{Ii}(\Re\theta, \Im\theta)$ at every root θ of h.
3-3: If $h_{I1}(\Re\theta, \Im\theta) = \cdots = h_{In}(\Re\theta, \Im\theta) = 0$ then

$$R := R \cup \{(h_{R1}(\Re\theta, \Im\theta), \ldots, h_{Rn}(\Re\theta, \Im\theta))\}.$$

It is important to note the algorithm **Real_Roots** works correctly only if the system has finitely many common roots. In particular, the algorithm may not be able to find real points lying on some (complex) positive dimensional components.

4.2. Application to Boundary Evaluation. We now give a brief description of how the exact RUR can be applied to a specific geometric computation—boundary evaluation.

4.2.1. *Overview of Exact Boundary Evaluation.* Boundary evaluation is a key operation in computer aided design. It refers to the process of determining the boundary of a solid object produced as the result of an operation - usually a Boolean combination (union, intersection, or difference) of two input solids. It is the key element for conversion from a Constructive Solid Geometry (CSG) format to a Boundary Representation (B-rep) format. Achieving accuracy and robustness with reasonable efficiency remains a challenge in boundary evaluation.

Boundary evaluation involves several stages, but the key fundamental operations involve finding solutions to polynomial systems. The input objects are usually described by a series of rational parametric surfaces described as polynomials with rational coefficients. Implicit forms for these surfaces are often known, or else can be determined. Intersections of these surfaces form the edges of the solids, sometimes represented inside the parametric domain as algebraic plane curves. These curves represented in the patch domain and defined by the intersections of two surfaces are known as either *trimming curves* if they are specified in the input, or *intersection curves* if they arise during boundary evaluation. Intersection curves that are output become trimming curves when input to the next boundary evaluation operation.

Intersections of three or more surfaces form vertices of the solids. Such vertices may be represented in 3D space (as the common solution to three or more trivariate equations), within the parametric domain of the patches (as the common solution of two or more bivariate equations), or a combination of these. The coordinates of these vertices are thus tuples of real algebraic numbers. The accuracy, efficiency, and robustness of the entire boundary evaluation operation is usually a direct result of the accuracy, efficiency, and robustness of the computations used to find and work with these algebraic numbers. Determining the signs of algebraic expressions evaluated at algebraic numbers thus becomes key to performing the entire computation.

4.2.2. *ESOLID and Exact Boundary Evaluation.* The ESOLID system was created in order to perform exact boundary evaluation [**KCF$^+$02, KCF$^+$04**]. ESOLID uses exact representations and computations throughout to guarantee accuracy and eliminate robustness problems due to numerical error (e.g. roundoff error and its propagation). ESOLID finds and represents points in the 2D domain only, using the MAPC library [**KCMK00**]. This 2D representation is sufficient for performing all of boundary evaluation. Though significantly less efficient than an equivalent floating-point routine, it runs at "reasonable" speed—at most 1-2 orders of magnitude slower than an inexact approach on real-world data. Unfortunately, ESOLID is designed to work only for objects in general position where infinitesimal perturbations of the data would not change the nature of intersections. Overcoming this limitation has been a major motivator of the work presented here.

The 2D vertex representation (coupled with techniques that produce bounding intervals in 3D) is sufficient for performing the entire boundary evaluation computation. MAPC represents points (with real algebraic coordinates) using a set of bounding intervals, guaranteed to contain a unique root of the defining polynomials. These intervals are found by using the Sylvester resultant and Sturm sequences to determine potential x and y values of curve intersections. Then, a series of tests are performed to find which x coordinates belong with which other y coordinates, thus forming an isolating interval. These isolating intervals can be reduced in size on demand as necessary. One nice aspect of this approach is that only roots in a particular region of interest are found, eliminating work that might be performed for roots outside of that region. Generally, a lazy evaluation approach is used; intervals are reduced only as necessary to guarantee sign operations. Often they need to be reduced far less than worst-case root bounds would indicate.

4.2.3. *Incorporating the Exact RUR.* The exact RUR could be incorporated directly into a boundary evaluation approach by following the ESOLID framework, but replacing the MAPC point representations with RUR representations. Only 2D point representations would be necessary. The following information would then be kept at each point:

(1) The exact RUR: the univariate polynomials h, h_1 and h_2 with rational coefficients.
(2) A bound on a unique (complex) root of h.
(3) A bounding interval determined by evaluating the root of h within h_1 and h_2.
(4) The two original polynomials used to compute the exact RUR.

While only the first two items are necessary to exactly specify the root, the other information allows faster computation in typical operations.

Points are isolated using **Real Roots**. Whenever we need to use the points, the computation is phrased as an algebraic expression in which that point must be evaluated. From this, we can obtain root bounds for the algebraic numbers, and thus the precision to which we must determine the root of h. In doing so, we guarantee accurate sign evaluation at any point in the boundary evaluation process.

Note that in practice, due to the relatively slower performance of the RUR method compared to the MAPC method, it is likely that a hybrid representation should be used instead. In particular, one would like to use the MAPC approach where it is most applicable, and the RUR where it is most applicable. Specifically, the RUR seems better suited for degenerate situations.

4.3. Degeneracy Detection in Boundary Evaluation. As stated previously, degeneracies are a major remaining obstacle in our exact boundary evaluation implementation. The exact RUR is able to cope with degenerate situations, however, and thus is more well-suited for detecting when a degenerate situation is occurring.

We outline here how the exact RUR method can be applied to detect each of the common degenerate situations that arise in boundary evaluation. Space does not permit a detailed review of how these cases fit within the larger boundary evaluation framework, which requires significantly more computation than that outlined here. See [**Key00**] for a more thorough discussion of the ways that each of these degeneracies arise and manifest themselves within boundary evaluations. We focus here on degeneracies that can be easily detected using the exact RUR method. Certain other degenerate situations are easily detected by simpler, more direct tests. For example, overlapping surfaces are immediately found in the vanishing of an intersection curve.

We focus here only on degeneracy detection, as there is more than one way one might want to handle the degeneracy (e.g. numerical perturbation, symbolic perturbation, special-case treatment).

4.3.1. *General Degeneracy Consideration.* The exact RUR method is particularly well-suited for problems that arise in degenerate situations. Unlike methods created under genericity assumptions (e.g. [**KCMK00**]), the exact RUR will find roots for systems of equations, even when they are in highly degenerate configurations. This makes it highly useful for treating degeneracies, beyond just detecting them.

An example from boundary evaluation is when intersection curves contain singularities (e.g. self-intersections, isolated point components, cusps). There are some approaches that can deal with these curves [**Sak91, Hon96, GVN02, Wol03, EKSW04**]. However, methods such as that in [**KCMK00**] fail in such situations, while the exact RUR approach is perfectly capable of finding, e.g. the intersection between two curves at a self-intersection of one of the curves.

Note that in a practical sense, the exact RUR might not have been used to represent all points to begin with, due to its relatively poorer performance for generic cases (see section 5.2). Instead, the exact RUR might be used to check for coincidence only when it seems likely that a degenerate situation is occurring. The best approach for pursuing such a combination is a subject for future study, but a common operation that would be necessary in such cases would be the conversion of a point given in the existing system to one expressed as the exact RUR. We will limit our discussion to the 2D case, as this is what is necessary for boundary evaluation.

Degeneracies are detected during the boundary evaluation process by checking for irregular interactions. They occur when one of the following happens in 2D:

(1) Three or more curves intersect at a single point. This includes the cases of singularities (where the curve and its two derivative curves intersect at a common point).
(2) Two curves intersect tangentially.
(3) Two curves share a positive-dimensional component.

All of these cause problems for a generic system solver (potentially resulting in crashes), and ESOLID in particular.

The exact RUR will find roots for systems of equations, even when they are in highly irregular configurations. The exact RUR can be used to detect these situations:

(1) *Three or more curves meeting at a point.*
Algorithm **Real_Roots** can be used to compute intersections of the system of curves. Recall that the exact RUR method works for curves having singularities. Algorithm **Exact_Sign** can also be used if several of the curves have already been intersected (generically). In that case, the test is for whether that point lies on another curve.
(2) *Tangential intersections.*
Algorithm **Real_Roots** will again be able find these intersections.
(3) *Positive-dimensional components.*
Algorithm **Positive_Dimensional_Components** will be able to determine whether there is a positive dimensional component. In general, though, other methods may be more efficient for detecting this degeneracy, in particular, in 2D.

To detect that a particular degenerate situation has occurred, the exact RUR-based method is combined with knowledge gained from an attempt to locate a generic solution. For example, the generic approach used to find roots of a pair of curves in ESOLID can return a set of points that are either: 1) guaranteed to be transverse intersections of the curves, or 2) to be either a tangential intersection of the curves or no intersection at all [**KCMK00, KCF$^+$04**]. In the second case, the RUR approach can be used to determine whether or not there is indeed a tangential intersection. Similar approaches can be used for other degeneracies; the key is that the exact RUR approach allows us to compute the intersection points where a generic approach would fail (or return an ambiguous result).

5. Experimental Results

We have implemented the exact RUR method in GNU C++. We use the Gnu Multiple Precision (GMP) arithmetic library. All the experiments shown in this section are performed on a 3 GHz Intel Pentium CPU with 6 GB memory.

We present examples here in order to indicate the time requirements of various parts of the exact RUR implementation, as well as to understand the time requirement in comparison to competing methods. These are *not* meant to provide a comprehensive survey of all potential cases or to show the range of cases we can handle (e.g. higher degree curves can be dealt with, albeit somewhat inefficiently), but rather to give greater intuitive insight into the exact RUR procedure.

Concerning the worst case complexity, our toric resultant approach should have better performance than those based on Gröbner bases. Both have complexity exponential in n, but, the Gröbner basis method may have quadratic exponents in the worst case. However, informal comparisons with the fastest known Gröbner basis implementation, GBRS [**RZ04**], indicate that our implementation is significantly slower. There are many possible contributing factors (e.g. GBRS is highly optimized in memory management, while our implementation uses GMP straightforwardly), and optimizations to our code (we have almost none, currently) would narrow the gap. However, it must be kept in mind that alternative Gröbner basis approaches could provide a more efficient approach in all practical cases. One key

Input System	F_1	F_2	F_3	F_4	F_5	F_6	F_7	F_8
# polynomial	2	2	2	3	2	2	2	2
# variable	2	2	2	3	2	2	2	2
max deg of monomials	4	3	3	2	2	2	2	2
max len of coeff	4	2	2	2	309	402	54	97
# roots of system	∞	2	1	∞	2	2	2	4
# roots of h	4	2	1	4	2	4	2	2
# real roots of system	2	2	1	-	2	2	0	4
max len of coeff of h_i	111	5	6	19	364	311	103	451
total time (sec)	2.59	0.0865	0.103	269.0	0.0926	0.104	0.183	0.687
% res matrix	0.8	19.7	18.8	0.1	14.1	12.5	11.8	3.1
% compute h	26.4	15.3	16.2	15.1	14.6	15.3	17.4	5.5
% compute $q_i^{\pm}$	70.4	60.7	64.4	84.8	58.0	59.6	67.5	22.5
% compute h_i	2.2	2.0	0.0	0	10.0	9.8	1.8	67.9
% approx roots of h	0.1	0.5	0.3	0	0.9	0.8	0.3	0.4
% evaluate h_i	0.1	1.8	0.3	0	2.4	2.0	1.2	0.6

TABLE 1. **Timing breakdown for several examples.**

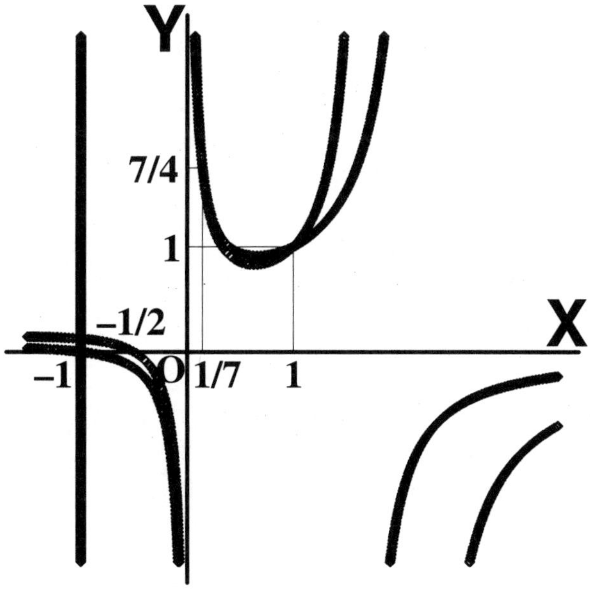

FIGURE 1. F1

advantage of our method, however, is that the Gröbner basis is discontinuous w.r.t. changes in the coefficients of the input polynomials.

5.1. Solving Systems in the Exact RUR. In this section, we show timing breakdowns for the application of the exact RUR to a few sample systems. We give a brief discussion of each case (a detailed description in one instance), and summarize the results.

5.1.1. *Details of Examples.* F_1: **Positive Dimensional Components**

Consider a system F_1 of two polynomials

$$\begin{aligned} f_1 &= 1 + 2X - 2X^2Y - 5XY + X^2 + 3X^3Y, \\ f_2 &= 2 + 6X - 6X^2Y - 11XY + 4X^2 + 5X^3Y \end{aligned}$$

The roots of the system F_1 are the points $\left\{(1,1), \left(\frac{1}{7}, \frac{7}{4}\right)\right\}$ and the line $X = -1$.

We give a detailed description of Algorithm **Positive_Dimensional_Components** here as illustration.

Since F_1 is a square system, Algorithm **RUR** immediately calls Algorithm **RUR_square**.

If we choose $u_1 = \frac{9}{2}$, $u_2 = \frac{17}{2}$,

$$f_1^* = 17X^3Y + 74X^2Y + 94, \quad \text{and} \quad f_2^* = 105X^3Y + 93X^2 + 98XY,$$

at step 2-1, then we obtain the exact RUR for some finite subset Z_1' of the roots of F_1 as

$$\begin{aligned} h^{(1)}(T) &= -284032T^4 - 4928704T^3 + 261419107T^2 + 4401863937T - 1411684417, \\ h_1^{(1)}(T) &= \frac{7689907930880}{5870227699098039}T^3 + \frac{1037528584832}{124898461682937}T^2 \\ &\quad - \frac{1313640082212220}{5870227699098039}T - \frac{851151731612679}{1956742566366013} \\ h_2^{(1)}(T) &= -\frac{23069723792640}{33264623628222221}T^3 - \frac{3112585754496}{707757949536643}T^2 \\ &\quad + \frac{27435113904634}{33264623628222221}T + \frac{7660365584514111}{33264623628222221}. \end{aligned}$$

If we choose $u_1 = \frac{9}{2}$, $u_2 = \frac{17}{2}$,

$$f_1^* = 17X^3Y + 110 + 112X^2Y, \quad \text{and} \quad f_2^* = 58X^3Y + 50X^2 + 63XY,$$

at step 2-1, then we obtain the exact RUR for some finite subset Z_2' of the roots of F_1 as

$$\begin{aligned} h^{(2)}(T) &= -278656T^4 - 4324928T^3 + 381956827T^2 + 4775312157T - 1790337263 \\ h_1^{(2)}(T) &= \frac{23827150611712}{20005826573410785}T^3 + \frac{21652317350528}{4001165314682157}T^2 \\ &\quad - \frac{4673278548948724}{20005826573410785}T - \frac{6694752830187523}{20005826573410785} \\ h_2^{(2)}(T) &= -\frac{23827150611712}{37788783527553705}T^3 - \frac{21652317350528}{7557756705510741}T^2 \\ &\quad + \frac{227539310412994}{37788783527553705}T + \frac{6694752830187523}{37788783527553705} \end{aligned}$$

We test whether or not the system $Z_1' \cap Z_2' = \emptyset$ using Algorithm **Exact_Sign**. For $i = 1, 2$, construct expressions r_{Ri} and r_{Ii} s.t.

$$h_i^{(1)}(\theta) - h_i^{(2)}(\theta) = r_{Ri}(\Re\theta, \Im\theta) + r_{Ii}(\Re\theta, \Im\theta).$$

The root bounds for r_{R1}, r_{I1}, r_{R2} and r_{I2} are all smaller than 64 bits. Now, we apply precision-driven computation to approximate the coordinates of the roots to the (absolute) precision 128 bits. We list the values of the real and imaginary parts of $h_i^{(1)}(\theta)$ for $i = 1, 2$:

((	$\Re h_1^{(1)}(\theta)$,	$\Im h_1^{(1)}(\theta)$	),	(	$\Re h_2^{(1)}(\theta)$,	$\Im h_2^{(1)}(\theta)$	))
((	-1,	$-7.329 * 10^{-54}$	),	(	-0.428,	$8.903 * 10^{-54}$	))
((	1,	$4.222 * 10^{-52}$	),	(	1,	$-4.449 * 10^{-52}$	))
((	0.1429,	$5.432 * 10^{-52}$	),	(	1.75,	$-4.250 * 10^{-52}$	))
((	-1,	$3.770 * 10^{-54}$	),	(	0.173,	$1.231 * 10^{-54}$	))

and the values of the real and imaginary parts of $h_i^{(2)}(\theta)$ for $i = 1, 2$:

((	$\Re h_1^{(2)}(\theta)$,	$\Im h_1^{(2)}(\theta)$	),	(	$\Re h_2^{(2)}(\theta)$,	$\Im h_2^{(2)}(\theta)$	))
((	-1,	$-1.299 * 10^{-51}$	),	(	-0.614,	$1.418 * 10^{-51}$	))
((	1,	$1.022 * 10^{-49}$	),	(	1,	$-1.065 * 10^{-49}$	))
((	0.1429,	$-4.076 * 10^{-50}$	),	(	1.75,	$3.203 * 10^{-50}$	))
((	-1,	$2.030 * 10^{-51}$	),	(	0.144,	$4.200 * 10^{-52}$	))

For each of these, the imaginary component is small enough to indicate that the root is real. Furthermore, points

$$\left(\left(1, 4.222 * 10^{-52}\right), \left(1, -4.449 * 10^{-52}\right)\right)$$

and

$$\left(\left(1, 1.022 * 10^{-49}\right), \left(1, -1.065 * 10^{-49}\right)\right)$$

are identical because their difference is smaller than the root bounds for r_{R1} and r_{I1} (and we know $(1, 1)$ is a root). Similary, points

$$\left(\left(0.1429, 5.432 * 10^{-52}\right), \left(1.75, -4.250 * 10^{-52}\right)\right)$$

and

$$\left(\left(0.1429, -4.076 * 10^{-50}\right), \left(1.75, 3.203 * 10^{-50}\right)\right)$$

are identical because their difference is smaller than the root bounds for r_{R2} and r_{I2}. On the other hand, the other pairs are distinct. Note also the other roots found in the two iterations, while different from iteration to iteration, are consistent with the positive dimensional component, $X = -1$ (although we cannot actually determine this component). Thus, we conclude that the zero set of F_1 consists of 2 isolated points and some positive dimensional components.

F_2 and F_3: Singularities

Consider the system F_2 of two polynomials

$$\begin{aligned} f_{21} &= X^3 - 3X^2 + 3X - Y^2 + 2Y - 2, \\ f_{22} &= 2X + Y - 3 \end{aligned}$$

An elliptic curve $f_{21} = 0$ has a cusp at $(1, 1)$ and intersects with the line $f_{22} = 0$ at this point.

We obtain the exact RUR for the zero set of F_2 as

$$\begin{aligned} h(T) &= -T^2 - 5424051198T - 6426597229148218232, \\ h_1(T) &= -\frac{2}{963579587}T - \frac{2533312437}{963579587}, \\ h_2(T) &= \frac{4}{963579587}T + \frac{7957363635}{963579587} \end{aligned}$$

which immediately implies that F_2 has 2 real roots. One of these is $(1, 1)$.

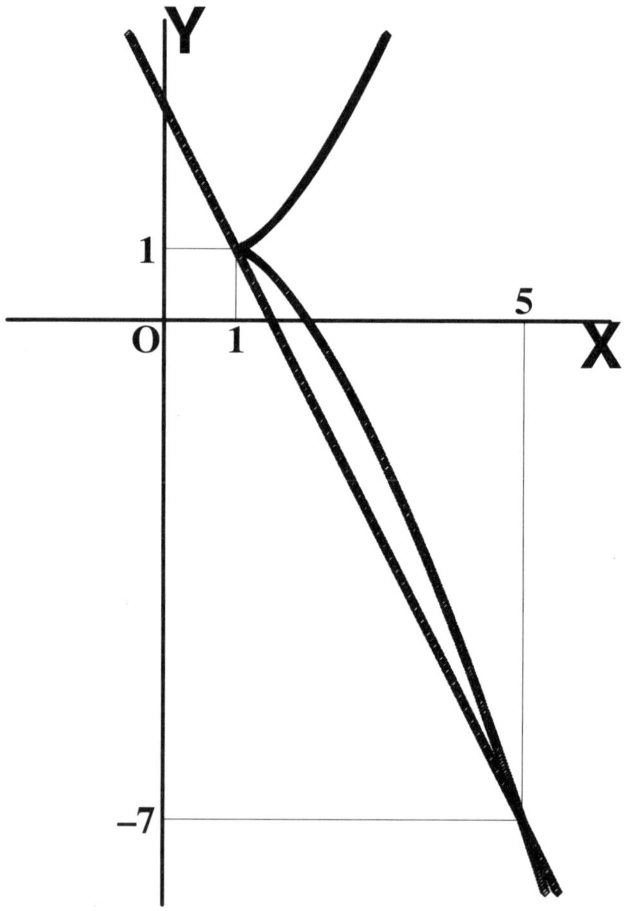

FIGURE 2. F2

Consider the system F_3 of two polynomials

$$\begin{aligned} f_{31} &= X^3 - 3X^2 - 3XY + 6X + Y^3 - 3Y^2 + 6Y - 5, \\ f_{32} &= X + Y - 2 \end{aligned}$$

Folium of Descartes $f_{31} = 0$ has a self-intersection at $(1, 1)$ and intersects with the line $f_{32} = 0$ at this point.

We obtain the exact RUR for the zero set of F_2 as

$$\begin{aligned} h(T) &= T + 3284000871, \\ h_1(T) &= -T - 3284000870, \\ h_2(T) &= -T - 3284000870 \end{aligned}$$

which immediately implies that F_3 has only 1 real root including $(1, 1)$.

Because the intersections of both systems are at singular points, we cannot apply methods, such as MAPC, that depend on general position.

F_4: Complex Positive Dimensional Components

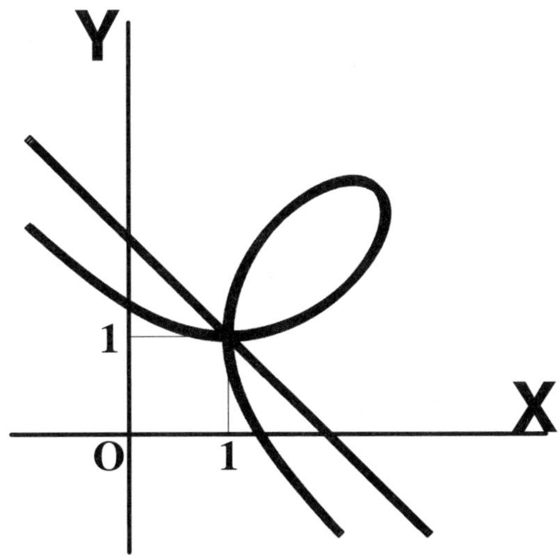

FIGURE 3. F3

Consider the system F_4 of three polynomials

$$\begin{aligned} f_{41} &= Z-1, \\ f_{42} &= X^2+Y^2+Z^2-4Z+3, \\ f_{43} &= X^2+Y^2+Z^2-1 \end{aligned}$$

It is easy to see that the only real point on the zero set of F_4 is $(0,0,1)$. However, the exact RUR for the zero set of F_4 consists of 4 complex points each of which lies on some complex positive dimensional component (satisfying $X^2+Y^2=0, Z=1$). In general, there is no clear way to extract finitely many real points that lie on these complex positive dimensional components.

F_5 through F_8

Cases F_5 through F_8 are all drawn from cases encountered in an actual boundary evaluation computation. The source data is real-world data provided from the BRL-CAD [**DM89**] solid modeling system.

F_5 and F_6

Both F_5 and F_6 consist of an intersection of a line with an ellipse. Both have 2 roots and all roots are real.

F_7: No Real Roots

The system F_7 consists of two ellipses. Rather than real intersections, these ellipses have 2 complex intersections. **Real_Roots** is used to distinguish real roots from the other roots.

F_8: Burst of Coefficients of RUR

The system F_8 consists of two ellipses. F_8 has 4 roots and all of them are real.

For this example, we spent the most time computing polynomials h_1 and h_2. This was because the coefficients of the polynomial h, h_1 and h_2 become huge. This

Input System	F_5	F_6	F_7	F_8
# real roots by RUR	2	2	0	4
# roots of h	2	2	2	4
# real roots by MAPC	2	1	0	2
total time by RUR (sec)	0.0926	0.104	0.183	0.687
total time by MAPC (sec)	0.017	0.015	0.024	0.017

TABLE 2. **Timings to Solve Systems F_5, F_6, F_7, F_8 by RUR and ESOLID/MAPC in seconds**

also slows down the following step which iteratively approximates the roots of the polynomial h.

5.1.2. *Summary of Timing Breakdowns.* While the examples shown above are not comprehensive, from these and other cases we have examined, we can draw the following conclusions:

- The performance of the method is reasonable for lower dimension/degree systems. However, for higher dimension/degree systems (e.g. F_4), the method is not practical.
- For lower dimension/degree systems, the most time consuming part of the algorithm is repeated computation of the determinant of the sparse resultant matrix. In fact, the size of the toric resultant matrix grows quite rapidly with respect to the dimension/degree of the system in its current implementation.
- For higher dimension/degree systems, the most time consuming part is computing univariate polynomials forming the exact RUR, mainly because of their huge coefficients.
- Constructing the toric resultant matrix, and finding and evaluating over the roots of the minimal polynomial takes up an insignificant portion of the time.
- Although the cost for performing exact computation over algebraic numbers expressed in the exact RUR is insignificant for the above examples, applying a zero-test is a "bad" application for exact computation using the root bound approach. For example, in F_1, in order to answer whether or not $h_i^{(1)}(\theta) = h_i^{(2)}(\theta)$, we need to evaluate the (real and imaginary parts of) the quantity to precision all the way up to the root bound.

5.2. RUR vs. MAPC. In this section, we show the timings for comparison of the exact RUR method with that used in the MAPC library within the ESOLID system. Note that the exact RUR method is able to find solutions in several cases where MAPC/ESOLID would fail (e.g. case F_1, F_2, and F_3), and thus we can only compare a limited subset of cases.

It is important to note that our current exact RUR implementation is a relatively straightforward implementation of the algorithm. For instance, no effort is made to control intermediate coefficient growth in several stages. The MAPC/ESOLID code, on the other hand, has been significantly optimized through the use of various speedup techniques, such as floating-point filters and lazy evaluation approaches. It is likely that by aggressively applying such speedups to the exact RUR implementation, for instance when we compute determinants of matrices, its performance can also be improved markedly. However, we believe it is unlikely that even such improvements would be enough to change the overall conclusions.

One of the basic routines in ESOLID / MAPC is finding real intersections of 2 monotone pieces of curves.

A real algebraic number ξ is specified by a pair polynomials (f_1, f_2) and the region R such that ξ is the only intersection of f_1 and f_2 in R. We could naively use the exact RUR here to specify algebraic numbers.

We compare only the last steps of root isolation in ESOLID/MAPC with the exact RUR method. Because ESOLID/MAPC only finds roots over a particular region, it does not necessarily find all the roots of the system. This is in contrast to the exact RUR, which finds all roots (and would then select only those in the region). This describes the difference in the number of roots found in F_6 and F_8.

In addition, the exact RUR loses efficiency by finding all the complex roots while ESOLID/MAPC finds only real roots. This phenomenon can be observed in the example of F_7. The size of the coefficients of polynomials in the exact RUR grow independent of the location of the roots.

We note that performance could be improved by isolating only real roots of the minimal polynomial, h, which would allow us to isolate the real roots of the system. In fact, this is the approach taken by the GBRS system [**RZ04**]. However, doing so would not allow us to determine the positive dimensional components, where the points isolated on the component could be complex.

From these cases, the clear conclusion is that for generic cases that a method such as those MAPC/ESOLID can handle, the exact RUR has an unacceptably high performance cost. For this reason, it will be best to use the exact RUR in implementations only in a hybrid fashion, when the other methods will fail. An important caveat should be considered, in that our exact RUR implementation has not been fully optimized. Such optimizations should increase the exact RUR performance significantly, though we still believe that fundamental limitations (such as finding all roots, including complex ones) will make the exact RUR less efficient in most generic cases encountered in practice.

6. Conclusion

We have presented the Rational Univariate Reduction as a method for achieving exact geometric computation in situations that require solutions to systems of polynomials. This is applicable to detecting situations where polynomials describe objects in degenerate configurations. We compute the exact RUR. In addition, we have presented an approach for exact sign determination with root bounds for complex algebraic numbers, extending previous approaches that apply to real numbers. This allows us to use the exact RUR for geometric operations, including determining whether or not there are positive dimensional components, and distinguishing real roots.

Our method is significant in the following sense:

- The method and generated answers are readily adapted to exact computation.
- The method finds complex roots and identifies real roots if necessary.
- The method works even if the set of roots has some positive dimensional component,
- The method is entirely factorization-free and Sturm-free.

Finally, we have implemented the exact RUR method described and presented the timing breakdown for various stages for a selected set of problems. These timings highlight the computational bottlenecks and give indications of useful avenues for future work.

6.1. Future Work. A primary avenue of future development will be in finding ways to optimize the computation and use of the exact RUR, particularly in reference to problems from specific domains. Experience has shown that by aggressively applying filtering and other speedup techniques, many exact implementations can be made far more efficient [**KCF$^+$04**]. There are numerous potential avenues for such speedups, and we describe a couple of them below.

Our experiments show that for low-dimensional cases, most of the computation time is dedicated to evaluating the determinant of the sparse resultant matrix. There are thus some immediately obvious issues to tackle:

- Use some algorithm which produces sparse resultant matrices of smaller size [**EC95, DE02, Emi02, CK03, Khe03**]. Although the matrix construction is not time-consuming, the size of the sparse resultant matrices directly affects the performance of the next step, namely, repeated computations of the determinant of the matrix.
- Computing the determinant of a sparse resultant matrix is one of the most time consuming steps. Currently, we use a standard Gaussian elimination method implemented in multi-precision arithmetic. Improvements might include taking advantage of the sparse structure of the matrix itself, or finding the determinant using a filtered approach, resulting in only partially determined, but still exact, polynomial coefficients. Such coefficients would allow faster computation in most cases, and could still be later refined to greater or complete accuracy in the few cases where it is needed.
- We would like to have some control over the size of coefficients of the univariate polynomials forming the exact RUR. Experiments show that, for higher dimensional cases, univariate polynomial ring operations tend to be very expensive, because of the growth of coefficients. To control this growth, we could choose generic values more carefully based on a sharper estimate for bit-complexity of the subroutines. Next, we could make better choices for subroutines, e.g., using a subresultant method instead of the Euclidean algorithm, etc., as this tends to dominate running time for higher degrees. Also, the use of rational functions instead of polynomials in the RUR might help.

Finally, we are in the process of integrating the exact RUR completely into a solid modeling system [**OK04**] to support fully exact and robust geometric computation. We have shown that (with the current implementation) the naive use of the exact RUR is not attractive in terms of performance. However, we have also shown that the exact RUR is able to handle degeneracies smoothly, and can thus be used to determine how to resolve problematic cases (where a generic algorithm fails). A practical solution would be to develop a hybrid system, incorporating both the exact RUR and a general-position system solver. Determining precisely how to put together such a hybrid system remains a challenge for future work.

References

[Abe73] O. Aberth, *Iteration methods for finding all zeros of a polynomial simultaneously*, Mathematics of Computation **27** (1973), no. 122, 339 – 344.

[ARD02] P. Aubry, F. Rouillier, and M. Safey El Din, *Real solving for positive dimensional systems*, Journal of Symbolic Computation **34** (2002), no. 6, 543 – 560.

[BCSS97] L. Blum, F. Cucker, M. Shub, and S. Smale, *Complexity and real computation*, Springer, 1997.

[Ber75] D. N. Bernstein, *The number of roots of a system of equations*, Functional Analysis and its Applications **9** (1975), no. 2, 183 – 185.

[BF00] D. A. Bini and G. Fiorentino, *Design, analysis, and implementation of a multiprecision polynomial rootfinder*, Numerical Algorithms **23** (2000), no. 2, 127 – 173.

[BFM+01] C. Burnikel, S. Funke, K. Mehlhorn, S. Schirra, and S. Schmitt, *A separation bound for real algebraic expressions*, Proc. of ESA '01, LNCS 2161, Springer, 2001, pp. 254 – 265.

[BFMS97] C. Burnikel, R. Fleischer, K. Mehlhorn, and S. Schirra, *A strong and easily computable separation bound for arithmetic expressions involving square roots*, Proc. of 8th ACM-SIAM Symposium on Discrete Algorithms, ACM, 1997, pp. 702 – 709.

[BFMS99] ———, *Exact efficient computational geometry made easy*, Proc. of 15th Annual Symposium on Computational Geometry, ACM, 1999, pp. 341 – 350.

[BFMS00] ———, *A strong and easily computable separation bound for arithmetic expressions involving radicals*, Algorithmica **27** (2000), no. 1, 87 – 99.

[Bin96] D. A. Bini, *Numerical computation of polynomial zeros by means of Aberth's method*, Numerical Algorithms **13** (1996), no. 3/4, 179 – 200.

[BMS96] C. Burnikel, K. Mehlhorn, and S. Schirra, *The LEDA class* **real** *number*, Tech. Report MPI-I-96-1-001, Max-Planck-Institut fur Informatik, Im Stadtwald, D-66123 Saarbrücken, Germany, 1996.

[BPR03] S. Basu, R. Pollack, and Marie-FranÇoise Roy, *Algorithms in real algebraic geometry*, Springer, 2003.

[Can90] J. F. Canny, *Generalized characteristic polynomials*, Journal of Symbolic Computation **9** (1990), no. 3, 241 – 250.

[CE00] J. F. Canny and I. Z. Emiris, *A subdivision-based algorithm for the sparse resultant*, Journal of ACM **47** (2000), no. 3, 417 – 451.

[CK03] A. D. Chtcherba and D. Kapur, *Exact resultants for corner-cut unmixed multivariate polynomial systems using the dixon formulation*, Journal of Symbolic Computation **36** (2003), no. 3/4, 289 – 315.

[CLO98] D. Cox, J. Little, and D. O'Shea, *Using algebraic geometry*, Springer, 1998.

[D'A02] C. D'Andrea, *Macaulay style formulas for sparse resultants*, Transactions of the AMS **354** (2002), no. 7, 2595 – 2629.

[DE01] C. D'Andrea and I. Z. Emiris, *Computing sparse projection operators*, Symbolic Computation: Solving equations in Algebra, Geometry and Engineering (E. L. Green et al., ed.), Contemporary Mathematics, Vol. 286, AMS, 2001, pp. 121 – 139.

[DE02] ———, *Hybrid sparse resultant matrices for bivariate polynomials*, Journal of Symbolic Computation **33** (2002), no. 5, 587 – 608.

[DE03] ———, *Sparse resultant perturbation*, Algebra, Geometry, and Software (M. Joswing and N. Takayama, eds.), Springer, 2003, pp. 93 – 107.

[DFMT02] O. Devillers, A. Fronville, B. Mourrain, and M. Teillaud, *Algebraic methods and arithmetic filtering for exact predicates on circle arcs*, Computational Geometry **22** (2002), no. 1-3, 119 – 142.

[DM89] P. C. Dykstra and M. J. Muuss, *The BRL-CAD package an overview*, Tech. report, Advanced Computer Systesms Team, Ballistics Research Laboratory, Aberdeen Proving Ground, MD, 1989, http://ftp.arl.mil/brlcad/.

[DY95] T. Dubé and C. Yap, *The exact computation paradigm*, Computing in Euclidean Geometry (D. Du and F. Hwang, eds.), Lecture Notes on Computing, World Scientific, 2nd ed., 1995, pp. 452 – 492.

[EC95] I. Z. Emiris and J. F. Canny, *Efficient incremental algorithm for the sparse resultant and the mixed volume*, Journal of Symbolic Computation **20** (1995), no. 2, 117 – 149.

[EKSW04] A. Eigenwillig, L. Kettner, E. Schömer, and N. Wolpert, *Complete, exact, and efficient computations with cubic curves*, Proc. of 20th Annual Symposium on Computational Geometry, ACM, 2004, pp. 409 – 418.

[Emi96] I. Z. Emiris, *On the complexity of sparse elimination*, Journal of Complexity **12** (1996), no. 2, 134 – 166.

[Emi02] ———, *Enumerating a subset of the integer points inside a Minkowski sum*, Computational Geometry **22** (2002), no. 1-3, 143 – 166.

[Emi03] ———, *Discrete geometry for algebraic elimination*, Algebra, Geometry, and Software (M. Joswing and N. Takayama, eds.), Springer, 2003, pp. 77 – 91.

[For97] S. Fortune, *Polyhedral modeling with multiprecision integer arithmetic*, Computer-Aided Design **29** (1997), no. 2, 123 – 133.

[GLS01] M. Giusti, G. Lecerf, and B. Salvy, *A Gröbner free alternative for polynomial system solving*, Journal of Complexity **17** (2001), no. 1, 154 – 211.

[GV91] L. Gonzáez-Vega, *A subresultant theory for multivariate polynomials*, Proc. of ISSAC '91, ACM, 1991, pp. 79 – 85.

[GV97] L. Gonzalez-Vega, *Implicitization of parametric curves and surfaces by using multidimensional newton formulae*, Journal of Symbolic Computation **23** (1997), no. 2/3, 137 – 151.

[GVN02] L. Gonzalez-Vega and I. Necula, *Efficient topology determination of implicitly defined algebraic plane curves*, Computer Aided Geometric Design **19** (2002), no. 9, 719 – 743.

[GVRR99] L. Gonzalez-Vega, F. Rouillier, and M-F. Roy, *Symbolic recipes for polynomial system solving*, Some tapas of computer algebra (A. M. Cohen, H. Cuypers, and H. Sterk, eds.), Algorithms and computation in mathematics, Vol. 4, Springer, 1999, pp. 34 – 65.

[HHK89] C. M. Hoffman, J. E. Hopcroft, and M. S. Karasick, *Robust set operations on polyhedral solids*, IEEE Computer Graphics and Applications **9** (1989), no. 6, 50 – 59.

[Hof89] C. M. Hoffman, *The problems of accuracy and robustness in geometric computation*, IEEE Computer **22** (1989), no. 3, 31 – 41.

[Hon96] H. Hong, *An efficient method for analyzing the topology of plane real algebraic curves*, Mathematics and Computers in Simulation **42** (1996), no. 4-6, 571 – 582.

[KCF+02] J. Keyser, T. Culver, M. Foskey, S. Krishnan, and D. Manocha, *ESOLID: A system for exact boundary evaluation*, Proc. of 7th ACM Symposium on Solid Modeling and Applications, ACM, 2002, pp. 23 – 34.

[KCF+04] ———, *ESOLID - a system for exact boundary evaluation*, Computer-Aided Design **36** (2004), no. 2, 175 – 193.

[KCMK00] J. Keyser, T. Culver, D. Manocha, and S. Krishnan, *Efficient and exact manipulation of algebraic points and curves*, Computer-Aided Design **32** (2000), no. 11, 649 – 662.

[Key00] John C. Keyser, *Exact boundary evaluation for curved solids*, Ph.D. thesis, Department of Computer Science, University of North Carolina, Chapel Hill, NC, 2000.

[Khe03] A. Khetan, *The resultant of an unmixed bivariate system*, Journal of Symbolic Computation **36** (2003), no. 3/4, 425 – 442.

[Kho78] A. G. Khovanskii, *Newton polyhedra and the genus of complete intersections*, Functional Analysis and its Applications **12** (1978), no. 1, 51 – 61.

[KLY99] V. Karamcheti, C. Li, and C. Yap, *A Core library for robust numerical and geometric computation*, Proc. of 15th Annual Symposium on Computational Geometry, ACM, 1999, pp. 351 – 359.

[Kro31] L. Kronecker, *Leopold kronecker's werke*, Teubner, 1895 - 1931.

[Kus76] D. N. Kushnirenko, *Newton polytopes and the Bezout theorem*, Functional Analysis and its Applications **10** (1976), no. 3, 82 – 83.

[Lec02] G. Lecerf, *Quadratic newton iteration for systems with multiplicity*, Journal of Foundations of Computational Mathematics **2** (2002), no. 3, 247 – 293.

[LY01] C. Li and C. Yap, *A new constructive root bound for algebraic expressions*, Proc. of 12th ACM-SIAM Symposium on Discrete Algorithms '01, ACM, 2001, pp. 476 – 505.

[MN99] S. Mehlhorn and M. Näher, *Leda - a platform for combinatorial and geometric computing*, Cambridge University Press, 1999.

[MOBP94] D. Michelucci M. O. Benouamer and B. Peroche, *Error-free boundary evaluation based on a lazy rational arithmetic: a detailed implementation*, Computer-Aided Design **26** (1994), no. 6, 403 – 416.

[MŞ99] N. Mignotte and D. Ştefănescu, *Polynomials: An algebraic approach*, Springer, 1999.

[MVY02] B. Mourrain, M. N. Vrahatis, and J. C. Yakoubsohn, *On the complexity of isolating real roots and computing with certainty the topological degree*, Journal of Complexity **18** (2002), no. 2, 612 – 640.

[OK04] K. Ouchi and J. Keyser, *Handling degeneracies in exact boundary evaluation (poster)*, Proc. of ACM Symposium on Solid Modeling and Applications, ACM, 2004, short paper, pp. 321 – 326.

[PS93] P. Pedersen and B. Sturmfels, *Product formulas for resultants and chow forms*, Mathematische Zeitschrift **214** (1993), no. 3, 377 – 396.

[PY03] S. Pion and C. Yap, *Constructive root bound for k-ary rational input numbers*, Proc. of 19th Annual Symposium on Computational Geometry, ACM, 2003, pp. 256 – 263.

[Roj94] J. M. Rojas, *A convex geometric approach to counting the roots of a polynomial system*, Theoretical Computer Science **133** (1994), no. 1, 105 – 140.

[Roj97] ______, *Toric laminations, sparse generalized characteristic polynomials, and a refinement of Hilbert's tenth problem*, Foundations of Computational Mathematics (F. Cucker and M. Shub, eds.), Springer, 1997, pp. 369 – 381.

[Roj99] ______, *Solving degenerate sparse polynomial systems faster*, Journal of Symbolic Computation **28** (1999), no. 1/2, 155 – 186.

[Roj00] ______, *Algebraic geometry over four rings and the frontier to tractability*, Hilbert's tenth problem : relations with arithmetic and algebraic geometry (J. Denef et al., ed.), Contemporary Mathematics, Vol. 270, AMS, 2000, pp. 275 – 321.

[Roj03] ______, *Why polyhedra matter in non-linear equation solving*, Topics in Algebraic Geometry and Geometric Modeling (R. Goldman and R. Krasauskas, eds.), Contemporary Mathematics, Vol. 334, AMS, 2003, pp. 293 – 320.

[Roj04] ______, *Arithmetic multivariate Descartes' rule*, American Journal of Mathematics **126** (2004), no. 1, 1 – 30.

[Rou99] F. Rouillier, *Solving zero-dimensional systems through the rational univariate representation*, Applicable Algebra in Engineering, Communication and Computing **9** (1999), no. 5, 433 – 461.

[RZ04] F. Rouillier and P. Zimmermann, *Efficient isolation of polynomial's real roots*, Journal of Computational and Applied Mathematics **162** (2004), no. 1, 33 – 50.

[Sak91] T. Sakkalis, *The topological configuration of a real algebraic curve*, The Bulletin of the Australian Mathematical Society **43** (1991), no. 1, 37 – 50.

[SI89] K. Sugihara and M. Iri, *A solid modelling system free from topological inconsistency*, Journal of Information Processing **12** (1989), no. 4, 380 – 393.

[Stu94] B. Sturmfels, *On the Newton polytope of the resultant*, Journal of Algebraic Combinatorics **3** (1994), no. 2, 207 – 236.

[Stu02] ______, *Solving systems of polynomial equations*, AMS, 2002.

[SV00] A. J. Sommese and J. Verschelde, *Numerical homotopies to compute generic points on positive dimensional components*, Journal of Complexity **16** (2000), no. 3, 572 – 602.

[SVW01] A. J. Sommese, J. Verschelde, and C. W. Wampler, *Numerical decomposition of the solution sets of polynomial systems into irreducible components*, SIAM Journal of Numerical Analysis **38** (2001), no. 6, 2022 – 2046.

[SVW03] ______, *Numerical irreducible decomposition using phcpack*, Algebra, Geometry, and Software (M. Joswing and N. Takayama, eds.), Springer, 2003, pp. 109 – 129.

[SW96] A. J. Sommse and C. W. Wampler, *Numerical algebraic geometry*, The Mathematics of Numerical Analysis (J. Renegar et al., ed.), Lectures in Applied Mathematics, Vol. 32, AMS, 1996, pp. 749 – 763.

[Wol03] N. Wolpert, *Jacobi curves: Computing the exact topology of arrangements of non-singular algebraic curves*, Proc. of ESA '03, LNCS 2832, Springer, 2003, pp. 532 – 543.

[Yap97] C. Yap, *Towards exact geometric computation*, Computational Geometry **7** (1997), no. 1, 3 – 23.

[Yap00] ______, *Fundamental problems in algorithmic algebra*, Oxford University Press, 2000.

[Yu91] J. Yu, *Exact arithmetic solid modeling*, Ph.D. thesis, Department of Computer Science, Purdue University, West Lafayette, IN, 1991.

JOHN KEYSER, DEPARTMENT OF COMPUTER SCIENCE, 3112 TEXAS A&M UNIVERSITY, COLLEGE STATION, TX 77843-3112
E-mail address: keyser@cs.tamu.edu

KOJI OUCHI, DEPARTMENT OF COMPUTER SCIENCE, 3112 TEXAS A&M UNIVERSITY, COLLEGE STATION, TX 77843-3112
E-mail address: kouchi@cs.tamu.edu

J. MAURICE ROJAS, DEPARTMENT OF MATHEMATICS, 3368 TEXAS A&M UNIVERSITY, COLLEGE STATION, TX 77843-3368
E-mail address: rojas@math.tamu.edu

DIMACS Series in Discrete Mathematics
and Theoretical Computer Science
Volume **67**, 2005

Mass Properties of the Union of Millions of Identical Cubes

W. Randolph Franklin

ABSTRACT. We present an algorithm and implementation, UNION3, for computing mass properties of the union of many identical axis-aligned cubes. This implementation has processed 20,000,000 cubes on a dual processor Pentium Xeon workstation in about one elapsed hour. The ideas would extend to the union of general polyhedra.

UNION3 works by combining three techniques. The first is a set of local topological formulae for computing polyhedral mass properties. No global topology is used, such as complete face or edge information, or edge loops and face shells. For example, the volume of a multiple component, concave, rectilinear object is $\sum s_i x_i y_i z_i$, summed over the vertices (x_i, y_i, z_i), where each s_i is a function of the directions of the incident edges. The second technique is a 3-D grid of uniform cells, which is overlaid on the data. Each cell has a list of edges, faces, and cubes that overlap it. Only pairs of objects in the same cell are tested for intersection. Experimentally, the uniform grid is quite tolerant of varying densities in the input. The third technique, building on the second, is the covered-cell concept. When a cell is completely contained within a cube, its overlap list is cleared, and objects in it are not intersected against each other (unless they both also overlap the same other, noncovered, cell). These techniques combine to reduce the expected computation time to linear in the number of cubes, regardless of the number of intersections.

In the slowest step of UNION3, which is testing pairs of objects in each cell for intersection, no inter-cell communication is required. Therefore UNION3 parallelizes quite well.

1. Introduction

3-D geometric applications sometimes create an object as the union of a large set of polyhedra. For example, in CAD/CAM, the volume swept by a cutting tool may be approximated by one polyhedron for each segment of the tool path. On occasion we don't need the resulting polyhedron itself, but only certain mass properties. This presents several opportunities for optimization.

2000 *Mathematics Subject Classification.* Primary 65D32; Secondary 68U07, 65D18, 49Q15, 26B15, 28A75, 51M25.

Key words and phrases. boolean operation, union, polyhedron, volume, mass property, uniform grid, parallel geometry computation.

This research was supported by NSF grant CCR 03-06502.

First, the usual method for computing the volume of the union of many polyhedra proceeds by forming a $(\log N)$ level computation tree that combines successively larger intermediate polyhedra pair by pair. The execution time is probably $N(\log N)^2$ to $N^2(\log N)^2$ depending on the size of the intermediate polyhedra and the efficiency of the polyhedron intersections. There may be many intermediate vertices even when the output polyhedron has few.

Next, efficiently combining polyhedra is more difficult than combining polygons. Sweeping space with a plane, while keeping order among all the active faces, is more intricate than sweeping a plane with a line that keeps track of active edges.

Third, if we need only the volume, or similar mass property, of the computed polyhedron, then it is unnecessary completely to compute the union polyhedron first. Indeed, we need only its vertices with their neighborhood geometries.

2. Prior Art

There appear to be few results for object combination in 3-D, except for the following special cases: efficiently intersecting convex polyhedra (CD87; Cha89; MP78; Sug94), intersecting a convex polyhedron with a general one (DMY93; MS85), fast detection of polyhedral intersection (DK83), and uniting convex polyhedra (AST97). Constructive Solid Geometry (CSG) is well documented, (Cam91; Epp92; LTH86), though its efficient methods tend to use bounding boxes.

(ZE02) presents a fast algorithm for intersecting boxes in 3D. Various heuristics are combined to optimize the implementation, for which tests on 500,000 boxes were reported. Although the general optimization techniques are very useful, extending this particular algorithm to compute properties of a union is not at all obvious since the characteristics of intersections and unions of boxes are quite different. E.g., intersections are convex, while unions are often not even connected. (Eri03) computes any Boolean combo of two *local* polyhedra in $O(N \ \log N)$ time. However since the faces of a union are often not simplices, our polyhedra are not local. (Vig) gives a new algorithm to report intersecting pairs among n objects in a fixed dimension d. Again, this is considerably easier than computing unions.

In 2-D, the well-known plane sweep methods for line segment intersection take from $T = \theta(K + N \log N)$ for a complex topological sweep up to $T = \theta((N + K) \log N)$ for a simple plane sweep. N is the number of input line segments, and K the number of output intersections. For the large values of N in this paper, the $(\log N)$ is significant. Of course, these methods' worst-case time is much better than ours, since an adversary can hurt the performance of our "input-sensitive" method. Also our computation model is more powerful than that used by the plane sweep methods; we can compute *modulo* in constant time.

The prior art on mass properties of boolean combinations of polyhedra is similarly sparse, (BN83; Mir96). Some of these local topological formulae could well date to the discovery of analytic geometry, (Des44), but actual citations earlier than

(Fra87; NF91b; FCK$^+$90) are elusive. Large geometric and cartographic data sets are now being processed, perhaps with external algorithms, (TWA$^+$01; Cen01).

Our big problem is to compute mass properties, including second order moments, center of gravity, volume, area, and edge length, of a very complex polyhedron defined by a constructive solid geometry tree. The immediate goal is to demonstrate a solution to a restricted subproblem, where the only operation is *union*, and the primitive objects are identical cubes, and to demonstrate it on as large an input example as possible.

We have produced a prototype implementation, UNION3, of the restricted case, which finds mass properties of the union of many identical cubes. This implementation has processed 20,000,000 cubes, containing half a billion subfacets, on a dual processor Pentium Xeon workstation in about one hour. Smaller examples take proportionally less time. E.g., processing 1000 cubes with edge length 0.1 takes 2 elapsed seconds on a 1600 MHz laptop.

UNION3 works by combining three techniques. The first is a set of local topological formulae for computing polyhedral mass properties. Let $\{v_i\}$ be some polyhedron $\mathbb{P}$'s set of vertices. Let M be any one of the mass properties listed earlier. Then there exists a function f_M such that $M = \sum_i f_M(v_i)$. (The center of mass is the ratio of two such formulae.) $f_M(v)$ depends on only v's location and local geometry and topology. That includes the directions in which any edges or faces are adjacent on v, and which adjacent sector is inside. No global topology is used, such as complete face or edge information, or edge loops and face shells. Provided that the polyhedron is valid, these formulae are valid regardless of global connectivity such as the number of components and their containment hierarchy.

The first advantage of such formulae is that determining vertices and local information of a boolean combination is much easier than determining global information, such as complete output edges and faces. The second is that our data structures are much more compact, so larger examples are feasible.

The second technique is a 3-D grid of uniform cells. Each cell records the overlaps of itself with any edge, face, or polyhedron. Intersection tests are performed only between objects overlapping the same cell. Therefore, we can quickly cull out pairs of groups of objects that cannot possibly intersect. This simple technique is effective because testing two line segments (or other primitives) for intersection is much faster than updating a complicated topological data structure. Therefore, testing, say, ten pairs to find one intersection is faster than performing a plane sweep operation while maintaining ordering information. This is also why the uniform grid also performs quite well in practice on even quite irregular actual data. For example, we describe intersecting all the edges in two anisotropic meshes, whose edges' lengths have a spread of 100:1.

The third technique, building on the second, is the covered-cell concept. It solves the problem that, for fixed-size input objects, the number of intersections of subfacets grows as $O(N^3)$. Thus, no initial cull can suffice. However, although the total number of intersections may be $O(N^3)$, the number of *visible* intersections, that is, intersections that are not contained in some input object, grows only linearly, under broad conditions. Only these visible intersections contribute to the result. However, if the grid cell size is smaller than the object size, then as the set of objects gets denser, with ever higher probability a particular cell will fall completely inside some object. Thus, in that cell, no intersections will visible, and so no intersections need be determined in that cell. The number of tests that need to be performed, i.e., in cells not completely covered by objects, is linear, not cubic.

UNION3's simple flat data structures permit it to fork copies of itself to utilize multi-processor machines. This contrasts favorably to topological sweep methods, which are harder to parallelize. The parallelizability of the uniform grid is presented in more detail in (KF95; FK93; NF92a; NF92b; NF91a; FK90; Kan90; Nar91).

UNION3 processes all the polyhedra in one pass instead of repeatedly combining them pair by pair. The first step finds the candidate output vertices. These are the 3-face intersections, edge-face intersections, and input vertices. Next, the candidates are culled by deleting those inside any polyhedron. The volume is the sum of a function of each survivor. Input degeneracies are processed with Simulation Of Simplicity. Since UNION3 never explicitly determines the output polyhedron, messy non-manifold cases become irrelevant. No complicated topological structures are computed. The expected time is linear in the number of input, even when the number of intersections is superlinear. That the time is indeed linear is supported by our implementation's ability to process 20M small cubes.

3. What UNION3 is Not

(1) We do not propose that objects be modeled as the union of identical cubes. The point of this paper is to demonstrate the feasibility of the concept by means of a prototype implementation, using an easily implementable subset of all possible polyhedra.
(2) Although this prototype implementation is for identical cubes, the concepts are not restricted to this, but extend to general polyhedra.
(3) This is not a Monte Carlo or other statistical sampling technique. The output is exact to the arithmetic precision of the computations.
(4) UNION3 is also not a voxelization or octree method, where the universe is partitioned into voxels, and each object is represented as the union of a set of voxels, and everything is only as accurate as the voxel size. Although superficially similar, our uniform grid is quite different. Different grid sizes affect only the execution time and space, but have no effect on the result. Input cubes generally overlap grid cells only partially.
(5) The general concept is also not limited to unions. Processing general constructive solid geometry trees appears feasible. However, efficiently culling the impossible output vertices in the general case is still future research, which would build on the foundation established in (Til84). Special cases, such as two level sum-of-products formulae, appear quite doable, efficiently.

4. Volume Determination

The volume of a cube with vertices (x, y, z) may be expressed as $\mathbb{V} = \sum_i s_i x_i y_i z_i$, where $s_i = \pm 1$. For any particular vertex, $s = +1$ iff an odd number of the three faces adjacent to that vertex are on the high side of the cube. The volume of any rectilinear polyhedron, whose edges and faces are all aligned with the coordinate axes, is still $\mathbb{V} = \sum_i s_i x_i y_i z_i$, if the definition of s is generalized for every possible local vertex geometry. Our implementation uses this formula. The above formula easily extends to other mass properties, such as mass moments of any order. These formulae are related to Greene's theorem. (LR82a; LR82b) is an excellent summary of Greene's theorem and many related formulae for different representations.

Likewise, lower dimensional mass properties, such as face area and edge length may be computed as $\mathbb{A} = \sum s_i x_i y_i$ and $\mathbb{L} = \sum s_i x_i$ with the s_i in each case a function of the local geometry of the vertex. Throughout this paper "volume" includes any mass property.

All the above extends to general polyhedra, as described in (NF91a; Fra87), although the implementation is considerably harder. Indeed, there are varying classes of formula, depending on how much topology is available. (Mir96) also describes efficient polyhedron formulae.

Suppose that we have only the set of incidences of output vertices, edges lines, and face planes, together with which side is inside. For instance, for a cube, each vertex would induce six such incidences. Then if P is the vertex position, $\hat{T}$ is a unit tangent vector along the edge incident on it, $\hat{N}$ is a unit vector normal to $\hat{T}$ in the plane of the face, and $\hat{B}$ is a unit binormal vector, normal to both $\hat{T}$ and $\hat{N}$ pointing into the polyhedron, then the volume is $\mathbb{V} = -\frac{1}{6} \sum P.\hat{T}\, P.\hat{N}\, P.\hat{B}$. Similar formulae obtain for other mass properties. Figure 1 illustrates this for a cube. Each visible *(vertex, edge, face)* tuple is check marked, except for one that is starred. For that one, the vectors P, $\hat{T}$, $\hat{N}$, and $\hat{B}$ are shown.

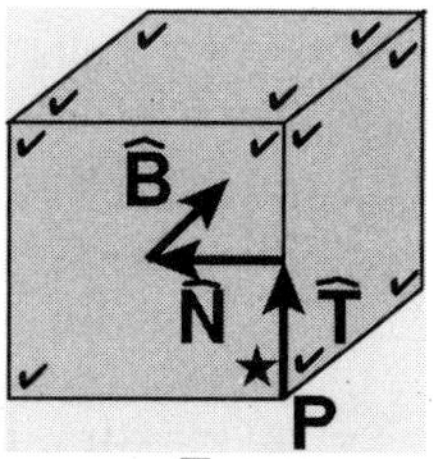

FIGURE 1. $P, \hat{T}\hat{N}, \hat{B}$

There are three classes of output vertices resulting from uniting a set of polyhedra: (1) a vertex of one of the input polyhedra, (2) an intersection of an edge with a face, and (3) an intersection of 3 faces.

An output vertex must not be contained in any polyhedron. Therefore, the process goes as follows. (1) Generate a candidate output vertex of one of the above types. (2) Test to see if it's outside all the polyhedra. (3) If so, then compute its sign based on its neighborhood, and add another term into the running total of the volume.

We define "intersection" to exclude components of the same polyhedron. That is why edge-face and 3-face intersections are different cases, even though every edge might be considered as the intersection of two adjacent faces.

The naive algorithm would test all triples of faces for intersection, requiring $T = \theta(N^3)$ time. However we use a uniform grid data structure, introduced by (Fra78), with applications, i.a., in (Nar91), (Kan90), (AFKN89), (HH90), (FK90), (KF95), and (FSS$^+$94), to reduce that, as follows.

(1) Superimpose a 3-D grid with $G \times G \times G$ cells on the universe. A reasonable cell size is half the average polyhedron size; however a wide range of G is acceptable
(2) For each cell, initialize an empty list, to be populated later, of items overlapping it.
(3) Iterate through the polyhedra. For each polyhedron, $\mathbb{P}$:
  (a) Determine which grid cells $\mathbb{P}$ *completely encloses.* Mark those cells as *covered.*
  For a cube, a range test is done in each dimension separately, and a cartesian product is taken of the results. For instance, let $G = 10$, and let the x-range of a test cube be $[.25, .57]$. Then, in the x-dimension it will cover cells 3 and 4, counting from 0.
  (b) For steps 3c and 3d below, whenever an item would be inserted into a cell, do not insert it if that cell is covered.
  (c) Determine which cells $\mathbb{P}$ *overlaps.* Add the polyhedron to those cells' lists. Continuing the example from item 3a above, that cube overlaps cells 2, 3, 4, and 5, in the x-dimension. Since it covers cells 3 and 4, the cube is added to the overlap lists of only cells 2 and 5.
  (d) Ditto for each face and edge.
(4) Iterate through the cells. Process the contents of each cell's overlap list, $\mathbb{L}$, as follows.
  (a) Test all triples of faces from $\mathbb{L}$, which are from three different polyhedra, for intersection. Three faces intersect if the intersection point of their three planes is contained in each face polygon. For each triple that does intersect, say at point $\mathbb{X}$, test if $\mathbb{X}$ is outside all polyhedra. If it is, then look up its sign in a table, based on the directions of its three faces, and add a term to the running total for the volume.
  (b) Test all pairs of edges and faces from $\mathbb{L}$, from two different polyhedra, for intersection, and process the intersections as before. A face and edge intersect if (a_0, a_1, a_2), the intersection point of the face plane and the edge line, is inside both the face and edge. The contribution to the volume of the union is computed as follows.

$$
\begin{aligned}
f &\triangleq \text{index of this face in its cube, from 0 to 5} \\
e &\triangleq \text{index of this edge in its cube, from 0 to 11} \qquad (1)\\
\Delta v &= (2(f\&1)-1)(2(e\&1)-1)(2((e \gg 1(-1))))a_0a_1a_2
\end{aligned}
$$

where & means *bitwise and* and $\gg$ means to shift the left argument right by the number of bits specified by the right argument. The

other operations are arithmetic. This assumes that the faces of each cube are ordered: lo X, hi X, lo Y, hi Y, lo Z, hi Z, and that the edges likewise have a fixed precise order.
The contributions of this intersection to the surface area and edge length of the union are two other similar formulae.

(5) Iterate through the vertices, testing whether each is outside all the polyhedra. For each outside vertex, then determine its s, and add $s \cdot x \cdot y \cdot z$ to the volume running total.

Determining whether point $\mathbb{X}$ is contained in any polyhedron proceeds as follows. (The precise identity of the containing polygon is unnecessary.)

(1) Compute which cell, $\mathbb{C}$, contains $\mathbb{X}$.
(2) If $\mathbb{C}$ is covered, then $\mathbb{X}$ is contained in some polyhedron.
(3) Otherwise, test whether any polyhedron in $\mathbb{C}$'s list contains $\mathbb{X}$.

Our implementation handles only identical cubes, although the theory extends to general polyhedra. Allowing only cubes removed numerical roundoff problems and vastly simplified the implementation, while still demonstrating the algorithm.

This implementation handles degeneracies via Simulation of Simplicity (SoS), (EM88) for volume computation. For instance, whenever a point is tested against a face, the three polyhedra whose faces intersected to create that point are brought along. In cases of a coordinate equality, the polyhedron numbers are compared to break the tie. However, then the computed area and length differ from the result of a regularized set union. Say that two polyhedra share a face. With SoS, either both faces will be counted, or neither will, depending on which polyhedron has the lower number.

The importance of the covering cell concept requires emphasis. The bad case for any intersection algorithm is when the objects are large, as shown in Figure 2, so that there are a cubic number of 3-face intersections. However we don't need all the intersections, but only those that are outside all polyhedra, i.e., *visible*. Figure 3 shades the covered cells, and X's the covered intersections. The number of visible intersections grows much more slowly than the total number of intersections. Indeed, under some reasonable assumptions analyzed in (Fra04), it grows only linearly. Therefore, the covering cell concept can reduce the intersection time from cubic to linear. It also reduces the query time to test point containment from linear time per point down to constant time per point. Indeed, the average number of polyhedra contained in each cell is constant, with reasonable cell sizes.

A major objection to the uniform grid is that it will fail when presented with real world data, since the real world is not uniform. This point is obvious, but *wrong*, as described in (AFKN89). Although it is counterintuitive, it can be shown that, in these applications, the grid structure tolerates nonuniform input as well as a hierarchical structure such as the quadtree, and the implementation is simpler. Consider, for example, the simpler case of finding 2D edge intersections. Let the number of edges in the i-th grid cell be k_i, and the total number of edges be N,

FIGURE 2. Seriously Overlapping Objects

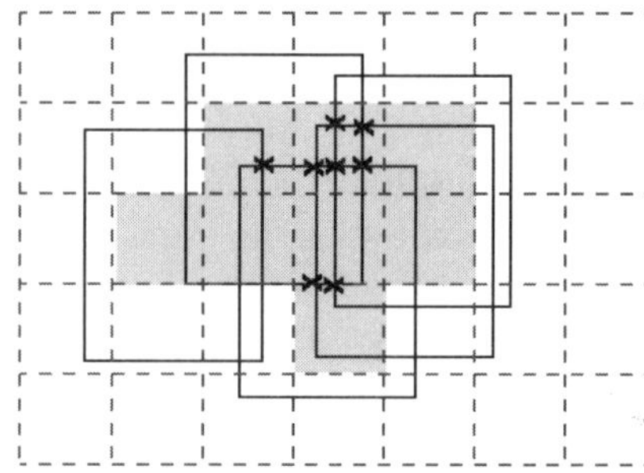

FIGURE 3. Covered Cells

and the grid be $G \times G$ cells. A reasonable value is $G = 1/\sqrt{N}$. $\sum_i k_i = cN$ where c is the average number of cells that an edge is incident on. The average number of edges per cell is $\overline{k} = cN/G^2 = c$. If the edges are uniformly, independently, and identically distributed, then the number of edges per cell follows a Poisson distribution with $\lambda = \overline{k} = c$, and so $\overline{k^2} = \overline{k}^2 = c^2$. The total number of edge pairs tested by pairwise comparing all the edges in each cell to find the intersections is

$$\begin{aligned} T &= \sum_i k_i(k_1 - 1)/2 \\ &= \frac{1}{2}G^2\left(\overline{k^2} - \overline{k}\right) \\ &= c_2 N \end{aligned} \tag{2}$$

An analysis of the asymptotic execution time for the full algorithm is given in (Fra04).

The bad case for both a grid and a quadtree would be many close, correlated, edges, which do not intersect. Imagine $N/2$ edges uniformly distributed, and the other $N/2$ edges uniformly distributed in a square of size $2^{-N} \times 2^{-N}$ in one corner. One grid size is inadequate for both groups of edges. However, a quadtree would require so many levels that it would also be too slow. A topological sweep could handle this case, but at the cost of a complicated data structure that is difficult to parallelize and takes $O(N \log N)$ time. However, first, $\log N$ is nontrivial for current N. Second, topological sweeps are much harder in 3D. Finally, and most important, the sweep finds all intersection vertices, including all $\Omega(N^3)$ hidden ones.

The parallel implementation of the algorithm goes as follows.

(1) Do steps 1 to 3 in one process (the main process); they are only a small fraction of the total time.
(2) Let M be the desired number of parallel processes.

(3) Partition the G^3 cells into M groups, one per process, preferably so that the groups are not too uneven.
(4) Repeat the following M times.
  (a) The main process creates a fifo pipe.
  (b) It forks a subprocess.
  (c) It gives each subprocess the data on all the input polyhedra, and the overlap lists of the G^3/M cells in its group.
  (d) Each subprocess totals up the contributions to the total volume, area, and length caused by the cells in its group. This is the compute-bound step.
  (e) Each subprocess returns these three numbers to the main process via the pipe.
(5) The main process waits until all M processes have completed and reads the volume, area, and length subtotals from each pipe.
(6) Finally, it sums the M volume subtotals, etc, and reports them.

5. Why Not Use Global Topology?

The intersection process directly produces only local topological information about the output polyhedron. It may look easy to link up this local topological info to find the output polyhedron's boundary. In practice, it is not easy, even in 2-D, and is much worse in 3-D. Our estimate is that doing this would double the number of lines of code, more than double the execution time (since exact computation would now be required), and introduce assorted nasty special cases, with the variety of possible manifold and non-manifold instances that would need correct handling. Nevertheless, if this is desired, a sketch of the process goes as follows, to combine a set of isolated vertex neighborhoods into edges and faces, then into loops and shells. The result might be in the form of the very useful data structures in (GS83; Bri89), except that variable-length arrays are more efficient than linked lists.

(1) Each output vertex neighborhood forms the *end* and *direction* of one or more edges. Compute these edge ends, as a set $\{(P, \hat{D})\}$, where P is the endpoint of an edge, and $\hat{D}$ is the unit vector along it.
(2) Since the two ends of one edge must be on the same infinite line, group those ends according to whether they are on the same infinite line.
(3) For each infinite line: if there are exactly two ends on it, and they point to each other, then we have an edge. If there are an even number greater than two ends, then sort them along the infinite line, and group them into adjacent pairs, each forming one edge. Otherwise, if there are an odd number of ends, then report an inconsistency.
(4) Each output vertex neighborhood also forms a vertex of one or more output faces. Compute those vertices.
(5) Group the vertices according to the infinite planes that they are incident on. This is well defined since each vertex knows the directions of its incident edges.

(6) For each infinite plane found in the previous step, identify any edges that are incident on it.
(7) In each infinite plane, connect up the vertices and edges on it into faces, while checking for inconsistencies.
(8) If there is more than one face per plane, then determine the inclusion relations among them.
(9) Group the faces into shells.

The actual process is worse than is apparent from the above brief summary. Our experience is that the much simpler case of combining only two (2-D) polygons requires perhaps 1000 lines of code, of quality such that testing point inclusion in a polygon is only 10 lines.

When combining only two polyhedra, there is an alternative to computing an unorganized set of vertices and then determining the global topology. That is, to trace out the boundary of the resulting polyhedron by following along the boundaries of the two input polyhedra. As always, nonmanifold cases cause complications. In addition, it is completely unclear how to extend this to trace out the union of many polyhedra in one step, without forming a tree of unions of smaller sets of input polyhedra.

6. Implementation Tests

The HW is a dual 2.4 GHz Xeon with 4GiB of real memory. The SW is SuSE 8.2 linux and the Intel C++ compiler. The program is about 1000 lines of code, excluding debugging lines, comments, and blank lines. The input cubes are generated with a combination of three Tausworth random number generators, which is much better than the widely used class of linear congruential generators. One problem with even the best linear congruential generators is that, if we let the generated numbers be x_i, then the 3-D points (x_i, x_{i+1}, x_{i+2}) fall on a relatively small number of parallel planes.

Figure 4 shows some sample runs, varying the number of cubes, edge length, grid resolution, and number of processors. The grid resolution and number of processors affect the time for a given input. For each run, the CPU time is reported, conservatively, as the sum over all the processes. However, since 4 threads can execute in parallel, the elapsed time is usually much less.

The added line has a slope of one, which would be $T = \Theta(N)$. The observed time performance appears to be slightly worse than linear. This is principally caused by memory limitations, which force a suboptimally small grid resolution to be used for large datasets.

Table 1 shows some statistics from the largest case. Lengths are scaled so that the universe is $1 \times 1 \times 1$. The actual edge length is 32768/(approx inverse edge length), rounded to a multiple of 10 (to make the numbers easier to read when debugging).

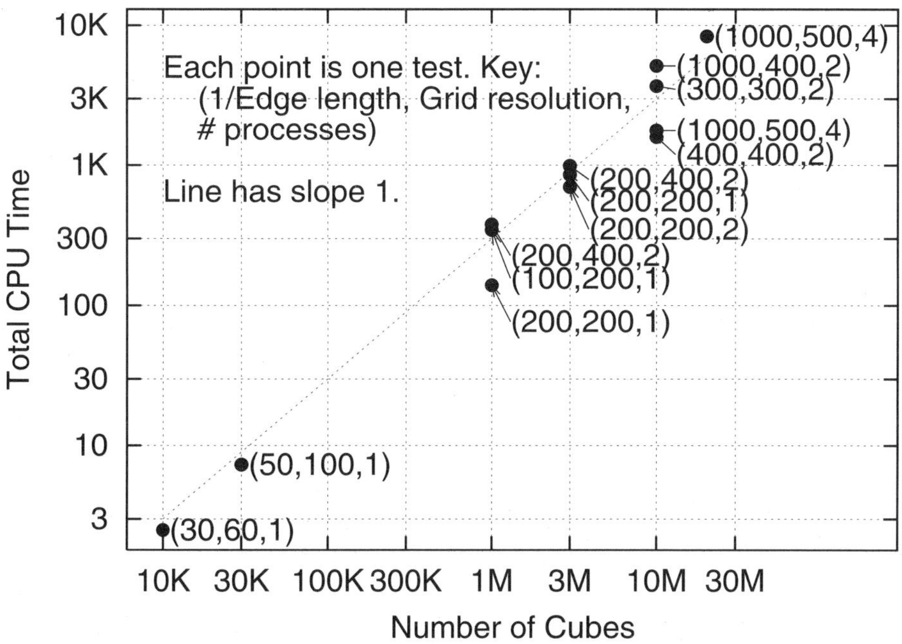

FIGURE 4. Execution Time vs Number of Cubes

Number of cubes	20,000,000
Edge length	0.00916
Number of processes	4
Number of grid cells	125, 000, 000
Number of cell – cube overlaps	61,890,769
Number of cell – face overlaps	254,823,296
Number of cell – edge overlaps	349,733,052
Sum of the cubes' volumes	0.015
Volume of the union	0.009
Sum of all the cubes' surface areas	100.583
Surface area of the union	99.053
Sum of all the cubes' edges	219,727
Edge length of the union	219,708
Number of input vertices	160,000,000
Number of those vertices that were outside all cubes	157,561,445
Number of 3-face intersections	37,170
Number face–edge intersections	7,259,255

TABLE 1. Statistics of the 20M Cube Test

The hard limit affecting the implementation is the available real memory. (If the total virtual memory of all the processes greatly exceeds the real memory then the amount of paging can hurt performance by more than an order of magnitude.) This limit also inhibits the use of a finer grid, which would reduce the time in some cases. In this largest case, that limitation meant that there were no covered cells. One solution to the memory problem is to subdivide the problem, then execute the parts sequentially, not in parallel.

How do we know that the above numbers are correct? Although errors are always possible, several indicators give us confidence. First, summing the addends mostly cancels out so that the final volume is in the range $[0, 1]$. Any erroneous addend is likely to produce a total that is out of range. Second, we constructed nasty test cases involving many cubes exactly overlapping or aligning along faces. (This also tested the SoS code.) The computed volume was always correct.

Third, for random input, we can estimate the volume as follows. For N independent and uniformly distributed cubes, each of volume v, the expected union volume is $V_{est} = 1 - (1 - v)^N \approx 1 - e^{-vN}$. The agreement between the predicted and computed volumes is excellent.

7. Non-Uniform Data and the Uniform Grid

It's not a priori obvious that the uniform grid can handle non-uniform data, at least without adapting into a quadtree. Formally, since the execution time contains terms that both increase and decrease with G, the best G is a broad optimum, and therefore if some cells contain, say, an order of magnitude more objects than the average, there's no problem. However, the best demonstration is by using real data, from diverse application domains.

Figure 5 shows the intersections of the edges of two anisotropic meshes from Mark Shephard and Xiangrong Li. Figure 6 is a detail from it. Together these two meshes have 54057 edges, whose lengths are quite variable, ranging in length from about 0.002 to 0.23, a factor of 100. These figures show the edges' extremely uneven density and orientation, which is a bad case for the uniform grid. Nevertheless, we can easily process this data. Before these experimental results were available, we considered using a multilevel grid, but that was unnecessary.

With a 90×90 uniform grid, we found all 64273 intersections of these edges in under 2 CPU seconds on a 1600 MHz laptop. That time includes reading the edge file and writing the intersections. No drastic optimization steps were necessary. Serious profiling and optimizing would probably reduce that time considerably. The average edge crossed 3 grid cells. In each of the 8100 cells, we tested all pairs of edges for intersection; 2,650,725 tests in all. Those tests are quite fast, so it's better not to filter the edges more thoroughly. There resulted 127,606 intersections, or 63,273 after duplicates were removed.

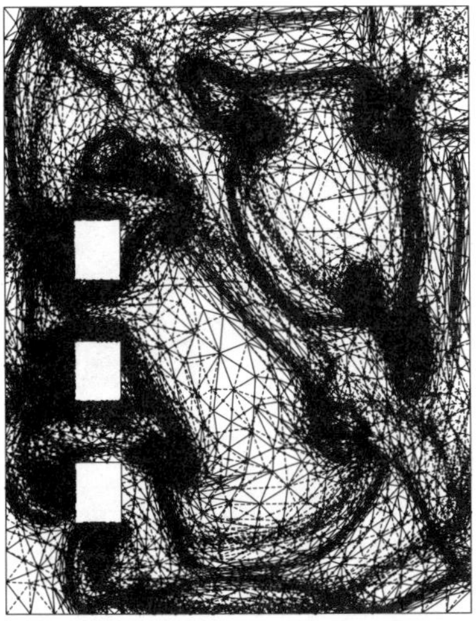

FIGURE 5. Two Anisotropic Meshes Intersected

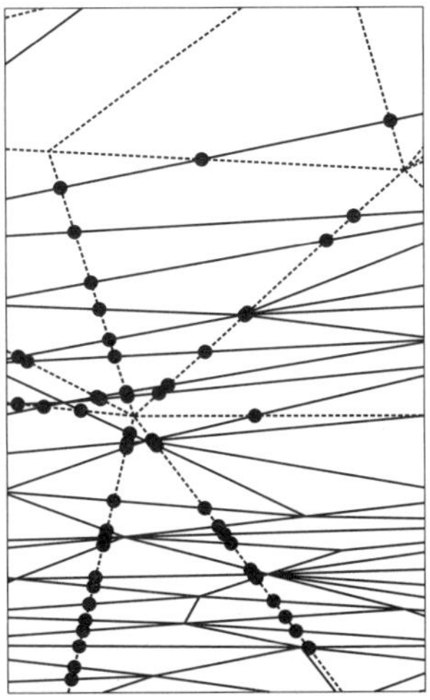

FIGURE 6. Detail from the Previous Figure

Figure 7 shows a cartographic application, with data from Denis White. The input is two planar graphs or maps. The first map is coterminous US counties, with 55068 vertices, 46116 edges, and 2985 polygons. The second map shows hydrography polygons, with 76215 vertices, 69835 edges, and 2075 polygons. The density variation between urban and rural regions is clear. Intersecting the two maps and reporting the areas of all the nonempty polygon intersections took 1.58 CPU seconds on the laptop, plus I/O time.

Finally, Figure 8 shows an integrated circuit designed by Jim Guilford, again with a wide variation in line segment density. With a uniform grid, we easily found all the intersections. Since these three examples are 2-D, we are searching for large 3-D examples to test.

8. Conclusion

This prototype implementation demonstrates that simple data structures, with no global topology, are an excellent method for processing large geometric datasets in 3D. The resulting implementations are simpler, faster, can process very large datasets, and can be parallelized.

Our method performs particularly well for very dense sets of objects, where most cells are covered, which is the worst case for sweep-line methods. It also is not merely a toy demonstration, but scales up to very large datasets.

FIGURE 7. Counties Overlaid on Hydrography Polygons

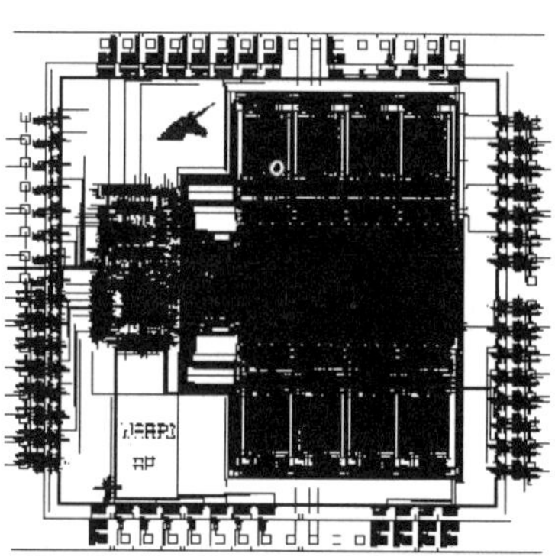

FIGURE 8. VLSI Design

In addition, this technique would also support other operations, not yet implemented, such as online computation of mass properties, as polyhedra are inserted and deleted. Insertion is easy. Deletion requires identifying the candidate output vertices that are now visible since they were hidden only by the deleted polyhedron. Computing properties of more complicated boolean operations is feasible. Producing the explicit output polyhedron is also possible, though considerably more complicated. This could even be extended to collision detection of moving objects, each composed of the union of many simple polyhedra.

References

[AFKN89] Varol Akman, Wm Randolph Franklin, Mohan Kankanhalli, and Chandrasekhar Narayanaswami, *Geometric computing and the uniform grid data technique*, Computer Aided Design **21** (1989), no. 7, 410–420.

[AST97] Boris Aronov, Micha Sharir, and Boaz Tagansky, *The union of convex polyhedra in three dimensions*, SIAM J. Comput. **26** (1997), 1670–1688.

[BN83] H. Bieri and W. Nef, *A sweep-plane algorithm for computing the volume of polyhedra represented in boolean form*, Linear Algebra and its Applications **52–53** (1983), 69–97.

[Bri89] E. Brisson, *Representing geometric structures in d dimensions: Topology and order*, Proc. 5th Annu. ACM Sympos. Comput. Geom., 1989, pp. 218–227.

[Cam91] S. Cameron, *Efficient bounds in constructive solid geometry*, IEEE Comput. Graph. Appl. **11** (1991), no. 3, 68–74.

[CD87] Bernard Chazelle and D. P. Dobkin, *Intersection of convex objects in two and three dimensions*, J. ACM **34** (1987), no. 1, 1–27.

[Cen01] Center for Geometric and Biological Computing, *External memory algorithms and data structures*, `http://www.cs.duke.edu/CGC/subjects/external.html`, 2001.

[Cha89] Bernard Chazelle, *An optimal algorithm for intersecting three-dimensional convex polyhedra*, Proc. 30th Annu. IEEE Sympos. Found. Comput. Sci., 1989, pp. 586–591.

[Des44] R. Descartes, *Principia philosophiae*, Ludovicus Elzevirius, Amsterdam, 1644.

[DK83] D. P. Dobkin and D. G. Kirkpatrick, *Fast detection of polyhedral intersection*, Theoret. Comput. Sci. **27** (1983), no. 3, 241–253.

[DMY93] Katrin Dobrindt, Kurt Mehlhorn, and Mariette Yvinec, *A complete and efficient algorithm for the intersection of a general and a convex polyhedron*, Proc. 3rd Workshop Algorithms Data Struct., Lecture Notes Comput. Sci., vol. 709, 1993, pp. 314–324.

[EM88] H. Edelsbrunner and E. P. Mücke, *Simulation of simplicity: a technique to cope with degenerate cases in geometric algorithms*, Proc. 4th Annu. ACM Sympos. Comput. Geom., 1988, pp. 118–133.

[Epp92] D. Eppstein, *Speed-ups in constructive solid geometry*, Report 92-87, Dept. Inform. Comput. Sci., Univ. California, Irvine, CA, 1992.

[Eri03] J. Erickson, *Local polyhedra and geometric graphs*, 2003.

[FCK+90] Wm Randolph Franklin, Narayanaswami Chandrasekhar, Mohan Kankanhalli, Varol Akman, and Peter YF Wu, *Efficient geometric operations for CAD*, Geometric Modeling for Product Engineering (Michael J. Wozny, Joshua U. Turner, and K. Preiss, eds.), Elsevier Science Publishers B.V. (North-Holland), 1990, pp. 485–498.

[FK90] Wm Randolph Franklin and Mohan Kankanhalli, *Parallel object-space hidden surface removal*, Proceedings of SIGGRAPH'90, vol. 24, August 1990, pp. 87–94.

[FK93] Wm Randolph Franklin and Mohan S. Kankanhalli, *Volumes from overlaying 3-D triangulations in parallel*, Advances in Spatial Databases: Third Intl. Symp., SSD'93 (D. Abel and B.C. Ooi, eds.), Lecture Notes in Computer Science, vol. 692, Springer-Verlag, June 1993, pp. 477–489.

[Fra78] Wm Randolph Franklin, *Combinatorics of hidden surface algorithms*, Technical Report, Center Res. Comput., Harvard Univ., Cambridge, MA, June 1978.

[Fra87] ———, *Polygon properties calculated from the vertex neighborhoods*, Proc. 3rd Annu. ACM Sympos. Comput. Geom., 1987, pp. 110–118.

[Fra04] W. Randolph Franklin, *Analysis of mass properties of the union of millions of polygedra*, 2004, (to appear).

[FSS+94] Wm Randolph Franklin, Venkateshkumar Sivaswami, David Sun, Mohan Kankanhalli, and Chandrasekhar Narayanaswami, *Calculating the area of overlaid polygons without constructing the overlay*, Cartography and Geographic Information Systems (1994), 81–89.

[GS83] Leonidas J. Guibas and J. Stolfi, *Primitives for the manipulation of general subdivisions and the computation of Voronoi diagrams*, Proc. 15th Annu. ACM Sympos. Theory Comput., 1983, pp. 221–234.

[HH90] Sara Hopkins and Richard G. Healey, *A parallel implementation of Franklin's uniform grid technique for line intersection detection on a large transputer array*, 4th International Symposium on Spatial Data Handling (Zürich) (Kurt Brassel and H. Kishimoto, eds.), 23-27 July 1990, pp. 95–104.

[Kan90] Mohan Kankanhalli, *Techniques for parallel geometric computations*, Ph.D. thesis, Electrical, Computer, and Systems Engineering Dept., Rensselaer Polytechnic Institute, October 1990.

[KF95] Mohan Kankanhalli and Wm Randolph Franklin, *Area and perimeter computation of the union of a set of iso-rectangles in parallel*, J. Parallel Distrib. Comput. **27** (1995), 107–117.

[LR82a] Y. T. Lee and A. A. G. Requicha, *Algorithms for computing the volume and other integral properties of solids. I. Known methods and open issues*, Commun. ACM **25** (1982), 635–641.

[LR82b] ———, *Algorithms for computing the volume and other integral properties of solids. II. A family of algorithms based on representation conversion and cellular approximation*, Commun. ACM **25** (1982), 642–650.

[LTH86] D. H. Laidlaw, W. B. Trumbore, and J. F. Hughes, *Constructive solid geometry for polyhedral objects*, Comput. Graph. **20** (1986), no. 4, 161–170, Proc. SIGGRAPH '86.

[Mir96] B. Mirtich, *Fast and accurate computation of polyhedral mass properties*, J. Graphics Tools **1** (1996), no. 2, 31–50.

[MP78] D. E. Muller and F. P. Preparata, *Finding the intersection of two convex polyhedra*, Theoret. Comput. Sci. **7** (1978), 217–236.

[MS85] K. Mehlhorn and K. Simon, *Intersecting two polyhedra one of which is convex*, Proc. Found. Comput. Theory (L. Budach, ed.), Lecture Notes Comput. Sci., vol. 199, Springer-Verlag, 1985, pp. 534–542.

[Nar91] Chandrasekhar Narayanaswami, *Parallel processing for geometric applications*, Ph.D. thesis, Electrical, Computer, and Systems Engineering Dept., Rensselaer Polytechnic Institute, 1991, UMI no. 92-02201.

[NF91a] C. Narayanaswami and Wm Randolph Franklin, *Determination of mass properties of polygonal CSG objects in parallel*, Internat. J. Comput. Geom. Appl. **1** (1991), no. 4, 381–403.

[NF91b] Chandrasekhar Narayanaswami and Wm. Randolph Franklin, *Determination of mass properties of polygonal CSG objects in parallel*, Proc. Symposium on Solid Modeling Foundations and CAD/CAM Applications (Joshua Turner, ed.), ACM/SIGGRAPH, June 1991, pp. 279–288.

[NF92a] Chandrasekhar Narayanaswami and Wm Randolph Franklin, *Boolean combinations of polygons in parallel*, Proceedings of the 1992 International Conference on Parallel Processing, August 1992.

[NF92b] ———, *Edge intersection on the Hypercube computer*, Information Processing Letters **41** (1992), no. 5, 257–262.

[Sug94] Kokichi Sugihara, *A robust and consistent algorithm for intersecting convex polyhedra*, Comput. Graph. Forum **13** (1994), no. 3, 45–54, Proc. EUROGRAPHICS '94.

[Til84] R. B. Tilove, *A null-object detection algorithm for constructive solid geometry*, Commun. ACM **27** (1984), 684–694.

[TWA+01] Laura Toma, Rajiv Wickremesinghe, Lars Arge, Jeffrey S. Chase, Jeffrey Scott Vitter, Patrick N. Halpin, and Dean Urban, *Flow computation on massive grids*, Proc. ACM Symposium on Advances in Geographic Information Systems, 2001, see also `http://www.cs.duke.edu/geo*/terraflow/papers/bib.html`.

[Vig] Antoine Vigneron, *Reporting intersections among thick objects.*

[ZE02] Afra Zomorodian and Herbert Edelsbrunner, *Fast software for box intersections*, International Journal of Computational Geometry and Applications **12** (2002), no. 1-2, 143–172.

ECSE Dept, 6003 JEC,, Rensselaer Polytechnic Institute, 110 Eighth St, Troy NY 12180, USA

E-mail address: `mail@wrfranklin.org`

URL: `http://wrfranklin.org/`

Titles in This Series

68 **A. Ashikhmin and A. Barg, Editors,** Algebraic Coding Theory and Information Theory

67 **Ravi Janardan, Michiel Smid, and Debasish Dutta, Editors,** Geometric and Algorithmic Aspects of Computer-Aided Design and Manufacturing

66 **Piyush Gupta, Gerhard Kramer, and Adriaan J. van Wijngaarden, Editors,** Advances in Network Information Theory

65 **Santosh S. Vempala,** The Random Projection Method

64 **Melvyn B. Nathanson, Editor,** Unusual Applications of Number Theory

63 **J. Nešetřil and P. Winkler, Editors,** Graphs, Morphisms and Statistical Physics

62 **Gerard J. Foschini and Sergio Verdú, Editors,** Multiantenna Channels: Capacity, Coding and Signal Processing

61 **M. F. Janowitz, F.-J. Lapointe, F. R. McMorris, B. Mirkin, and F. S. Roberts, Editors,** Bioconsensus

60 **Saugata Basu and Laureano Gonzalez-Vega, Editors,** Algorithmic and Quantitative Real Algebraic Geometry

59 **Michael H. Goldwasser, David S. Johnson, and Catherine C. McGeoch, Editors,** Data Structures, Near Neighbor Searches, and Methodology: Fifth and Sixth DIMACS Implementation Challenges

58 **Simon Thomas, Editor,** Set Theory: The Hajnal Conference

57 **Eugene C. Freuder and Richard J. Wallace, Editors,** Constraint Programming and Large Scale Discrete Optimization

56 **Alexander Barg and Simon Litsyn, Editors,** Codes and Association Schemes

55 **Ding-Zhu Du, Panos M. Pardalos, and Jie Wang, Editors,** Discrete Mathematical Problems with Medical Applications

54 **Erik Winfree and David K. Gifford, Editors,** DNA Based Computers V

53 **Nathaniel Dean, D. Frank Hsu, and R. Ravi, Editors,** Robust Communication Networks: Interconnection and Survivability

52 **Sanguthevar Rajasekaran, Panos Pardalos, and D. Frank Hsu, Editors,** Mobile Networks and Computing

51 **Pierre Hansen, Patrick Fowler, and Maolin Zheng, Editors,** Discrete Mathematical Chemistry

50 **James M. Abello and Jeffrey Scott Vitter, Editors,** External Memory Algorithms

49 **Ronald L. Graham, Jan Kratochvíl, Jaroslav Nešetřil, and Fred S. Roberts, Editors,** Contemporary Trends in Discrete Mathematics

48 **Harvey Rubin and David Harlan Wood, Editors,** DNA Based Computers III

47 **Martin Farach-Colton, Fred S. Roberts, Martin Vingron, and Michael Waterman, Editors,** Mathematical Support for Molecular Biology

46 **Peng-Jun Wan, Ding-Zhu Du, and Panos M. Pardalos, Editors,** Multichannel Optical Networks: Theory and Practice

45 **Marios Mavronicolas, Michael Merritt, and Nir Shavit, Editors,** Networks in Distributed Computing

44 **Laura F. Landweber and Eric B. Baum, Editors,** DNA Based Computers II

43 **Panos Pardalos, Sanguthevar Rajasekaran, and José Rolim, Editors,** Randomization Methods in Algorithm Design

42 **Ding-Zhu Du and Frank K. Hwang, Editors,** Advances in Switching Networks

41 **David Aldous and James Propp, Editors,** Microsurveys in Discrete Probability

40 **Panos M. Pardalos and Dingzhu Du, Editors,** Network Design: Connectivity and Facilities Location

39 **Paul W. Beame and Samuel R Buss, Editors,** Proof Complexity and Feasible Arithmetics